Library of
Davidson College

Gray's Monitor Lizard

The author holds an extremely shy Gray's monitor lizard (*Varanus olivaceus*). Only three specimens were known to scientists, despite the fact that the species was described in the last century.

Gray's Monitor Lizard

Walter Auffenberg

University Presses of Florida
University of Florida Press
Gainesville

Copyright © 1988 by the Board of Regents of the State of Florida

All rights reserved

Printed in the U.S.A. on acid-free paper

Library of Congress Cataloging-in-Publication Data

Auffenberg, Walter.
  Gray's monitor lizard.

  Bibliography: p.
  Includes index.
  1. Gray's monitor lizard. I. Title.
QL666.L29A933   1988   597.95   86-15894
ISBN 0-8130-0841-7

UNIVERSITY PRESSES OF FLORIDA is the central agency for scholarly publishing of the State of Florida's university system. Its offices are located at 15 NW 15th Street, Gainesville, FL 32603. Works published by University Presses of Florida are evaluated and selected for publication by a faculty editorial committee of any one of Florida's nine public universities: Florida A&M University (Tallahassee), Florida Atlantic University (Boca Raton), Florida International University (Miami), Florida State University (Tallahassee), University of Central Florida (Orlando), University of Florida (Gainesville), University of North Florida (Jacksonville), University of South Florida (Tampa), University of West Florida (Pensacola).

ORDERS for books published by all member presses should be addressed to University Presses of Florida, 15 NW 15th Street, Gainesville FL 32603.

Contents

Acknowledgments	vii
Preface	xi
1. The Study Area and Methodology History. Geography. Plants. Butaans. Other Study Methods.	1
2. External Morphology and Color Size, Form, and Mass. Color and Pattern.	16
3. Internal Morphology The Head. Postcranial Skeleton. Gastrointestinal Tract. Fat.	33
4. Environment Macroecology. Microecology.	68
5. Movement The Daily Pattern. Factors Affecting Activity.	96
6. Spatial Relations Horizontal Distribution. Vertical Distribution. Home Range. Shelter Characteristics and Use.	122
7. Individuals and Populations Individual Characteristics. The Senses. Locomotion. Population Characteristics.	147
8. Reproduction Sexual Maturity. The Female Reproductive Cycle. The Male Reproductive Cycle. Behavior and Reproduction. Egg Morphology and Complement. Nest Location. Incubation Time. Control of Annual Reproductive Cycle. Reproductive Strategy.	159

9.	**Food and Feeding**	178
	General Utilization Patterns. Food Types Taken in Captivity. Seasonality. Characteristics of Important Foods. Toxicity and Other Chemical Properties. Food Location. Consumption Techniques. Consumption Rate and Amount. Digestion. Fecal Pellets. Gastric Pellets. The Feeding Niche.	
10.	**Food Resources**	216
	Animal Foods. Plant Foods.	
11.	**Foraging Strategy**	235
	Optimum Diet. Optimum Foraging Space. Optimum Time. Competition and Fruit. Frugivory as a Feeding Strategy in Lizards. Conclusions.	
12.	**Interspecific Interactions**	269
	Potential Competition. Parasitism. Enemies. Plants and the Butaan.	
13.	**Intraspecific Interactions**	292
	Behavioral Acts. Combat. Courtship. Releasers. Priming Mechanisms.	
14.	**Systematics and Evolution**	326
	Systematics. Phylogenetic Relations. Ecology and Phylogeny. Geology.	
15.	**Conservation**	357
	Refuges. Predation by Man. Habitat Destruction. Recommendations.	

Appendixes

1. Annotated List of Plants Mentioned in Text   373
2. Annotated List of Birds and Mammals Whose Food Resources Overlap Those of *Varanus olivaceus* in the Caramoan Area   379
3. Abbreviations Used in Text   383

Literature Cited   385

Index   409

Acknowledgments

THE support my work received during the course of seven years cannot be acknowledged in a few words. The work in this book was financed by several sources, including the Auffenberg family. My first thanks must go to the University of Florida and the Florida State Museum, who encouraged me in every way to carry out my plans to study this animal. Additionally I must thank the Animal Research Council of the New York Zoological Society, for they supplied financial assistance throughout the work. During the early phases the World Wildlife Fund (American Appeal) also supported the investigations, and during the last year additional funding was provided by the United States Fish and Wildlife Service (Office of Endangered Species). Pena de Francia School (Naga, Philippines) and the Dallas Zoological Society made certain of their facilities available for the study of live specimens. My wife, Elinor, deserves much credit for her willingness to spend much of our own money on an enterprise whose outcome, while never in doubt, was often partially obscured by the trials and tribulations of living in a camp far from home, friends, and relations.

A work like this always draws heavily on the kindness of friends and of the people one meets in the course of gathering and analyzing material during a long-term scientific study. But this particular study involved special demands, and I should like to express my gratitude to the people who taught me what they knew of the Caramoan wilderness, made available their homes and lands, their servants and trucks, who read and commented upon draft versions of the manuscript, whole or in part, shared their knowledge regarding the *butaan*, and offered their hospitality and support to myself and my family in the course of our travels.

Acknowledgments

First of all, a study like this would not have been possible without the official assistance and cooperation of, particularly, Dr. Jesus B. Alvares, Jr., assistant director, Philippine Bureau of Forestry Development. Additionally, I wish to thank Sirilo B. Serna, (former) chief, Parks, Range and Wildlife Division, and his able assistant, Luz C. Gonzales. Benjamin Echano of the same office worked part-time on the project and provided additional information on the range of this species on Luzon. Locally, official assistance was given by Faustinad Perfecto, chief district forester, and Epifanio R. Lopo, senior conservation officer, both of the Bureau of Forestry Development Office (Naga). I thank Robert Z. Villarino of the bureau's education staff for expert help regarding certain photographic problems. Of the bureau's Naga staff, I extend my special appreciation to Stephen C. A. Alba (then forestry extension officer) for his unfailing interest in the problems attending my field work and his untiring efforts to make our life in the Caramoan Peninsula as pleasant as possible.

During the first two years of the work Rene E. Mendoza, head, External Affairs Office of the Philippines Center for Advanced Studies, was instrumental in making necessary governmental arrangements. Dr. Godofredo L. Alcasid, former director of the Philippine National Museum, assisted with logistic problems during the early phases of the project. Finally, the work proceeded more efficiently because of the interest and attention provided by the offices and personal attention of José Fuentabella, chief, Provincial Development Staff, and Dolores H. Sisson, Assemblywoman and Office of the Ministry of Tourism.

The attention of Romeo Gonzales (Manila) to our needs in the earliest parts of the work deserves mention. But my warmest thanks are extended to Dr. and Dra. Heracleo and Naida Guballa and Nilo and Tiseng Roa for their sincere friendship, hospitality, and steadfast belief in the importance of the work. Without the constant help of the Guballa family the study would never have been successfully initiated or completed. The Roa family was particularly gracious in the Caramoan area, where they consistently offered their assistance and properties for the benefit of the project. The base camp was constructed on their land at their invitation. For their hospitality and many kindnesses extended to me, I also wish to acknowledge the International Rotary Clubs of Manila, Davao, Puerta Princesa, Tacloban, Naga, and Calapan.

Dr. Harry Hoogstraal and the staff of the United States Naval Medical Research Unit No. 3 (Cairo) identified all of our tick samples; Dr. Gerald D. Schmidt, University of Northern Colorado (Greeley), studied the nematodes. My son Kurt, biologist, Florida State Museum, University of Florida (Gainesville), was extremely helpful regarding our studies of the molluscs eaten by Gray's monitor. Dr. Sam R. Telford, Danish International Development Agency, Morogoro, Tanzania, contributed much time in study of the blood smears sent to him. Dr. Elliot Jacobson, School of Veterinary Medicine, University of Florida, arranged for histological preparation in his laboratory; I also thank Mr. Kent Vliet for assistance in regard to this and related matters. John G. Fiskell and Mary McCloud graciously allowed us the use of the University of Florida Forest Soils Laboratory for protein analyses. Mr. Efron P. Cantor, assistant manager, Radio Mindanao Network (Naga), deserves special thanks for his help in constructing some of the radio transmitters.

The personnel associated with several zoological parks have been particularly helpful and have maintained a strong interest in the project: James Murphy, Dallas Zoological Garden, John Behler, New York Zoological Society, and Dr. James Bacon, Zoological Garden of San Diego. Ardell Mitchell, Dallas Zoo, has generously provided all information possible on captives of this species.

Dr. George Zug (U.S. National Museum of Natural History, Washington), Drs. Alicia Grandison and Nick Arnold (British Museum of Natural History, London), Dr. Hymen Marx (Field Museum of Natural History, Chicago), and Pedro Gonzales (Philippine National Museum, Manila) assisted in my study of preserved specimens of Gray's monitor lizard in their respective institutions. Dr. Edmond Malnate (Academy Natural Sciences, Philadelphia) provided important information on type material of Philippine varanids in his care.

For field assistance I must acknowledge the often long and arduous hours, sometimes under rather uncomfortable situations, contributed to the project without complaint by William Holzmark (Sarasota, Florida), Ardell Mitchell (Dallas Zoological Garden), Arthur Strange (Toronto, Canada), and Laurie Wilkins (Florida State Museum, Gainesville).

Though I am completely responsible for errors of commission or omission, I wish to thank the following people for the time and energy they spent and the constructive criticism they provided in greatly improving the manuscript: John Eisenberg, Stephen Humphrey, J. C. Dickinson, and the two reviewers selected by the University of Florida Press.

In addition to these acknowledgments, I have several very special debts of affection. This is the second book I have written on varanid lizards, and both have been typed, and retyped a number of times by Rhoda J. Bryant, editor, Florida State Museum. Her patience, reliability, and experience, as well as that of my editor at University Presses of Florida, have made the final product possible. Mr. Peter A. Meylan, biologist, Florida State Museum, provided research support of exemplary thoroughness and energy. He has coped graciously with hundreds of requests for information and copies of library and other materials; much of the completeness of the literature cited section is due to his labor. The chemical analyses of fruits are entirely the result of his diligence and care. He has been considerate but thorough in his reading of sections of the manuscript, for which I am very thankful. Like Rhoda, he has done much to make the impossible possible. Those who have read selected sections of the complete manuscript are Drs. Kent Redford and Stephen R. Humphrey of the Florida State Museum and Kent Vliet of the Department of Zoology, University of Florida. I also owe a debt of gratitude to the two unidentified referees selected by the University of Florida Press for noting a number of my early errors and omissions.

Additionally, these acknowledgments would be incomplete without mention of the incalculable hours devoted to the project in so many ways by my family. My sons, Garth and Troy, are both due my most sincere thanks for jobs well done. Both were eager and dependable field assistants of the highest possible caliber, spending innumerable hours laboring and traveling in the local forests, often far removed from the luxury of the family base camp at Terogo. Both shared in the important job of camp management.

I also wish to thank the people of the Caramoan Municipality, for without their interest, hospitality, and cooperation this work never could have been completed. Particular thanks are extended to Keekoi and Aning Delloro, Wilson and Marilynn Co, and Leo and Nebis Morilla.

I also want to extend my sincere appreciation to the local hunters and other forest-wise folk, such as the Aeta tribals in Camarines Norte and hunters in the Caramoan area, who provided so much information so freely, so many specimen materials, and so many clues to then unanswered questions regarding the biology of Gray's monitor. While many were helpful, particular thanks are extended to Felizardo Teoxin (Barrio Ilawod) and Artemio Tolero (Kasine Mountain).

My greatest debt is to Elinor, my wife. She has twice suffered great uprooting so she could accompany me to environments in which few women would freely choose to live. During these trips she has remained my most valued camp assistant and my best friend. She made the good times better, and in the bad times it was her faith that sustained me. Without the help and encouragement of the entire family this study would have been impossible, and it is to them that this book is dedicated.

Preface

GRAY'S MONITOR lizard has been surrounded with more ignorance, misconception, and confusion within the scientific community than any other large lizard species living in the world today. Given a scientific name that has been in use for 150 years but can no longer be used, known only from a preserved juvenile and the cleaned skull of an adult until 1976, considered exceedingly rare for many years, and probably restricted to some remote Philippine island, it actually is rather common within a short drive of Manila—one of the larger cities of the world. In spite of the fact that male Gray's monitors reach a length greater than the height of most of the humans with whom it shares the forests, it has been mostly overlooked by biologists. These facts, plus the recent discovery that it represents the only fruit-eating species in an otherwise carnivorous family of lizards, are just a few of the many features of this animal that make it one of the world's most interesting reptiles.

Given the opportunity to study this unique animal in a beautiful tropical forest setting, among people not only cooperative but kind, it is no wonder that I consider this the crowning experience of my thirty years of study and play in some of the most fascinating parts of the world.

1
The Study Area and Methodology

THE contents of this report result from a 22-month study of the ecology and behavior of the Monitor lizard *Varanus grayi,* locally called *butaan* (sometimes called *batua* in and near Quezon Province, Luzon; *baneas* by the Aeta Negritos of Camarines Norte). The study was conducted primarily in the Republic of the Philippines, on the islands of Luzon (1976, 1978, 1981, 1982–83), Samar (1978), and Catanduanes (1983). The primary intention was to accumulate data considered important for comparing this forest-dwelling, largely frugivorous varanid lizard to the equally specialized, savanna-inhabiting, totally carnivorous Komodo dragon, *Varanus komodoensis* (Auffenberg, 1981a).

While parts of the Komodo dragon project generated important and interesting data, other sections, such as those concerned with physiology, demography, and especially reproduction, were all rather meager. On the other hand, information on its feeding biology and social behavior was especially enlightening and reasonably complete. Perhaps the most significant factor in the comparison of these two *Varanus* species is the great size and highly predaceous habits of the Komodo monitor. Hatchlings of that species are considerably larger than adults of almost all other lizard species in the world. This great size, coupled with certain morphological and behavioral specializations, reflect a feeding strategy for adults that is often based on the search for dangerous prey frequently larger than the predator itself. Social interactions in that species are often aggressive, sometimes leading to fierce fights, death, and even cannibalism. The overall picture of the general life-style of the Komodo dragon drawn by that long-term study is one of a highly active predator with a pattern of behavior that can be expected to take aggressive turns during both intra- and interspecific interactions.

NOTE: The abbreviations used in this text are defined in appendix 3.

This study is concerned with an entirely different life-style for a monitor lizard. Its largely frugivorous diet in tropical evergreen forests represents a feeding strategy in which search patterns tend to be geographically restricted compared to those of the Komodo monitor. Aggressive behavior toward conspecifics is much reduced and usually highly stereotyped. Individuals of this species are secretive, and an observer may spend months in forests in which butaans are reasonably common without seeing one. Even when observed, their escape pattern is circumscribed and slow. They are cryptically colored during all stages of their lives and are thus difficult to see in the luxuriant vegetation in which they most commonly occur.

As a result of these behavioral and morphological attributes, some aspects of the biology of the butaan remain poorly understood. Demographic data remain woefully inadequate. However, data accumulated on reproductive cycles and feeding are of good quantity and quality.

Varanus grayi is the only regularly frugivorous lizard species in the world. Relatively few extensive field studies have been conducted on any herbivorous reptile—and these have been concentrated on species with largely folivorous rather than frugivorous habits (see especially Burghardt and Rand, 1982). The only other extensive field study of a varanid lizard is that on *Varanus komodoensis* (Auffenberg, 1981a), though many excellent shorter papers dealing with various aspects of physiology, behavior, and field ecology have been published (see especially Vogel, 1979). No other field study has ever been conducted on this species. A progress report on the research conducted from 1976 to 1979 (Auffenberg, 1979a) is the only relevant publication.

History

In 1845, John E. Gray described *Varanus ornatus* on the basis of a single juvenile specimen in the British Museum of Natural History (BMNH 1946.8.30.98) from an unknown locality in the Philippine Islands, collected by H. Cuming. (Hugh Cuming was a major collector for the British Museum in the Philippines from 1836 to 1840. Many of the specimens he collected during that time were accompanied by imprecise, incorrect, or no locality data, including the holotype of *Varanus grayi*.) Later, Boulenger (1885) realized that *Uaranus ornatus* Gray was preoccupied by *Varanus ornatus* (Daudin) (a synonym of *Varanus niloticus*) and proposed instead *Varanus grayi* for the distinctive Philippine species.

During this study I discovered an important paper showing that the valid name of this species should be *Varanus olivaceus*. In 1856, Edward Hallowell described a number of new species of reptiles in the collection of the Academy of Natural Sciences, Philadelphia, including a *Varanus olivaceus* (ANSP 9916, a skin, has not been located in a recent search of that collection), which was based on a large specimen of monitor purchased some time earlier in Manila by "Dr. Kane, U.S. Navy." The diagnosis Hallowell provided makes it clear that his specimen was identical with the type of Gray's *Uaranus ornatus*. Therefore, according to the rules of priority, Hallowell's *olivaceus*, not Boulenger's *grayi*, replaces Gray's *ornatus*, and the Philippine species should be called *Varanus olivaceus*. Boulenger (1885) placed Hallowell's *olivaceus* in the synonymy of *Varanus bengalensis* without comment, but with a question mark. The most recent generic reviser (Mertens, 1942) followed Boulenger, also without comment, though he dropped the question mark (see chap. 14 for full discussion).

Though other specimens were evidently taken subsequently (de Elera, 1895–96, but see below), none was available for study until another forty years had passed, in spite of the extensive herpetological studies of Edward H. Taylor in the Philippines during the early part of the twentieth century (1912–19). When he completed his work, all Taylor (1922) could say of this species was that "it must be restricted to some remote island, because [he was] unable to secure any specimens of it in [his] travels."

Twenty-one years later, Robert Mertens (1942) published his monumental work on the Family Varanidae. In that publication he provided a more detailed account of the juvenile holotype of *Varanus grayi* in the British Museum. More important, he had located the skull of a much larger varanid from "Luzon" in the collection of the Munich Museum (ZMS 2640/0) that he thought represented this species. His study of this specimen suggested to him that its structural similarity to the skulls of the Nile Monitor (*Varanus niloticus*) was due to convergence and that *Varanus grayi* was, in fact, more closely related to *Varanus salvator*. Though he erected a number of new subgenera and placed all the known *Varanus* species into his new system, he remained uncertain regarding the peculiar skull which he believed represented *Varanus grayi* (= *olivaceus*). A number of years later he faced the problem of still not having located any additional material and published a separate paper (1962) in which he reemphasized the peculiarities of the Munich skull, reiterated that he thought it represented an adult of *Varanus grayi* (= *olivaceus*), and established a new subgenus—*Philippinosaurus*—to receive it as its only member species.

During the decades after World War II, attention was again directed to study of the herpetofauna of the Philippine Islands. However, the extensive collecting trips of Walter C. Brown, Dioscoro Rabor, and A. C. Alcala made during that time brought no additional material of this species to light, in spite of the fact that these competent biologists were particularly interested in lizards and that they had made a point of visiting the areas not touched by Taylor in his earlier travels. As a result, some conservationists and biologists thought that *Varanus olivaceus* was probably extinct.

A chance discovery in 1974 reopened the question of the ecological and geographic distribution of the species. In that year I found a stuffed skin, including a skull, of an adult of this species in the United States National Museum of Natural History (USNM 27776)—the third individual to come light in 130 years (Auffenberg, 1976). The specimen had escaped earlier notice because of its poor condition and the fact that it was catalogued under *Varanus olivaceus*. It had three important features. First, the scalation and color pattern were identical to those of the juvenile type in the British Museum, and the teeth were identical to those of the distinctive Munich specimen described by Mertens; thus, this specimen confirmed Mertens's earlier suggestion that the Munich skull was that of an adult *Varanus grayi* (= *olivaceus*). Second, the United States National Museum individual was the first to have precise locality information ("Pasacao, Luzon"), thereby pinpointing an area in which additional specimens might be sought. Third, it had been collected in the 1930s, documenting that the species was extant at that time.

Shortly after that discovery I obtained support from both the World Wildlife Fund (U. S. Appeal) and the New York Zoological Society to explore southern Luzon and nearby islands for evidence of its existence. In the resulting preliminary publications (Auffenberg, 1978a, 1979a) I described the habitat of the species, listed localities at which individuals had been seen, captured, or reported by locals, pointed out the peculiar frugivorous habits of the species, and outlined what I believed to be the status of the species at that time. (In these publications *V. dumerilii* was erroneously listed from the Philippines in the identification key provided; *V. rudicollis* may be part of the Philippine fauna in spite of statements to the contrary [preserved adult specimen, FSM].) I also suggested that *V. olivaceus* was probably restricted to the forested mountains of central to southern Luzon in the area of most consistent rainfall throughout the year. In 1978 I returned to continue my investigations of the geographic range of this species (Auffenberg, 1979a) and found that it seemed to be absent from Mindanao, Leyte, and Samar and certainly from Mindoro and Palawan. However, unconfirmed reports of it were obtained from the island of Catanduanes. Additional information was also obtained on feeding habits and particularly on ecology. It was during this trip that the Caramoan Peninsula, Camarines Sur, Luzon, was chosen as the site for more intensive study. The base camp location was selected, and radio telemetry techniques and equipment were tested during a three-month visit to this area in the summer of 1981. The final study was begun in June 1982 and ended in October 1983.

Perhaps the most interesting part in the entire story of the rediscovery of this species in the Philippines is that only fifty years after its description *Varanus olivaceus* (as *grayi*) had been listed from "Laguna Province, Luzon" by de Elera (1895–96). Unfortunately, Mertens (1942) rejected the locality record as highly unlikely. Evidence gathered during the course of this study makes it clear that *V. olivaceus* is indeed found in at least the eastern part of that province and extends even farther north. We now know that the butaan, practically unknown for 130 years following its discovery, exists within an hour's drive of the largest city in the Philippines. That it remained undiscovered in that area is due largely to its secretive habits, which this study was designed to examine.

Geography

Almost all the fieldwork was conducted in the Caramoan "township," Camarines Sur Province, southern Luzon Island (123°51'E, 13°55'N, fig. 1-1) during 1976 (2 weeks), 1978 (2 months), 1981 (3 months), and 1982–83 (17 months; total 22 months). The Caramoan area was selected because of good local cooperation, the ease with which suitable habitat could be reached, and the apparent high density of resident butaans.

The base camp was located at Barrio Terogo, about 2 km north of Caramoan. From this point field trips of varying lengths were made to forested hills and mountains east and southeast, covering an area of about 15 km^2, both within and outside the Caramoan National Park. All localities mentioned in the text are shown in figure 1-1.

The area in which the study was located has a basement complex of ancient metamorphic rocks, though they are exposed at the surface in the valley of the Manipot River. The remainder of the area is covered with a highly eroded Tertiary limestone, and caves, sinks, fissures, and block fields are common features. Slopes tend to be steep (limestone slopes average 38.2°, SD = 19.5°, N = 30; igneous hill slopes average 31.9°, SD = 17.3°, N = 30); cliffs are a dominant feature of most local limestone mountainsides. Rivers regularly disappear into sinks and caves in mountains, to reappear as springs on the other side. Lowland areas are covered with loamy clays; mountain soils are calcareous organics, usually restricted to small pockets in the surface rock, or lateritic clays in deeper fissures and sinks. Mountaintops are usually jumbles of fallen limestone boulders, among which extensive labyrinthine passages are formed. The highest mountain in the study area is approximately 300 m; most are lower ($\bar{X}$ of 16 highest local mountains = 210.8 m). However, just outside the study area there are a few mountains reaching 450 m.

All of the flatter valley areas were long ago converted to agricultural use, primarily rice culture. Even some of the lower slopes, particularly in the igneous areas, were cleared and converted to sweet potato, coconut, and upland rice production, but they also contain extensive areas of grass and brush. With the exception of somewhat scattered agricultural areas in small valleys, most of the Caramoan Peninsula east of Caramoan is covered with secondary and

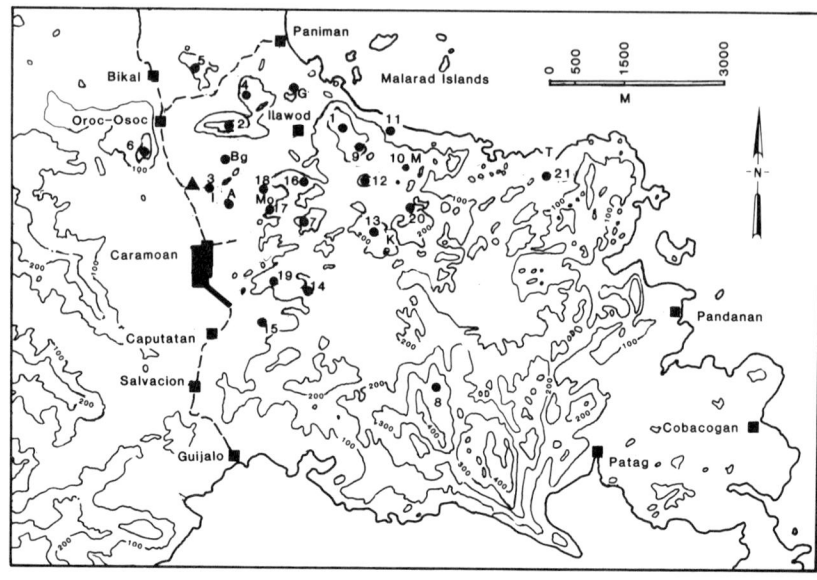

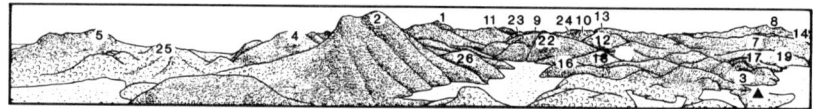

Fig. 1-1. Study area location. *Top:* Outline map of tip of Caramoan Peninsula (see figs. 4-3, 6-1 for location on Luzon), showing general landform and local place-names used in text. Contours are 100 m, except 50-m base. *Letters* show location of plant study blocks: *A*, Ahop Mountain; *Bg*, Batungan; *G*, Gota Beach; *I*, Ibu I, II, and III; *K*, Katiagawan; *M*, Mabho I and II; *Mo*, Matingapo I and II; *T*, Tiak Lake. *Numbers* locate major mountain peaks: 1, Damulog; 2, Taison; 3, Ahop; 4, Batlayan; 5, Malipot-na-tubig; 6, Pagtanawan; 7, Matingapo; 8, Kabangan; 9, Kasini; 10, Mabo; 11, Kabutanan; 12, Anuling; 13, Katiagawan; 14, Ubus; 15, Kaputatan; 16, Balanac; 17, Burukan; 18, Libo; 19, Sorak; 20, Patag; 21, Tiak Lake; 22, Kalibung; 23, Sulok; 24, Tipolo; 25, Inangkauhon; 26, Laog. *Bottom:* Same area but view east from Pagtanawan (6); numbers as above. The base camp is shown as a solid triangle.

primary forest; the dominant trees in relatively undisturbed and well-soiled sites are members of the family Dipterocarpaceae. This dipterocarp forest is quite heterogeneous, though *Ficus* and *Sterculia* are common and often dominant floral components. In igneous mountains to the west, these forests are now more or less restricted to the higher ridges and peaks. Though several streams traverse the study areas, most drainages are subterranean due to extensive exposures of limestone at the surface. The environment of the major study site is discussed in detail in chapter 4.

Plants

The plants of the area were investigated from several different standpoints, with the primary goals of floral density descriptions and tree phenology. The fruits of some tree species were studied in detail, with particular emphasis on chemical composition, seasonal and spatial distribution on the forest floor, density, and biomass. Fruit and seed morphology was examined closely to allow taxonomic identification of seeds found in the intestinal tracts of the butaans (see chap. 9, 10).

Although relationships among the dominant plants and animals in plant communities have received considerable attention in many parts of the world, this has not occurred in most of the Philippine Islands. In fact, few plant communities have even been defined, and few have been intensively studied. Therefore one of the primary goals of this work was to characterize the plant communities of which the butaan is an integral part. Plant identifications were based on the publications of Whitford (1911), Merrill (1912, 1926), Brown and Mathews (1914), Brown and Merrill (1920), Brown and Fischer (1920a, b, 1921), West and Brown (1921), and Salvoza and Lagrimas (1940). Botanical names used are those currently recognized, but it is likely that many will change as more work is conducted on local floras.

Density.—While much has been written on methods of determining density of plants in temperate areas (see Oosting, 1958, for review), relatively little attention has been given to the significant problems of determining plant species densities in diverse tropical forests. Recommended plot sizes for various plant types in temperate regions do not work well in tropical forests, especially for trees. Therefore, a plotless method was used in this study—specifically the "quarter method" (see Cottam and Curtis, 1956; Phillips, 1964; Mueller, Dombois, and Ellenberg, 1974). Using this method requires slightly more time to acquire the same amount of data as the "Bitterlich" and similar sighting methods (see Grosenbaugh, 1952), but it requires no equipment. More important, it is more widely applicable to vegetation other than trees. It is based on a measure that reflects the amount of space surrounding a given plant (= distance between plants). Each sampling point (determined randomly) is considered the origin of four quadrants of a Cartesian grid. The tree (in this study, trees >10 cm bhd) in each quadrant nearest to the origin is identified. The sum of the distances between these trees (in m) and the origin divided by four gives the average distance from the origin. Sample origins are selected so that a quadrant of one grid never overlaps that of another. Each tree tabulated by the plotless system was also classified in relation to one of three height classes (small, <10 m; medium, 10–25 m; tall, >25 m), leaf function (deciduous or evergreen), leaf shape (grasslike, medium or small, broad and larger, and compound or simple), leaf texture (papery or succulent), and whether the plants touched one another. At each study site, at least 20 sample areas were randomly chosen for the grid origins (= 80 measurements per sample area). Six study sites were examined in each plant community. In addition, five other sample areas were selected in each community, and 20 trees, each over 10 cm bhd (= 100 data points per site, a total of 500 data points per community) were randomly selected in each area and tabulated as to the three height classes. The shrubs, herbs, and substrate were all described for each community and a sketch made of each study area. Locations of these sample areas are shown in figure 1-1. Collectively they offer a good picture of the phenology of the trees in the local forests. Similar data were collected from the vicinity of Panaquasan, Isarog National Park, el. 500 m, to obtain information on community structure near the upper altitudinal limits of the butaan.

Cover.—Vegetative cover was measured in respect to its obstruction of vision at the level of butaan activity (see Wight, 1938; deVos and Mosby, 1963) and by the amount of light present at different levels in the forest (see Allard, 1947; Auffenberg, 1981a, for details).

Plant foods.—Early in the study it was thought that the factor most critical to the behavior and distribution of the butaan was the abundance of food and its precise pattern of distribution. I therefore concentrated on the feeding niche of the butaan, which could be accomplished only through reasonably long-term studies; only long periods of observation and data recording offered prospects for successfully unraveling the details of its ecology and feeding strategy. For this reason use was made of chemical studies of food sources, for it has become increasingly clear that as the behaviorist depends on the ecologist for a full interpretation of results, so the ecologist is dependent on the nutritionist.

The density of food plants, biomass of fruit fall per food species per hectare, periodicity of fruit flushes, and a number of other parameters important in understanding the availability of the butaan's plant foods were studied by means of belt transects. Twenty-seven transects of 1000 m^2 each (10 × 100 m) were established and checked weekly throughout an entire year, beginning in September 1982. Of these, 21 were located in the exact area in which butaans had been found during the early part of the study, 3 in typical habitats of the species, 2 in second growth forests on a low ridge, and another at the base of a mountain slope in mesic secondary forest in the floodplain of a small river. Figure 1-1 shows the location of all transects. All trees with edible fruit in each transect were plotted on a transect map, numbered individually for future use, identified, and the canopy height, bhd, and crown width measured. Though monthly sampling intervals are commonly used to establish seasonal patterns of floral phenology, seasons do not begin and end at monthly intervals; therefore, weekly visits were made to each transect. All fruits found on the ground under each tree were collected and kept in separate numbered sacks. At the base camp the fruits were later counted and weighed (to closest 0.1 g), their greatest diameter measured (to closest 1.0 mm), and the volume determined (to closest 1.0 ml). Seeds from all fruit types were collected for identification of stomach and intestinal contents of the butaans.

To reduce the number of observations and simplify the subsequent analysis of the data, the results from all 27 vegetational transects were combined to give values for each plant food species each week. Each species was given two values: *relative frequency,* or the number of fruit species within each plot that were represented by at least one fruit that week, expressed as a percent of the fruit per plot, and *absolute density,* or the total number of fruits of that species falling in all plots combined and converted to square meters, i.e., absolute density. This information was later compared to data on tree height and crown width to characterize the trees that produce the fruits upon which the butaan feeds. The physical characteristics of the fruits eaten were determined by computing the mean and standard deviation of fruit size (greatest diameter), weight, and volume for all food species per plot per week, the summed frequencies for all species together per plot per week, and the number of fruits per m^2 for all species together per plot per week. In addition, the total number of species over all plots each week was also tabulated for all species and for each ecological category of fruit (cauliferous v. axillary terminal fruiting, crown width and height, etc.). Ecological categories used in various comparisons were (1) canopy trees greater than 10 m and requiring almost full sunlight for reproductive maturity, (2) understory shrubs and smaller trees less than 10 m tall and reproducing in shaded conditions, (3) vines and climbers rooted in the ground and not self-supporting, and (4) epiphytes perched or climbing on trunks and branches with no ground contact.

Only the fruits that had fallen to the ground were measured and weighed, for only these are eaten by the butaan (chap. 9). There are some possible sources of bias in counts. (1) Small fruits might be more widely dispersed after hitting the ground than larger ones, and they clearly are more difficult to find in the surface leaf litter; thus they may be underrepresented compared to larger ones. (2) Fruits may be removed from below the tree by frugivorous vertebrates, possibly the most important biasing factor. (Foster [1973] reported that at least half of the seeds collected from traps on the floor of Panamanian forests had already either been removed from the edible portions, regurgitated, or passed through the digestive tracts of animals before reaching his traps.)

Differences in the representation of fruit species per unit of forest floor are clearly important to discovery by the butaan, so that the samples provide a reasonable index of the phenological behavior of the trees important in the feeding biology of the butaan (see Charles-Dominique et al., 1981, for similar methodology).

Fruit chemistry.—Caloric determinations were made on a Parr model 1411 adiabatic standard bomb calorimeter, corrected for nitric acid, water, and ash. All weighing errors were less than 1%. Crude proteins (nitrogen) analyses were based on a semi-micro Keldahl method (Fleck and Munroe, 1965), using a mean conversion factor of 6.25 and standards of 20% aliquats.

Butaans

Capture.—Because of their frugivorous feeding habits and disinterest in carrion, butaans cannot be trapped, nor can they be excavated from the rock cavities in which they are found. With a few exceptions, all were captured by trained dogs owned by three local families (1–4 dogs per pack per hunt). Parties were usually in the field by 8:00 A.M., and hunts continued until about 3:00 P.M. Only twice were two individuals ever taken on the same day by a single party. Our collecting permit (BF/PWR #11) allowed us to take no more than five individuals each month. Once located by the dogs, butaans ran into either a nearby tree or nearby cavities in the rock. They were then usually captured by means of a noose attached to the end of a bamboo pole and were transported to the base camp in large sacks. Unlike *Varanus komodoensis* or similarly aggressive varanid species, even the largest butaans were easily held and otherwise manipulated by one person.

Individual characters.—Total weight was obtained with spring balances accurate to 1 g; all measurements were made to the closest mm. In addition to standard scale counts of midbody longitudinal scale rows and transverse rows from gular fold to hind leg insertion, other nonstandard counts were made on the head. Each specimen was examined externally for ectoparasites, shedding pattern, scars, and injuries.

Some individuals were sacrificed for internal studies. These were killed by injecting 90% alcohol into the brain through the orbital fissure.

Anatomical dissection.—Gastrointestinal observations and measurements follow those outlined in Hladik (1976) and Chivers and Hladik (1980). The proportional weights of muscle, skin, and bone of various parts of adult individuals (chap. 2, fig. 2-6) were based on the following generalities: tail, transversely severed at the level of the cloaca; upper hind leg, the entire femur, with all associated muscles and skin severed transversely at the femoral junction with the tibia distally and the pelvis proximally; lower leg, all tissue included between transverse cuts at the distal and proximal parts of the tibia; hind foot, all skin, muscle, and bone distal to the tibio-astragulus articulation; front limb parts dissected in same pattern as the rear elements; pelvic girdle, all tissues from base of tail to anterior end of cranial first sacral vertebrae; all tissues associated with the pectoral girdle were dissected from the underlying vertebral column; the body constituted all tissues (except the viscera) from the first to last thoracic vertebrae; neck, all cervical vertebrae and associated tissues, including the atlas-axis complex, dissected transversely at the base of the skull; head, all tissues as far posterior as the occipital condyle.

The viscera (lungs, liver, stomach, and intestines) were examined for endoparasites, and any collected were preserved in alcohol or formalin. Blood smears were made and fixed in 90% alcohol at this time. Various visceral parts were measured, weighed, and examined, depending on the part being studied (gall bladder, liver, stomach, caecum, and large and small intestines). Skin samples (4 × 4 cm) were excised at the lateral fold immediately in front of the hind leg insertion and preserved in formalin for later sectioning. Sometimes the entire skin of the body and neck was removed, slightly stretched, and dried for later studies of pattern and color. Testes were measured for diameter to the closest 0.1 mm, weighed to the closest 0.1 g, and preserved in formalin for sectioning; ovaries were examined for corpora lutea (which were counted and weighed to the closest 0.1 g) and ova (which were counted, measured, weighed, and examined for extent of yolking).

Stomach contents were identified and counted, as were intestinal contents. Skeletal material was sometimes prepared for later study.

One of the testes of each male dissected was prepared for microscopic examination. Tissues were fixed for several days and later rinsed in water and transferred to 70% ethyl alcohol for storage. As time permitted, tissues were prepared for light microscopy examination using standard histological techniques (Luna, 1968; Mamason, 1979). Paraffin-embedded tissues were sectioned at 5–7 microns and stained with hematoxylin and eosin. Prepared microscopic testes sections were examined using techniques outlined in Underwood (1970).

The process of spermatogenesis was divided into four stages based upon the diameter of the seminiferous tubules, thickness of and number of cell layers in the wall of the tubules, relative proportions of cell types in the walls, and relative abundance of spermatozoa within the tubules and/or epididymis of each testes. Each male was categorized by having testes in one of these four stages.

In addition to the weight of ovaries obtained upon dissection of females, detailed analyses of ovarian follicles, corpora lutea, and oviductal eggs were made from preserved tissues. The diameter of each ovarian follicle greater than 3 mm was determined with vernier calipers. Follicles were grouped into 5 mm size categories for analysis of the data. Regardless of season, all females had numerous undeveloped follicles less than 3 mm in diameter, but these were not counted. The number of corpora lutea and oviductal eggs present was determined, and their diameters measured to the closest 0.1 mm and weights to the closest 0.01 g.

Captive specimens.—Butaan captives were maintained at the University of Florida and the Dallas Zoological Gardens and in a large pen built and used during 1982–83 at the Caramoan base camp. In each case, caging size was more than adequate to assure normal behaviors and feeding. This size was most important at the base camp where most behavioral observations were made. Here a hardware cloth and bamboo structure approximately 2.5 m on each side (ca 15.5 m^3) was constructed adjacent to the camp to facilitate observation. Cover was provided by coconut fronds and several large tubes made of rolled sections of woven bamboo laid on the floor. Because it was constructed outside, the light and climate factors were those that would normally be experienced in the nearby forest in which they occurred naturally. An adult male and female (originally captured together in a large hollow tree) were permanent residents, though others were introduced from time to time to study certain types of interactions (combat between adult males, for example).

These captives were observed periodically throughout each day, particularly their haul-out and haul-in periods, walking and climbing speeds, and behaviors related to feeding, courtship, and combat. Emphasis was on recording and analysis of adjacent transitional behavioral acts (see Carpenter and Ferguson, 1977). Analyses techniques follow those of Auffenberg (1983b), based on Oden (1977), in which stationarity between acts is not assumed and assessment of the significance of dependence between acts separated by an arbitrary number of steps is allowed.

Telemetry.—Several radiotelemetric techniques were used during the course of the study. Frequencies used were in the 150- and 27-MHz ranges: the former for long-distance work and more precise tracking and the latter for telemetry of ambient and physiological variables over shorter distances. Receivers used were Dav-Tron 12-channel types (Models MS 254 with automatic scanning features and RS 227 with all manual controls), the former in the base camp and the latter for fieldwork. The 27-MHz receivers were handheld Lafayette Model HA-420 types, with whip antennas modified to include a beat-frequency oscillator for improved signal reception.

The base camp location was selected partly with a view to the best possible radio signal reception. It was located on top of a 30-m hill cleared of all large trees and shrubs and with an unobstructed line-of-sight reception possibility in the directions of nearby mountains in which telemetered butaans were located.

The base antenna was a 7-element Yagi type, mounted on a movable swivel in the ceiling of the camp and provided with a small compass rose to determine bearing when necessary. Maximum reception distance from butaans was

1.0–1.5 km when they were on or slightly under the surface of the ground and almost 3 km when they were in a tree. This system was mainly used to obtain information on activity periods and intensity, ambient forest temperatures, and general location of the telemetered butaans.

The 150-MHz field receivers were provided with 3-element Yagi antennas for users' ease in moving through the forest. The field system was almost exclusively used in precise location of individuals. Reception distance was usually about 400 m.

The 27-MHz receivers were provided with whip antennas and used only for fieldwork. Maximum receiving range was usually 200 m.

Transmitters were of several types, though all circuits were blocking oscillators based on early designs of Cochran and Lord (1963) (for the external units) and MacKay (1970) (for gastrointestinal units). With appropriate modifications the following parameters were investigated with basically similar devices: ambient light intensity striking the dorsal surface of the butaan, ambient temperature of the dorsal region, gastrointestinal temperatures, axial orientation (horizontal or vertical) of the butaan, location in the field, and intensity of activity. Kimmich (1980) provides an excellent summary and bibliography of the pertinent literature for measuring these and other parameters.

Transmitters used were purchased (Dav-Tron and AVM) or constructed by Efron R. Cantor (Radio Mindanao Network, Inc., Naga City, Philippines) or me, based on the outline provided by Bradbury et al. (1979).

After construction, all transmitters were dipped twice in warm paraffin and once in a thick liquid rubbery vinyl compound (Plasti-dip, used for tool handles, etc.), then covered with a 2–3 mm coating of a 2-part epoxy (Marine-Tex). The 150-MHz units were powered with one or more Sanyo lithium batteries (CR-2N, 3 v, 1000 hma) and the 27-MHz units with Mallory mercury cells (variable sizes, depending on application). Average field life for the lithium cells was 6 months (none was used till dead), for the mercury cells about 3 months. The few units that stopped working in the field were later found either to be improperly sealed against moisture or to suffer antenna breakage (rare). All antennas were constructed of 35 kg stainless steel braided fish line leader, soldered to an electrical coupling that could then be fastened to the transmitter with a small bolt and nut. Entire weight of the largest and most powerful units was 61 g, size about 2×5 cm. Though other types of harnessing were tried (leather, nylon straps, etc.), the only method that ensured long service, because of the sharp rocks in the environment, was plastic-coated stainless steel banding (min. 50 kg breaking strength). Harness components were held together with nickel steel press connectors applied with a special tool for attaching steel bands. The most comprehensive recent basic work covering all phases of construction and use of radio telemetric units is that by MacDonald and Amlaner (1980).

All radio telemetry of activity and physiological parameters exhibiting continuous variation (temperature, activity, and light intensity) were measured by first calibrating the variable (measured as click rates produced by the oscillating radio circuitry) against known levels of each parameter. Thus variation in each parameter was recorded as a function of either the number of clicks produced per unit of time or the length of time that elapsed in 50 clicks. Time was measured with a Micronata stopwatch (accurate to 0.01 sec). For activity intensity records, readings were taken from each of the telemetered animals (up to four at a time) every five minutes during the observation period (usually 12 hrs/day). Thus for a one-day observation period, 144 contacts were made with each individual. Some records were made during the night for comparative purposes, though these sessions never lasted as long as those during the day (max 4 hrs at a time). Some individuals were monitored in this manner for up to 5 months, producing a total of up to 21,500 contacts per individual.

For tracking and positional plotting most individuals were checked at least once each day. Positions were plotted on a base map of the study area, and directions and distances of movements important in the analyses were calculated from that record.

Other Study Methods

Weather and climate.—Two weather stations were established about 0.5 km from one another, one on the cleared hill on which the base camp was located and the other in the forest at the base of the closest mountain in which butaans were regularly studied. At the base camp we recorded wind velocity (in 0.5 km/hr, hand-held Dwyer meter), rainfall (in mm), relative humidity (% RH with a Barach sling psychrometer), barometric pressure (100 m Thommen pocket altimeter-barometer), and shaded air temperature (Yellow Springs telethermometer with shielded air probes) at 10 cm and 1 m above the surface, as well as a shaded black bulb temperature probe at 1 m height. All recordings were taken at or close to noon each day throughout the study.

At the forest locality, temperatures were taken at least once each day, sometimes more frequently, with an attempt to sample the environment at all hours. We recorded shaded air temperatures from 2 m below the surface in the rocks (= fissure in the following accounts), at the forest floor (= surface), at 10 cm height (= 10 cm), at 1 m (= 1 m), and at 10 m (= 10 m) in the crown of a *Ficus nota* understory tree (all with an Atkins telethermometer, accurate to 0.1°C, and appropriately placed air probes). Percent relative humidity at 1 m height was taken with a Barach Sling Psychrometer (accurate to 0.5%) during each visit to the forest station.

The same equipment was also used periodically at various places on the same mountain, measuring surface and 10-cm temperatures and % RH each hour of the day for one-week periods to obtain information on diurnal variations at and near ground level in different parts of the community.

Trail systems.—A series of trails was established on Base Camp Mountains A and B. These passed through the slopes of the mountains, over their tops, and along their bases in ways that made almost the entire area available for observation and study.

Mapping.—The entire study area was mapped (Brunton compass and plane table method) during our work at Caramoan, with particular emphasis placed on the area within 2 km of the base camp, near and including Ahop Mountain. The most recent topographic map was used as a mapping base (Anon., 1979). General topographic features used as major reference points were taken from available United States and Philippine sources (U.S. Army Map Service, 1944, 1:25,000, Camarines Sur Province, Caramoan Peninsula, Series 711; Philippine Board Technical Surveys and Maps, 1979, Caramoan Sheet 3761 II, 1:50,000; and Camarines Sur Provincial Map, Provincial Planning Office, Area III, Partido [unpublished]). Data on size of province and forested tracts are taken from Hendry, 1959.

In almost all instances the common local usage has been followed for names not found on official maps. With a few exceptions, place-names used in the Caramoan karst area are not as localized as is usual in general cartographic usage; rather place-names usually refer to a valley (or plain) and the adjacent mountain slope. Thus, most mountains will have one name for a slope and valley on one side and another name for a slope and valley on the other side. In general, I have tried to avoid the resulting confusion by adding descriptive words, such as "mountain," "valley," etc.

Unsuccessful techniques.—Early in the planning stages it was presumed that blinds could be set up in the forests and telemetered free butaans observed from reasonably close distances. It was immediately obvious that this plan was unfeasible, for two reasons: the secretiveness of this species and the dense cover of the local forest and the extremely fissured rocky substrate, both of which offered so much cover that it was impossible to watch the monitors even when the general area in which they were located could be observed from behind an appropriate and effective blind.

Abbreviations.—All abbreviations used are defined in Appendix 3.

2
External Morphology and Color

ACCORDING to Mertens (1942), the diagnostic characters separating *Varanus olivaceus* from all other described varanid species are a slitlike nasal opening, located about one-half the distance between the eye and snout tip; broad, supraocular scales smooth; nuchal scales smooth and somewhat larger than the middorsal scales and about the same size as the posterior head scales; belly scales keeled, in about 100 transverse rows (from the gular fold to the insertion of the hind limb at its anterior edge); tail laterally compressed, with a double row of keeled scales dorsally; and posterior teeth blunt and barrel-shaped in the adult. His diagnosis relied entirely on external morphological and scale characters. It is now possible to add a number of other characters, including internal ones, that greatly extend knowledge of the anatomy of this species. The emphasis in the descriptions and discussions that follow is on the characters that are important in understanding the behavioral ecology of this species in the wild and on those that are of value to systematists in defining this and other species of monitor lizards. Many comparisons are made with *Varanus salvator* because *V. salvator* is believed to be the most primitive extant species of the genus (Mertens, 1942; King and King, 1975), it occurs sympatrically with *V. olivaceus* in the study area, and it belongs to an entirely different feeding guild.

Size, Form, and Mass

Length.—One of the most surprising discoveries about this lizard is that it was overlooked for such a long time in spite of its very large size; butaan males may reach lengths of over 1.5 m and weights exceeding 9 kg. (Native reports of individuals to 3 m and a weight of 40 kg are considered exaggerations.)

Only 8 of 42 *Varanus* species attain greater lengths or weights; thus *V. olivaceus* is one of the largest lizards in the world.

We measured 99 adult specimens, providing an average snout-vent length (SVL) of 51.2 cm. Of these, 90 were sexed by dissection. Like most (perhaps all) varanids, males reach a greater length than females do; the 10 largest males have an average SVL of 65.43 cm and the 10 largest females an average of 50.88 cm. There is no overlap in the SVLs of the adult males (59.0–73.0, range = 14 cm, SD = 3.71) and females (48.5–56.8, range = 8.3 cm, SD = 2.36). Total lengths reflect the same pattern as SVL lengths, since there is no appreciable difference in tail proportion in the two sexes (see below).

The largest individual examined was a male with a total length of 175.5 cm (SVL 73.0 cm, tail 102.5 cm); the largest female was 144.0 cm (SVL 56.8 cm, tail 87.2 cm).

Weight.— Weight distributions show essentially the same pattern as SVL and total lengths. Males are heavier (heaviest 20, $\bar{X}$ = 6.69 kg, SD = 1.30; all males N = 50, $\bar{X}$ = 4.15 kg, SD = 2.46), females considerably lighter (heaviest 20, $\bar{X}$ = 2.59 kg, SD = 0.41; all females N = 40, $\bar{X}$ = 2.09 kg, SD = 0.69) (fig. 2-1). The average weight of all individuals combined (N = 97) is 3.13 kg. The relationship between weight and total length is shown in figure 2-2 and can be expressed by the equation $Y = 1.47^{-8}X^{3.91}$. Figure 2-3 shows the distribution for weight classes in the population.

Hatchlings.—Hatchlings of *V. olivaceus* (with open umbilical scars) are rarely found because of their habits, habitat, and coloration. The few available (N = 3) have an overall range in total length of 34.5–41.8 cm ($\bar{X}$ total length 40.7 cm, $\bar{X}$ SVL of 16 cm, and $\bar{X}$ tail length of 24.7 cm). Local hatchlings of *V. salvator* (N = 23) have an average total length of 32.5 cm, SVL

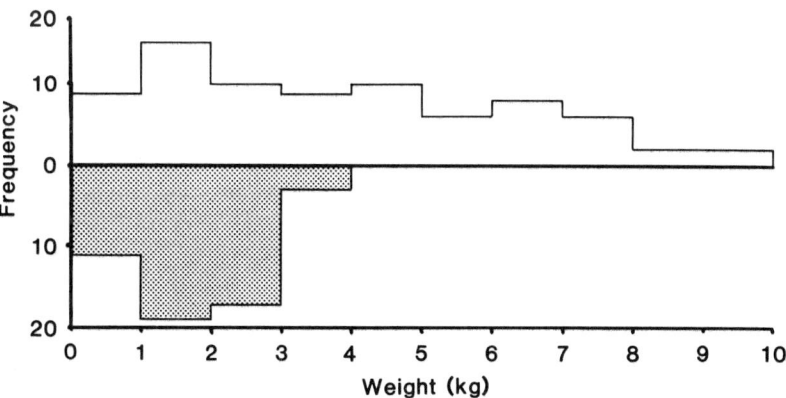

Fig. 2-1. Frequency distribution for weight classes of adult *V. olivaceus*, males (*above*) and females (*below*), showing the greater size range of males.

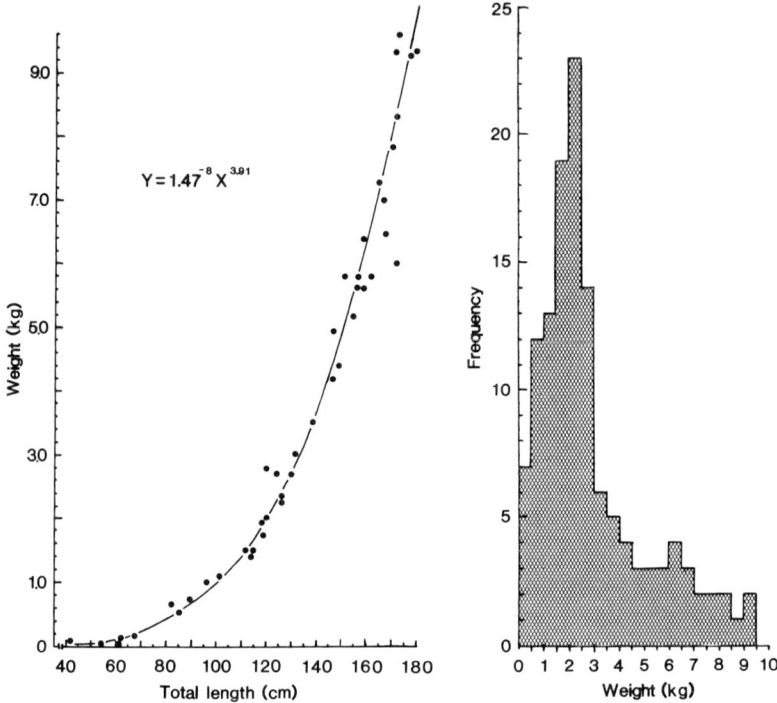

Fig. 2-2. Weight–total length curvilinear relationship in *V. olivaceus* ($N = 106$; some dots represent duplicate values).

Fig. 2-3. Population structure of *V. olivaceus* (both sexes), based on total weight, showing a modal value of 2.00–2.50 kg ($N = 123$, all populations included).

13.4 cm, and tail 19.1 cm. The average weight of the *V. olivaceus* hatchlings is 25 g and of *V. salvator* 21.4.

Proportions.—Body proportions usually change greatly throughout life in all varanid lizards (Auffenberg, 1980a, 1981a). Early workers sometimes made misidentifications due to unrecognized ontogenetic changes in head proportions, form and position of nostril, skull bones, and form of teeth. Thus Gray (1845) placed young and adult *V. b. bengalensis* in two separate genera (*Monitor* and *Uaranus*) on the basis of the form of the nostril. Werner (1893) was the first to recognize the great changes from juvenile to adult in monitor lizards. In general, juvenile monitors usually have more rounded nostrils than adults and more pointed teeth (if those of the adults are blunted). The head is larger in proportion to the body in adults, and the tail is longer in juveniles.

Table 2-1 provides data on the proportional size of component bodily parts of adult *V. olivaceus* and *V. salvator*. It shows that the body-tail proportions are about the same. It also suggests that *V. olivaceus* has a proportionally longer femoral part of the leg, while *V. salvator* has longer SVL, neck, and ulnar portion of the leg. However, none of these differences is statistically significant. When added together, they do indicate that *V. olivaceus* has a significantly longer hind leg (and is longer and heavier; see below) than *V. salvator*, placing the former within the category Mertens (1942) termed "long-legged" arboreal varanids (SVL/hind leg L = 2.01–2.27). *V. salvator* has somewhat shorter legs (but not nearly as short as some species, such as *V. flavescens, V. exanthematicus*, and others, in which the same ratio is 2.60–3.00). The tail of *V. olivaceus* is significantly greater in circumference at its base ($P \leq 0.05$) and is laterally more compressed than that of *V. salvator*.

Not only is the tail of *V. olivaceus* proportionally greater in cross section at its base than that of *V. salvator*, but the two species also differ in the cross-sectional shapes of their tails (see below).

In conclusion, both species are similar in respect to their body components. Both are rather slender species compared to the heavy-bodied, short-tailed

TABLE 2-1. Comparison of body component lengths of *V. olivaceus* and *V. salvator* (both are Caramoan populations).

Body part	*V. olivaceus*[a] $\bar{X} \pm SE$ (%)	(%)	OR (cm)	*V. salvator*[b] $\bar{X} \pm SE$ (%)	(%)	OR (cm)
Snout-vent length	40.6 ± 2.67		38.7–45.4	41.7 ± 3.01		38.5–46.3
Tail length	59.1 ± 4.78		57.5–61.4	58.2 ± 3.77		53.7–61.5
Body length	26.2 ± 1.68		17.5–28.2	26.4 ± 1.91		18.9–28.0
Neck length	10.7 ± 0.90		9.0–12.2	11.4 ± 0.79		10.8–12.3
Head length	4.0 ± 0.31		3.6– 4.3	4.0 ± 0.41		3.5– 4.6
Femoral length	6.1 ± 1.01		4.7– 7.0	5.6 ± 1.10		4.9– 6.5
Tibia length	6.5 ± 1.11		5.4– 7.8	6.2 ± 1.13		4.9– 8.0
Hind food length	8.4 ± 1.00		7.0– 9.3	8.5 ± 0.89		7.7– 9.5
Humeral length	4.0 ± 0.78		3.8– 5.4	4.5 ± 0.69		4.2– 5.0
Ulnar length	4.9 ± 0.81		4.6– 5.8	5.1 ± 0.91		4.4– 6.0
Front foot length	6.3 ± 1.15		4.3– 7.6	6.2 ± 1.00		6.0– 6.4
Tail base circumference	11.8 ± 2.45		9.3–13.2	10.0 ± 2.31		5.9–14.4
Greatest tongue length	12.3 ± 2.05		10.8–14.5	13.4 ± 1.89		12.6–15.3
Normal tongue length	7.9 ± 1.86		5.9– 9.8	7.6 ± 1.45		5.9–11.1

a. $N = 8$; OR total length = 61.3–131.0 cm.
b. $N = 11$; OR total length = 71.5–115.0 cm.

adults of *V. komodoensis*, or the short-necked, large-headed *V. exanthematicus*. They are not, however, as slender as some monitors, such as the apparently more arboreal *V. prasinus* and *V. rudicollis*.

Table 2-2 compares the weights of different anatomical parts of *V. olivaceus* and *V. salvator* (see chap. 1 for methods); some measurements are shown diagrammatically in figure 2-4. Total head weight represents a larger percentage of total weight of *V. olivaceus* than of *V. salvator*. Relative neck weight is not significantly different in the two species (though the mean neck length is greater in the latter). The pectoral and particularly the pelvic girdles of *V. olivaceus* are significantly heavier than those of *V. salvator*, while the body of *V. olivaceus* is somewhat lighter proportionally. While front leg weight is greater in *V. salvator*, it is the hind leg that is heavier in *V. olivaceus*. The tail of *V. salvator* is considerably heavier proportionally than that of *V. olivaceus*, though this may be due to an inverse correlation between tail length and over-

TABLE 2-2. Percent of total weights of different anatomical parts of *V. olivaceus* and *V. salvator* ($N = 3$ adults of each).

	V. olivaceus		V. salvator	
	(g)	(%)	(g)	(%)
Total weight	2031.0		511.0	
Total muscle, bone, and skin	1606.8		444.7	
Muscle		56.5		63.3
Bone		10.2		10.3
Skin		12.4		13.6
Viscera		11.1		11.5
Body fat		9.8		1.3
Head		13.1		8.2
Neck		6.1		4.0
Pectoral		4.0		3.0
Body		27.4		30.3
Pelvic		5.9		3.8
Tail		17.5		23.8
All 4 legs		24.8		26.8
Front legs		6.4		9.8
Humeral		2.5		5.1
Ulnar		2.2		3.1
Manes		1.7		1.6
Hind legs		17.8		17.1
Femoral		11.1		8.7
Tibial		4.1		5.0
Pes		2.3		3.3

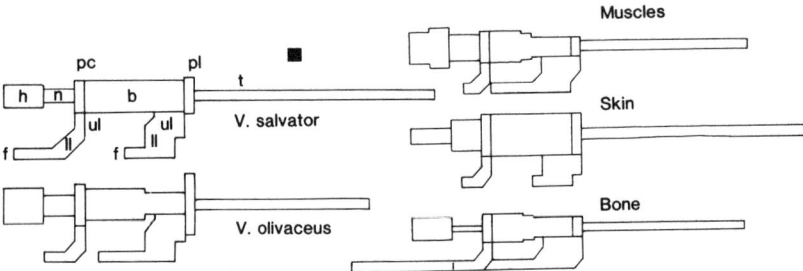

Fig. 2-4. Proportional weight of different parts of the body in *V. olivaceus* and *V. salvator*. *Left:* Two figures show total weights of parts considered. *Right:* three figures show proportional weights of muscles, skin, and bone of *V. olivaceus* only. Abbreviations: *b*, body; *f*, feet; *h*, head; *ll*, lower leg (front and back separate); *n*, neck; *pc*, pectoral girdle; *pl*, pelvic girdle; *ul*, upper leg (front and back separate). Area of each box equals the proportion of each body part. The black square represents 1% of the total weight of the individuals dissected (see tables 2-2 and 2-3 for details).

all size (as has been indicated for other species studied [Mertens, 1942; Auffenberg, 1981a]), coupled with the smaller total weight (TW) of local *V. salvator*.

Comparable weights from a large *Iguana iguana* (TW = 2070 g) are most like those of *V. olivaceus* or intermediate between those of *V. olivaceus* and *V. salvator*. The chief divergences from *V. olivaceus* are the heavier body, neck, and head of *Iguana iguana* (about 67.7% of TW, compared to 56.5% for *V. olivaceus*) and its lighter limbs (13.5% of TW, compared to 24.8% for *V. olivaceus*). In general, *Iguana iguana* seems to spend more time in the trees than *V. olivaceus* (see Rand, 1978). In mammals, arboreality involves less use of muscles and a general reduction in the total scale of motor activity, particularly in the limbs, resulting in a lower total limb weight compared with that of terrestrial forms. Thus the difference in limb weights of *Iguana iguana* and *V. olivaceus* are as expected.

Figure 2-4 and table 2-3 compare the proportional wet weights of the skin, muscles, and bones associated within different functional units of the body of *V. olivaceus*. These data show that the head is comprised of a larger proportional weight of muscle than of skin; bone of the head is of intermediate weight. In the neck and the body (pelvic and pectoral girdles excluded) the skin weight is greatest in proportion to total weight. Skin and muscle are approximately equal in their relative contribution to the total weight of the front legs, but bone weight is proportionally much greater. Muscles of the hind limbs weigh less than the bones but considerably more than the skin. The weights of the bones and muscles of the tail are nearly identical, but the skin constitutes a greater part of the total.

TABLE 2-3. Percent of total weight of muscle, bone, and skin in different anatomical parts of *V. olivaceus*.

	Muscle[a]	Bone[b]	Skin[c]
Head	14.0	9.8	6.1
Neck	7.8	2.5	9.3
Pectoral	5.0		
Body	24.9	25.8	40.0
Pelvic	7.6		
Tail	16.7	16.1	22.5
All 4 legs	22.9	38.0	22.2
2 front legs	4.8	16.1	5.2
2 humeral	2.3	4.1	2.3
2 ulnar	2.1	2.7	1.6
2 manes	0.4	9.2	1.6
2 hind legs	18.1	21.9	12.8
2 femoral	13.1	5.5	6.2
2 tibial	4.5	3.2	2.2
2 pes	0.6	3.9	2.0

a. Total weight = 1146.9 g.
b. Total weight = 206.9 g.
c. Total weight = 251.0 g.

A comparison between the two species casts a somewhat different light on the relative importance of these parts. The data show that the shoulder and pelvic girdle weights are greater in *V. olivaceus*. The heavy musculature of the pelvic girdle is associated with the proportionally heavy weight of the hind leg (mostly muscular). The differences illustrated are undoubtedly related to differences in their behavior and locomotion. While these seem logical, functional explanations are not obvious. The heavy neck of *V. olivaceus* is certainly related to the necessity of holding up a proportionally heavier head, which results from extreme development of bone and muscle needed to prepare some foods before swallowing (see below).

Mertens (1942) noted correctly that the cross section of the tail of *V. olivaceus* is similar to that of *V. salvator*; it is laterally compressed and provided with a double row of strongly keeled scales dorsally. Closer study reveals that the cross sections of the tails in these two species differ in several regards. The study of these differences is based on examination of cross sections of tails made at one, three, and five head lengths behind the vent. These sections (fig. 2-5) show that there are two major differences in the shapes of the tails of these two species: the tail of *V. olivaceus* is even more compressed that that of *V. salvator*, for the latter is more flattened on the ventral surface, and the dorsal keel of *V. salvator* is comprised of a strip of dense connective tissue that is missing in *V. olivaceus*. Thus the apparent lateral compression of *V. salvator* seems due more to a dorsal addition than to simple compression, as found in *V. olivaceus*.

External Morphology and Color   23

It is usually assumed that tail compression is an adaptation to a semiaquatic life, but this interpretation is clearly incorrect for *V. olivaceus*, because it rarely swims. Its tail compression is related to the function of both ventral and lateral surfaces of the tail during climbing. The strongly keeled scales found on the ventral tail surfaces of both *V. salvator* and *V. olivaceus* are used as frictional surfaces for climbing vertical trunks, etc. In this case the tail is usually held in a sigmoid position with the ventral tail surface pressed against the substrate (see chap. 7). The tail is also used as a prop in climbing, with the flat, *lateral* surface pushing against the substrate (see chap. 7).

Mertens (1942) emphasized the SVL/tail length ratio in distinguishing monitor species. While it is useful in certain species, there is clearly an ontogenetic change in proportion in many others (considerable in *V. komodoensis*; see Auffenberg, 1981a). As in other species, the tail in *V. olivaceus* is longest (relatively) in juveniles. Even among adults the variation in this ratio is 1:1.18–1.62 ($N = 86$), with the difference of 0.46 equaling 30% of the entire range in ratios for all species (Mertens, 1942).

Claws and feet.—All monitor lizards do a great deal of climbing and digging. Thus the shape and mechanics of the feet and claws are important. Mertens (1942) noted that arboreal species of *Varanus*, such as *V. prasinus* and *V.*

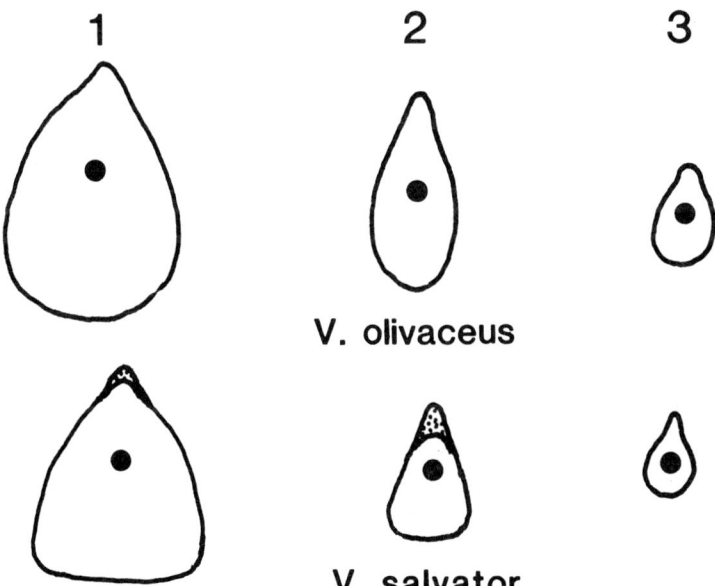

Fig. 2-5. Tail sections of *V. olivaceus* and *V. salvator*, showing shape differences at (*1*) one head length behind vent, (*2*) three head lengths behind vent, and (*3*) five head lengths behind vent. Dots represent position of vertebral column.

rudicollis, have shorter toes, with larger, more curved claws. These he called "climbing specializations." Though deserticolous species (*V. griseus* and *V. eremius*) also have short toes, their claws are small and straight. The claws of *V. olivaceus* are also large and curved. Those of *V. salvator* are proportionally longer, less curved, and with finer tips. The smallest claw of the butaan is that on toe V of the hind foot (2.6 times into toe length), the longest of that foot on toe I (1.3 × toe length). The claws of the front feet are generally longer, ranging from 1.0 times the length of toe I to 1.1 times its length on toes II and V. The claw depth at the base is 0.75 times its length (*V. salvator* 0.50, showing its generally more slender nature).

Though the large, curled claws of the butaan are logically expected to be important in clambering over rocks and tree trunks, *V. salvator* does the same, and apparently with similar ease. The large hooked claws of *V. olivaceus* may be more important in moving about in the terminal branches of trees (which *V. salvator* does infrequently). Several varanids are known to turn their feet posteriorly (with normal dorso-ventral orientation) when descending through terminal twigs and leaves, hooking their claws into the vegetation to slow or stop their descent (Auffenberg, 1981a; Auffenberg, field notes), including *V. olivaceus*, which seems particularly adept at it. This use is possible not only because of the claw curvature but, in the case of *V. olivaceus*, because the claws can be flexed more completely than those of *V. salvator*. Each of the claws of *V. olivaceus* can be turned at a right angle at its junction with the penultimate phalange. Furthermore, there is a mechanism by which this position is maintained with a sliding lock type of anatomical arrangement. When the *flexor profundus* tendon is retracted (see Landsmeer, 1981, for digital morphology of *Varanus*), the claw is sharply flexed, rotating around the distal end of the penultimate phalange. Further tension pulls the claw proximally, changing its position on the articular surface of the penultimate phalange to such a degree that the claw is locked into a flexed position by a loss of contact between articular surfaces. Fascial connections keep the *flexor profundus* tendon from moving laterally. These mechanisms make it possible for large adults (up to 10 kg) to hang from a single claw.

Ear and narial openings.—Though potentially useful, there are no studies of variation in the shape of ear openings of varanids. The ear opening of *V. olivaceus* is small compared to that of some species (e.g., *V. dumerilii*, *V. salvator*) but clearly larger than that of some other species (e.g., *V. exanthematicus*). In both size and shape it seems most like *V. komodoensis*, *V. bengalensis*, and *V. rudicollis*. There is no clear association between ear opening size and shape and current phylogenetic arrangement, or with ecology or behavior. In the case of the nostrils there is some evidence of correlation between shape, size, and foraging technique.

While the external narial opening of *V. olivaceus* differs in shape from the other species, such differences are minor compared with species in which the opening is slitlike. Internally the differences are greater. In *V. bengalensis* the choanate passageway bends anteriorly just inside the external nares, passing anteriorly to the premaxillary-maxillary suture, where it makes a sharp U-bend, passes the external narial opening again, and then reaches the internal nares. Mertens (1942) stated that this arrangement is probably found in all varanids with slitlike external nares. However, the narial passageway of the butaan is different, in spite of the fact that this species also has a slitlike opening. In this species the internal passage is quite short and not particularly tubelike. From the opening the passageway is directed ventroposteriorly to the internal nares. Immediately inside the external nares there is a small pocket beneath a flat sagittally placed shelf, with a similar but larger sagittally placed shelf deeper and more posterior in position. Near the posterior edge of the narial opening in the skull there is a vertical wall of tissue that forms the posterior limit of the narial chamber. On the basis of Wegner's (1922) figure, a similar system of baffles occurs in *V. bengalensis*. Thus the major difference between these two species is the absence of a long, folded nasal tube in *V. olivaceus*. A thorough investigation of both the macro- and microstructure of the narial tube in all of the species with slitlike exterior nares would be particularly enlightening as an extension of recent analyses of phylogeny based on karyotype and immunological studies.

To my knowledge, all slit-nosed monitors engage in much "dabbling" behavior when foraging, using their snout to push and turn over debris much as a duck uses its bill. The fact that the nasal passage is so simple suggests that to the butaan the position of the external nares is the only (or major) matter of functional importance. However, the complex nasal passage of *V. bengalensis*, and presumably other species, suggests that the long, tubelike passage, sharply bent to accommodate its length in a shorter snout, is somehow of importance in another context, as yet unknown.

Scalation.—The following account of head scalation supplements that of Mertens (1942), which was based on the only specimen known at that time—the holotype. The data below were obtained from 106 butaans, most of which originated in the vicinity of Caramoan.

The scales of the dorsal head surface (fig. 2-6) are larger than those on the lateral surface and largest in the interorbital region, where there are usually 10–21 ($\bar{X} = 15.1$) scales between the inner edges of the supraorbital scales. Although the holotype has supraoculars nearly as broad as long, there is considerable variation among the additional material, most having supraoculars about one-third as broad as long. The large supraoculars are rarely divided (one to all) so that the remaining small "supraoculars" are no larger than the surrounding scales. The large supraoculars vary from 0 to 14 (mode 3) on each side (fig. 2-7). The number of scales from the eye to the nostril varies

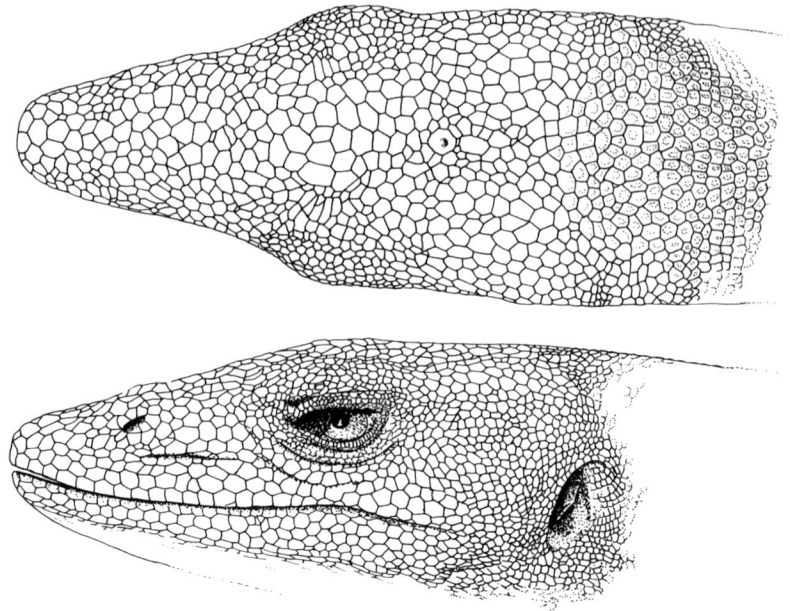

Fig. 2-6. Head scalation of adult *V. olivaceus,* top and side, from photo; individual was released.

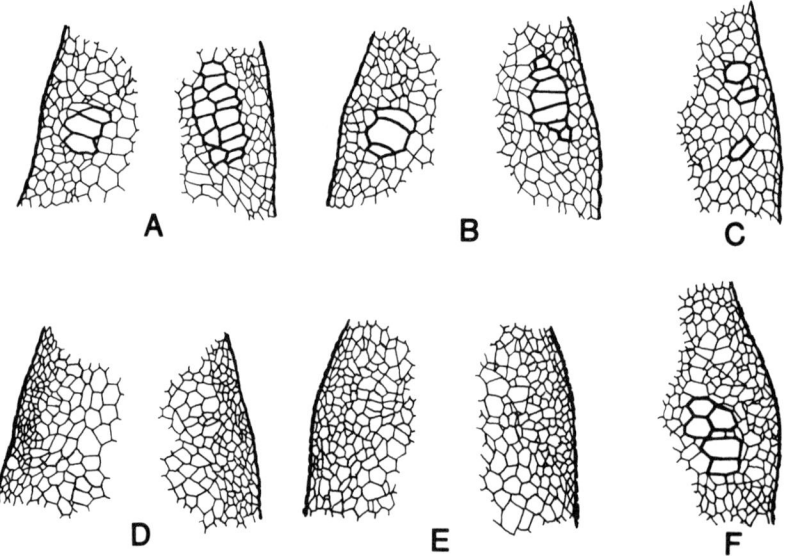

Fig. 2-7. Variation in supraocular scales of *V. olivaceus;* heavy borders emphasize largest scales in supraocular groups. *A and B,* left and right supraorbital area of wild individual, Caramoan; *C,* right supraorbital area of wild individual, Caramoan; *D, V. grayi* holotype, BMNH 1946.8.30.98, left and right; *E,* left and right sides, wild individual, Caramoan; *F,* right side, wild individual, Caramoan (see also fig. 2-6). All wild individuals are drawn from slides.

from 9 to 13 (usually 10), supralabials 27–34 ($\bar{X} = 28.8$). The scale rows from the angle of the mouth on one side to the angle on the other are 50–61 ($\bar{X} = 58.4$). The lamellae under the fourth toe vary from 27 to 31 ($\bar{X} = 29.2$). The anterior rows of nuchal scales are smooth and about as large as the posterior scales of the head and are as wide as long. Those scales behind the ninth to tenth rows are noticeably smaller and longer than wide. The dorsal body scales are smaller yet and strongly keeled. Those on the upper limb surfaces are generally larger and strongly keeled. The scales of the dorsal part of the front feet are smooth; those of the hind feet are usually keeled. Ventral scales are feebly keeled. Scales from the gular fold to the anterior edge of the insertion of the hind limb vary from 101 to 121 ($\bar{X} = 109.0$). Midbody scale rows vary from 169 to 214 ($\bar{X} = 186.1$). There are no preanal pores present in either sex.

Color and Pattern

In general, the butaan is a greenish gray lizard, its neck, back, and tail crossed by a number of darker transverse bands (pl. 1). Ontogenetic color and pattern changes are slight compared to changes in other varanid species. There is little significant variation correlated with the geography (specimens examined on Luzon from as far north as Rizal to Sorsogon provinces, as well as the island of Catanduanes). Nor is there is any difference in color or pattern that can be correlated with sex or season, except that the largest males are often suffused with melanin dorsally and frequently so badly scarred (from male-male combat; see chap. 13) that patterns are sometimes difficult to discern—particularly in the scapular region. Figure 2-8 shows the normal extremes of patterns found in adults within the Caramoan population. A more detailed description of color and pattern in *live* individuals follows.

The ventral surfaces of the head, neck, and tail are generally grayish, grayish green, or yellow-gray. The throat is marked with 3–4 longitudinal brownish black to black stripes, 3–7 scales wide. The ventral body surface is lightest medially—rarely with any obvious markings. When present, markings are faint, usually dark gray, becoming more distinct laterally. In shape they tend to form juxtaposed, somewhat circular figures, forming a reticular pattern, or they may form faint extensions of the dorsal body bands. The underside of the tail is generally faintly banded with gray-brown, those bands being extensions to the dorsolateral patterns.

The neck is usually darkest dorsomedially, provided with brown to (usually) black speckles and dots, covering 2–6 scales each. These are generally most numerous immediately behind the head, becoming more scattered posterolaterally. Three darker V-shaped bands cross the neck, with their open ends facing anteriorly (fig. 2-8). Ventrally some or all of these are confluent with narrow longitudinal stripes of the ventral part of the throat. The edges of the

PLATE 1. A half-grown *V. olivaceus*, showing the bright color pattern of subadults. The light ground color will darken with maturity.

V-bands are often darker (brownish black to black) than the centers (brownish to greenish gray). In extreme examples the edges may remain completely joined and rather widely separated from one another, allowing the ground color to fill completely the space between (fig. 2-8).

On the body, there are five bands that vary from nearly solid brownish black or black to dark-edged with much lighter centers. The anterior bands have a greater tendency to be less solid than the more posterior ones, which tend to be wider and more diffuse along their edges. The lateral edges of the scapular band are always divided laterally, with the one anterior arm extending anterior to the insertion of the front limb, while the posterior arm extends posterior to the limb. The anterior and posterior parts of the band may be variously divided along their entire lengths, and each of these separated parts may have lighter centers, forming a rather vivid pattern extreme that I term "open type" (fig. 2-8A). More often the anterior and posterior parts of the scapular band are each provided with a middle row of lighter dots and dashes (fig. 2-8B), or even completely closed and the two parts fused to one another along their entire transverse lengths (fig. 2-8C, "closed type"). Together each band (and space between, when band divided) is about 15–19 scales wide. The remaining four similar body bands are usually fused along most or all of their median lengths, though the most posteriorly located bands are less obviously separated than the more anterior ones. Laterally these bands are not split, as is the scapular band, but are narrower ventrolaterally, extending onto the larger ventral scales as variable, rather reticulate, darker patterns. The most posterior band is the least well defined, with the anterior ones generally more clearly separated from the ground color and its lateral brownish gray reticular pattern.

Behind the pelvis the bands become progressively wider and less distinct on the tail. There are usually 12 bands (11–12) on the tail, becoming indistinct (with the addition of progressively more light and dark spots). A lighter central area is common in each tail band, though it is somewhat difficult to see.

The limbs are generally darker than the rest of the body, being grayish to blackish brown, with lighter spots liberally sprinkled at random over the entire upper surface of all four limbs. The feet are dark brown to black, with nearly every scale having a gray, green, or yellow spot, often fused into larger dots; each dorsal scale of the fingers is similarly spotted. The claws are uniformly black.

The head is often the lightest part of the entire animal, particularly anteriorly. Here the lightest hues are yellows and grays, particularly near the external nares and along the lips of both the upper and lower jaws. There are no obvious dark marks on the lateral surfaces of either the head or the lower jaws. In general, the top of the head is darker but usually dotted and irregularly spotted with darker and lighter hues ranging from greenish yellow to yellow, yellowish gray, gray, grayish brown, and nearly black. The posterior part of

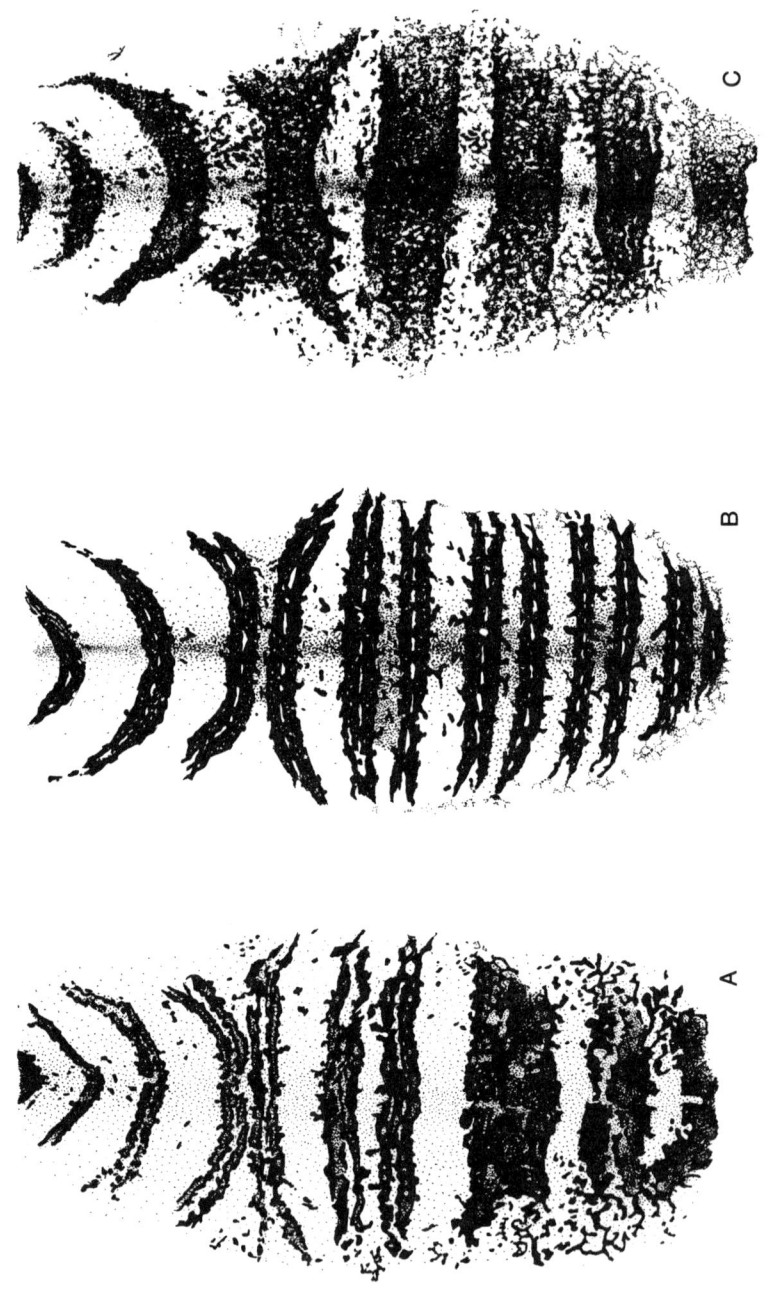

Fig. 2-8. Extreme dorsal body color patterns in *V. olivaceus* from the Caramoan area: *A*, uncatalogued, released adult male, "open type"; *B*, same, more "open"; *C*, UF 55024, adult male, "closed type." See also fig. 14-1.

Fig. 2-9. Color pattern of near hatchling (UF 50915); SVL 15.5 cm.

the head is darkest, especially the temporal area (mistakenly overemphasized in the illustration in Auffenberg, 1976) which is more or less uniformly grayish brown or greenish gray. Bright yellow (sometimes lemon yellow) appears only on the head, most frequently around the nostrils, in front of the eyes, on the dorsal surface of the most anterior part of the snout, and along the anterior parts of the lips of the lower jaws. The chin is usually uniformly grayish yellow to greenish yellow, without any obvious markings, though small dots and dark smudges may occur.

In general, the juveniles have the coloring of adults; there are fewer dark spots and reticulations between the bands, so the bands of young individuals are better defined, including those of the tail (fig. 2-9). In addition, there is more green in the ground color over the entire body so that the general impression is of a bright green lizard with dark bands. The head has no yellow markings and tends to be more or less uniformly green on the sides and top. On the other hand, adults give the impression of a greenish brown to grayish green lizard, liberally speckled with both light and dark dots and having dark bands. The eyes of both juveniles and adults are medium brown.

The color and pattern of this species blends well with both the rocks and tree trunks with which it is often associated. The black, yellow, and cream spots over the body and limbs are remarkably similar to the large lenticles frequently found on the smooth bark of the many forest trees. The darker hues of the body and neck blend with the shadows and darker lichen and liverwort patches on the larger tree trunks and limbs; the subtle gray-greens of the ground color are reflected in colors of the bark of forest trees, particularly near the base. The same colors also match the rocky substrates of the environment, particularly the limestones on which mosses, lichens, liverworts, and the natural colors of the weathered rocks are similar to the hues and intensities of those found on the dorsal surfaces of the lizards. Thus the colors and patterns of the butaan produce a very disruptive camouflage. Even after one is located in a particular tree (by radio telemetry), this pattern, plus the tendency for the species to remain still for such long periods of time after discovery, makes it difficult to see (as Henne an Rhyne [1903] reported for a treed *V. salvator* during a hunt with dogs on Sumatra). Other largely arboreal varanids of tropical evergreen forests frequently have similar colors and patterns (green, olive-green, or brown, with darker transverse bands, as in *V. rudicollis, V. prasinus, V. dumerilii*). Even *V. bengalensis* from the more humid and more heavily forested part of the range (= *V. b. nebulosus*) is more obviously banded and more gray than populations of the nominal subspecies in the arid western part of the range.

The most obvious, and possibly nondisruptive, color and pattern is the bright yellow found only on the heads of adults. The bright head color may serve as an important visual releaser in social encounters in the dark gray-green forests in which this species lives, but as yet there is no evidence that head movements are important in social contexts other than in displaying social status at close distances (usually less than 1 m; see chap. 13). Bright head colors are found in other evergreen forest species (*V. karlschmidti*, for example) but are sometimes restricted to or more obvious in juveniles (*V. dumerilli*, for example).

Though no seasonal color changes have been demonstrated for *V. olivaceus*, they are known for males of both *V. griseus* and *V. flavescens*, in which the colors tend to become more intense and vivid during the rainy season and may somehow be associated with courtship (Mertens, 1942). Thilenius (1897) reported color changes in *V. griseus* associated with temperature changes as well, as did Muller (1905) and Werner (1908) for *V. exanthematicus* and Lesson (1830) for *V. indicus*.

3

Internal Morphology

The Head

THE cranial anatomy of varanid lizards is fairly well known, some studies having been done as long ago as 1812 (Camper). However, for *V. olivaceus* the knowledge is incomplete, for only one skull has been described (Mertens, 1942). In this work it has been shown to be that of an adult male. The skull anatomy of juveniles and of adult females has never been studied, and there is no description of the muscles. The following notes on head anatomy of the butaan are not intended as an exhaustive description but as a synopsis of major cranial features. They are further intended to provide background for understanding the feeding adaptations that characterize this species. The features of the skull important in determining evolutionary relationships are discussed in chapter 14.

Skull.—A total of 24 skulls, representing an equal number of males and females, was selected for study. The measurements taken were the same as those made by Mertens (1942) to facilitate comparison with the species for which he gives skull measurement data. Figure 3-1 shows the measurements taken on all skulls and used in the following ratios (numbers refer to points on the skull in fig. 3-1):

Head Measurement Ratios Defined

$$\text{I:} \quad \frac{\text{Snout tip–posterior edge of parietal at midline } (1-3) \times 100}{\text{Snout tip–occipital condyle } (1-2)}$$

II: $\dfrac{\text{Squamosal width (5–5)} \times 100}{\text{Snout tip–occipital condyle (1–2)}}$

III: $\dfrac{\text{Parietal–basisphenoid (7–8)} \times 100}{\text{Squamosal width (5–5)}}$

IV: $\dfrac{\text{Snout tip–posterior edge of nasals at midline (1–4)} \times 100}{\text{Snout tip–occipital condyle (1–2)}}$

V: $\dfrac{\text{Maxillary length (9–10)} \times 100}{\text{Snout tip–occipital condyle (1–2)}}$

VI: $\dfrac{\text{Maxillary height (11–12)} \times 100}{\text{Maxillary length (9–10)}}$

VII: $\dfrac{\text{Least width parietal (14–14)} \times 100}{\text{Greatest width parietal (13–13)}}$

VIII: $\dfrac{\text{Mandibular length (15–16)} \times 100}{\text{Snout tip–occipital condyle (1–2)}}$

IX: $\dfrac{\text{Dentary length (15–17)} \times 100}{\text{Mandibular length (15–16)}}$

X: $\dfrac{\text{Dentary height (18–19)} \times 100}{\text{Dentary length (15–17)}}$

XI: $\dfrac{\text{Dentary height at fifth tooth (19–19)} \times 100}{\text{Dentary length (15–17)}}$

Note: Numerals in parentheses are mapped on figure 3.1.

In general, the skull of *V. olivaceus* is similar to that of other varanids, though unquestionably rather specialized in certain anatomical features—apparently all related to feeding technique. In form it is between the slender skull of *V. rudicollis* and the wide, spoon-shaped skull of adult *V. komodoensis*. The general impression is of a lightly built skull, with numerous fenestrae. However, closer inspection shows that *V. olivaceus* is one of the few species (i.e., *V. komodoensis*, *V. niloticus*, and *V. exanthematicus*) in which the bones of the skull exhibit unusual thickening and strengthening. Thus individual bones tend to be more massive than those of other varanids, in spite of the

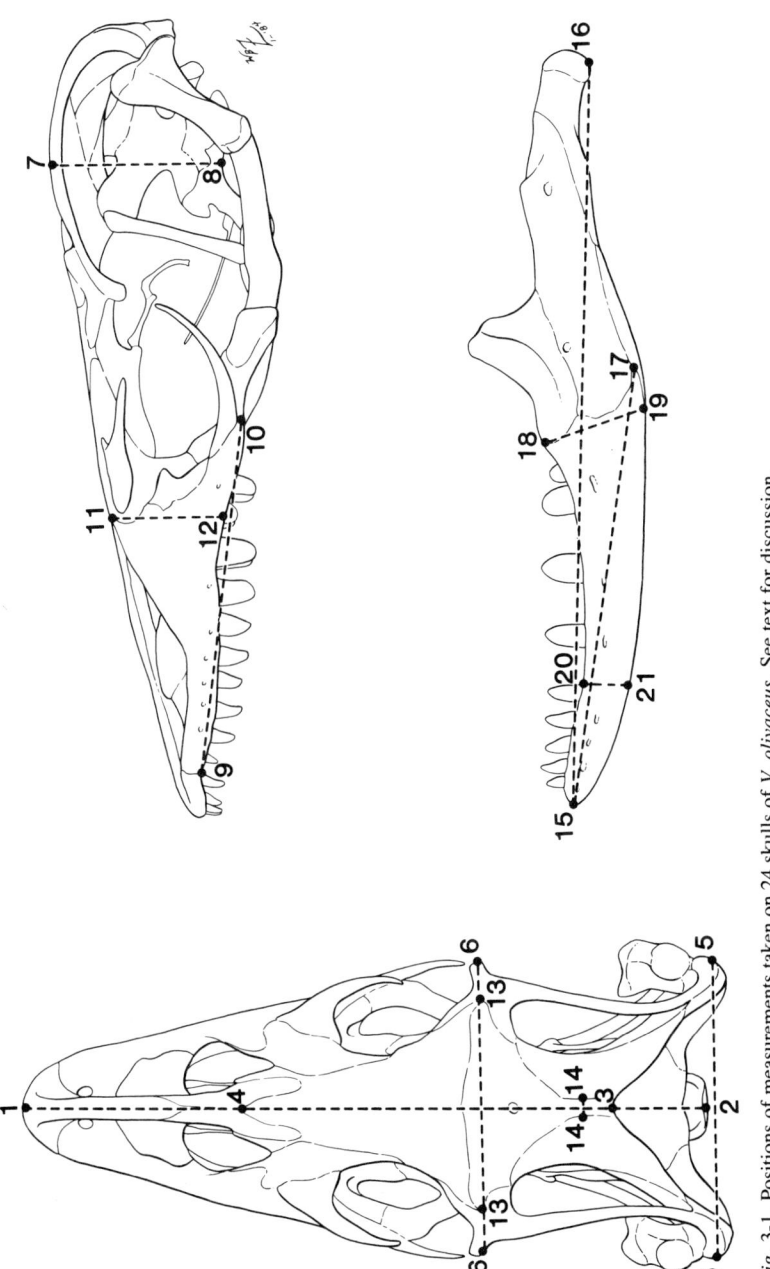

Fig. 3-1. Positions of measurements taken on 24 skulls of *V. olivaceus*. See text for discussion.

TABLE 3-1. Cranial indices of *V. olivaceus* ($N = 12$ each females and males).

Index[a]	Females			Males			Males and females	
	OR (cm)	X̄ (cm)	SD	OR (cm)	X̄ (cm)	SD	X̄ (cm)	SD
Skull length	87.5– 95.8	89.5	3.2	88.5– 98.3	92.3	4.1	91.2	3.9
Length-width	38.9– 52.8	41.9	5.4	42.9– 56.3	44.3	1.6	43.1	4.0
Length-height	21.8– 25.1	23.9	1.2	25.0– 29.2	27.0	1.7	25.5	2.2
Width-height	41.9– 63.8	58.0	8.1	58.0– 63.1	61.0	1.3	59.5	5.8
Snout length	36.2– 39.2	37.9	1.2	36.9– 43.9	39.6	3.4	41.8	10.2
Maxillary length	44.8– 57.2	50.1	4.5	47.0– 54.4	49.6	2.8	49.8	3.5
Maxillary length-height	26.0– 37.8	30.9	4.4	32.7– 41.5	37.6	3.3	34.5	5.1
Parietal width	12.5– 22.8	17.0	4.0	4.8– 10.3	6.8	1.9	11.9	6.1
Mandibular length	99.3–101.0	100.0	10.6	99.2–104.2	102.4	9.0	100.8	1.8
Dentary length	57.5– 61.1	59.7	1.7	58.9– 62.4	61.0	1.4	60.3	1.6

a. See text and figure 3-1 for definitions and measurements.

large fenestrae. The skull also has certain parts ossified that are cartilaginous in other species, such as the nasal capsule.

The skull-length-to-skull-width indices (index I, table 3-1) of most species are similar and can be classed as "intermediate." This grouping includes *V. olivaceus* (skull-length-to-skull-width index I, 43.1), whose skull is narrower than wide-headed forms such as *V. flavescens* and *V. exanthematicus* (index I, 48.9–49.0) and broader than narrow-headed forms such as *V. prasinus* (index I, 33.8–35.4). In general, the widest part of the skull is at the outer squamosal width (5–5, fig. 3-1); the width across the postorbitals (6–6) is about the same. Thus the outer edges of the cranial table in the butaan are somewhat square. It is from the eyes forward that the skull is narrowed (fig. 3-2).

Table 3-1 provides those cranial indices and measurements believed important in comparisons with other species. Some are clearly related to overall size, explaining some of the sexual dimorphism noted. The most obvious of these is the parietal width (index VII), which becomes progressively less with increasing lizard size because as the masseter muscle complex becomes larger it creeps up onto the dorsal surface of the parietal (see below).

The lower jaw of the butaan is noticeably increased in vertical dimension compared to that of all other varanids except *V. niloticus* and *V. exanthematicus*. The cranial portion itself in *V. olivaceus* is not as high as in *V. niloticus* and *V. exanthematicus,* so that the cranial part of the lower jaw of *V. olivaceus* is intermediate within the genus.

Among the various skull proportions showing important differences among the monitor lizard species is the shape of the snout (i.e., straight along the dorsal surface, curved, or even with an obtuse angle, as well as either long or short). Within varanids, snout lengths vary from very short (*V. brachyurus,* index V, 27.57) to long (*V. rudicollis,* index V, 50.29, and *V. komodoensis,* index V, 54.08). *V. olivaceus* has an intermediate-length skull (index V, 41.8).

The position and shape of the nasal fenestrae are also of considerable taxonomic and functional importance. Each is bordered by the premaxillary nasal process, the nasal, frontal, prefrontal, maxillary, and the posterior edge of the septomaxillary. They are longest in *V. salvator* and shortest in species of the subgenera *Ondatria* and *Empagusia.* Among varanids, fenestra length is not correlated with snout length (i.e., the long-snouted *V. rudicollis* has short openings). In *V. olivaceus* they are shorter. Their pointed posterior margin is similar to those in most other *Varanus* species (see Mertens, 1942).

Supratemporal fenestra size correlates with age in most varanids (Mertens, 1942), as it does in *V. olivaceus.* Their size in the butaan is probably greater than in any other varanid, with the increase clearly demonstrated by the changing ratio of the narrowest part of the parietal plate (14–14, fig. 3-1). Narrowing of the parietal plate at this point is due to the proportional increase in the size of the masseter muscle complex (see below).

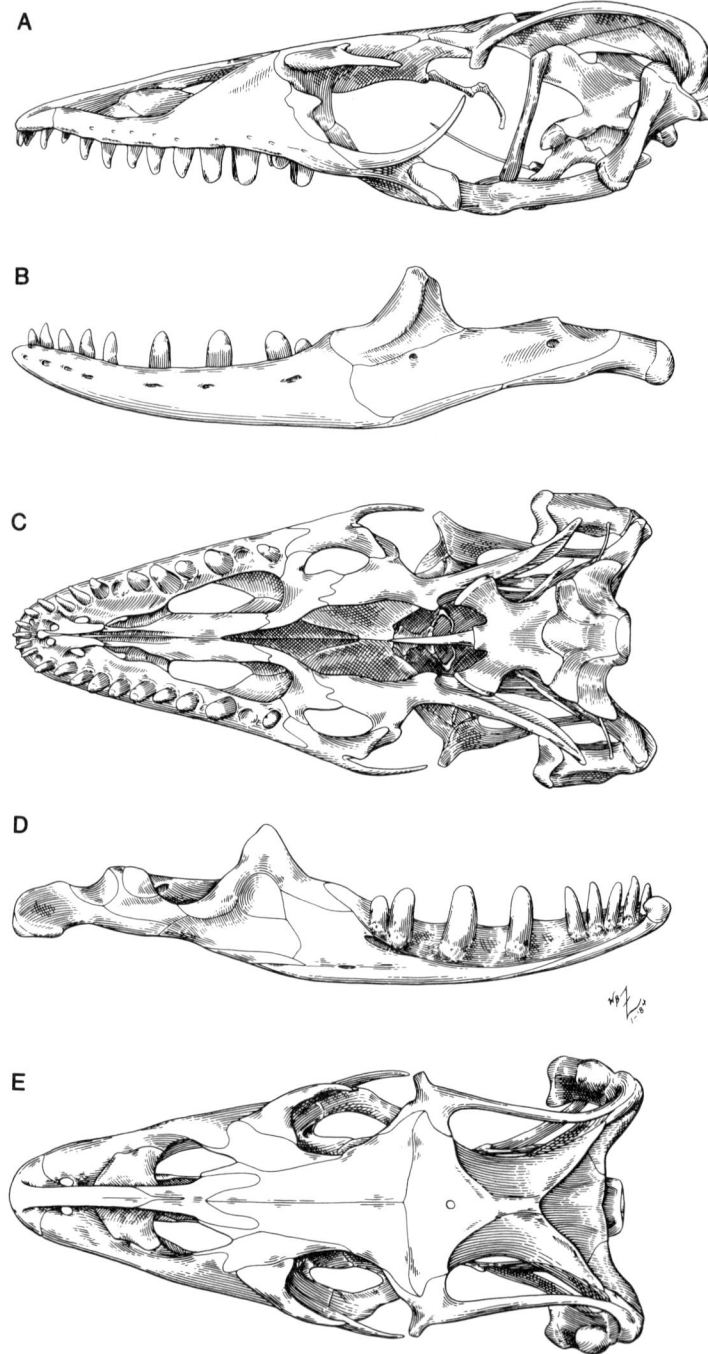

Fig. 3-2. Views of skull and left lower jaw of adult male *V. olivaceus* (UF 56586). *A*, lateral skull; *B*, exterior lower jaw; *C*, ventral skull; *D*, interior lower jaw; *E*, dorsal skull.

Mertens (1942) stated that though the skull of the butaan is similar to that of *V. niloticus*, closer scrutiny suggests that this is due to convergence (short, rounded *posterior* teeth, heavy jaw elements, narrowed parietal plate, with a median keel on the *posterior* cranial surface). *V. olivaceus* differs from *V. niloticus* in having posteriorly pointed (not rounded) nasal fenestrae, lower posttemporal fenestrae, septomaxilla anterior edge not covering palatine fenestra, more posteriorly placed dorsal foramina of the premaxillary, more posterior position of the vomer, more or less triangular shape of the parietal plate, much longer frontonasal suture, stronger hypsiprosopic maxillary with an edge that is considerably bent, long and well-developed postfrontal parietal processes, nonconvex supratemporal process of the parietal, less ventrally bent quadrate process of the pterygoids, less obviously bent basipterygoid process of the basisphenoid, smaller splenioangular tubercle, very short dentary, and relatively tall but few maxillary and mandibular teeth (10 maxillary v. 10–11 in *V. niloticus*, and 13 mandibular teeth v. 11–12 in *V. niloticus*). The mandibular-dentary length index of *V. olivaceus* ($\bar{X} = 60.3$) is not approached by any other varanid species, except the juveniles of *V. niloticus* (60.6). Though the similarity between *V. olivaceus* and *V. salvator* is thought due to convergence, Mertens (1942) also believed that both likely derive from a *salvator*-like ancestor. Their close relationship is suggested by similarities in the shape of the septomaxilla, which he suggests is of considerable taxonomic importance. The palatine shape is also similar, as is that of the prootic and the posttemporal fenestra, and the shape and position of the dorsal premaxillary foramina. These similarities become even more pronounced when the skulls of juvenile *V. olivaceus* are included in the analysis. Thus the only significant differences between *V. salvator* and *V. olivaceus* are parietal shape, general bone robustness, construction of the tooth-bearing bones, and the teeth themselves.

Skull musculature.—Of prime importance in the feeding biology of *V. olivaceus* is the arrangement and size of the muscles of the skull—particularly those of the masseter complex responsible for closing the jaw. Though the head muscles of varanids are reasonably well known, only a few species have been studied in detail. Those of *V. olivaceus* have not been studied previously. Furthermore, they differ from those of other varanids in several important *functional* details, a short description of which follows (based on three adults, with no appreciable differences between them).

On removing the skin of the head, the *levator annulis oris* (MLAO, fig. 3-3) is the first muscle encountered; it covers the lateral temporal area. As in *V. bengalensis* and *V. varius* (Lakjer, 1926; Haas, 1973), this thin muscle inserts on a large tendinous sheet (lateral rictal plate) that arises from the coronoid process of the lower jaw. The muscle originates from a wide area across the entire upper temporal arch (fig. 3-3A) from four narrow tendons and a broader connection to the quadrate. There is no strong interlacing of the fibers

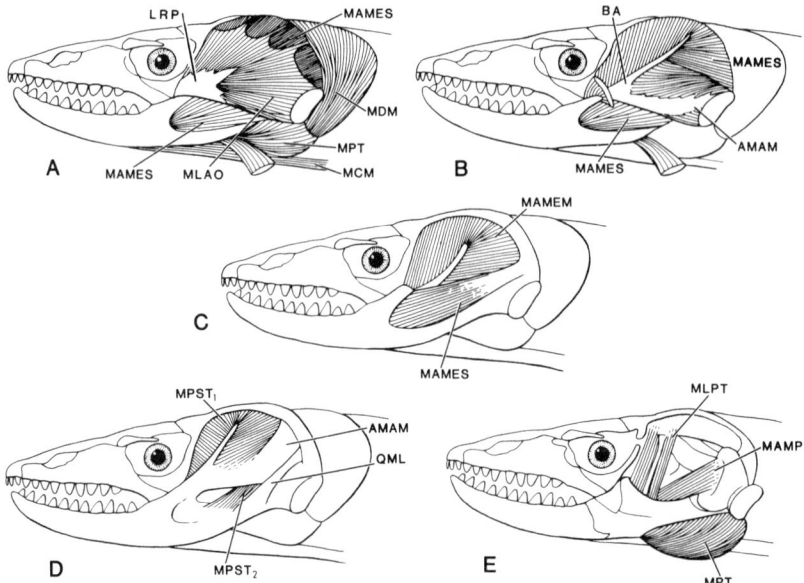

Fig. 3-3. Muscles of the masseter complex in *V. olivaceus*, from superficial (*A*) through the deepest members of the group (*E*). AMAM, *aponeurosis adductor mandibularis;* BA, *basal aponeurosis;* LRP, lateral rictal plate; MAMEM, *adductor mandibularis externus,* medial portion; MAMES, same, superficial portion; MAMP, *adductor mandibularis profundus;* MCM, *cervicomandibularis;* MDM, *depressor mandibulae;* MLAO, *levator annulis oris;* MLPT, *levator pterygoidei;* MPT, *pterygoideus;* MPST$_1$ and MPST$_2$, *pseudotemporalis* (superficial and deep); QML, quadratomandibular ligament.

of this muscle with those of the superficial layers of the *adductor mandibularis externus* (MAME), as in *V. varius* (Haas, 1973). There is also a strong ventrally located bundle below the *posterior* attachment to the quadratomaxillary ligament, running horizontally to the inner angle of the mouth (*rictus oris*). Thus, as in *V. varius,* this muscle serves as a levator as well as a retractor.

The adjacent deeper muscle is the *adductor mandibularis externus* (MAME), which is divided into three parts (*superficialis, medialis, profundus,* fig. 3-3). Together these constitute the largest muscle mass in the mandibular complex and are most important in bringing about the great biting pressure characteristic of this species. The superior portion (MAMES) is similar to that described for other varanids. It originates from the postfrontal, squamosal, and lateral quadrate surfaces and inserts on the lateral surfaces of the coronoid and surangular, primarily through a tendinous sheet that extends posteriodorsally into the muscle (fig. 3-3B) of the lateral rictal plate. Some of the most posterior fibers tend to form a separate muscle mass inserting on the exterior surface of the surangular. Most of the fibers making up both this part of the *adductor mandibularis* and the parts closer to the bones of the skull are disposed in a pinnate pattern, many of them inserting on a basal aponeurosis that runs to the coronoid process. This aponeurosis is longer and better developed than in other varanids studied. There is, in addition, another well-developed aponeurosis (MAME) that runs from the coronoid process to the quadrate at the anterodorsal corner of the tympanum to which most of the fibers of the posterior part of the main muscle mass are attached.

The medial portion of the adductor mandibularis complex lies still deeper, with fibers more or less parallel to those of the superior part. In the butaan this muscle is not divisible into superficial and deeper layers as it is in *V. bengalensis* (Lakjer, 1926). However, tendinous arches and the basal aponeurosis are larger and more clearly defined than in other varanids reported upon. In fact, muscle fibers do not run from bone to bone but join sheets of connective tissue which are then attached to the bones. This type of system allows a great increase in the number of muscle fibers that can apply a force to any one insertional area. The system is important in *V. olivaceus,* for it allows the force generated by the masseter complex to be much greater than would otherwise be the case.

The deepest (*profundus*) portion of the mandibularis complex is split into three parts, each with a different origin. The lower, posterior part originates from the medial surface of the quadrate (fig. 3-3B), as in *V. varius.* The middle part (which makes up most of the mass of the entire mandibular complex) originates from the medial surface of the supratemporal. Its fibers form the major muscle mass filling the posttemporal fossa and are directed anteriorly to insert on the medial surface of the basal aponeurosis. The deepest portion originates from the prootic bone.

Another important set of muscles used by varanids in closing the jaw is the pseudotemporalis complex (Smith, 1982). As in *V. varius,* a superficial part originates from the lateral surface of the parietal and bulges dorsally into the temporal fossa. Its fibers converge in a pinnate pattern on the dorsal tendinous extension of the basal aponeurosis; they insert on the medial side of the coronoid process. The deeper part of this muscle arises on the wall of the braincase and inserts on the prearticular.

The pterygoideus muscle also plays a major role in jaw closure in varanids (Smith, 1982). Though it can sometimes be separated into superficial and deeper portions in some varanids and is thought to be always weakly divided (Haas, 1973), it seems to be a single structure in *V. olivaceus.* It arises from the ectopterygoid and adjacent parts of the pterygoid, forming the large, fleshy masticatory cushion that can be seen so easily when the mouth is opened. This cushion plays a major role in guiding and moving fruits into a position appropriate for swallowing. Its fibers insert chiefly on the dorsolateral part of the retroarticular process and a considerable part of the centromedial surface of the posterior process dorsal to the region just anterior to the articular fossa.

As in most lizards, the *adductor mandibularis profundus* originates from the body of the quadrate and inserts on the mandibular (= Meckelian) fossa.

Those muscles changing the position of the palatal bone complex (levators) include the *M. levator pterygoideus,* originating from the descending process of the parietal and inserting on the pterygoid. It raises the pterygoid anteromedially in respect to the prootic. The *M. protractor pterygoideus* originates from the anteroventral prootic process and the basisphenoid, inserting on the entire medial surface of the quadrate process of the pterygoid.

In general, the head muscles of *V. olivaceus* are similar to those described for other varanids (particularly *V. exanthematicus,* Smith, 1982). But their one outstanding characteristic in *V. olivaceus* is the massiveness of the masseter muscle complex, surpassing (proportionally and at least in general appearance) that of *V. komodoensis*. The complex is 14% larger than the corresponding mass in *V. salvator* (*V. olivaceus* TW/masseter complex wt = 260.4; *V. salvator* = 300.0).

Skull kinetics.—Together, the massiveness of the head muscles and bones with which they are associated in *V. olivaceus* enables this species to bring great pressure on food particles at the level of the posterior teeth (details in chap. 9). This pressure is used to crush the shells of the land snails upon which it regularly feeds (the fruits upon which it also feeds are almost always swallowed without being crushed; see below). The only other varanids with similar blunted teeth and heavily constructed skull (*V. niloticus* and *V. exanthematicus*) also feed on molluscs (Mertens, 1942; Cisse, 1972), though this habit may vary geographically. The fossil *V. hooijeri* (Pleistocene–Post-Pleistocene, Flores Island, Indonesia) has similar teeth (Brongersma, 1958). I presume it too fed on molluscs. Another lizard with blunted teeth—a specialized molluscivore—is the large teiid, *Dracaena,* of tropical South America. Other varanids that occasionally feed on molluscs (*V. indicus, V. bengalensis;* see chap. 9) have nonspecialized teeth. An amphisbaenid known to feed on hard-covered prey is also known to possess somewhat blunted teeth and a large masseter complex (Pregill, 1984).

Other differences exist in this muscle-bone system between *V. olivaceus* and *V. salvator;* the former has reduced ability for movement at the frontonasal and mandibular hinges, a broader and more bony palatal area involving primarily the vomers and pterygoids (both wider and closer together), and basipterygoid processes more ventrally deflected and thus closer together. These differences are probably correlated with the need for more muscle fibers to pass to the coronoid process of the lower jaw—made possible by the narrowing of the palatal complex rather than a widening of the posterior part of the mandibles (which is already narrower than that in *V. salvator*).

In general, these differences suggest that the skull of the butaan is less flexible than that of *V. salvator* (as it is in the similarly structured *V. exanthematicus;* Smith, 1982). These differences, in addition to those of the shape of the teeth and the larger masseter complex, also are probably related to a molluscivorous diet, for the same osteological adaptations are noted in *V. niloticus* and *V. exanthematicus* (in spite of the fact that they are not particularly closely related to one another, or to *V. olivaceus;* see chap. 14).

All three of these partially molluscivorous varanids have these ten characteristics in common: short snouts; thick, deep lower jaws with short postarticular processes; teeth blunted posteriorly; strong muscle scars for the attachment of particularly heavy adductor externus and pterygoideus muscle complexes (both used primarily for closing the lower jaw); a small median palatopterygoid fossa; broad, short vomers; small vomeropalatine and ectopterygoid-palatine fenestrae; a narrow space between the quadratomaxillary ligament and the pterygoid; narrow, more downturned basipterygoid processes; and a shortened distance between the coronoid and the major crushing site of the lower jaw (see below).

In all three molluscivorous varanids the skull is a functional compromise between one adapted completely for crushing the shells of land molluscs, on the one hand, and capturing, killing, and swallowing a variety of completely different animals, on the other. In *V. olivaceus* there is additional accommodation for swallowing sometimes remarkably large, hard fruits that are resistant to compression. Most students of cranial morphology believe that for such foods to be swallowed the skull must be flexible to at least a certain degree. It does not seem to be the case for *V. olivaceus*.

The earliest study of reptilian skull kinetics was by Versluys (1910), who characterized the forms that he studied as kinetic or akinetic, depending on the presence or absence of a movable joint(s) in the skull roof. Species possessing kinetic skulls have been classified according to the position of the cranial joint: metakinetic (between supraoccipital and parietal bones), prokinetic (between the nasal and frontal bones), mesokinetic (between the parietals and frontals), and amphikinetic (if both of the last two conditions apply). Amphikinesis is the usual condition in varanids, most of which have a reasonably flexible skull (see Bahl, 1937; Hofer, 1960; Frazetta, 1962, 1983; Impey, 1967; Auffenberg, 1981a). In comparison, the skull of the adult butaan is quite rigid. There is no functional prokinetic joint, and the metakinetic one is limited. Thus the skull is more mesokinetic than amphikinetic. In addition, there is less movement possible between the nasal and maxillary bones than in other varanids studied (e.g., *V. komodoensis*, Auffenberg, 1981a). (The juveniles of all *Varanus* species have more flexible skulls than adults, so that the various benefits ascribed to varanids because of skull kinesis may actually apply mainly [perhaps only] to juveniles.)

Of the various interconnections and interactions that have been discussed by different workers, those that tend to widen the gullet and allow the passage of the largest food particles are the most important in the present context. Opening the mouth (protraction) is more important than closing it (retraction), for several movements of the components (muscles and bones) bring about changes in the gullet shape. These changes are particularly important when the butaan is eating large fruits, for, unlike the soft bodies of rodents, etc., fruits do not respond to skull movements designed to change the shape of the food particle.

Studies of varanid feeding (Loop, 1974; Smith, 1982) and skull anatomy (Frazetta, 1962, 1983; Haas, 1973) suggest that it is the least diameter of the prey (food) relative to the least diameter of the gullet (not gape) that determines the size of the particle that can be swallowed. Gape is important only in determining the efficiency with which varanids are able to grab a living animal and manipulate it. Gape is not related to skull kinesis (Frazetta, 1983; Smith and Hylander, 1985). Head diameter and length are important only as relative measures of food size. The most important measurement is the diameter of the narrowest, nonexpandable portion of the gullet. In varanids this portion is located just posterior to the mandibular hinge (fig. 3-4A). In this study the gullet diameter was recorded as the widest distance between the flesh covering the bones at this point when the mouth was completely opened in a freshly killed individual. The chief differences between *V. olivaceus* and *V. salvator* are that the former has a slightly wider head and a significantly smaller gape.[1] The gullet diameter relative to total length is the same. The mandibular hinge is less well developed in *V. olivaceus*, allowing less lateral bending of the mandible at the junction of the dentary and postdentary elements. However, a ventral view of the mandible shows that it is appreciably wider just posterior to the mandibular hinge in *V. olivaceus* due to outward flaring of the postdentary elements (fig. 3-4A). Another important difference between this species and *V. salvator* in this part of their cranial anatomy is that the palatopterygoid complex is not only considerably more robust in the butaan but is projected ventrally near the articulation of these two bones, forming an obvious obtuse angle rather than a convex bow as in *V. salvator* (fig. 3-4D, E).

1. *V. olivaceus*—head length/head width: 1.54–1.94, $\bar{X} = 1.79 \pm 0.41$; head length/gape: 1.54–2.00, $\bar{X} = 1.79 \pm 0.37$; head length/gullet width: 2.50–3.07, $\bar{X} = 2.77 \pm 0.31$.
V. salvator—head length/head width: 1.84–2.08, $\bar{X} = 1.94 \pm 0.33$; head length/gape: 0.92–1.57, $\bar{X} = 1.20 \pm 0.33$; head length/gullet width: 2.45–2.94, $\bar{X} = 2.71 \pm 0.52$.

Internal Morphology 45

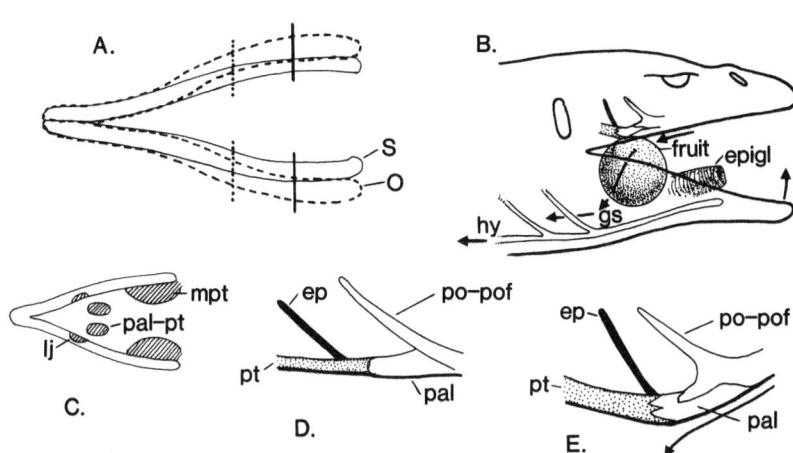

Fig. 3-4. Important features of the skull related to the swallowing of large fruits by *V. olivaceus:* A, ventral view of lower jaw of *V. olivaceus* (O) and *V. salvator* (S), showing the wider jaw of *V. olivaceus* from the middle to the posterior part; dotted vertical line shows level of mandibular hinge; and solid line shows level of jaw fulcrum. B, diagrammatic view of *V. olivaceus* individual swallowing a large round fruit; arrows show generalized directions of movements of the fruit and several skeletal elements involved (see text). C, diagrammatic ventral view of buccal area, showing structures (shaded) that tend to force large fruits ventroposteriorly into that part of the esophagus that is in the gular sack. D, palatopterygoid complex in *V. salvator*. E, same, *V. olivaceus*. Abbreviations: *gs*, gular sack; *hy*, hyoid; *ep*, epipterygoid; *epigl*, epiglottis; *pal*, palatine; *pal-pt*, palatopterygoid complex; *po-pof*, fused postorbital and postfrontal bones; *pt*, pterygoid; *mpt*, pterygoideus muscle; *lj*, lower jaw.

To maintain this angle in *V. olivaceus*, the epipterygoid forms a strut that is not only noticeably more robust but also has a ventral end that joins the lower elements at their articulation (the most ventral part of the complex). In *V. salvator* this junction is much posterior to the palatopterygoid juncture. Furthermore, in *V. olivaceus* the support function of the ectopterygoid is enhanced by its more vertical position. The function of this entire palatal complex is to serve as a strong wedge to force large, hard-walled fruit downward and backward into the gular sack below the widest part of the mandibles. Thus the food does not pass posteriorly from the palatopterygoid complex but largely *ventrally* into the elastic capacious gular sack, where there are few muscles and bones to impede its progress. (A maximum food particle diameter of 4.8 cm can pass into the gullet of the largest *V. olivaceus*—considerably more than the average food particle size [see chap. 9].) During this maneuver the lower jaws play an important role. While the mandibles of the butaan are robust and incapable of much bending, they can and do rotate independently of one another on their longitudinal axes. By this means they tend to conform to the radius of the fruit on its ventrolateral surface. More important, when the lower jaw is closed on a large fruit, its inner surface tends to clamp the food particle, and further closure causes the palatopterygoid wedge to push or roll the fruit posteroventrally into the gular sack (see fig. 3-4). Meanwhile the genohyoid muscle has pulled the entire hyoid apparatus anteriorly, including the epiglottal projection, so that this projection and the anterior end of the hyoid apparatus take the rough form of a spoon located ventroanteriorly to the fruit. Retraction of the hyoid apparatus tends to pull the fruit back into the gullet and move it even more posteriorly into the smaller diameter part of the esophagus (as it also does in *V. exanthematicus;* Smith, 1982), where other muscles take over the responsibility of pushing the food posteriorly by peristalsis.

Hyoid apparatus.—The structure of the hyoid apparatus is poorly known in varanids, though potentially of value in phylogenetic comparisons and probably worthy of careful study. This apparatus has been described in a number of varanids (*V. niloticus, V. salvator, V. griseus, V. bengalensis,* and *V. exanthematicus;* Zavattari, 1908; Furbringer, 1922; Linz, 1939; Mertens, 1942), but it has not been studied previously in *V. olivaceus.* The following is a preliminary investigation, pending future studies, of all varanid hyoid apparati. In addition to serving as a base for the larynx, supporting the tongue and the floor of the mouth, the hyoid also supports the floor of the pharynx, which in varanids can be much enlarged by moving the entire hyoid apparatus posteriorly.

The main body is an osseous plate, triangular and slightly curved dorsally, with lateral processes. In the shallow groove between these processes lies the trachea. Two successive pairs of visceral arches articulate with the body. Both pairs arise from the lateral articular process. The lingual process is also bony, being long and somewhat filiform, tapering to a fine point anteriorly. The cornu hyale is well developed and bony. It curves dorsally as it extends laterally from the hyoid body, forming a broad U-shape when viewed in anterior aspect (fig. 14-2). Though nearly round in cross section throughout, it is flattened at its anterior end, which is oriented vertically. The flattened area has a shallow notch at the terminus and may have an elongate opening of variable shape near the middle of the generally subtriangular piece. The epihyale is a long, flattened band of bone that extends from connective tissue at the ventroanterior process of the cornu hyale directly posterior and most medial to the cornu hyale and then bows outward to connect with the columellar apparatus, as in most lizards (Schumacher, 1973). At about one-third of its length there is a first small dorsal process, and behind it there is a second less ossified and smaller one that is sometimes absent.

There is no second cornu branchiale in any varanid as far as I am aware. However, cornu branchiale I is bony and well developed. It is somewhat shorter than the epihyale and longer than the cornu hyale. In cross section it is rounded. Each of the pair members extends posteriorly, diverging from the midline to near the posterior tip of the epihyale. Here there is a bony epibranchiale 1, which is as long as the cornu branchiale 1 and attaches to the skin of the lateral neck surface.

This hyoid apparatus differs from that of all other varanids I have investigated (*V. salvator, V. bengalensis, V. dumerilii, V. flavescens, V. exanthematicus, V. timorensis,* and *V. prasinus*), mainly in the shape of the cornu hyale. In all those mentioned, it is a flat, flexible element that lies parallel to the lower jaws but in a horizontal plane. The flattening allows the element to be bent in a vertical plane to accommodate the prey. In *V. olivaceus* the cornu hyale pair members cannot be bent because of their rounded shape. Instead they form a shallow U-shaped channel through which the rounded food particles (fruits and snails; see chap. 9) can pass without restriction. This shape brings the proximal ends of the cornu hyale and the entire epihyale into a vertical orientation, allowing horizontal but no vertical bending because of their flattened shape (fig. 14-3). In all other varanids examined the epihyale is flattened in the horizontal plane as the cornu hyale. Thus the hyoid of *V. olivaceus* seems highly modified in a special way—presumably as an adaptation to swallowing hard, spherical objects of standard size. The lateral curvature of the cornu hyale is particularly important in this regard, for it allows the passage of large spheres without vertical stress on that part of the hyoid located in the area of the narrowest passageway (between the posterior part of the lower jaw and the palatopterygoid complex; see chap. 9).

Thus the bones of the mouth, the hyoid, and associated muscles of the butaan are highly adapted to manipulate the unique diet upon which it subsists—large, hard fruits and hard-shelled land snails. The latter are crushed by a battery of stout, blunted teeth and a massive muscle mass; the large fruits cannot be crushed or shaped prior to swallowing so that a combination of bone shapes in the palatal surface and muscle actions in the anterior gular area assist in rolling and forcing them into the esophagus at the level of the gular sack.

Teeth.—The teeth of varanid lizards have, unfortunately, not received as much attention as have the skull and head muscles (see review in Edmund, 1969). The teeth of all adult butaans examined ($N = 99$) are characterized by being relatively barrel-shaped, not laterally compressed in cross section (fig. 3-5B, C) and blunt. This general tooth type was first described in varanids by Owen (1840–45), who briefly compared those of adult *V. niloticus* with somewhat similar teeth of the large teiid *Dracaena*. Lonnberg (1903–4) was the first to point out the probable relation between diet and the stumpy teeth of *V. niloticus*, showing that the major food of this species was land snails of the family Achatinidae. Other aspects of the teeth of varanids and their development were worked out by a series of German authors, particu-

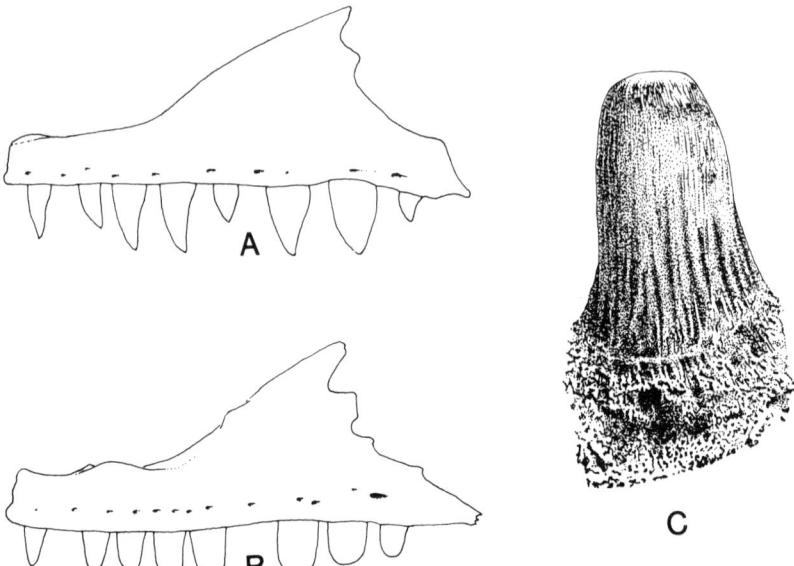

Fig. 3-5. Maxillary teeth of *V. olivaceus: A*, UF 53607, a juvenile (SVL 61 cm), showing the pointed teeth found throughout the entire series. *B*, UF 55055, an adult (SVL ca 70 cm, skeleton found in field). *C*, posterior maxillary tooth of adult, greatly enlarged, showing fine enamel striations on distal half of tooth and large, deep furrows at base.

larly during the early part of the twentieth century (see Bullet, 1942, for review). A 1920 manuscript by Peyer (a Festschrift for Prof. Dr. O. Stoll) also dealt with the blunt teeth of *V. niloticus*. Finally published in 1929, Peyer's work concentrated on the structure of the teeth and the relationship of this structure to that of other pleurodont lizards. Reinholz (1923) studied thin sections of the teeth of *V. niloticus*, emphasizing sectional views and including descriptions of the development of replacement teeth. Estes and Williams (1984) summarize all the literature on fossil and living lizards that have blunted teeth as adults.

As in other varanid species whose dental anatomy has been studied, the teeth of *V. olivaceus* rest on a broad base on the inner, somewhat vertically concave surface of the jaw bones with the lingual part of the tooth base much longer than the labial part. The differences that occur between these teeth and those of other varanids are in shape and number. A description of the teeth of an adult male follows, with comments based on observation of other adults ($N = 27$).

The premaxillary teeth are usually nine, one of which is median (but often missing). All teeth on each side are small and of about the same size, except the last one, which is about one-third larger and slightly more compressed mesiolabially; the anterior ones are not compressed but are almost perfectly round at their bases. All are conical and have a very sharp tip that is directed slightly posteriorly only in the large last pair.

The maxillary tooth row is almost straight, with a slight mesial bend near the middle of the series. It always consists of ten teeth, with considerable variation in size and shape from the anterior to the posterior part of the row. None is noticeably compressed (the anterior ones slightly, if at all, in some individuals), nor are there any cutting edges. All are conical, though the last six (or seven) are blunted to varying degrees, depending partly on tooth position and partly on age. The most blunted are the last two (or three); the largest is always the third from the last, and the smallest is always the first. Close inspection shows that teeth surfaces, particularly the largest teeth, are covered with fine longitudinal striations that run the entire height of the teeth, often anastomosing (fig. 3-5C). The tip is often also blunted by wear, with small, thin pieces of enamel broken off, especially on the largest posterior teeth. The pleurodontal angle is from 35° to 45° in the maxillary teeth, less on the premaxillary series.

There are always 13 teeth on the dentary, with the tooth row inside that of the maxillary and following it closely. The smallest are the most anterior, the largest in diameter the second through the fourth from the posterior end of the series. The first five (or six) teeth are never blunted but are simple cones with their tips bent outward and backward. From this level posteriorly all are more or less blunted, though the last four are particularly low and broadly rounded. Posterior replacement teeth of both the upper and lower jaw are always blunted to the same degree as functional teeth. Surface decorations (if any) are as in the maxillary teeth. The anteriormost six (five to seven) teeth are the most laterally directed; the posterior ones become progressively more vertically placed.[2]

The teeth of juveniles are different from those of adults in shape (fig. 3-5) though not in number. Proportionate basal lengths are the same, but the teeth of adults are rounded in cross section and in general proportionally shorter. This same barrel-like shape is seen in replacement teeth as well. The most significant differences are that juveniles (fig. 3-5A) lack the blunted posterior teeth of adults (fig. 3-5B) and that their teeth are mesiolaterally compressed, are provided with a more or less obvious cutting edge on at least the posterior

---

2. Measurements (in mm) of tooth height above external surface dentigiferous bones in 21 adult butaan skulls (condylobasilar length 40–120.5 mm): tooth no. 4 (premx) 1.7–4.0, $\bar{X} = 3.2 \pm 0.1$, SD =0.6; tooth no. 8 (max) 1.2–3.4, $\bar{X} = 2.6 \pm 0.1$, SD = 0.6; tooth no. 12 (max) 0.8–1.4, $\bar{X} = 1.8 \pm 0.1$, SD =0.4; tooth no. 14 1.0–4.8, $\bar{X} = 2.8 \pm 0.2$, SD =0.8.

Maximum diameter of tooth at level of external surface of the maxillary bone: tooth no. 8 1.8–4.8, $\bar{X} = 3.5 \pm 0.2$, SD =0.7; tooth no. 14 1.5–3.3, $\bar{X} = 2.5 \pm 0.1$, SD = 0.4.

surface (but often on the anterior one as well), and are pointed, with tips directed slightly posteriorly. In both juveniles and adults there are fine striations on the parts of the teeth above the gum line. Though the teeth become progressively thicker with increasing size of the individual (<42 cm SVL), the major changes (extreme blunting of posterior teeth and dramatic increase in relative tooth diameter) occur when individuals reach an SVL between 42.0 cm and 52.0 cm. That this change is not due to wear alone is proven by the fact that newly erupting teeth of adults are already blunted. However, minute chipping of the surface does occur on old teeth ready to be shed.

The teeth of the butaan become progressively shorter with age, being proportionally shortest in the largest (and in general the oldest) individuals. Some of the largest specimens have misshapen teeth in different parts of the jaws, particularly posteriorly where they apparently get the most wear. In some, a few teeth are missing that do not seem to have any replacement members forming below them, so that the hiatus is probably permanent. Bellairs and Miles (1960, 1961) also reported missing teeth that probably were not replaced in a particularly old individual of *V. niloticus*.

The proportional size differential in the upper teeth with respect to condylobasilar length remains the same for all teeth through the maturing process so that, other than the shape changes mentioned, there are no significant differences in allometric growth of teeth in different parts of the dental series.

Replacement teeth move dorsoanteriorly on the dentary and ventroanteriorly in the upper jaw, as they do in all other varanids studied so far (see Edmund, 1960, 1969). Bullet (1942) described replacement and tooth development in *V. salvator; V. olivaceus* appears to be similar. Edmund (1969) reported four changes per year for *V. bengalensis*. The same rate in *V. olivaceus* would lead to 200 new teeth annually (53 × 4). Earlier I estimated an equivalent rate for *V. komodoensis* (Auffenberg, 1981a). *V. salvator* replaces more teeth per year because it has a greater total number of functional teeth at any one time. Dissection of the gum in *V. olivaceus* reveals only 1 or 2 developing replacement teeth for each functional tooth. There are 2 or 3 in *V. salvator*, suggesting that the replacement rate is probably slower in *V. olivaceus*.

In summary, the teeth of the butaan are generalized as juveniles. Later in life the teeth at the posterior end of both the upper and lower jaw become highly modified to crush snail shells. This modification involves both a shortening of the crown height as well as a significant enlargement in basal diameter of only the posterior teeth (correlated with durophagy in several families of fossil and modern lizards; see Estes and Williams, 1984, for summary, though Dessam, 1985, has shown that for at least one teiid lizard this is not the case). The size at which this occurs in the butaan is approximately that at which there is a significant change in the diet (see chap. 9). The modification is clearly diet-related. The broad, blunted posterior teeth of adults are used primarily for crushing the shells of land snails; to do this effectively, blunt tooth shape is coupled with a massive muscle system for closing the jaw, as well as an osteomuscular system that provides great mechanical advantage (see chap. 9).

Postcranial Skeleton

Unlike other varanid species (Mertens, 1942), *V. olivaceus* has no dermal ossicles. Skeletal adaptations for climbing are insignificant compared to those of other reptiles. Because the species is so rare in collections, reference drawings of major postcranial skeletal elements are provided in figure 3-6. Appropriate description and comparative illustrations for other varanids can be found in Lecuru (1968, 1969) and Hoffstetter and Gasc (1969).

Gastrointestinal Tract

While the morphology of the skull and teeth are commonly related to reptilian diets, it is only recently that attention has been given to the morphology and physiology of the digestive organs; such studies have a long history in the field of mammalogy. Physiology of digestion is beyond the scope of this study, but morphology is described here in some detail to identify specializations related to the peculiar feeding habits of *V. olivaceus*.

Recent studies of reptiles have provided major advances in our understanding of their feeding behavior and diet (see Rand, 1978; Iverson, 1982; VanDevender, 1982, for reviews and discussion). However, all of these have been conducted on lizards and emphasized the folivorous part of the diet. They have also made the error of assuming all herbivory to be the same. Actually, herbivory includes folivory, frugivory, granivory, and nectivory, each of which is a highly specialized feeding strategy in its own right and has relatively little relation to the other types in regard to nutrition, digestive physiology, or morphological adaptations. The butaan is a frugivore, not a folivore.

In the following descriptions and discussions, three different digestive morphologies of lizards found in the study area will be compared and contrasted: the folivorous *Hydrosaurus pustulosus* (Agamidae), the carnivorous *Varanus salvator,* and the frugivorous *V. olivaceus*. (Further comparisons in regard to feeding strategies are presented in chapter 11.) Only the description of the gut morphology of *V. olivaceus* is considered more or less complete; the others are included only for comparison to certain aspects of butaan morphology.

The structure of the wall of the gastrointestinal tract follows a pattern common to all vertebrates: the inner lining of mucosal membranes is separated by connective tissue from an outer cylinder of at least two layers of muscle. Variation in histological structure reflects divisions into stomach, small intestine,

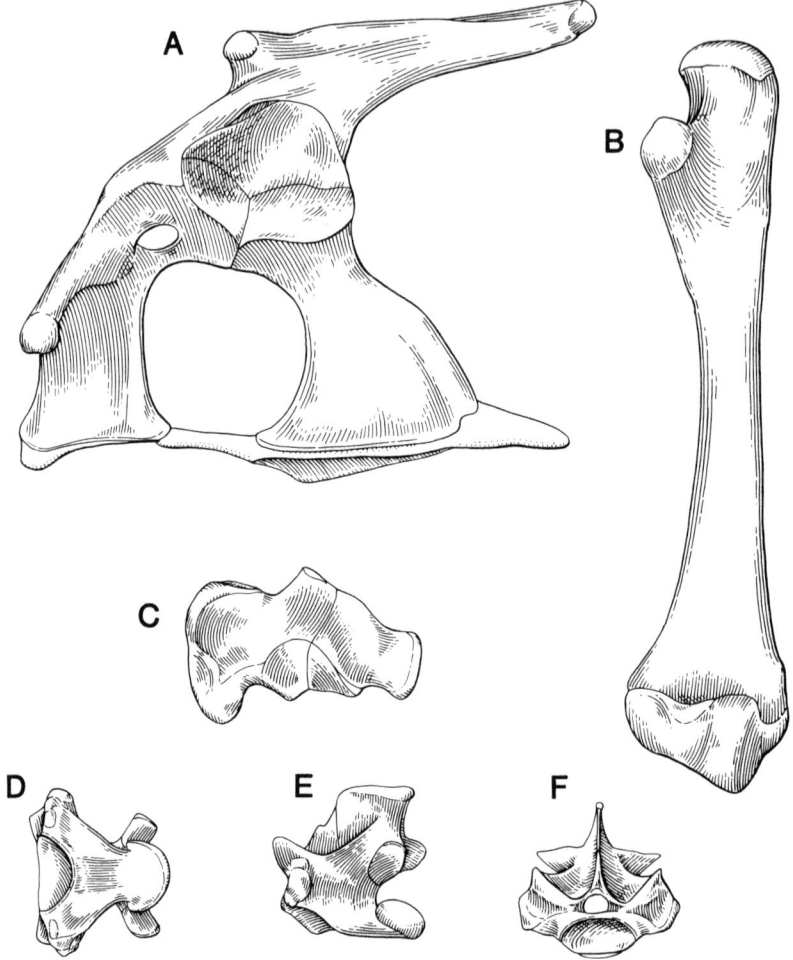

Fig. 3-6. Postcranial skeletal elements of *V. olivaceus* adult: *A*, pelvis; *B*, femur; *C*, astragalus and calcaneum; *D*, ventral view meddorsal vertebra; *E*, lateral view, same; *F*, anterior view, same; *G*, pectoral girdle; *H*, ulna; *I*, radius; *J*, fibula; *K*, humerus.

and large intestine (including a caecum in *V. olivaceus*). In mammals having different diets, these components vary in three ways: length, type and degree of mucosal foldings, and histology (recent reviews in Hladik, 1976; Chivers and Hladik, 1980). The same types of differences occur in reptiles (Iverson, 1982), although less is known about them.

Gastrointestinal tracts were taken from 14 *V. olivaceus*, 23 *V. salvator*, and 2 *Hydrosaurus pustulosus*. These were measured while still fresh (not in preservatives, which cause significant contraction). Digestive tracts were exam-

Internal Morphology

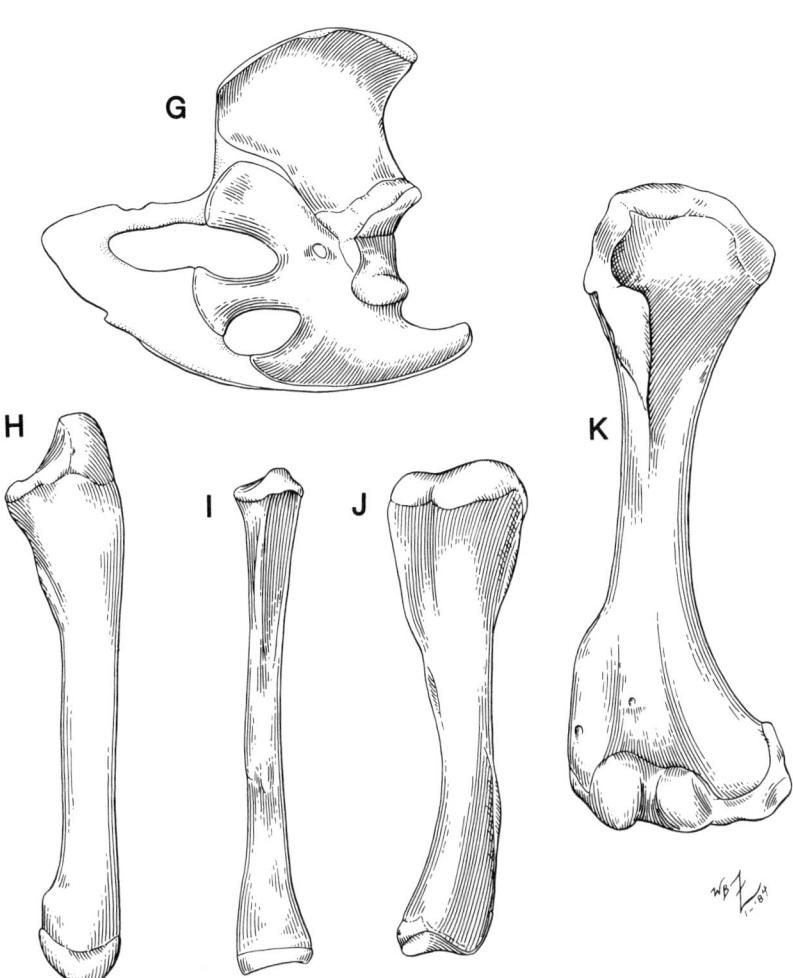

Fig. 3-6. (*continued*)

ined, drawn, or otherwise manipulated while underwater in a shallow dish. The dimensions of each region were then noted for calculations of area and volume (see Chivers and Hladik, 1980, for technique, formulas, etc.) and weighed. Mucosal linings were examined underwater after the tracts were split, cleaned, and spread out.

The following description of the gastrointestinal lining of *V. olivaceus* is based on an adult male, SVL 67.5 cm, 7.2 kg. No appreciable differences were found in the mucosal lining in any other individuals of this species examined.

The esophagus is lined with thick, generally nonbranching, closely packed, longitudinal, but slightly wrinkled folds. These are about 3 mm high and fairly straight, without sharp bends or obvious flat spaces between.

At the cardiac end of the stomach there is an abrupt change in the folds so that they are generally transverse, smooth (without wrinkles), larger in diameter than those of the esophagus, and branching, with little or no space between them. On the dorsal part of the stomach the folds are lower and more anastomosing, and they have more spaces between them, a pebbly surface, and larger wrinkles and folds. At the pyloric end the folds smooth out almost entirely, so that few (sometimes none) persist.

At the anterior end of the small intestines there are large, wavy folds that are diagonally wrinkled and nonbranching. They are more or less uniform in diameter and are covered with low villi. From the middle of the small intestine posteriorly the folds are obscured by long, flexible villi. Though the folds in this area have the same degree of obliqueness, they are lower and less wrinkled.

There is a caecum in the butaan just posterior to the beginning of the large intestine (fig. 3-7). Its lining lacks any large folds; rather, the inner surface is pebbly and slightly wrinkled with small crypts (pits) scattered over its surface. The anterior part of the large intestine is covered with low anastomosing, wrinkled folds that tend to circle the opening of the caecum and become more transverse posteriorly. In the middle of the large intestine these folds become higher, thinner, and completely smooth-surfaced. The more or less circular orientation of these folds suggests high absorptive potential. I conclude that the folds around the caecal orifice may somehow be associated with the movements of particles from the large intestine into the caecum. If there was any food at all in the gut, some was always pushed into the caecum. The most interesting morphological feature in the rest of the large intestine is the platelike structure of the folds in the posterior section, which tend to form partial divisions running across the lumen, though not as well developed as those in some folivorous lizards (see Iverson, 1982). I do not mean to imply that they are valvular but that they have the same general structure. Iverson's conclusion that all herbivorous lizards have a partitioned colon should be reworded to read "all folivorous lizards possess" one.

The following account of the gastrointestinal lining of *V. salvator* is based on a male 39.0 cm SVL and 0.95 kg. Other individuals of the same species were identical in all important aspects.

The esophagus is lined with closely packed, longitudinal, nonbranching, low (about 2 mm high), straight folds, with no flat spaces between, thus essentially identical to that in *V. olivaceus*.

The cardiac portion of the stomach has an abrupt change in the folds whereby they become highly convoluted and sometimes branching. They are narrower at the cardiac end and wider toward the middle of the stomach. At

Internal Morphology 55

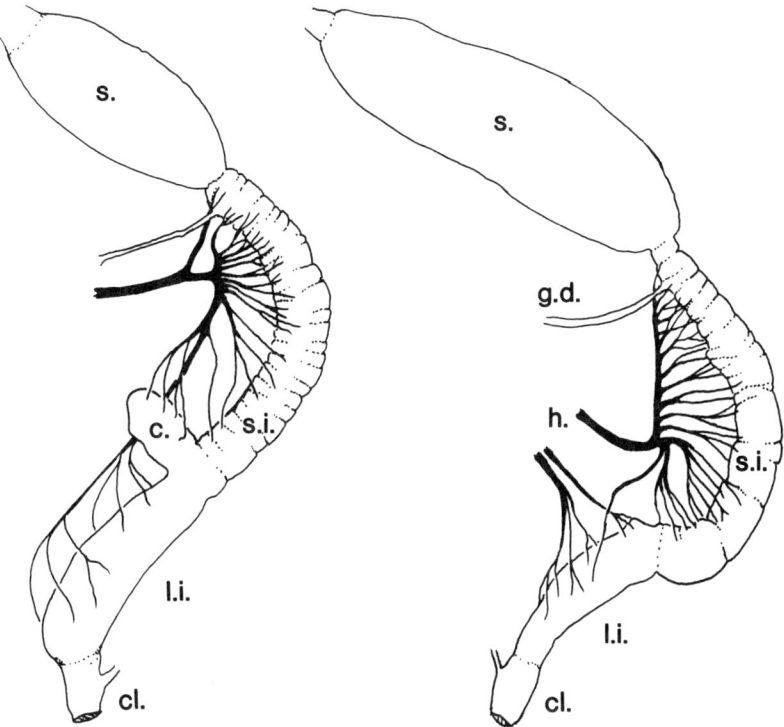

Fig. 3-7. Gastrointestinal tract of *V. olivaceus* (*left*) and *V. salvator* (*right*). Abbreviations: *c*, caecum; *cl*, cloaca; *g.d.*, gallbladder duct; *h*, hepatic portal system; *l.i.*, large intestine; *s*, stomach; *s.i.*, small intestine. Note that the large intestine is primarily supplied by a separate main branch of the hepatic portal system in *V. salvator* and that the same species has a small outpocketing of the large intestine in place of the large caecum found in *V. olivaceus*.

the pyloric end they become tall, narrow, few, and widely separated, with interfold areas that are relatively smooth. Thus the linings of the stomach in these two species differ in several important regards, particularly at the pyloric end where there are no or few folds in *V. olivaceus* and tall, separated ones in *V. salvator*.

The lining of the small intestine has large, wavy, oblique folds at the anterior end that are not anastomosing and are almost constant in diameter. They are covered with villi similar to those in the same area in *V. olivaceus*. The interfold areas are smooth. Posteriorly the folds become lower, less wavy, and more transverse. They lack the long villi found in *V. olivaceus*. Hladik (1976) has shown that primates ingesting more insects than fruit often have the shortest villi. Thus the long villi of *V. olivaceus* suggest a more highly specialized surface, with much greater absorptive potential in the small intestine than *V. salvator* has.

There is no caecum in *V. salvator*, though in perfectly prepared individuals one sometimes sees a *slight* triangular rise at the area where the caecum occurs in *V. olivaceus*. This area is not particularly well supplied with blood vessels, as is the caecum in *V. olivaceus*. In fact, the entire small intestine of *V. olivaceus* seems more copiously supplied with blood vessels than that of *V. salvator*. Thus the intestinal wall of *V. salvator* is less specialized for "heavy duty" absorption than in *V. olivaceus* and is, furthermore, a more "clean" bore, allowing the food probably to pass through more rapidly, particularly in the lower part of the intestine. The more extensive folding of the stomach in *V. salvator* may be related to greater holding capacity (see below). As will be shown (chap. 12), there is no appreciable difference in the *number* of nematode parasites in the guts of these two species, so that mechanical manipulation and chemical destruction by intestinal parasites seem to be at the same level in both species.

The *Hydrosaurus pustulosus* on which the following account is based was a male, 26.0 SVL, 525 g. In this part of the Philippines this species appears to be a complete folivore (occasional insects are probably eaten). The leaves in the gut are almost completely chewed; they are *not* clipped and swallowed whole as in *Cyclura* and *Iguana* (Iguanidae) (Rand, 1978; Auffenberg, 1982a). Furthermore, other features of this lizard suggest a different folivore strategy than the iguanines studied so far and would undoubtedly repay intensive study handsomely. The following description is included as an introduction to the peculiarities of this genus, as well as to serve as a basis of comparison with a sympatric frugivore.

The esophagus is lined with low, wide, anastomosing folds that become higher and more interlaced in the cardiac portion of the stomach. Then there is an abrupt change to a rather finely wrinkled surface over most of the remaining part. These folds become so low and fine at the pyloric end that on casual inspection the surface appears smooth.

The lining of the small intestine (which has a *much* smaller diameter than that of either *Varanus* species) is provided with fine, low, anastomosing longitudinal striations that are more wrinkled anteriorly. The entire wall of the small intestine is remarkably thin compared to that of the varanids.

As in *V. salvator,* there is no caecum. The lining of the large intestine is thrown into fine longitudinal striations. There are a *few* large folds that tend to be short, longitudinal, and rather variable in both size and length. There are no major anatomical changes along its length, although distally there is a small transverse partition nearly identical to that described by Iverson (1982) for *Hydrosaurus amboinensis*.

The gastrointestinal lining of *Hydrosaurus pustulosus* is remarkably simple. With the exception of the distal colic valve the entire surface remains relatively smooth throughout a thin-walled small and large intestine. However, the large capacity of both the stomach and large intestine are in keeping with a highly folivorous diet (Chivers and Hladik, 1980).

Component sizes.—*H. pustulosus* has no definitive gall bladder, though the liver produces bile and sends it to the small intestine through a bile duct. This lack shows that fats are not a major constituent of its diet, or at least that fats are not ingested in large amounts over short periods of time. On the other hand, the gall bladder of both *V. olivaceus* and *V. salvator* is well developed. There is no appreciable difference between the proportional lengths of their gall bladders.[3]

As far as is known, the function of the liver in reptiles is probably nearly identical to that in mammals. Its major roles are concerned with protein synthesis, glycogen storage, cholesterol and bile salt synthesis, and metabolism of steroid hormones. From a dietary standpoint, the liver is most important in fat degradation and detoxification of multiple proteinaceous substances. In general, larger livers suggest greater systemic levels of hepatotoxins and a greater number of different kinds of chemicals metabolized (see Freeland and Janzen, 1974; Swain, 1976, for general reviews).

The liver is a large organ in all three lizards, though smallest in *Hydrosaurus pustulosus* (1.6% of total wt), suggesting that this species probably ingests the smallest amount of oils and toxic proteins. In both varanids studied, liver weight is inversely related to total lizard weight (fig. 3-8). However, *V. salvator* has a more rapid rate of liver weight decrease per unit body weight increase than does *V. olivaceus*. Furthermore, data for the largest butaans form an almost straight line, suggesting that after the total lizard weight of 2500–3000 g is reached, changes in proportional liver size no longer occur. The reasons for this cessation are currently obscure, for there is no evidence of dietary change at this size in *V. olivaceus*. Local populations of *V. salvator* do not attain the large size of *V. olivaceus*, so it is not known if a change in rate occurs in the former at greater body size. In general, the liver of *V. salvator* juveniles is proportionally larger than that of juvenile *V. olivaceus*, but the change with increasing size is more rapid; the result is that the average

---

3. Gall bladder size and weight in *Varanus* adults (juveniles in parentheses, $N = 10$ and 7, respectively):
   For *V. olivaceus:* gall bladder wt $\bar{X} = 11.0$ (1.0) g; TW/(gall bladder wt $\times$ 100) = 6.5 (5.2); greatest length $\bar{X} = 36.5$ (16.0); SVL (cm)/(length $\times$ 100) = 1.8 (2.1).
   For *V. salvator:* gall bladder wt $\bar{X} = 2.8$ (0.3) g; TW/(gall bladder wt $\times$ 100) = 4.1 (9.6); greatest length $\bar{X} = 20.0$ (10.1); SVL/(length $\times$ 100) = 1.8 (2.1).

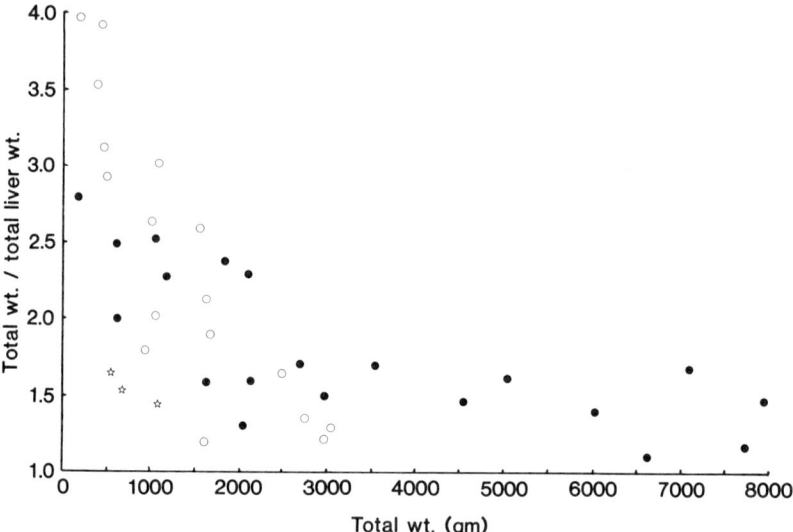

Fig. 3-8. Relation between liver weight and total weight in large lizards of the Caramoan area: open circle, *V. salvator;* solid circle, *V. olivaceus;* star, *Hydrosaurus.*

proportional liver weight is about the same in individuals of both species weighing from 1750 g to 2500 g. Average liver weight in all *V. olivaceus* is 1.89% of total weight ($N = 13$) and in *V. salvator* 2.64% ($N = 11$).

The allometric change in proportional liver weight in both species at smaller classes is apparently not entirely related to diet; even 1000- to 3000-gram individuals of *V. olivaceus* are clearly molluscivorous-frugivorous, yet they show the same rate of change with liver size as the completely carnivorous *V. salvator* in the same general size range (fig. 3-9). The allometric change in proportion may have some relationship with total body mass, so that the liver becomes less important in detoxification as the mass of the individual increases.

The lengths of the visceral parts concerned with digestion were measured for 9 *V. olivaceus,* 10 *V. salvator,* and 3 *Hydrosaurus pustulosus,* representing a variety of sizes for each species. While the viscera of the *Varanus* species are similar, those of *H. pustulosus* are different.[4]

4. Lengths of visceral organs (cm; as $\bar{X}$ % SVL) in three species of lizards:
   Hydrosaurus pustulosus ($N = 3$), stomach 51.9, small intestine 69.2, large intestine 33.8;
   V. salvator ($N = 10$), 19.3, 24.4, and 17.1, respectively;
   V. olivaceus ($N = 9$), 19.2, 27.0, and 23.6, respectively.

Among the chief differences in component lengths, all three parts are proportionally longer in *Hydrosaurus* than in either *Varanus* species; all three parts added together are longer than the SVL in *Hydrosaurus* (155% SVL) and less than the SVL in *Varanus* species (61% SVL). Thus relative length of the digestive tube of *Hydrosaurus* is over twice that of the two varanids. This difference is true not only of the entire tract but of each component part as well. The two *Varanus* species differ significantly from each other only in respect to the length of the large intestine, which is significantly longer in *V. olivaceus* than *V. salvator*, and in the fact that *V. olivaceus* has a well-developed caecum. This caecum, a small, rather thin-walled sack with a more or less uniform lumen size, about equals the size of the proximal end of the large intestine (see below for details), or about 0.005% of the SVL of adult individuals.

Early studies comparing gut morphology with diet (Cuvier, 1805) produced no obvious correlations, because they were concerned only with lengths of the component parts. Later demonstrations of useful correlations between structure and diet were based on surface areas rather than lengths. Recently these relationships have been made more clear by emphasis on quantitative methods and the inclusion of volume (derived from measurements of size rather than of quantity of fluids held, which is inaccurate). In the following section the approach and techniques of Chivers and Hladik (1980) have been employed.

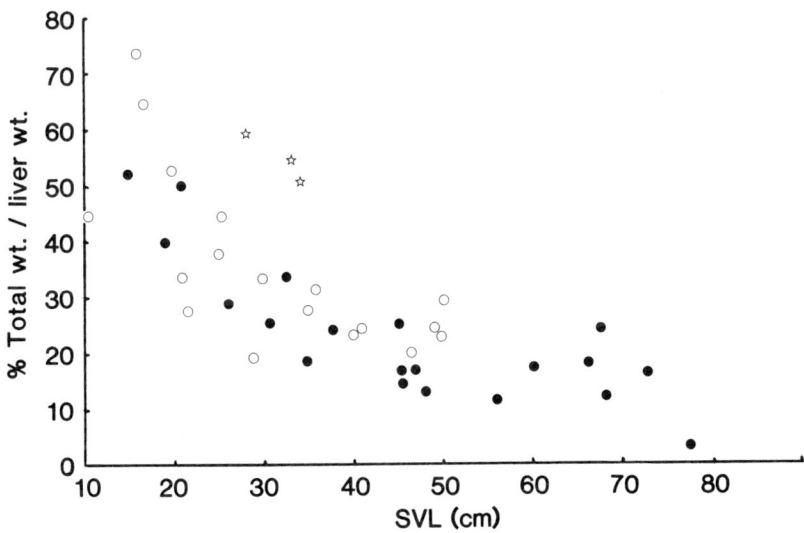

Fig. 3-9. Relation between liver weight and SVL in large lizards of the Caramoan area: open circle, *V. salvator;* solid circle, *V. olivaceus;* star, *Hydrosaurus.*

Though digestive tract lengths are nearly identical for *V. olivaceus* and *V. salvator,* important differences become apparent with consideration of internal surface area. *V. salvator* has a significantly greater surface area of the stomach and significantly smaller areas of both the small and large intestines.[5] Because of the large stomach, the total surface area of the gastrointestinal tract approaches that of *V. olivaceus.* Calculating the volumes of all parts shows similar results, except that the differences in the stomach, small intestine and large intestine are even greater.[6] The total volume is greater in *V. salvator* due only to the capacious stomach, which is probably related to the infrequent capture of relatively large prey by this species. On the other hand, the mature fruit food of *V. olivaceus* comes in evenly packaged sizes and is significantly common (chap. 10) so some selection is possible (if not necessary). Most important is that single prey items often constitute a meal for *V. salvator* (regardless of size), whereas *V. olivaceus* feeds on a *number* of items (fruits) at each meal so that the total volume ingested is more evenly distributed.

The absence of a caecum is, of course, common in faunivorous vertebrates, but it occurs frequently in folivores and frugivores (Hladik, 1976). Its absence in the totally carnivorous *V. salvator* is thus no surprise, but its exceptional development in the presumably closely related *V. olivaceus* is unusual. The caecum in most animals is apparently important in microbial fermentation and degradation of cellulose (Hladik, 1976). Thus its presence in *V. olivaceus* is undoubtedly related to the ingestion of large quantities of plant material. However, not all food passes through the caecum, for when it is full other food particles pass by it to the large intestine. Therefore, not all food is treated to whatever microbial process is normally restricted to the caecum. What food does lodge there tends to remain longer than food in the large intestine. Normally the food in the caecum is limited to two to three fruits at most (see chap. 9).

Though *Hydrosaurus* lacks a caecum, it is provided with a much longer large intestine than even *V. olivaceus.* The long-carbon-chain carbohydrates present in tree leaves consumed by *Hydrosaurus* require considerable degra-

---

5. Surface areas (cm^2) of gastrointestinal components:
   adult *V. olivaceus* ($N = 5$), stomach 76.8, small intestine 80.0, caecum 18.3, large intestine 90.0, total 265.1;
   V. salvator ($N = 3$), stomach 131.8, 65.3, caecum 0, large intestine 26.5, total 222.9 (all corrected for SVL differences).
6. Volumes (cm^3) of gastrointestinal components:
   adult *V. olivaceus* ($N = 5$), stomach 46.9, small intestine 50.8, caecum 2.0, large intestine 64.4, total 164.8;
   V. salvator ($N = 3$), stomach 138.8, small intestine 34.0, caecum 0, large intestine 5.6, total 178.4 (all corrected for SVL differences).

dation by symbiotic microorganisms (as in folivorous iguanines; see Iverson, 1982). The most conspicuous adaptation to folivory in herbivores is the large colon, with caecal valve and several species of nematodes (see chap. 12 for nematode loads in *V. olivaceus*), both of which are required for the bacterial fermentation of cellulose and the mechanical breakdown of the plant tissues and absorption of fatty acids. Little digested plant material is available for absorption in the small intestine of herbivores, hence its small size. It is not known if *Hydrosaurus* ingests its own feces after fermentation in the colon so that metabolites of the herbivorous diet can be absorbed in the small intestine (as apparently occurs in the largely folivorous iguanids *Cyclura* and *Iguana*). The butaan does not eat any reptile feces, whether containing partially digested plant food or not.

V. olivaceus, compared to *V. salvator*, has not only a different stomach lining (see above) but a much larger small intestine, where most of the absorption takes place in both species on the basis of blood vessel supply. Also, the villi of *V. olivaceus* are much larger. Such characteristics are common in herbivores, especially frugivores.

The sizes of stomachs and large intestines (with caecum) relative to small intestines in terms of surface area provide a quantitative index of gut differentiation (Coef. Gut Differentiation: Area stomach + caecum + colon/Area small intestine; see Chivers and Hladik, 1980, for complete analysis techniques and justifications). In *V. olivaceus* the index for area is 2.24 and in *V. salvator* 2.42. These differences are not as great as expected on the basis of their dissimilar diets (especially when compared to indices obtained in mammals). However, when the stomach data are ignored (due to accommodation to food particle size and shape in particularly *V. salvator*), the indices are 1.35 for *V. olivaceus* and 0.41 for *V. salvator*. This difference is undoubtedly a reflection of the largely frugivorous diet of *V. olivaceus* and is in agreement with the gut differentiation coefficients obtained by Chivers and Hladik (1980) for mammalian frugivores. The larger value for *V. salvator* is due to the smooth area of the large intestine and no caecum compared to the area of the small intestine. The smaller large intestine is undoubtedly associated with the nutritionally rich and easily digested normal diet of animal prey.

All mammalian frugivores supplement their diets with varying amounts of animal prey and leaves, and they generally lack distinctive structural specializations in the gut (Hladik, 1976, for review). The butaan is the same, retaining many features of the completely faunivorous gut from which it was derived (near *V. salvator, fide* Mertens, 1942). However, like other frugivores, it has a well-developed small intestine, amply provided with long villi, and a large caecum. These two areas are most important in absorption (Hladik, 1976) and are most developed in *V. olivaceus*, which is the species feeding on foods most difficult to digest (plant material).

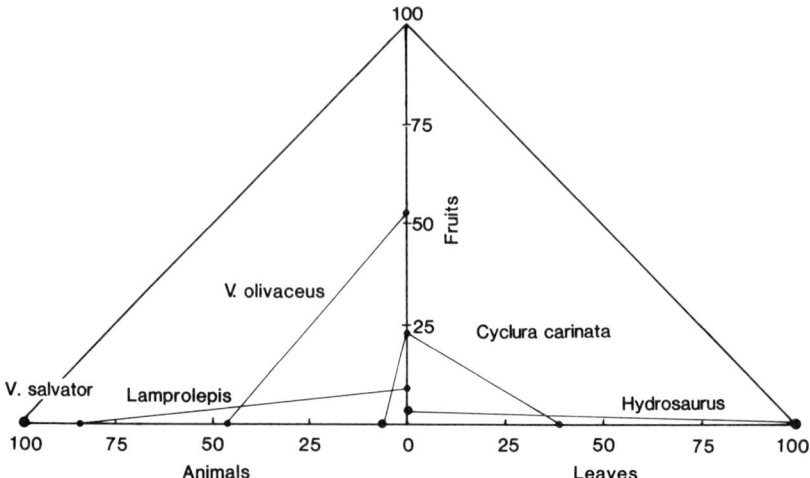

Fig. 3-10. Comparison of the annual mean percent of the diet (by item type) of five selected lizards, all but *Cyclura carinata* in the Caramoan area (data from this study, Auffenberg, 1982a, and Auffenberg and Auffenberg, 1987). *Lamprolepis smardigina* is a scincid, *Hydrosaurus pustulosus* an agamid, and *Cyclura carinata* a West Indian iguanid.

Figure 3-10 shows the dietary relations among the iguanine *Cyclura carinata,* the agamid *Hydrosaurus pustulosus,* the scincid *Lamprolepis,* and the varanids *V. olivaceus* and *V. salvator. Hydrosaurus* is almost exclusively folivorous, though a few fruits are eaten. Though *Cyclura carinata* is primarily a folivore, its diet is a mixture of animal matter, leaves, and fruits. *Lamprolepis* and *V. olivaceus* feed only on animals and fruits; *V. salvator* feeds only on animal food. Hladik et al. (1972) have shown that, while fruit is an adequate source of carbohydrate, it is generally inadequate in protein. This need probably explains why the butaan feeds on both fruits and animals, in spite of the fact that its gut is primarily modified for a frugivorous diet. Those animals eaten are almost always land snails (a secondary food, <20% dependence; fruits >45% dependence). As a protein source land snails are undoubtedly more easily obtained than any other animal in the environment in which the butaan is generally found; they are also a rich source of calcium and are important because of the low calcium content of most local plant foods (see chap. 9).

In conclusion, several features in the butaan gastrointestinal tract are apparently adapted for a predominantly frugivorous diet. In the faunivorous *V. salvator,* stomach volume is related to feeding strategy, while large intestine volume is correlated with body size. Both are voluminous in folivorous lizards (with caecal valves and rich nematode fauna as well; see Iverson, 1982). They are intermediate in *V. olivaceus*. Surface area for absorption is large in *V. olivaceus* compared to the faunivorous *V. salvator* and is probably related to metabolic body size. These features are all directly related to the evolutionary dietary shift from insectivory (or faunivory) to frugivory, on the one hand, and from faunivory to folivory on the other (though intermediate steps in the latter demand a mixture of fruit, animals, and leaves).

Fat

Fat deposition in reptiles is generally considered an indication of good health. Fat occurs in the tail and the visceral cavity. Visceral fat usually comprises almost all of the total body fat.

In all lizards in which fat deposits have been studied (see Fox, 1977), deposits in the tail are always indicative of food surfeit. In varanids tail fat deposits are particularly evident in smaller species of arid regions, where food resources may be highly seasonal. Thus the small Australian species *V. brevicauda* and *V. acanthurus* are notable for the fluctuation of fat deposits throughout the year. Both species may lose up to two-thirds of the tail circumference at its base when on a starvation diet in captivity (Mertens, 1942). Such deposits are also present in *V. olivaceus*, though the tail circumference is not nearly so variable seasonally. When present, the tail fat in the butaan is restricted to the anterior half of the tail, with the greatest concentration near the base. Here the fat is found in the interfacial spaces between the muscles rather than within them. The heaviest deposits are always found on the lateral surfaces of the neural arches, in a thin band in the sagittal plane between, and often dorsal to, the transverse processes of the caudal vertebrae. Smaller, ribbonlike deposits may also be found between major muscle complexes of the epaxial system, particularly the multifidus and spinalis groups. The maximum fat encountered in the tail of a *V. olivaceus* was 14.4 g (individual 2.9 kg, female, SVL 53.6 cm, May 1983). A mature female collected in July had no fat in the tail though visceral fat weighed 403 g, suggesting that tail fat is probably metabolized before visceral fat in time of need.

Visceral fat bodies have been described in a number of different varanid species (*V. exanthematicus*, Gunther, 1861; *V. giganticus*, Stirling, 1912; *V. niloticus*, *V. salvator*, and *V. komodoensis*, Mertens, 1942) and probably occur in all members of the genus. They also occur in *V. olivaceus*, in which they show considerable ontogenetic, seasonal, and sexual variation.

It is not clear how fat bodies are related to food and reproduction, for correlations have been noted with both. In several varanid species it has been noted that fat bodies are reduced in direct relation to food deprivation (Mertens, 1942). During this study it was noted that adult individuals of *V. olivaceus* with the proportionally smallest fat bodies were always skinny and demonstrably ill—sometimes heavily infested with cestodes. In any case, these individuals were not eating regularly. Presst (1971) obtained similar results in a large study of fat deposits in the snake *Vipera berus*. Working with the same species, Volsoe (1944) suggested that visceral fat bodies served as a reserved food storage, and Bellairs (1970) suggested that body fat was used in times of food shortage. Following initial work by Hahn and Tinkle (1965), others have shown that, at least in female lizards, visceral fat is important in follicular development. In his review of the subject, Fox (1977) demonstrated that seasonal-breeding lizards have fat bodies, year-round breeders do not. Some snakes show a similar relationship with reproductive cycles; Gibbons (1972) showed that fat bodies were reduced after parturition in the viper *Crotalus*. Burrage (1973) suggested that in *Chamaeleo pumilus* some fat may be used as sustenance for the embryos in the later stages of pregnancy. The significance of visceral fat bodies and their seasonal changes in males has not been well studied. Licht and Gorman (1970), as well as Fox (1977), suggested that these may be needed for higher energy demands during the breeding season. Fat body excision resulted in a decline of testicular activity of *Chamaeleo* (Burrage, 1973) and retarded yolk deposition and follicular development in female *Uta* (Hahn and Tinkle, 1965). Fox (1977) suggested that ovarian hormones may regulate lipid metabolism from the fat bodies. Bellairs (1970) has shown that body fat passes to the liver, where it is processed to form yolk in the developing oocytes. Thus there are currently a number of explanations for visceral fat deposition in reptiles, none of which satisfactorily explains all situations (at least in the males)—indeed the facts can, or probably should, be covered with several explanations. It appears that in many female lizards, visceral fat storage is primarily associated with reproduction. Deserticulous forms of both sexes often store caudal fat to sustain themselves during lean periods. It is highly likely that varanids use their fat deposits in a number of different ways.

All butaans have visceral fat. However, in juveniles (SVL about 40 cm) the fat bodies are proportionally small (19.5–20.0 g; body fat [g]/SVL [cm] about 0.50). Furthermore, there is no evidence in juveniles of seasonal variation in the proportional weight of these structures, nor is there any sexual correlation. Thus massive fat deposition (which may vary seasonally) is a characteristic of older, larger individuals only.

In adults these deposits may comprise as much as 12% of the total weight during at least part of the year. With the exception of June through September, males have significantly more abdominal fat than females. However, the seasonal pattern is similar in both sexes (fig. 3-11). While the greatest amount of body fat occurs in both sexes during the cooler months, there is more amplitude in the curve represented by the males, with a clear peak in November. The peak is less defined in females. In males the least fat occurs from June through August, whereas in females there is a more definite minimum point in

July. Thus male *V. olivaceus* accumulate proportionally more fat than females during the second (winter) monsoon but drop to approximately the same level as females during the first (summer) monsoon. Additional data on reproduction (fig. 8-1) show that in females the fat is not lost only after egg-laying but rather in two major steps, one during the period of heaviest monsoon rainfall (November to January) and the other in the first monsoon (June through July), *prior* to egg-laying. Thus the eggs are laid during a time when the abdominal fat bodies are near their smallest sizes during the year. For the males, the least fat (June through August) occurs during courtship and breeding. A comparison of figures 3-11 and 8-4 shows that the fat cycle of males is inversely related to seasonal changes in testes diameter; i.e., testes diameters are smallest when fat deposits are greatest. While this relationship might suggest that breeding is related to fat degradation, it is probably important to note that fat reduction actually begins in both sexes several months before the breeding season.

In my opinion the best explanation for demonstrable seasonal variation in size of abdominal fat masses in both male and female butaans is that this variation is related to food abundance and quality. Data presented in chapter 10 show that the greatest variety of food is available during July through September, and the October peak in fat reserves is probably a reflection of food abundance. As expected, fat accumulation is not precisely correlated with food availability but lags somewhat behind it.

Further evidence of the importance of food availability in relation to fat accumulation is suggested by the fact that the seasonal body fat of male and female *V. salvator*, living in the same climatic regime and with almost identical breeding and egg-laying seasons, is totally different from that of *V. olivaceus* (fig. 3-12). This difference is related to seasonal availability of the food of *V. salvator*, with surfeits occurring in the drier months of August and March through May. The least fat occurs during the months when food is most difficult to obtain for this semiaquatic species (September through December), when the freshwater crabs (most common prey in Caramoan area) are widely scattered because of the generally wet conditions.

The steepness of the fat accumulation curve of male *V. olivaceus* compared to that of the female is probably related to the higher feeding rate of males compared to females (see chap. 9 and Auffenberg, 1979c for the same pattern in *V. bengalensis*). Thus the basic similarity of both curves (in spite of the differences in slopes) is best explained by seasonal abundance and sexual differences in feeding.

When body fat is particularly abundant in butaan adults (October through January), it is a rich yellow-white in color and lobulated in appearance. Prior to and after this period the total mass is smaller, decidedly more flaccid, and whiter in color. If there is further reduction, the fat becomes pinkish white and the blood vessels are easily seen. At this reduced stage the fat is flexible and smooth to moderately wrinkled, without lobules. Even further reduction

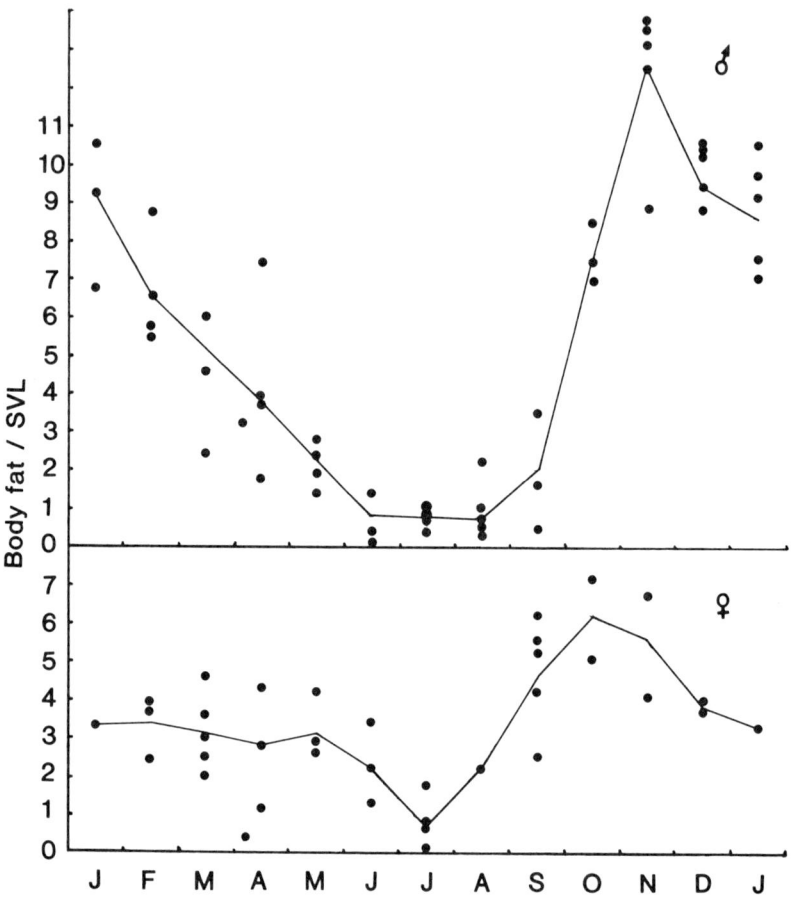

Fig. 3-11. Seasonal variation in abdominal body fat of adult *V. olivaceus* (SVL > 40 cm). Line indicates monthly averages.

results in small masses devoid of any noticeable fatty tissue, mottled with whitish pink and red—the former in the nonvascularized areas. This stage is never found in healthy adults, no matter what season of the year, but only in diseased, skinny individuals. Every adult we obtained in which the body fat was in this stage ($N = 3$) was particularly heavily infested with cestodes. While rare in *V. olivaceus*, atrophied abdominal fat masses are fairly common in *V. salvator* during periods of food shortages (i.e., severe drought during the early part of 1983). This difference suggests that food shortage is a less serious problem for *V. olivaceus* than for *V. salvator*—a conclusion reached on the basis of other observations as well.

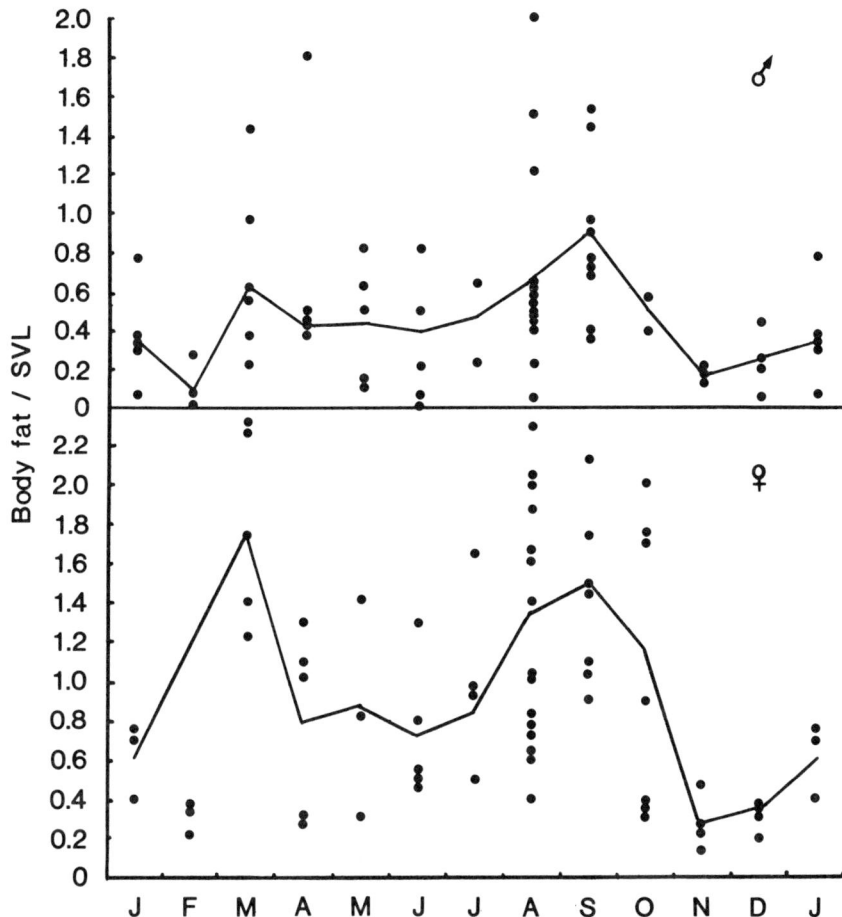

Fig. 3-12. Seasonal variation in abdominal body fat of adult *V. salvator* (SVL > 40 cm). Line indicates monthly average.

When present, the body fat deposits of all varanids are located within the ventrolateral part of the body cavity, from just anterior to the cloacal region to near the diaphragm when particularly well developed. There are two masses present, one on each side of the midline. The body fat of *V. olivaceus* is more yellow and firm than that of *V. salvator*. When rendered, it produces proportionally more oil than that of *V. salvator* and does not break down as readily at high heat (remains "sweet," as locals say), the major reason why butaan fat is preferred for cooking.

4

Environment

Macroecology

THE REGIONS in which the butaan is found are all heavily forested, hilly to mountainous country (pl. 2, 3). Because the mountains' uplift has been recent, they tend to be rather steep, though rapid runoff has filled valleys with a deep alluvium. Vegetational communities are complex; the high annual rainfall and equable climate produce deeply shaded forest associations of thousands of species within relatively small areas. Drought and fire are not major factors in the evolution of the butaan or the forest they occupy.

Most of the ecological data presented were gathered from the Caramoan Peninsula (chap. 1), though additional data were obtained from sources published on other parts of the butaan's range.

Edaphic Factors

Geologic.—Though altitude plays a major role in the vegetational variation of the Philippine Islands, so also does geology. The driest habitats in which the butaan is found are all in lowland coastal regions where the surface is very porous because of basement limestone rocks. In such areas (i.e., the tip of the Caramoan Peninsula) soils are thin, often practically nonexistent. Most of the soil that is being or has been formed quickly washes deep into the bedrock via fissures and sinkholes. There is little or no standing surface water, and many rivers drop into caverns to emerge elsewhere, sometimes several kilometers away. Mountaintops, ridges, and knolls are frequently a jumble of giant limestone blocks, with solid basement material often tens of meters below the surface (pl. 2, fig. 4-1). Rainwater percolates rapidly through such material, and though precipitation may be high the surface environment is often dry.

Forests growing on these limestones reflect the dry conditions, for the trees

PLATE 2. Close view of mixed dipterocarp forest, showing rough, rocky substrate typical of the areas in which *V. olivaceus* is most common.

PLATE 3. *Top*: Taison Mountain (site 2, fig. 1-1), showing the heavily forested karst topography typical of the habitat of *V. olivaceus*. *Bottom*: Dense tropical evergreen forest clothes the mountainsides. The tallest emergent trees are mostly dipterocarps.

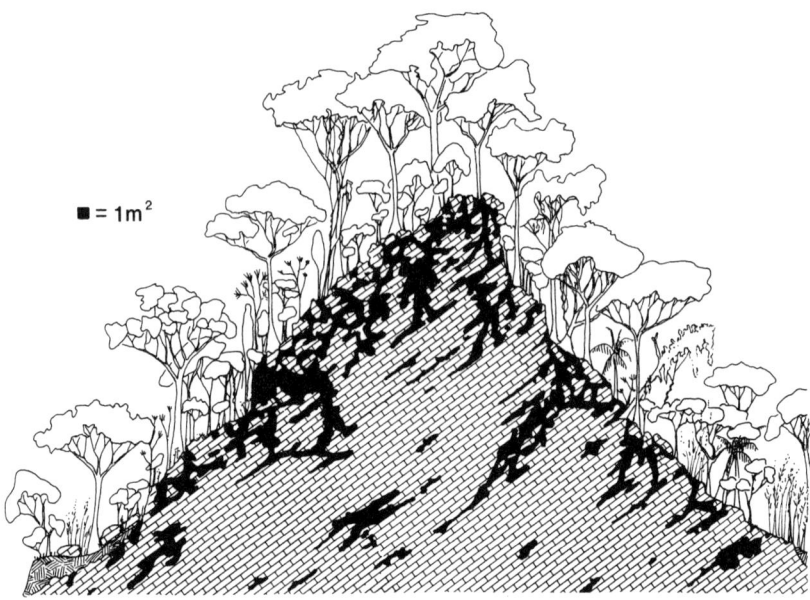

Fig. 4-1. Idealized cross section of a limestone hill (Mt. A) in butaan habitat near Caramoan.

tend to be smaller than those in more mesic situations. The generally lower and more open canopy results in a rich and dense surface herbaceous layer, though on rocky ridges this layer is often composed of only a few species capable of tolerating the drier conditions. The limestone soils support a forest rich not only in plant species but also in land and tree snails, some species of which are important food for the butaan.

The butaan is also found in mountains of igneous origin; in fact, igneous substrates compose a large part of its range. In general, these substrates form slopes less steep than those in limestone areas (angle of mountain slopes in Caramoan area: igneous $\bar{X} = 31.9°$ SD $\pm$ 3.0, $N = 30$; limestone $\bar{X} = 38.4°$ SD $= \pm 3.5$, $N = 30$). More important, subsurface and surface water is generally common on igneous substrates. Streams flow for longer distances, forming dendritic erosional hillside gulleys and ravines. In general, the plants here are adapted to more mesic conditions. Surface soils are usually deeper and richer. These soils and the generally higher rainfall produce a rich tree flora in terms of species, usually composed of taller and more widely spaced trees. In general, molluscs are less common. Finally, the igneous substrate usually offers fewer rock crevices as refuges for the butaan (see below).

At present, most relatively flat valleys, lower terraces, and plateaus on Luzon have been converted to rice culture and rarely have any extensive forests. But they did in the past, as they still do in parts of Samar and Mindanao islands. In such lowland areas the alluvial soils are deep and continually moist. Rock outcrops are almost completely absent. In general they produce the tallest of the Philippine forests, though current evidence suggests that these were probably not as varied as those of the drier limestone areas. So, while butaans certainly occurred in valley forests at one time, their density there was probably never higher than in karst areas at present.

Table 4-1 provides data on the extent of various soil types within the range of the butaan and the landforms with which they are normally associated.

Most surface rocks within the habitat of the butaan are of fairly recent geological origin (most of the limestones are Pliocene-Miocene, most volcanics the same or slightly older). Vulcanism does not play an important role in the life of the butaan, in spite of the fact that volcanos (some presently active) are liberally sprinkled through its range. Though these volcanos reach heights of 2400 m, most Luzon mountains within the butaan's range vary from 100 to 900 m and, on Catanduanes Island, from 100 to about 750 m.

Older, early Tertiary volcanos and metamorphic rocks are found in central Catanduanes, Camarines Norte, the Caramoan Peninsula, the eastern flanks of the Sierra Madre, and about half of the Bicol Peninsula from Camarines Sur to Sorsogon. The substrate in the remainder of the butaan's range is all Tertiary limestone (see chap. 14 for details).

Groundwater.—Because of the difference in surface materials, free water is available in surface pools only on the impermeable clays of valleys and igneous hills; standing water in the limestone hills is found only in small pools in hollow trees, limbs, and knotholes and, less frequently, in small basins perched on the sides or tops of particularly dense boulders. In this karst area,

TABLE 4-1. Soil type and landform correlates in the habitats of *V. olivaceus*.

Soil type	Total area (ha)	Land form
Alaminos clay	58,668	Generally upland forests
Annam clay loam	87,206	Mostly forested, some coconuts
Bulusan loam	6,770	Rolling to mountainous[a]
Casiguran clay loam	38,120	Usually hilly[a]
Guinobatan sandy loam	9,270	Riverine
Lusiiana clay	140,379	Rolling to mountainous[a]
Macolod sandy loam	26,793	Hilly forest
Mayon sandy loam	14,023	Mountain base areas[a]
Sevilla clay	98,708	Usually hill forest
Tigaon clay	59,006	Rolling to mountainous, some forest
Undifferentiated mountain soils	108,931	Mountain forest

SOURCES: Soil types are based on soil maps of Lucas, Salazar, and Engle (1947) and Marfori (1947, 1948a, b).
a. Now mostly in agriculture.

standing water is usually found only in rock fissures deep below the surface, where clays eroded from above have accumulated as perched impermeable lenses. Where such clay plugs are absent, groundwater may be up to 40 m below the general land surface. Groundwater is often close to the surface in alluvium-filled valleys. Streams flowing across valley floors often disappear below the surface at the edge of limestone mountains, where groundwater is lower. In any event, drinking water is generally available to *V. olivaceus* in all habitats in which they occur, though in highly karsted limestone hills they may have to climb down into rock crevices to get it.

Topography.—Local irregularities of the rocky surfaces are important, for they provide shelter both for these lizards and for the molluscs upon which they feed (chaps. 9, 10; fig. 4-1). Deep fissures are often the only water sources in these mountains during the dry season. Furthermore, fruits regularly eaten by *V. olivaceus* accumulate in these crevices. Figure 4-2 shows the surface irregularities in respect to the general form of Mountain A—one of the major study sites near the base camp—and the activity level of *V. olivaceus* in different parts of the same mountain.

In a large valley system, like that in the Caramoan area, the downward movement of cooler air develops a characteristic down-valley wind by evening and an upslope wind by day. (Both are of local occurrence.) Because of the vegetational density these daily winds do not greatly affect the microecology of the butaan on the hillsides, but they do in open areas such as the base camp.

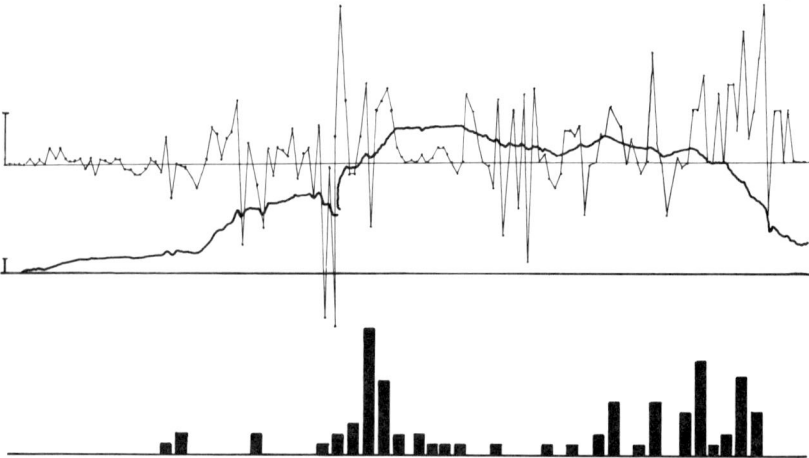

Fig. 4-2. Top: line shows surface configuration every meter across surface of mountain, 3 m shown by vertical bar on left. *Middle:* heavy line shows profile of Ahop Mountain on which surface was measured; the most irregular surface is found on both slopes and in a section near the middle of the mountain. *Bottom:* histogram shows the frequency with which butaans were located in different parts of the mountain profile; most butaan activity is located at places where the surface is irregular.

Thus the effect of local topography on wind direction, etc., is not as important for the butaan as it is for *Varanus* species that range more widely, feed on carrion (*V. salvator* and *V. komodoensis;* Auffenberg, 1980a, 1981a), and live in more open places.

Climatic Factors

There are few good climatological data for parts of Luzon in which *V. olivaceus* is found. For example, no weather data are available for the entire Caramoan Peninsula. Some stations do exist within all the provinces in which this species occurs, but they are often remote in respect to the actual ecological setting of *V. olivaceus* and bear little relation to the conditions under which it lives.

To compare butaan behavior with local conditions, climatic data were obtained at two sites—at and near the base camp. One station was located on the top of a low, exposed hill which tended to have extreme conditions; another was within the forest at the base of Mountain A, 0.25 km from the base camp. The same climatic data classes were obtained from other sites selected as the occasion demanded. Only data from the base camp station are considered here, as they are more or less comparable to those obtained from other macroecological weather stations throughout the Philippines. The data from other stations are discussed under microecology below, because they tend to emphasize differences of smaller magnitude over shorter distances.

Precipitation.—The geographic range of the butaan falls between the wettest and driest areas in the Philippines: the highest annual total rainfall tends to occur in the mountainous areas in the eastern part of the islands and the lowest in a north-south central strip through the archipelago and in the south and west (fig. 4-3). These areas are reflected in the distribution of the climatic zones Coronos established (1918) for these islands—four zones, of which the butaan is only found in two (II and IV; fig. 4-3).

Zone II lacks a definite dry season. It rains throughout the year, but there is a pronounced season of heavy precipitation from July to February, with frequent cyclonic storms contributing to the generally high annual rainfall. On Luzon, the zone is restricted to a narrow strip along the east coast, extending inland to the crests of the Sierra Madre Mountains in the northern and central parts of the islands and the eastern half of the Bicol Region of the southern part.

Zone IV, restricted to the western part of southern Luzon, has no pronounced rainy season. Most of the rest of Luzon falls in Zone I, with a generally lower total annual rainfall (except in the northern mountains) and alternating and distinct dry and wet seasons. Zone III on Luzon is restricted to the Cagayan River in the far northeast. As far as is known, the butaan does not occur within

Fig. 4-3. *Left:* mean annual maximum rainfall (mm) for the eastern Philippine islands. *Right:* climatic zones (slightly modified from Coronos, 1918, and Anon., 1975): Zone I, pronounced seasons, dry in winter and spring, wet in summer and autumn; Zone II, no dry season, maximum rainfall in winter; Zone III, no pronounced maximum rainy period, no dry season; Zone IV, no pronounced maximum rainy period, short dry season. Solid dot, location of Mt. Maquiling; triangle, Caramoan.

Zone I; its occurrence is unlikely in Zone III, though it can be expected throughout Zones II and IV on Luzon.

Within the known range of this species, average annual rainfall is about 3000 mm (OR 2540–4064 mm). For Zone II most rainfall occurs from October through January, with a minor peak in July; April–May and August–September are generally the driest months. For Zone IV the pattern is a little different, with most of the annual total rainfall from August through November and the lowest from February through June. The average annual rainfall for the rest of the Philippines shows that the entire monthly rainfall pattern is shifted earlier than in Zones II and IV so that the peak falls during July through September.

The precipitation data available at the base camp show a pattern similar to that of Zone II, with which it is geographically associated (fig. 4-4). The peak shown in July corresponds to the passing of two typhoons, as does the noticeably high peak in September, when two even stronger cyclonic storms passed close to the camp. February rainfall was less than normal. When these discrepancies are removed, the resulting curve is similar to that of Zone II (Tigaon curve; fig. 4-5). An annual precipitation curve for Mt. Maquiling, near Los Banos (el. 1500 m; fig. 4-5), shows that most of the rain falls from July through September—more like the climate of western Luzon (Zone III). The butaan is unknown in this mountain mass, in spite of the existence of likely forest situations. The fact that it does not occur there suggests that minor weather differences may be important in its distribution—probably because they affect seasonal patterns of fruit production (see chap. 10).

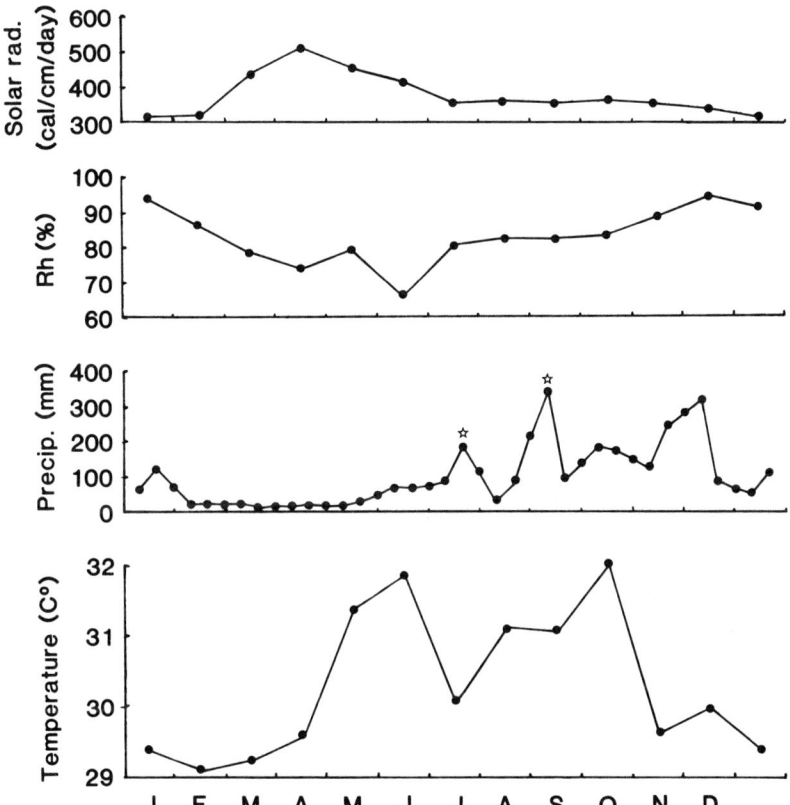

Fig. 4-4. Mean monthly climatic factors at the Caramoan study site and their annual variation (1982). All data are from base camp, except solar radiation, which originates from the Manila area (Lof et al., 1966). Stars mark major typhoons. Mean monthly temperature is shaded air at 1 m above surface; total annual precipitation is 2858 mm; annual mean is 383 cal/cm/day.

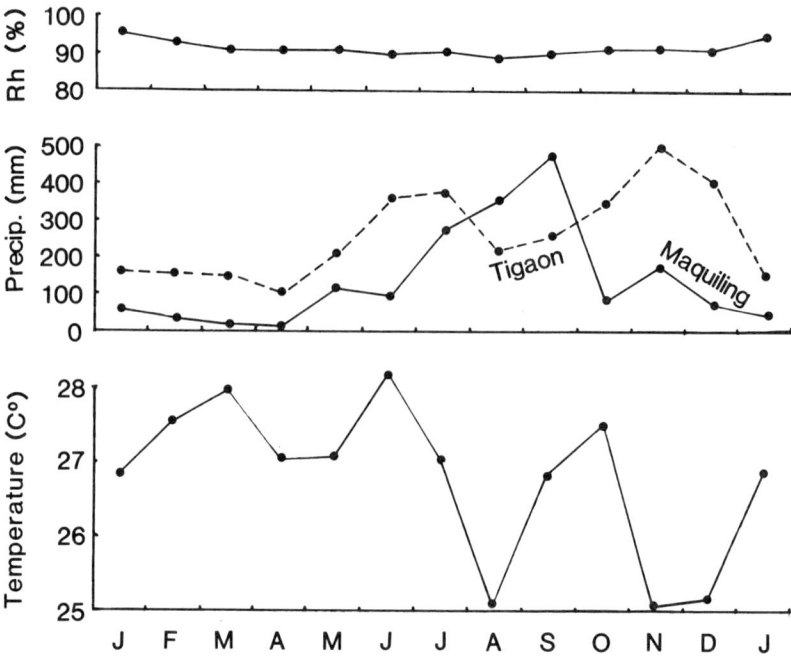

Fig. 4-5. Mean monthly climatic factors above the upper limit (750 m) of butaan distribution (*fide* Brown and Mathews, 1914 for Mt. Maquiling, 1500 m, climatic zone III). Precipitation is compared with that in Zone II at Tigaon, a lowland area near the base of the Caramoan Peninsula, well within the range of the butaan.

Temperature.—Tropical evergreen forests are known to show little seasonal change in average temperatures, and the forest around the Caramoan study site is no exception (fig. 4-4). Its generally low elevations and proximity to the sea tend to ameliorate even the slightest differences. Within the butaan's range average temperatures for January are 29.4°C and for July 30.2°C. Minimum temperatures of 16°C occur during the northeast monsoon (October-January), and a maximum of 35°C may be reached before the southeast monsoon (May). Considerably more detail concerning daytime temperatures at the study site is given below. In general, diurnal variation in temperature is almost always greater than the variation in averages between seasons.

Wind.—Wind velocity varies greatly within the habitat of the butaan. In general, the forest interior, especially at and below ground level, experiences little wind, even during rather violent storms. However, in exposed places such as rocky, nearly treeless hilltops and in the canopy of the emergent forest trees, the wind is an important environmental factor. In such places it was found to vary from 0 to about 35 km/hr, except during typhoons, when it sometimes reached velocities above 150 km/hr. In general, wind velocity is lower in the early morning and early evening and through the night than during the day. Temperature inversions due to insolation usually produce significant surface winds (in open places only) beginning about 0700 hr to 0900 hr each morning, and these are sustained for varying periods but often fall off by about 1700 hr. Except for stormy weather, wind velocities from July through November are the lowest for the entire year (monthly $\bar{X} = 6.8-8.7$ km/hr). In November average velocity increases to 9.4 km/hr, and from December through June monthly averages are from 11.2 to 14.7 km/hr, with the three highest months February, March, and April.

Wind direction changes drastically, depending on the season, and is generally dominated by the direction of low pressure areas associated with either the northeast or the southwest monsoon season. Wind direction is not as important for this species as for *V. komodoensis* and *V. salvator,* which frequently locate their carrion foods by means of wind-carried scents (Auffenberg, 1982). The effect of wind on activity patterns of *V. olivaceus* is discussed in chapter 5. In general, sustained wind reduces butaan activity, though some may initially move to shelter to escape the wind. Heavy gusts at the beginning of storms often lead to increased foraging.

Barometric pressure.—Barometric pressure at the near-sea-level base camp remained relatively high, with little variation throughout the year (755–760 mm Hg) except during the monsoon periods, when it sometimes dropped as low as 744 mm during cyclonic storms. There is no evidence that barometric pressure affects the behavior of any species of *Varanus* (Auffenberg, 1981a), and no data obtained during this study suggest that it does. However, the accompanying rain and wind certainly do affect the behavior of the butaan.

Relative humidity.—Monthly average relative humidity at the base camp varied from 72.2% (April) to 96% (December). These values are sometimes a little lower than those in dense forest (fig. 4-4). There is no evidence suggesting that relative humidity is an important factor in butaan behavior.

Solar radiation.—In the Philippines solar radiation is fairly low compared to open, relatively cloudless regions, such as the Lesser Sunda Islands. Data from Quezon City, Manila area (no closer station available), show that solar radiation varies from a low monthly value of 302 cal/cm^2/day in February to 518 cal/cm^2/day during April (fig. 4-4, all data from Lof et al., 1966). The first few months of the dry season usually have the most hours of sunshine, the rainy months (June–July, October–November) the least, but all depends on the cloudiness of the southwest monsoon.

Figure 4-6 summarizes the macroclimatic data of the base camp at Caramoan in the style of Walter (1973) so that comparisons can be made with other parts of the world.

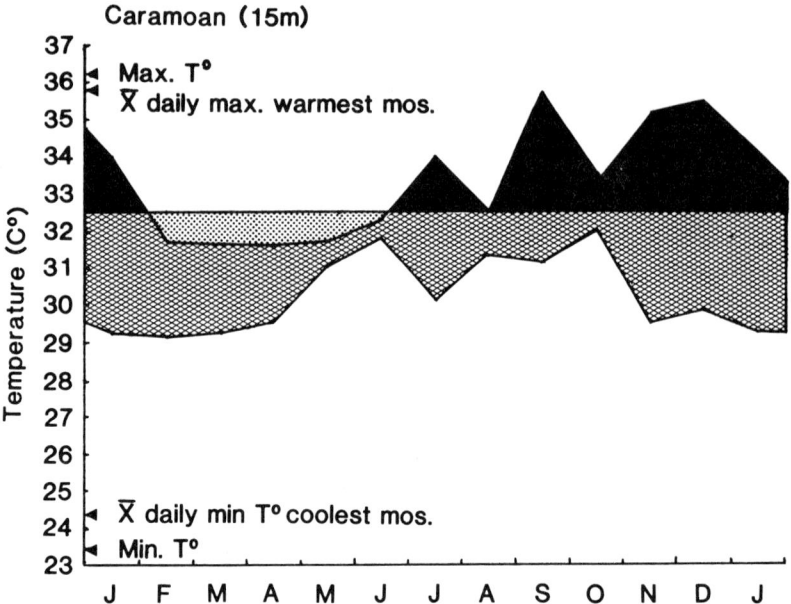

Fig. 4-6. Mean monthly climatic factors, Caramoan study site, plotted in the manner of Walter (1973) for comparison with data from other areas. *Lower curve:* temperature C°. *Upper curve:* total monthly precipitation; *solid,* over 100mm/mo; *dotted,* a deficit for tropical evergreen forest (Walter, 1973; McClure, 1966) and thus a relatively dry period (though humidity may still be high for these months). Mean annual T = 30.2; mean daily T variation = 3.7; mean annual precipitation = 2858 mm. Record duration = 1.75 yrs.

Light.—Daylight duration varied from 11 hrs 24 min in December to 12 hrs 52 min in June. Sunrise varied from 0534 hr in June to 0618 hr (local standard time) in December, and sunset was from 1742 hr in December to 1820 hr in June. In general, the diurnal activities of *V. olivaceus* individuals are geared to sunset and sunrise, so that these differences are of some significance in their activity cycles and thus in their energy expenditure. Light levels within the habitat, those actually experienced by the butaan in its movements through the forest, are discussed below.

In general, the macroclimatic factors important in the daily life of the butaan show little significant variation throughout the seasons. One corollary of these small seasonal changes, compared with temperate climates, is that several years may pass without a critical period in which there is a shortage of resources. It is at such times that competition between coexisting animal species often becomes apparent and consequential to their survival. The major drought conditions prevailing over much of the archipelago during the spring of 1982 obviously produced a year of such critical shortages. Still, there was no evidence that the lack of water or food, or both, affected the local *V. olivaceus* populations—though there are considerable data to show that the same conditions greatly stressed the local populations of smaller lizards, especially the skinks. The potential competitors of the butaan are discussed in greater detail in chapter 13.

Vegetational Factors

The principal habitats of the butaan are dipterocarp and secondary forests (pl. 3). The former is thought to be the climatic vegetational climax throughout most of Southeast Asia, including the Philippines. It extends (or did extend) over most of the lowlands to heights of about 1000 m. It receives its name from the tree family Dipterocarpaceae, species of which dominate the association. It is a tall, tropical forest that occupies localities most favorable to tree growth. In the Holdridge (1964) system it would represent a major Asian association of the tropical lowland moist life zone. Structure in dipterocarp forest varies greatly, so generalities are difficult. However, most trees have slender trunks with thin bark, and the crowns tend to be high and small due to crowding. Another characteristic is high floristic diversity; it is certainly one of the richest plant associations in the world. (Woody plant species in dipterocarp forests in Borneo number 4000 in 8–9 ha. Ashton [in Fogden, 1972] reported nearly 550 tree species with a girth of a half meter or more; in the Malay Peninsula, Whitmore [1975] stated there are 2500 species of trees.) It has only a few widely distributed tree species, most of which are rather restricted both ecologically and geographically. This well-developed local endemism (up to 85%; Fogden, 1972) makes it difficult to provide a detailed description of the dipterocarp communities of which the butaan is a part. It means that thousands of species could be included in descriptions of the primary forests of Luzon and Catanduanes in which the butaan occurs. It is obvious that only the most general descriptions can be included here.

The secondary forest is more open and is characterized by many more shrubs and herbaceous plants. It is less uniform in temperature and humidity than dipterocarp forest. With the advent of man on these islands and the accompanying forest clearing, this habitat has been greatly extended.

Lowland and hill dipterocarp forest types.—Dipterocarp forests extend from northern India through Sri Lanka, Burma, the Malay Peninsula, western Indonesia, and into the Philippine Islands. Dipterocarp species usually occur below 1000 m and generally show the greatest development below 100 m. In the Philippines these forests (9 dipterocarp genera, 50 species) originally covered over 75% of the islands (ca 48,000 km^2) and contained 95% of all standing timber.

While some dipterocarp forests are composed almost entirely of a single dipterocarp species (i.e., *Shorea robusta* forests of India, *Dipterocarpus tuberculosus* forests of Burma; Brandis, 1895), most have several to many species (as do the Philippine types). But, in most cases, the proportion of trees belonging to this family is usually small. Therefore they are often referred to as mixed dipterocarp forests. However, owing to the large size of most dipterocarps, they give a characteristic appearance to the vegetation and usually comprise a large proportion of the timber. In a Hill Dipterocarp Forest on Luzon, Brown and Mathews (1914) reported that of 92 large tree species only 2 were dipterocarps; Whitford (1911) reported 7 different dipterocarp species out of 120 tree species in a Luzon Lowland dipterocarp type.

Whitford (1906) divided the Philippine dipterocarp forests into five different types, more or less corresponding to those recognized more recently in Malaysia (Wyatt-Smith, 1953). Whitford's major divisions were Lowland (to 300 m), Hill (to 750 m), and Upper (ca 1200 m) Dipterocarp Forests. The first two belong to the Lowland Evergreen Formation (Whitmore, 1975) and the Upper to the widely distributed Lower Montane Formation of many other parts of the world (fig. 4-7). The water monitor's range extends to about the upper limit of the Upper Dipterocarp Forest, but the butaan is restricted to the Lowland Evergreen Formation (Lowland and Hill Dipterocarp Forests). The Lowland Dipterocarp Forest can be further divided into a number of communities, of which two—*Vitex-Ficus* and *Pentacne-Hopea*—are the most important habitats of the butaan in and near Caramoan. Of the Hill Dipterocarp Forests, the most extensive community is the *Shorea-Hopea* type.

Though mixed dipterocarp forest is often described as structurally layered, the layers are not always well developed (Fogden, 1972). However, layering

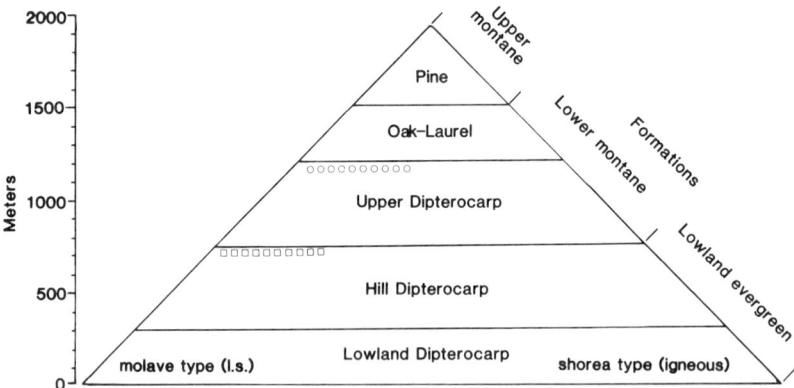

Fig. 4-7. Plant formation and forest type in the range of the butaan. Hollow squares represent the approximate vertical limit of *V. olivaceus;* circles represent the limit of *V. salvator.*

Fig. 4-8. Mixed dipterocarp forest of Caramoan Peninsula, generalized, showing its major structural features. Forest layers are *1*, shrub; *2*, main canopy; *3*, emergent.

does occur in the Caramoan area, where three tree strata can be recognized, in addition to shrub and herbaceous layers (fig. 4-8).

In the Caramoan *Vitex-Ficus* (Molave) Association the top story is formed of trees with slender trunks that often remain unbranched for as much as 20 m above the ground. (This is so because low branches are a competitive liability in vine- and scandant-rich vegetation [Walter, 1973]. In spite of this, butaans can and do frequently climb to the tops of these trees.)

Though trees of this strata are the tallest in the forest (30–40 m high locally), their crowns do not touch. Thus each of these trees projects above the general forest level, and they are usually called the sentinel or emergent trees of the association. These sentinels are considered the dominant trees of the forest and in this case are usually members of the Sterculiaceae and Dipterocarpaceae.

The next lower level (fig. 4-8) is composed of many more species of smaller trees whose canopies approximate the height of the lowest branches of the emergents. In this layer (5–20 m) the crowns of the trees tend to fuse, forming a more or less solid canopy. Beneath this layer is yet another composed of still smaller tree species and shrubs, mixed with seedlings and saplings of larger trees, rattan palms, pandans, and bamboos (sometimes over 100 sp/ha) (fig. 4-8). The lowest layer is near ground level and consists of herbs, including ferns and seedlings; grasses and sedges are found only in the well-lighted parts.

The *Vitex-Ficus* Community near Caramoan is characteristic of drier soils and limestone hills throughout the Philippines. It is dominated by molave (*Vitex parvifolia*), a dipterocarp that grows most luxuriantly near the coast. Figs (*Ficus*) are particularly rich in species represented, as well as form, size, and fruiting characteristics. Other common large trees are *Tarrietia sylvatica, Kingiodendron alternifolium, Aglaia harmsiana, Zizyphus talanai, Wallaceo-*

dendron celebicum, Sterculia spp., and the dipterocarps *Pterocarpus vidalianus* and *P. indicum.* Common smaller trees are *Mallotus* sp., *Cassia javanica, Maba buxifolia,* and *Canarium* (several species). Appendix 1 provides additional information on the common plants of this community. It is characterized to some extent by the large number of trees producing edible fruits (some of which are eaten by the butaan; chap. 9). It also has a wealth of vines and scandant palms, shrubs, and bamboos. Epiphytes are not a dominant feature; when they occur, they tend to grow in the canopy layer.

The butaan regularly moves through all these levels in the local forest, climbing to the top branches of the emergents, sheltering in the vines and scandant shrub layer of the canopy at night, and foraging primarily in the herbaceous layer at the foot of the trees (see below). (In a progress report of my work with this species [1979a] I mistakenly overemphasized the arboreal habits of this species, failing to recognize the importance of movement at and below ground surface.) Figures 4-9 and 4-10 illustrate typical forest situations in which *V. olivaceus* is found in the Caramoan area.

Fig. 4-9. Cross section (120 m, somewhat idealized) of typical butaan habitat (Mabho I transect area), showing a number of emergents, a dense canopy layer producing heavy shade, and a shrub layer in which species are few and widely scattered. However, ferns and other low-light-adapted plants may be abundant in the lowest forest layer. Individual trees of the only species whose fruits are eaten by the butaan are C, *Canarium hirsutum;* Cc, *Caryota cumingii;* and Fb, *Ficus benjamina.* The solid diamond indicates where the resident butaan was first captured.

Fig. 4-10. Cross section (120 m, somewhat idealized) of typical open butaan habitat (Matingapo I transect area) on extremely rocky hillside, showing few, relatively low emergents and high light levels near the surface, resulting in a varied and dense shrub layer. Individual trees of the only species whose fruits are eaten by the butaan are A, *Aglaia harmsiana;* C, *Canarium hirsutum;* and Fn, *Ficus nota*. The solid diamond indicates where the resident butaan was first captured.

Though the *Vitex-Ficus* Community is certainly the most widespread in the karst limestone areas, other lowland types probably occurred on the flat valleys before they were largely converted to rice fields many years ago. Much of the *Vitex-Ficus* Community has also been destroyed, but because it is not suitable for agriculture much also remains in a more or less undisturbed state.

The butaan is also found locally in other communities. The most important of these is the *Pentacne-Hopea* Community, which is probably best preserved near Katchigawan in the Caramoan National Park. Though developed on a limestone-derived substrate, the soil is richer and deeper than in the *Vitex-Ficus* community. The emergents are mainly *Pentacne contorta* and *Dipterocarpus gigantifolia*—both dipterocarps. Even the major canopy species, *Hopea basilanica*, is a member of this family. This forest is most typical of the main body of Whitford's Lowland Dipterocarp Forest. In general it is similar to the *Vitex-Ficus* Community, and the butaan is commonly found in it. Details of the vegetational differences are described below.

In the Caramoan area the Hill Dipterocarp Forest is represented by the

Shorea-Hopea Community, which is usually found at elevations over 300 m. The soil is less organic than in the molave type, being a poorly drained red clay. The emergents are composed almost entirely of dipterocarps (80%), with species of *Shorea* (*S. squamata* and *S. teysmanniana*) the most prominent. The dipterocarp *Hopea pierri* is the dominant canopy tree. There is also a well-developed herbaceous layer, though less rich than in the molave type because of the somewhat more shaded conditions. In some places where butaans occur, *Shorea polysperma* and *Anisoptera thurifera* are the dominant dipterocarp emergents.

As indicated, the butaan is rarely found above 650 m. It is at this elevation that the Hill Dipterocarp Forest changes its character in several important ways: it becomes essentially two-layered; the large trees that dominate the lower forests are replaced with different, usually smaller, nonbuttressed species; tree ferns become more abundant; dipterocarps generally become rare; and, most important, many trees with large edible fruits dispersed primarily by animals are replaced with nonanimal-dispersed types. Most of the food fruits of the butaan are therefore restricted to lowlands. The line of demarcation between hill and upland forest associations may be sharp or gradual, depending on water availability, slope, and exposure. In the Caramoan Peninsula area Upper Dipterocarp Forest occurs only on the tops of the ridges making up the spine of the peninsula and on the upper slopes of some of the volcanic cones dotting the area (Isarog, Mayon, etc.).

Secondary forests.—In addition to mixed dipterocarp forests, the butaan also occurs in certain kinds of secondary forest. These contain few or no dipterocarp species. Secondary forests occur wherever man has exploited the soil and has felled the larger forest trees.

The secondary forest, sometimes covering extensive areas, is different from the primary forest in species and in genera. Older secondary forests are composed of a great variety of small trees and shrubs, often with a considerable number of vines. Many of the trees have relatively short trunks and often branch freely like shrubs, with the result that ordinary secondary forest is difficult to walk through. There may also be more herbaceous species and grasses present than in undisturbed primary forest, particularly in more or less open places.

It is almost impossible to consider all the species that may occur in secondary forests of southern Luzon or, for that matter, even all the common ones. In addition, in regions where commercial contacts have been of long standing, one often finds a number of introduced and naturalized forms. In the Caramoan area these include *Sandoricum koetjape, Psidium guajava, Leucaena glauca, Acacia farnesiana,* and *Lantana camara.* With the exception of the almost completely naturalized *Sandoricum koetjape, V. olivaceus* is unknown to feed on any of these introduced fruits.

While endemism in primary forests may be as high as 75% even mature secondary forests rarely contain more than 10% endemics. Thus the primary dipterocarp forests must represent the original vegetation of the Philippines (see chap. 14). These great forests were undoubtedly more extensive before the advent of man, or at least before agriculture became established. Thus a high percentage of the secondary forest constituents—presently so widespread—represents species that originally were restricted to small isolated patches but were spread by wind, lightning, landslips caused by heavy rainfall, and river floods. Species of the secondary successional stages colonize and grow in these natural gaps in the forests. These species, unable to regenerate in their own shade, live only a single generation and are gradually replaced by species of primary forest, as the latter revert over a long period to the local climax type. Secondary forest species must, therefore, have efficient means of long-range dispersal if they are to survive, particularly as the forest openings they colonize are usually small and isolated compared to the large areas of primary forest that surround them. In fact, many of the species bear edible fruit and so are dispersed by animals, including *V. olivaceus* (see chap. 12). Furthermore, they tend to fruit at regular intervals, often several times a year (see chap. 10).

During the past 100 years the secondary forest area of the Philippines has been increased greatly, so that much of what was originally primary has now been converted to this secondary type. Lately the destruction of primary forest has been significantly increased, mainly because of human population pressure in areas that had been settled long ago and colonization of forested lands where population pressures were low.

Local detailed vegetational studies.—Several vegetational factors of the Caramoan area were studied in detail, both in places where butaans were captured and in places where they were not. Twelve sites were selected for study (see fig. 1-1 for locations). At each of these sites six blocks, each 10×100 m ($= 1000$ m^2), were marked out. All potential food trees of *V. olivaceus* were plotted in each block (fig. 10-3) and measured (bhd, h, and crown w); substrate, herbs, and shrubs were noted in terms of their general characteristics; 80 trees larger than 10 cm bhd were classified in regard to three height classes (<10 m, $10-25$ m, >25 m); and the general leaf shape and texture (grasslike, medium, small, broad, and large; compound or simple; etc.) were noted for the closest trees in each of four quadrants (plotless method) in 20 extra sample plots at each of the 12 sites ($4 \times 20 = 80$ samples/site $\times$ 12 sites $= 960$ data sets). Techniques are outlined in chapter 1, and results are shown in tables 4-2 and 4-3.

Table 4-2 shows that emergent trees at butaan home sites constitute from 0 to 19% of the larger trees, that those making up the canopy comprise from 14 to 65% of the total larger trees, and that the subcanopy trees are the most

TABLE 4-2. Forest characteristics in *Vitex-Ficus* dipterocarp forest near Caramoan. Values in percent of totals at each site. Number of data sets per site = 80, except for Ahop, where number of data sets = 134.

Place	Tree height			Leaf shape[a]				Crown cover	
	<10 m	10–25 m	>25 m	G	M	B	V	Incom-plete	Com-plete
Kabutanan[b]	86	14	0	1	35	58	6	100	0
Ahop II[b]	29	65	6	1	12	76	11	72	28
Gota	40	52	8	4	14	58	16	86	14
Mabho I[b]	48	33	11	0	10	77	13	70	30
Mabho II[b]	32	49	19	10	3	86	7	50	50
Matingapo[b]	43	44	16	0	10	78	12	70	30
Ibu I[c]	92	8	0	2	3	79	16	59	41
Ibu II and III[c]	94	4	2	0	61	16	24	24	76
Batungan[b]	87	13	0	0	13	32	8	85	15
Bulang-bugang[b,c]	50	50	0	0	7	75	14	67	33
Tiak Lake	78	28	0	0	55	35	6	59	41

a. G = grasslike; M = medium-small; B = broad, large; V = compound.
b. Sites at which butaans were taken.
c. Secondary forest; others are mixed dipterocarp.

common (from 29 to 94% of the total). There is no significant difference between sites at which butaans were taken and those where they were not; the only major difference was between secondary and mixed dipterocarp forests (primarily reflecting the lack of tall trees or a well-developed canopy in secondary forest). Leaf shapes varied considerably and showed no obvious pattern. All forests are dominated by broad-leafed trees. Crown continuity is variable with no correlation with presence or absence of butaans; the only correlation is that crowns tend to be more complete in secondary forest (butaans occur in secondary forest, though not as commonly as in the primary dipterocarp types).

Table 4-3 shows that butaans most frequently occur in forests with trees an average of 1.5-3.0 m apart ($\bar{X}$ = 2.5 m); few occur in forest sites with more widely spaced trees (4.33-4.5 m apart, $\bar{X}$ = 4.41 m). There was no difference in modal tree height in blocks with or without butaans (mode with butaans 2.5 m bdh, without 2.0).

Ecological distribution.—Perhaps one of the most important ecological discoveries about *V. olivaceus* is the fact that it is restricted to mountainous areas, below 750 m (fig. 4-7). It is also found (though perhaps somewhat less abundantly) in well-developed secondary forest. It is not found in the extensive *Imperata*-dominated grasslands of Luzon (= Gogon lands) or in any intermediate seral stages between this and secondary forest, probably because the fruit trees on which it depends, and perhaps many of the land snails that make

TABLE 4-3. Mean distance between trees 10 cm bhd and modal tree height class (see table 4-2) in *Vitex-Ficus* dipterocarp forest near Caramoan.

Place	N	Distance between trees (in m) X̄	SD	Modal tree height class (in m)
Kabutanan[a]	80	2.54	2.05	<10
Ahop II[a]	85	4.33	2.96	10–25
Gota	86	4.50	4.27	10–25
Mabho I[a]	81	2.12	1.53	<10
Mabho II[a]	81	2.85	1.95	10–25
Matingapo I[a]	80	2.50	1.99	10–25
Batungan[a]	104	2.32	1.50	<10
Bulang-bugang[a]	84	2.65	1.90	10–25
Tiak Lake	80	2.31	1.93	<10

a. Butaan sites. For all butaan sites, X̄ = 2.50, SD = 2.9; for all nonbutaan sites, X̄ = 4.41, SD = 4.3.

up its diet as well, are absent from these pioneer, more open associations. Nor are the trees and snails found in the extensive nipa swamp and mangrove forest associations in the area. Thus this species is unlike the sympatric *V. salvator,* which is found in all of these habitats plus riverine swamp forests of several types, marshes, lake borders, and coastal scrub. Even the geographically highly restricted *V. komodoensis,* which has a smaller total range than that of the butaan, lives in most habitats within its small range (Auffenberg, 1981a).

Within its ecological setting, the butaan regularly spends as much time under the surface of the ground as in the tops of the tallest emergent trees. Thus, vertical utilization of the habitats among both young and adults is greater than for *V. salvator* (though the latter can climb the tallest trees, it does so infrequently).

As will be shown in chapter 5, the butaan is rarely stressed by environmental conditions within its habitat. Shelter is available almost everywhere; temperatures never reach lethal extremes, rarely varying more than 5°C more or less than 30°C—an ideal temperature from the standpoint of reptilian physiology; and water is almost always available nearby. This thermal uniformity and the general availability of shelter, water, and food throughout the year provide almost perfect conditions for the butaan. As will be shown, this "easy life" has so many benefits that it is difficult to understand why it has not evolved for other varanids living in tropical forests of Asia and Africa. Compared to the life-styles of *V. salvator* and *V. komodoensis,* the life-style of the butaan is reflected in a lowered reproductive rate and unusually low activity levels, including much hiding and slow locomotion (see later chapters).

The structurally and taxonomically diverse dipterocarp forests of Asia provide a large number of potential niches into which numerous vertebrates, including the molluscivorous-frugivorous *V. olivaceus*, have radiated. As shown, it inhabits a rich plant association and experiences an equable climate. However, it shares this environment with a wide variety of distantly related animal competitors (chap. 12). In these circumstances it seems probable that any selective forces of significance are exerted by the butaan on other organisms, and vice versa, rather than by the physical environment on the butaan. This situation is in distinct contrast to that of varanids of more open environments for which the physical environment is clearly very important as a major limiting factor (*V. komodoensis, V. bengalensis*, etc.; see especially Auffenberg, 1981a).

Microecology

Vegetation

Vegetation plays an important role in the life of the butaan. Its density limits visual perception and modifies temperature and humidity close to the ground. The canopy supports arboreal thickets that serve as important retreats for *V. olivaceus* (see chap. 13), and its shape and density determine to a large degree the pattern of fallen fruits on the forest floor (chap. 10). Different associations in an area contain differing densities of food plants, some producing more food than others. Plant cover is also important to the snails upon which the butaan regularly feeds.

Ecotonal zones are less important to the butaan during foraging than they are to other large monitors. These zones are, however, important barriers to dispersal of the butaan; they seem loath to cross them. On the other hand, these same zones are the ones most frequented by the sympatric *V. salvator*, for it is here that small animal densities are usually highest. The large carnivorous *V. komodoensis* uses ecotonal zones for launching ambush attacks against some of its prey (Auffenberg, 1981a).

The fact that large butaans use trees for basking, shelter, and feeding means that at least certain trees are of greater importance to the butaan than to varanids inhabiting more open situations. In most other large monitor species only the juveniles make much use of trees.

Arbohabitat.—In large trees the butaan is often immersed in thick vegetation that is not a part of the parent tree but results from a host of other plants that use the tree as a substrate on which to grow—various epiphytes, scandant shrubs, and vines. In some areas these plants may make up a significant part of the total biomass of the local forest.

TABLE 4-4. Vines, scandant shrubs, and epiphytes important in the biology of *V. olivaceus*.

Species	Maximum height above ground (in m)	Type	Importance
Bauhinia sp.	7	Broad-leafed vine	Retreat
Calamus mollis	20	Scandant palm	Retreat
Dendrobium sp.	20	Epiphytic orchid	Retreat
Dinochloa scandans	20	Scandant bamboo	Retreat
Lygodium japonicum and *L. flexuosum*	6	Scandant fern	Retreat
Ophioglossum pendulum	10	Epiphytic fern	Retreat
Rhaphidophora merrillii	6	Stout, broad-leafed vine	Retreat
Schizostachyum diffusum and *S. dielsianum*	20	Scandant bamboo	Retreat
Uvaria sorsogonensis	15	Vine	Food

Of the various plant groups that make up this arbohabitat the scandant shrubs, vines, and epiphytes are the most important to the butaan, for they are often used as shelter both day and night (chap. 13). Table 4-4 lists the major species in the Caramoan area and their importance to the biology of the butaan.

In general, the biology of tropical vines and scandant shrubs is poorly understood. However, it is important to *V. olivaceus*, for in these forests vines and shrubs may produce a crown as extensive as that of a large forest tree. But not all species are equally important as cover: some have diffuse crowns and some dense, some are evergreen and others completely and regularly deciduous. The rapid growth and high opportunism of vines and scandant shrubs mean that their crowns often change location. Such changes undoubtedly affect the biology of the organisms associated with these plants. The fruit of only one vine species is eaten by *V. olivaceus* (see chap. 9).

Rocks.—The nature of rock crevices and their use by butaans is discussed in chapter 5. Figure 4-11 shows that the crevices are consistently cooler than the surface of the forest during all months of the year (cooler 215 days, warmer 99 days, same 51 days), and that the average differential varies from 0.30 to 2.11°C, with the greatest differential in April and May. This large differential corresponds to the period of lowest rainfall, highest rate of insolation, and highest level of deciduousness of the forest trees, as well as fairly high general ambient temperatures. During this period butaan activity levels are high, and most movement is at the ground surface (chap. 5). In addition,

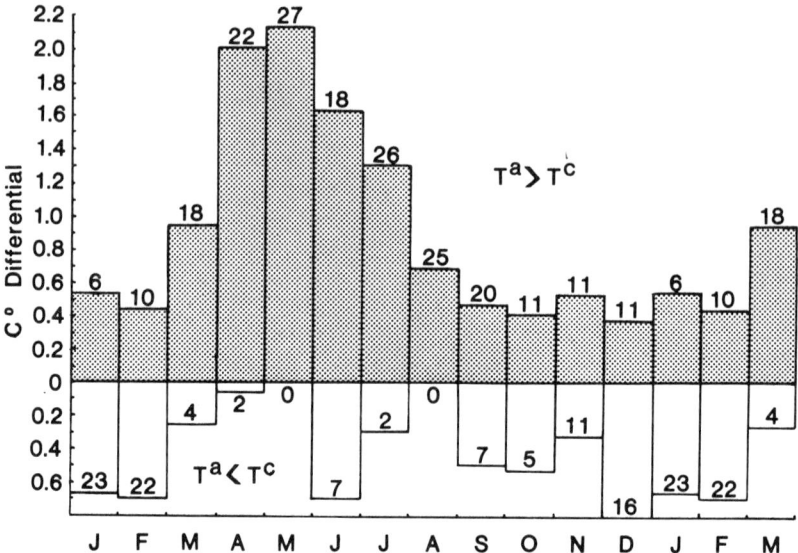

Fig. 4-11. Differentials (C°) between average daily surface (T^a) and rock crevice (T^c) temperatures for different months. Numbers show the total days each month when T^c was more (dotted) or less (open) than T^a. Days not accounted for in the total obtained by adding top and bottom monthly numbers represent days when mean surface and crevice temperatures were the same.

there are periods during most months when crevice temperatures are higher than those at the surface (though the differential is always less) and the annual pattern is less clear.

Temperature.—Though the temperature of the tropical forests in the Caramoan area is rather uniform, there is an important vertical thermal stratification—even during a single day (table 4-5). The data available show that the highest average temperatures are found at the highest recording station (10 m) at all times of the lighted part of the day, the highest averages are about 4°C higher than the lowest average recorded (subsurface), and the greatest differential temperatures are found from about 0900 to 1300 hrs each day. In general, the average temperatures of the subsurface, surface, and 10-cm levels are not significantly different from morning through afternoon and into night, nor are they different from one another at these levels. Thus, unlike the situation in open savanna habitats (Auffenberg, 1981a), all temperature levels close to the ground are essentially the same, so that the boundary effects at these levels are practically nonexistent. The average temperature at 1 m is intermediate between the lowest level and that at the 10-m height.

In general, higher levels show greater daily variations than progressively lower levels (OR of SD = 1.0–5.8 at 10 m, 0.7–1.8 at 1 m, 0.7–1.4 at 10 cm, 0.3–0.8 at surface, and 0.2–0.9 at subsurface). Thus the lowest temperature variability is found at subsurface and surface levels, with progressively higher levels showing progressively lower temperature reliability. When viewed on an hourly basis, morning temperatures are of lower variability at all above-surface levels than are temperatures at those levels at any other time of the day. Variability remains low throughout the rest of the day in the forest canopy but is reduced in all lower levels later in the afternoon. From the standpoint of overall temperature range at different times (measured by difference between highest and lowest), there is a shift in the daily pattern whereby the higher the station, the longer the period during the day where there is a significant difference between the highest and lowest readings made for those hours. In general, there is no noticeable effect in the daily patterns at any level that can be correlated with the change in incoming and outgoing radiation patterns (resulting in general thermal instability from 1000 to 1100 hrs), as is found in open savanna habitats and forest clearings (such as at the base camp). More important, the lack of both early morning thermal inversion and mid-morning instability results in no noticeable change in butaan activity patterns, such as has been shown to occur in species inhabiting savanna and deciduous forests (Auffenberg, 1981a).

What these variations mean to the butaan is that lower levels are reliably cooler at all hours of the day and that there seems to be little difference in temperatures at the subsurface, surface, and 10-cm levels. In general, relative humidities are expectedly higher at these levels as well. While moving upward into the trunk and canopy levels generally takes the butaan into slightly warmer conditions, this change can be reliably counted on only during the morning, for the afternoon pattern shows considerably more variation. The highest

TABLE 4-5. Vertical profiles of temperature (°C) in mixed dipterocarp forest (rainy season, July–November).

Clock intervals (hrs)	Subsurface		Surface		10 cm		1 m		10 m	
	$\bar{X}$	SD	$\bar{X}$	SD	$\bar{X}$	SD	$\bar{X}$	SD	$\bar{X}$	SD
01–07	25.5	0.1	25.5	0.4	26.1	0.8	27.3	0.9	27.5	1.1
07–09	25.6	0.2	25.5	0.6	25.7	1.0	27.3	1.8	28.2	1.0
09–11	26.1	0.7	25.8	0.8	26.3	1.4	27.2	1.7	29.9	2.1
11–13	25.9	0.6	26.1	0.8	26.3	1.2	27.4	1.6	30.2	2.4
13–15	25.8	0.5	26.4	0.5	26.7	1.3	28.4	1.8	29.3	5.8
15–17	26.1	0.4	26.1	0.8	25.7	0.7	25.6	0.7	27.2	2.4
17–19	25.4	0.2	25.8	0.3	25.3	0.9	26.4	1.4	26.1	5.1
19–24	25.7	0.9	25.5	0.6	25.1	1.1	26.2	1.4	25.6	5.0

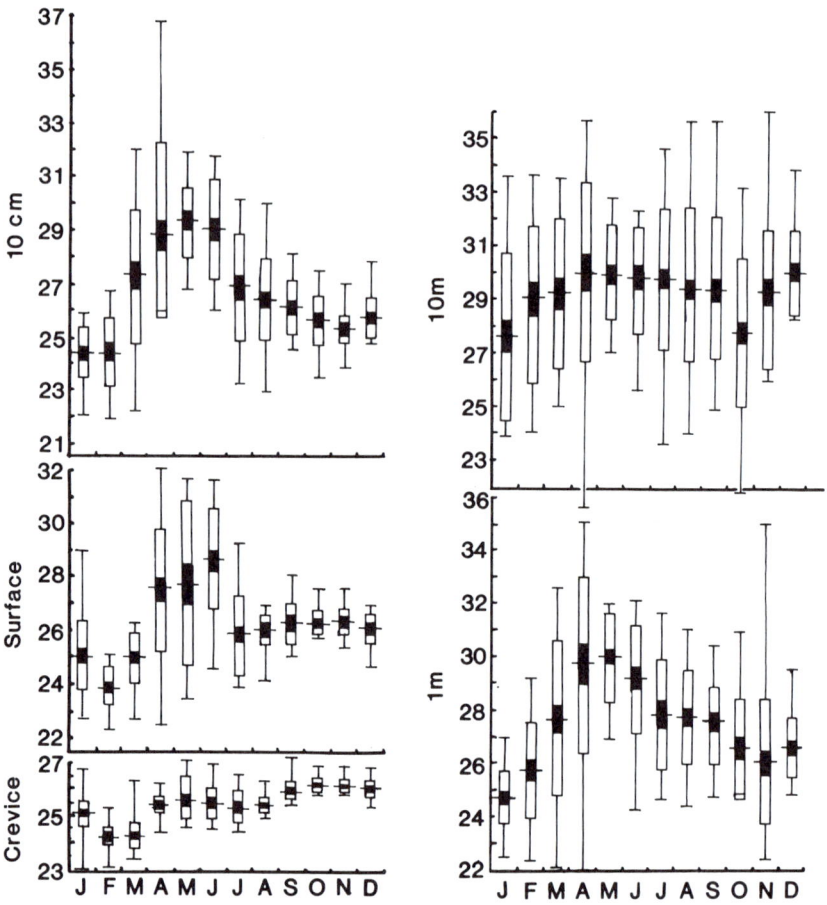

Fig. 4-12. Seasonal variation in temperatures (in C°) at five levels within the Caramoan forest.

shade temperatures are found from 1 m above the surface of the ground to the lowest part of the general forest canopy, and this temperature differential is most reliable from 0900 to 1300 hrs. Thus it is this period when basking would most effectively and most reliably raise butaan body temperatures—and this is exactly when most basking in the trees seems to occur (see chap. 5).

In general, temperatures within leafy arboreal retreats are somewhat ameliorated by the vegetative cover of the vines and scandant shrubs. The average differential maintained during the night is +1.75°C SE ± 0.28 (OR = 0.30–3.16°C). Ambient and retreat temperatures become equal at about 1000–1100 hrs; thereafter retreat temperatures remain in equilibrium with, or slightly below, ambient temperatures outside the retreat. The cool weather during typhoons would generally produce a greater differential during the daytime with warmer retreat temperatures, but during such periods butaans normally move down into rock crevices (see chap. 5).

Figure 4-12 shows the temperatures at crevice, surface, 10-cm, 1-m, and 10-m levels in the Caramoan molave forest for each month.

5

Movement

ONE OF the strengths of this study is the quality and quantity of data obtained through the use of radiotelemetry. The devices used were sensitive to temperature (for use both inside and outside the body), activity, light, and orientation (chap. 1). They were placed on 12 individuals and retained there for as long as 108 days, sending information back to the base camp. Table 5-1 provides data on the individuals telemetered, transmitter types used in each instance, type of record obtained, duration in days, and number of radio contacts established for each experiment. All individuals monitored were adults, representing varying sizes and both sexes. Movement data were collected through all seasons of the year. Certain facets of activity cycles were also studied through direct observation of both captives and free-ranging individuals without transmitters.

In the following discussion, telemetric data received at the base camp plays the major role. Telemetric devices were calibrated not only to the parameters mentioned but to the activity of the butaan when movement intensity was studied with devices sensitive to positional changes. Four categories of activity level were recognized: not significant, low, medium, and high. "Not significant" was based on a calculated standard deviation (SD) of less than 0.30 per 30 min. "Significant" movement was any higher than this, and it was arbitrarily divided into low (SD = 0.30–0.50), medium (SD = 0.51–1.00), and high (SD = 1.01 and above). Standard deviations were determined by transmitter click rates during each 30-min period, registered at 5-min intervals. Controls (placed in rock crevices) showed that background "noise" produced standard deviations of 0.22 to 0.28 ($\bar{X}$ = 0.23) per half hour periods. In addition, captives fitted with the same transmitters were noticeably active when SD of click rates was greater than 0.27. A probability calculation showed that in order to reach a p of 0.02 or less it would require an SD (per 30-min period measured every 5 min) of 0.30 or more.

The Daily Pattern

Local hunters say that some butaans go abroad at night, but I found no evidence that they do so in the wild. On the other hand, new captives are often restless and prowl about their quarters after dark, especially on bright moonlit nights. In fact, they sometimes remain hidden during all the daylight hours of the first few days of captivity. However, once adjusted to their new quarters, none moved after twilight, unless greatly disturbed. Telemetric data support these observations, for there is no single instance in over 20,000 contacts made during the night in which wild monitors could be said to be moving about.

Sleep.—The telemetric traces of sleep are identical to those produced by the nonmoving controls placed in crevices. An analysis of 20,151 contacts during periods when sleep was expected (1900–0430 hr) resulted in essen-

TABLE 5-1. Summary of radio telemetric data used as a basis for activity and movement conclusions.

Butaan no.	Sex	Wt (kg)	Transmitter type[a]	Type record[b]	Months	Total days	Total contacts
18	M	4.2	Tb, Ac	Continuous	Nov	23	5,990
			Tb, Ac	Periodic	Jul–Nov	108	18,900
5	M	1.6	Ac	Continuous	Nov	18	4,020
45	M	1.2	Ac	Periodic	Aug–Nov	78	17,100
6	M	2.6	Tb, Ac	Continuous	Nov	9	1,940
25	M	2.0	Tb, Ac	Periodic	Aug–Nov	55	9,900
35	F	2.3	Tb, Ac	Periodic	Aug–Nov	87	3,060
76	M	3.5	Tb, Ac	Periodic	Mar–May	49	12,600
76			Tb, Ac	Continuous	May	15	15,360
82	M	8.8	Tb, Ac, Li	Periodic	Apr	20	7,200
9	M	8.3	Tb, Ta, Li	Continuous	Jul–Aug	28	6,480
10	F	2.7	Tb, Ta, Po, Li	Continuous	Jul–Aug	22	4,320
11	F	2.8	Tb, Ta, Po, Li	Continuous	Jul–Aug	8	2,880
Totals						520	109,750

a. Ac = activity-sensitive and location-sensitive; Li = light intensity; Po = positional; Ta = ambient temperature; Tb = body temperature.
b. See chap. 1 for details.

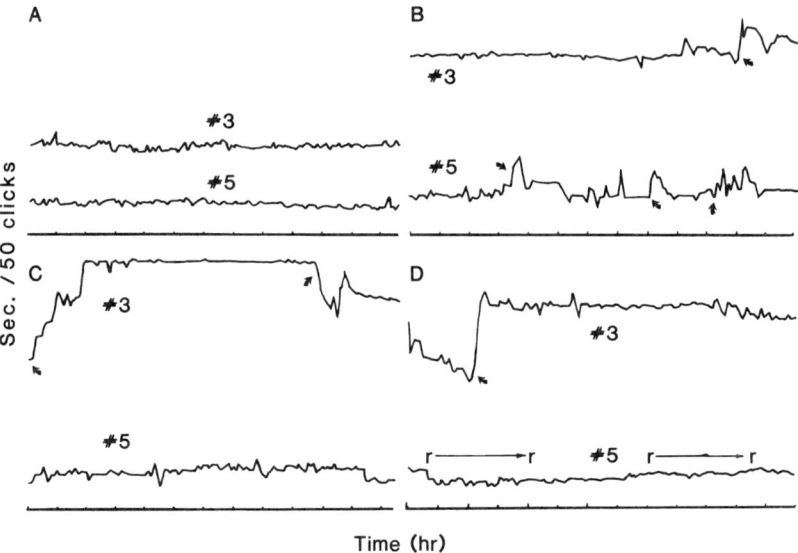

Fig. 5-1. Continuous (once/min) simultaneous telemetric actograms of individuals 3 and 5, showing different nighttime patterns. *A*, typical late night pattern of even, deep sleep for both individuals, with no significant body movements. *B*, typical early night patterns with minor body position changes in 5 from about 2000 to 2110 hr (arrows) and in 3 at 2100 hr. *C*, early night records, showing a normal sleep pattern with minor position changes for 5; 3 moved in a significant way (left arrow) from one site to another (1930–1958 hr); a period of deep sleep followed in the new place (2000–2053 hr) and later (*right arrow*) another significant site change. *D*, the difference in the movement patterns of the two butaans during a night rain (r); 5 slept through the two showers, but 3 moved extensively at 0015 hr after transmitter temperature suddenly dropped, indicating that this individual had been in a much more exposed position than 5. After this 4-min movement to a drier site, undoubtedly nearby, it slept during the rest of the record.

tially straight-line signal traces (fig. 5-1A) with relatively little deviation ($\bar{X}$ SD = 0.12, OR SD ± 0.0–0.16). Most butaans enter this pattern by about 1800 hr. By 2000 hr all individuals are inactive (mode 1930 hr).

However, there are periods in the night when positions are changed, and each of these movements shows clearly in the traces (fig. 5-1B). While most nighttime movements are completed within a few minutes, some involve as much as 43 min of activity. The most significant occur during rainstorms, when individuals are apparently awakened by the dropping water and seek other shelter (fig. 5-1D). Still other movements seem correlated with ambient temperature changes during the night.

In general, positional changes are most frequent during the early evening, from nightfall to about 2200 hr. There is, however, a slight increase in movements between 0400 and 0539 hr, just before they wake and start to move more significantly. The number of recorded positional changes during the night are 1800–2000 hr, total 53 movements; 2100–2300 hr, 12 movements; 2400–0200 hr, 4 movements; 0300–0500 hr, 31 movements ($N = 20$, 151 radio contacts).

Waking.—The butaan usually wakes early. Data from observations of both captives and telemetric records of wild ones ($N = 148$ "wake-ups") show that sleeping butaans often become restless about 0400, with movements that are always of low intensity and short duration and become more common until haul-out time ($\bar{X} = 0717$ hr). Figure 5-2 shows wake-up patterns of three individuals on three successive days. These and many other data show that wake-up time is not coordinated among individuals in the same area—not even when they are in the same retreat. Wake-up rate and time vary among individuals and for an individual on different days. Light intensity is an important factor, though within the shelter considerable activity sometimes takes place prior to sunrise. On overcast days there is often no well-defined morning activity pattern change but usually a continuation of the resting-sleep pattern.

Haul-out time.—The time when butaans leave the nighttime shelter varies from as early as 0522 hr (long before sunrise when the moon is still evident) to 1113 hr and sometimes not at all. Average haul-out time for all records ($N =$

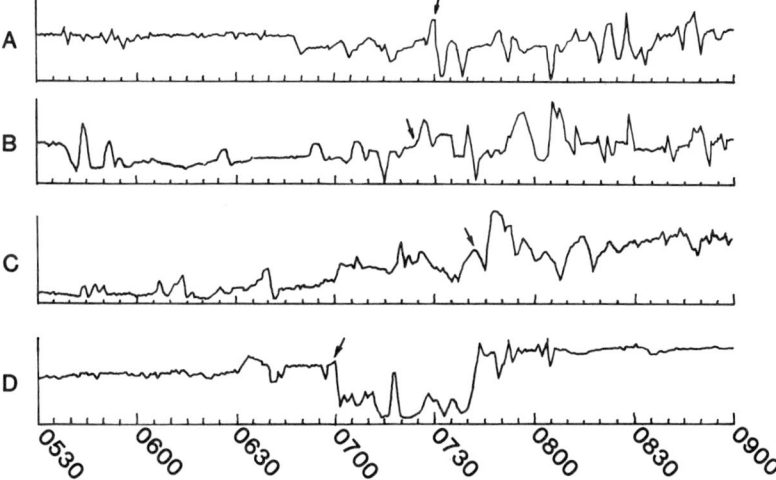

Fig. 5-2. Typical morning movement patterns of *V. olivaceus*, showing varying levels of restlessness before haul-out time (arrows). *A*, sleep continues to 0640 hr, followed by about 40 min of restlessness, then haul-out. *B*, scattered restlessness until haul-out. *C*, the same, though there are more waking movements before haul-out. *D*, relatively little movement before haul-out followed by moderate movement for 40 min, then back into shelter at 0745 hr with a rest pattern following.

416) is 0724 hr ± 6.7 min. There is a seasonal change in which summer haul-out time on sunny mornings (March–May; table 5-2) is 0703 ± 5.1 min ($N = 90$), while for sunny mornings in the fall-winter season (September–January; table 5-2) it is 0730 ± 5.7 min ($N = 140$), with the difference significant at the 0.5% level, which ordinarily would suggest that temperature plays a role in determining haul-out time (as has been suggested for *V. gouldii* by King, 1980). However, the temperatures at 0700 hr in summer and winter are identical (26°C) on sunny mornings so that temperature is not the determining factor for *V. olivaceus*.

Individuals leave their shelters progressively later from July through November (no December records) as the days get shorter (11.2 hr/day 21 December, 12.5 hr/day 21 June at this latitude). This change suggests that light level may be the primary triggering mechanism. Figure 5-2 shows that relatively few major movements are recorded before haul-out time.

Haul-in time.—Haul-in time is the time at which individuals return to their nighttime shelters and remain there for the evening. Unlike the varying behavior in haul-out time, there is no significant seasonal difference in the time at which butaans enter shelters at day's end, nor is the last significant movement of the day correlated with overcast or rainy days. However, haul-in time is much more variable than haul-out time (SD haul-out = 84.2, SD haul-in = 99.3, with difference significant at $p \leq 0.02$), though the difference is less marked in the rainy period. Average haul-in time varies from 1603 to 1805 hr (table 5-3). The average for all records ($N = 303$) is 1647 ± 5.8.

TABLE 5-2. Haul-out time for *V. olivaceus* during different morning weather conditions.

	Sunny			Overcast			Raining		
	N	$\bar{X}$ hr	SE (min)	N	$\bar{X}$ hr	SE (min)	N	$\bar{X}$ hr	SE (min)
January[a]									
February[a]									
March	8	0709	[b]						
April	36	0642	4.9	7	0730	[b]			
May	46	0719	3.1	—			3	0757	[b]
June[a]									
July	29	0728	7.1	—			13	0823	11.8
August	31	0727	5.6	10	0814	6.7	3	0805	8.2
September	49	0734	2.7	25	0803	18.8	11	0844	18.0
October	45	0719	3.1	25	0723	29.8	16	0812	8.9
November	46	0716	1.5	14	0809	3.2	2	0724	[b]
December[a]									
Overall	290	0717	4.5	81	0816	12.0	48	0834	11.6

a. No observations for January, February, June, or December.
b. SE not calculated for $N < 10$.

TABLE 5-3. Haul-in time (latest significant site changes) by month for *V. olivaceus*.

Month	Sunny			Overcast			Raining		
	N	$\bar{X}$ hr	SE (min)	N	$\bar{X}$ hr	SE (min)	N	$\bar{X}$ hr	SE (min)
January[a]									
February[a]									
March	19	1644	13.4						
April	28	1721	8.5				1	1650	[b]
May	13	1653	16.2	2	1632	[b]			
June[a]									
July	12	1748	13.2	3	1758	[b]	5	1650	[b]
August	31	1735	15.2	14	1651	39.3	10	1637	34.3
September	37	1632	29.1	21	1705	23.9	16	1603	14.9
October	45	1606	37.7	17	1631	19.2	21	1604	20.6
November	6	1805	[b]	1	1700	[b]	1	1713	[b]
December[a]									
Overall	191	1696	27.3	58	1679	27.5	54	1615	23.3

a. No observations for January, February, June, or December.
b. SE not calculated for $N < 10$.

Daylight movement patterns.—Most radio contacts with free-ranging butaans were obtained during daylight hours. Figure 5-3 shows the hourly pattern of all significant movements; it is clear that the overall pattern is essentially unimodal, whereas all other varanid species studied thus far are strongly bimodal (*V. komodoensis,* Auffenberg, 1981a; *V. gouldii,* King, 1980; and *V. bengalensis,* Auffenberg, field notes). In addition, all these species also show considerable seasonal variation in the degree and timing of bimodality, and the species found in xeric habitats exhibit considerable fluctuation in daily and seasonal body temperatures. Extreme environmental change is not found in the area in which the butaan occurs, and this study represents the first investigation of a varanid inhabiting mesic forest.

There is no significant hourly difference in movement intensity of the butaan from 0800 to 1600 hr—all daylight hours are almost evenly represented. However, both before 0700–0800 hr and following 1600–1800 hr (the rather uniform midday period), there is a significant decrease in activity ($p \leq 0.01$) which coincides with the periods after sunrise and before sunset. The even greater reduction in movement from 0600 to 0700 hr and 1800 to 1900 hr occurs at sunrise and sunset, regardless of season. The two lowest values at either end of the illustration represent movements prior to and following haul-out and haul-in times.

There are no significant seasonal changes in daily movement patterns when both sexes are analyzed separately ($p = 0.12$). But a difference appears and reaches a statistically significant level ($p < 0.01$) when the rainy months of September and October are calculated separately (females CV = 3%, males

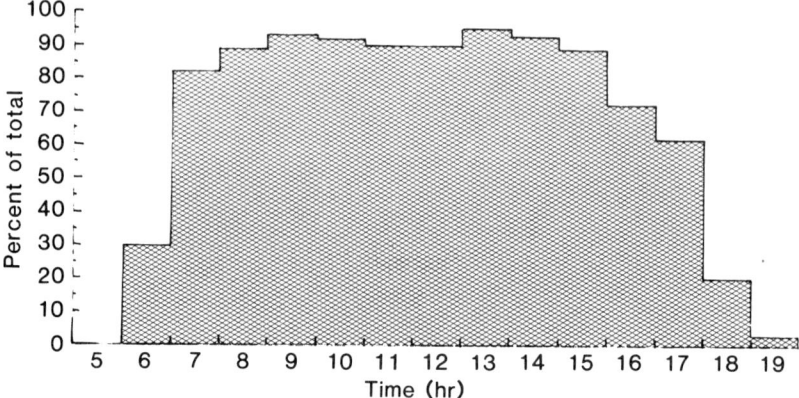

Fig. 5-3. Combined records of movements from all contacts (= 100,740) with telemetered individuals, showing diel activity pattern. All seasonal and sexual differences ignored.

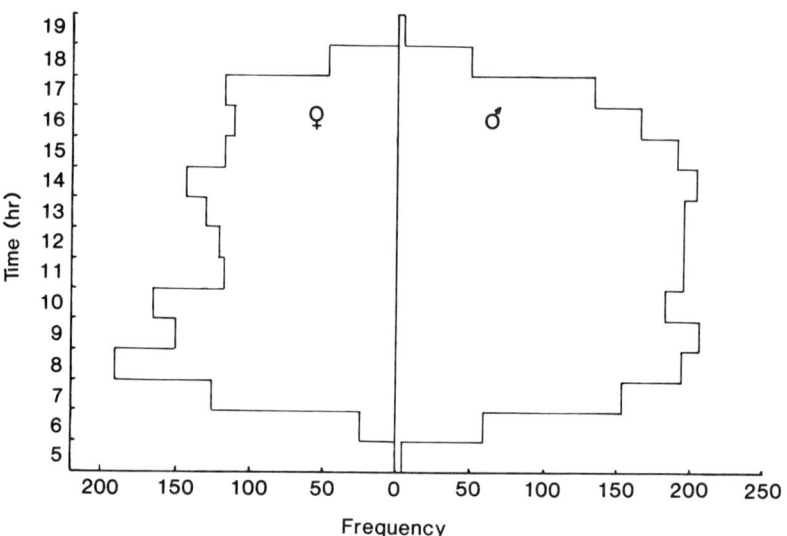

Fig. 5-4. Comparison of movements of female (*left*) and male (*right*) *V. olivaceus* during different times of the day (September–October), showing that males tend to move more uniformly throughout the day and females tend to move more in the morning.

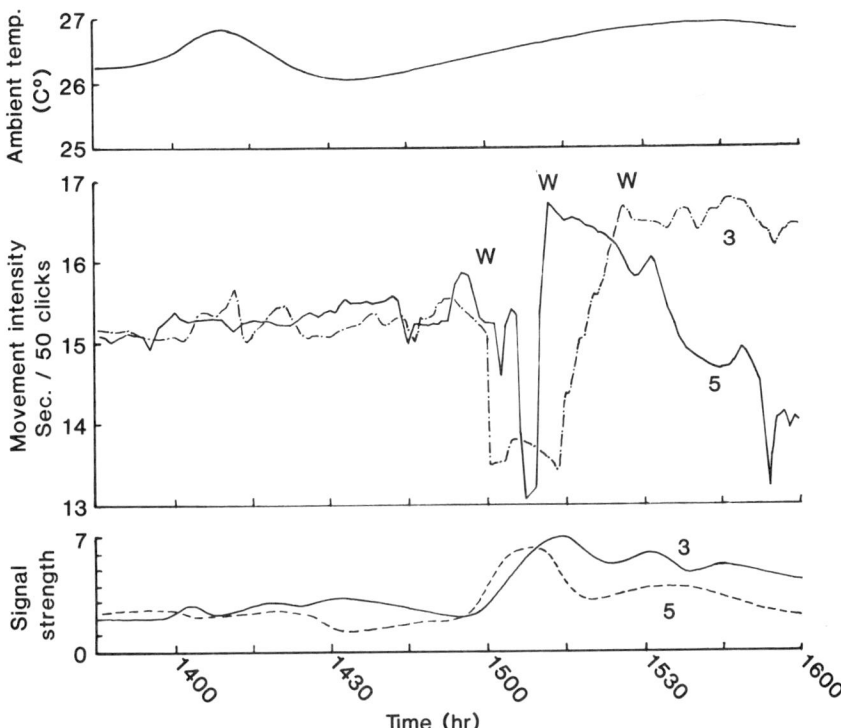

Fig. 5-5. Middle: Continuous telemetric records of two *V. olivaceus* (nos. 3 and 5) show simultaneous movement of both individuals in response to a sudden gusty period (*W*), resulting in different thermal environments, one a higher ambient temperature, the other a cooler one. Note also the similarity in movement intensity in the earlier parts of the records of both individuals, reflecting the simultaneous gradually increasing ambient forest temperature 10 cm above the surface (*top*). *Bottom:* curve of signal strength shows movement of the monitors into shelter as wind increases.

CV = 6.5%; see fig. 5-4). Furthermore, during this time of year females show a decided tendency for more activity in the earlier part of the day than during the remainder, whereas males remain consistently more active all day long (see below). I have shown elsewhere that captive male *V. bengalensis* also move and eat more than female siblings do under identical conditions (Auffenberg, 1979b).

Sometimes movement patterns are remarkably synchronized. Examples are shown in figure 5-5, where a sudden gusty period results in simultaneous movement of two individuals; other examples include simultaneous basking, and simultaneous movement inside the same shelter during the day. Simulta-

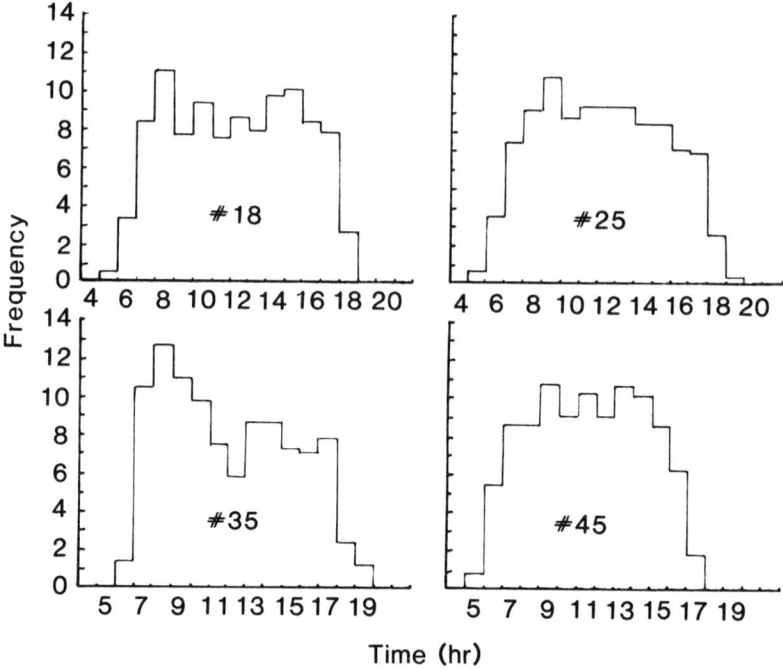

Fig. 5-6. Histograms of four adult individuals of *V. olivaceus,* showing diel activity patterns during October.

neous haul-out and haul-in activity is common. In general, hourly activity patterns are similar among adults living in the same area, as shown by numbers 18, 25, and 45 in figure 5-6. However, number 35 in the same figure exhibits a stronger bias for morning activity than do the remaining individuals (all more or less uniformly active throughout the day).

Differences in seasonal activity patterns are probably the rule in varanids, at least in those living under more temperate or xeric conditions (Corkhill, 1928; Cowles, 1930; Anderson, 1963; Haggag et al., 1965; Cloudsley-Thompson, 1966; Minton, 1966; Sokolov et al., 1975; Vernet, 1977; King, 1977, 1980).

Factors Affecting Activity

Light.—That butaans leave their shelters early in the morning is suggested not only by the telemetric actograms but by radiotelemetric studies of ambient light intensities falling on the backs of several free-ranging individuals (chap. 1). During July and August, 58 records of daily ambient light intensity levels were obtained via radio transmitter (table 5-1). The data obtained show that light intensity surrounding these individuals is generally lower than the maximum available in the area and about equal to the average light conditions in the forest (rank correlation of ambient v. maximum available light = 0.83). Thus butaans do not seek out or remain in light patches in the forest but pass through the light and dark patches of the mosaic throughout the day. Some of this movement is probably related to thermoregulation. There is a time after haul-out in the morning when the butaans remain in high light intensities. At the same time deep body temperatures are rising, and during this time these individuals are rather quiescent and in exposed local situations (on rocky pinnacles or in the trees); that is, they are basking.

As far as can be deduced, basking occurs only in the morning after haul-out (0804–1015 hr, $\bar{X}$ = 0837). It is not a regular, definite pattern but occurs commonly in all individuals of all sizes. However, males bask for longer periods than females (males $\bar{X}$ = 32.6 min; females $\bar{X}$ = 21.2 min, t-test significant at 0.02%, N = 126).

Figure 5-7 shows a typical curve for variation in ambient light intensity striking the back of the free-ranging butaans throughout the day. There is a sharp increase in incident light levels on emergence from the nighttime shelter. It is followed by a long period, lasting much of the day, when ambient light levels remain normal, with only slight variation as the butaans move through the light mosaic of the environment. The highest average values are usually recorded from 1000 to 1200 hr, coinciding with the often higher inten-

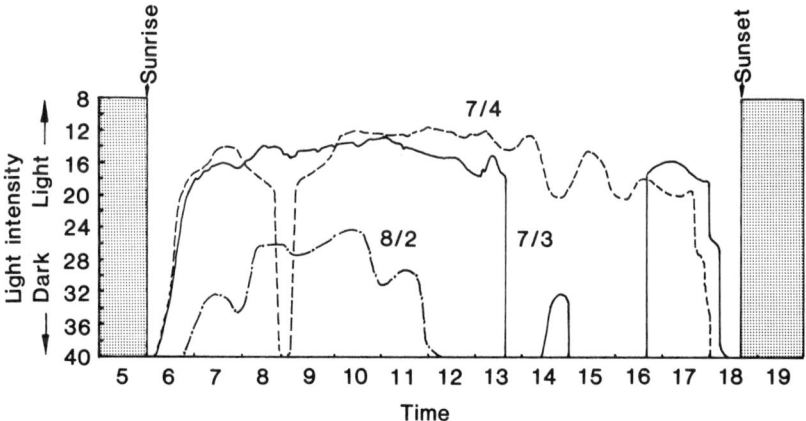

Fig. 5-7. Relative ambient light levels striking the back of a free-ranging *V. olivaceus* (no. 9) on three different days show major diel pattern variations. As an example, on 3 July the butaan was subjected to average forest ambient light from 0545 to 1300 hr; it then entered a crevice, emerged at 1405 hr, and reentered it at 1430 hr, remaining there until 1610 hr. From then until 1750 hr it received ambient forest light.

sity in activity levels during that time. Near the end of the day there is a sharp decline in light levels as the monitors remain on or near the substrate before entering shelters.

The data also show that, on the basis of light-sensitive transducers in the transmitter circuits, from 4.9% to 56.4% of the total available hours of daylight is actually spent in darkness ($\bar{X} = 18.0\%$). The demonstrated periods of low ambient light near the individuals coincide with periods when signal intensity from the same transmitter was extremely low ($r = 0.87$), suggesting that these periods are spent in rock crevices (chap. 6). This conclusion was corroborated on several occasions when individuals were located in crevices after radio signals received at the camp suggested that the monitor was in a nonexposed position under low light intensity.

Tables 5-2 and 5-3 show that available light affects the times at which haul-out and haul-in occur, the former more obviously affected than the latter. On overcast days haul-out time is almost one hour later than on sunny mornings, and the variation in haul-out time (as determined by CV values) is greater (CV = 6.2 v. 4%, respectively). Chi-square analysis for haul-out times and total available daylight hours shows a low probability that these factors are related ($p > 0.8$), suggesting that light is not the only factor determining haul-out time. Data presented earlier suggest that neither is temperature alone the major factor, though both obviously contribute to that combination of environmental conditions dictating haul-out time.

Temperature.—Three *V. olivaceus* (3.3–6.3 kg) were subjected to maximum radiation under the open sky and protected from wind in order to determine the temperature at which death occurred (the critical thermal maximum [CTM]) and to calculate the heat accumulation curve. Each was provided with telemetric probes placed in the cloaca to a distance of 3–5 cm. Ambient and deep-body temperatures were recorded simultaneously. During the experiment, ambient temperatures varied from 41° to 47°C.

Heating rates for the three individuals varied from 0.04 to 0.62°C/min (fig. 5-8), depending on butaan mass and the phase of the accumulation curve. As expected, smaller individuals heated faster than larger ones (Bartholomew, 1966), the mean heating curve for the 3.3-kg individual being 0.26°C/min and for the 6.3-kg one 0.18°C/min. These rates are identical to those obtained from telemetric devices placed deep within free-ranging individuals of the same approximate weights when other data suggested these individuals were basking in full sunlight.

King (1980) showed that the mean heating rates in basking *V. gouldii* varied seasonally from 0.15 to 0.28°C/min, reflecting differences in seasonal radiation levels. Such differences are not expected to show up for the butaan in view of the more or less uniform climatic conditions in the Philippines. However, the average heating rates for the butaan (0.18–0.26°C/min) fall within or

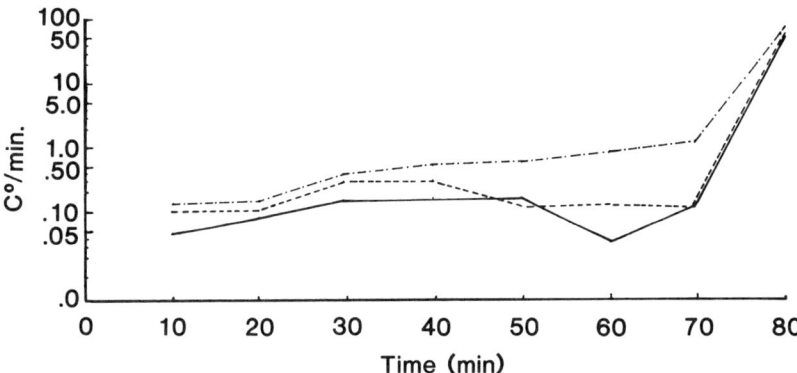

Fig. 5-8. Heat accumulation rates in three *V. olivaceus* adults (3.3–6.3 kg; see text), showing an ability to thermoregulate effectively up to approximately 70 min exposure to lethal temperatures.

near the ranges obtained for *V. gouldii* (0.15–0.28°C/min) by King (1980), *V. komodoensis* (0.11–0.14°C) by Auffenberg (1981a), and *V. varius* (0.14°C/min) by Stebbins and Barwick (1968). These data suggest that under field conditions all medium- and large-sized varanids accumulate heat at about the same rate. Size is obviously an important factor, as is incoming radiation level. Monitor lizards in temperate and xeric habitats are faced with a situation where most are required to bask after a cool evening to bring internal temperatures to operant levels. Such basking is, however, of less importance to *V. olivaceus* in view of the uniform conditions of the environment in which it lives.

Figure 5-9 shows the heat accumulation curve in two individuals leading to CTM with behavioral annotations. Strikingly, these individuals show few signs of heat stress as their internal temperatures approach CTM. There is no excess salivation, gular pumping, or excessive gaping, all of which have been reported in the behavioral syndrome accompanying high body temperature in *V. komodoensis* (Auffenberg, 1981a). The only obvious signs of heat distress in *V. olivaceus* are occasional tail twitching, air gulping, slight gaping, and gum vascularization. One gets the general impression that heat stress is not normally experienced by the butaan, and indeed this is true.

Critical thermal maximum temperature is from 41.6 to 42.4°C in the butaan, with no evident correlation with mass within the range of individuals available for study. These values are not significantly different from those of 42.7°C recorded for *V. komodoensis* (Auffenberg, 1981a), and it seems likely that most monitors have similar levels.

That reptiles can effectively change behavior to regulate their body temperatures has been well documented. It is also clear that while they can choose

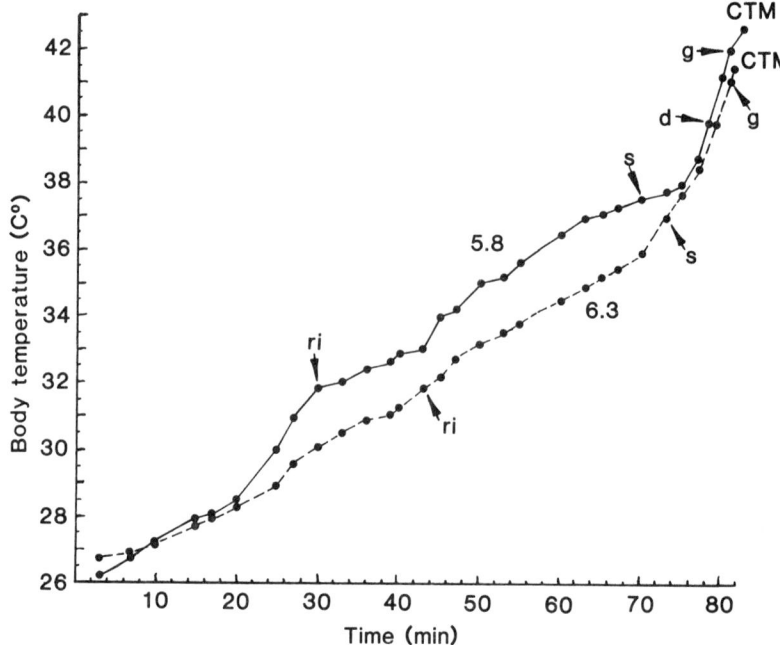

Fig. 5-9. Heat accumulation curve leading to CTM in two adult male *V. olivaceus* (wts. indicated) and behaviors associated with rising temperatures. Abbreviations: *d*, defecation; *g*, gaping; *ri*, respiration rate; *s*, stress noticeable.

the ceiling of their preferred body range through use of solar energy, they lack the cooling mechanisms of most birds and mammals and can cool themselves only to the level of the coolest segment of their environment.

Though varanid lizards generate more metabolic heat than other reptiles (Bartholomew and Tucker, 1964), it is unimportant in maintaining temperature differentials within the environment. Still, varanid lizards have elicited considerable research in regard to thermoregulation (Johnson, 1972; Sokolov et al., 1975; King, 1977, 1980, among others). Brattstrom (1968) has shown that heat retention in *V. varius* is controlled by a complex vasomotor shunting mechanism and is thus an important function in the endothermy reported in free-ranging varanids (Stebbins and Barwick, 1968; Francaz et al., 1976; McNab and Auffenberg, 1976). The skin of varanids is apparently not effective in reducing heat loss (Brattstrom, 1973). Though Auffenberg (1981a) reported gaping and panting at high temperatures in *V. komodoensis,* temperature lowering by evaporation, if effective at all, is probably too expensive in terms of water loss to varanid lizards in xeric habitats (Schmidt-Nielson,

1964). McNab and Auffenberg (1976) have shown that large size is an important factor in meliorating temperature changes within varanids living in tropical xeric habitats. Thus there are a number of mechanisms available at least to the larger members of the family for regulating temperature, though clearly the most important of these for most species is behavior. (A number of recent nonthermal physiological studies have provided much information on the capabilities of Australian varanids living in xeric habitats [Green, 1969, 1972; Braysher and Green, 1970, 1972; Bennett, 1971, 1972a, b, 1973; Webb et al., 1971], though all were laboratory studies dealing with water loss, oxygen consumption, and cardiac activity.)

Other studies on lizard activity and body temperature relations of conspecifics and congenerics have shown that one of two strategies is used in their thermal biology: (1) maintaining similar average activity temperatures through behavioral means, such as restricting activity to certain times of the day (Bogert, 1949; Spellerberg, 1971; Parker and Pianka, 1975); (2) maintaining similar diel activity patterns but tolerating activity temperature differences (Heatwole, 1970; Pianka, 1970). In an important study of the thermal biology of *V. gouldii,* King (1980) showed that this species utilizes the first strategy—changing its diel activity patterns in selecting the optimum activity temperatures to suit seasonal and habitat conditions. The nearly uniform environmental temperatures throughout the year in the range of *V. olivaceus* represent an entirely different thermal setting than that of any other varanid yet studied. As a result, this species is rarely, if ever, faced with any thermal conditions warranting major changes in its behavioral patterns. Therefore there is little variation in behavior other than what can be ascribed to changes in ambient temperatures during the year.

Rand (1967a) and others have shown that the body temperatures of many lizard species (generally small) are almost always close to, and usually above, air temperatures in their immediate vicinity. Radiotelemetric studies of temperature relations in *V. komodoensis* have shown that at least varanid daytime cloacal temperatures are intermediate between soil and air temperatures. The same pattern occurs in the butaan.

The following description of the diel temperature pattern is based on data gathered over a cumulative total of 215 days from three adult individuals, which ranged in weight from 2.2 to 8.3 kg (fig. 5-10). The pattern of juveniles would be similar but with wider fluctuations (none was telemetered in this study).

In general, early morning deep body temperatures are about 2°C higher than the shaded ambient temperature that accompanies the shift from outgoing to incoming radiation patterns. This difference usually occurs in Caramoan about 0630 to 0700 hr. Shortly after haul-out time and when ambient temperatures are rising quickly, butaan body temperatures also rise but at a slower rate

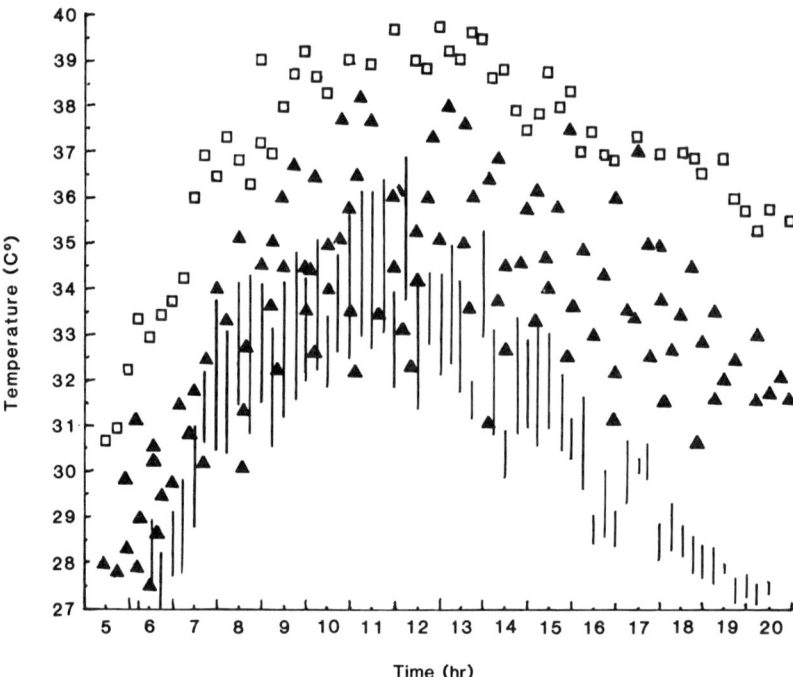

Fig. 5-10. Relationship between average shaded maximum ambient air temperatures available in the upper layers of the forest canopy (*squares*), range of maximum ambient air temperatures within 1 m of ground surface (*lines*), and deep body temperatures (*triangles*) of free-ranging adult butaans.

than those of the shaded immediate environment. The result is that by about 0800 hr deep body temperatures are equal to ambient temperatures. From 1000 to 1100 hr deep body temperature usually rises faster than shaded ambient temperatures, particularly when the butaan basks or passes through sunny patches on the forest floor. This rise results in a body temperature 3–8°C ($\bar{X}$ = 4.7°C) higher than ambient shaded temperatures. This differential may be maintained until 1600 to 1730 hr, when ambient temperature drops faster than deep body temperature. The average differential at 1800 hr is 1.4°C and at 2000 hr 1.8°C. Though both temperatures continue to fall through the night, the lizard's falls at a slower rate.

Varanus komodoensis has been shown to have a uniform body temperature through the day (McNab and Auffenberg, 1976; see Bartholomew and Tucker, 1964, and Stebbins and Barwick, 1968, for extensive diel fluctuations in an Australian varanid). This uniformity is due largely to its great mass (to 45 kg). However, figure 5-11 and table 5-4 show that the daily body temperature fluctuations in adult *V. olivaceus* are even less than in *V. komodoensis*. The more or less uniform diel temperature pattern in *V. olivaceus* is due to the thermal uniformity of the habitat and the generally much lower radiant energy load experienced by the species in the dense forest that is its habitat.

The seasonal differences in mean body temperatures reported for other varanids (all in xeric habitats; Vernet, 1977; King, 1980) could not be demonstrated for the butaan—undoubtedly because of the uniform environment in which it lives. Still, though the South Australian environment of *V. gouldii* is seasonally variable, individuals of this species expose themselves to only a narrow range of environmental conditions to facilitate thermoregulation (King, 1980). Though the end result in terms of thermoregulation is the same for both *V. olivaceus* and *V. gouldii,* the latter achieves it by spending much time either in the burrow or basking, both of which preempt time that otherwise might be spent in other behaviors. While data for xeric varanids are consistent with earlier work on other desert lizards, King (1980) suggested that the rate of water loss may be more important than temperature alone in determining details of behavior in xeric habitats. In the daily life of the butaan, however, water loss and high daily temperatures are not important factors.

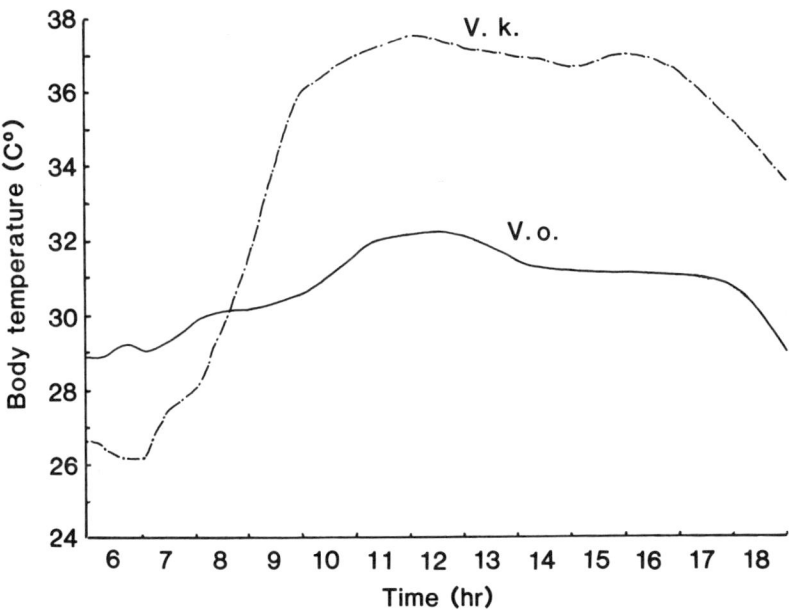

Fig. 5-11. Average daily deep body temperature curve for equal-sized wild *V. olivaceus* (V.o.) and *V. komodoensis* (V.k.), showing that the thermal pattern of the former is more uniform, presumably due to its more uniform environment (see text).

TABLE 5-4. Mean deep body temperatures of free-ranging *Varanus* species.

	OR of $\bar{X}$	CTM[a]	Authority
V. olivaceus	27.8–38.2	41.6–42.4	This study
V. komodoensis	26.1–37.0	41.7–42.3	McNab and Auffenberg 1976
V. gouldii	21.3–39.2	?	King 1980
V. caudolineatus	37.8		Pianka 1969a
V. eremius	37.3		Pianka and Pianka 1970
V. griseus	32.1–38.4 [b]		Sokolov et al. 1975; Francaz et al. 1976; Vernet 1977

a. CTM = critical thermal maximum (see text for definition).
b. Reflecting seasonal differences in mean cloacal temperatures in different environments.

TABLE 5-5. Temperature change in the body of *V. olivaceus* and its ambient shade forest environment throughout the day ($N = 834$).

	Average rate of change (°C per min)					
	04–06 hr	07–09 hr	10–12 hr	13–15 hr	16–18 hr	19–21 hr
Body temperature	+0.001	+0.018	+0.01	−0.005	0.000	−0.003
Environment temperature	+0.003	+0.01	+0.007	−0.006	−0.01	−0.007

In general, the body temperature–ambient temperature differentials are greater for *V. komodoensis* than for *V. olivaceus*. In the former the mean early morning differential at haul-out is 4.3°C (body higher); at 0900 to 1000 hr the body is 3.2°C cooler; at midday it is 7.2°C (body higher); and at 2000 hr the mean differential is about 4.0°C (body higher). However, the major difference in the thermal pattern of the two species is that *V. komodoensis* has a significantly higher body temperature in the middle of the day (32–40°C; fig. 5-11). This difference is undoubtedly due to the higher ambient air temperature in the open, high radiant habitat of *V. komodoensis* compared to the close, low radiation habitat of *V. olivaceus*. Table 5-5 shows the average variation in rate of deep body and ambient temperatures for *V. olivaceus*.

While shaded air temperatures are nearly identical in both habitats (to 32.0°C on Komodo, to 31.9°C in the Bicol area of Luzon), the sun-to-shade ratio is much higher on Komodo (the mean sunny patch size on Komodo is several hundred times larger than it is on the forest floor in the Philippines).

Figure 5-12 shows the relationship between body and ambient temperatures, exposure of the telemetered individual, and its movements during a typical day. The relation between the two temperatures taken has been discussed. The relationship between body temperature and exposure (lizard in or out of rocks, etc.) is clearly shown. Note particularly that the individual was in the most exposed position (basking) from about 0815 to 0930 hr, and that body temperature began to climb rapidly near the end of that period. This was followed by a period of considerable activity, into and out of shelter, until about 1200 hr. During this time a fairly constant relationship was maintained between ambient and body temperatures, the latter somewhat higher than the former. At 1300 hr it entered a shallow rock shelter and, except for body positional adjustments, remained there until 1635 hr, and body temperatures more or less followed the curve of the falling ambient temperatures. At 1635 hr the butaan entered a deeper rock shelter nearby, and its body temperatures remained more or less stable at 31.2°C. This activity-temperature tracing is typical for a bright sunny day.

Figure 5-13 shows a typical activity-temperature trace of the same kind for an overcast day. Note that ambient temperatures started to climb higher later in the morning, and at a slightly lower rate, and reached lower maximum at about the same time. A small shower at 1500 hr was preceded and followed by a cool period. The deep body temperature record remained uniform at about 30°C until 1100 hr. From 0530 to 0730 hr the monitor remained in the same vine-covered tree where it had spent the night. There was slight movement

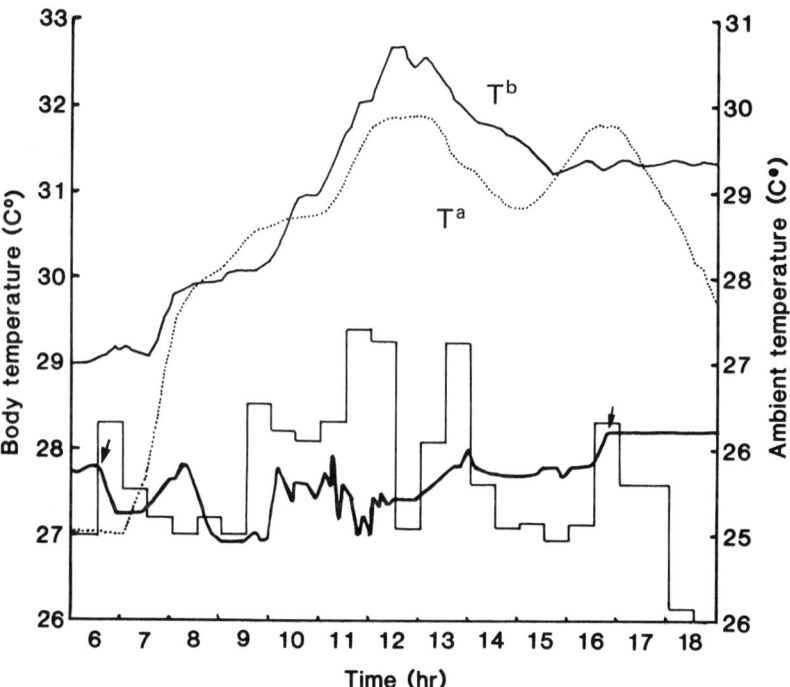

Fig. 5-12. Relationship of deep body temperature (T^b), ambient shaded air temperature (T^a), degree of exposure (signal strength, *heavy line*), haul-out and haul-in times (*arrows*), and movement intensity shown as histogram. Sunny day, 11 April 1983, male, 8.3 kg (see text).

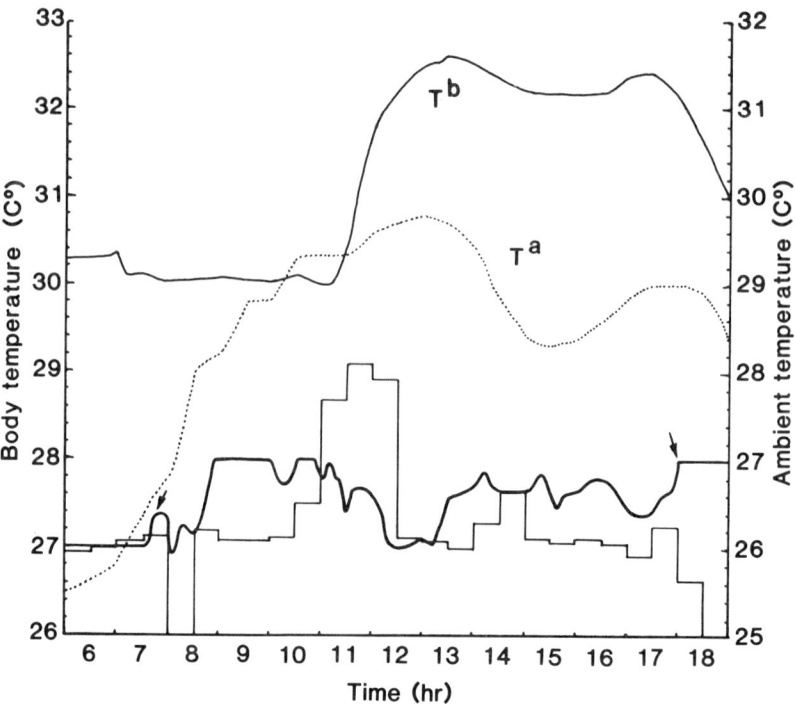

Fig. 5-13. Relationship of body and ambient temperatures, exposure, and activity intensity in *V. olivaceus* male, 8.3 kg, on overcast day, 13 April 1983. Symbols as in fig. 5-12 (see text).

(without change in body temperature) at 0830 hr. At this time it entered a deep crevice in the rocks and remained there until 1030 hr, when there was much movement, resulting in a general drift into more exposed situations. This exposure with higher ambient temperature led to a rapid increase in body temperature. For the most part, movements were few throughout the remainder of the day, except for a slight increase just before a rain storm, when movement tended to be in and out of sheltered situations. This movement lasted until 1730 hr, when it entered another deep crevice and settled down for the night. During this recording there was no close correspondence between ambient and deep body temperatures, and, except for a rapid rise in T^b on haul-out, temperatures were less directly dependent on changes in T^a.

Substrate.—Table 5-6 and Figure 5-14 show substrate use by several butaans throughout the year. In general, the ground surface is the most commonly used by all individuals; trees and crevices are only occasionally dominant substrate types, and no consistent pattern ever occurs among all individuals at the same time.

TABLE 5-6. Substrate use, movement intensity, and relationship between them (independence test) during the year (in 5-min observation periods; $N = 6,194$). Data excluded for butaans contacted less than 100 times per month.

Month	Butaan no.	Total no. contacts	Major substrate	Binomial tests (df = 3) X^2	p	Moving or still	X^2	p	Independence test (df = 3) Dependent	X^2	p
March	76	171	surface	49.69	<0.005	moving	13.90	>0.005	yes	9.83	<0.025
April	76	756	surface	14.41	<0.005	moving	42.00	>0.005	yes	7.96	>0.05
May	82	291	surface	9.55	<0.05	moving	9.16	≈0.025	yes	10.31	<0.025
July	76	217	trees	41.15	<0.005	equal	0.75	<0.50	no	4.61	<0.10
July	18	160	surface	24.69	<0.005	moving	14.40	>0.005	yes	9.91	<0.025
August	25	186	trees	12.00	<0.01	equal	1.36	<0.50	no	2.58	≥0.50
August	18	687	surface	23.05	<0.005	equal	1.48	<0.50	no	3.89	>0.10
September	35	516	trees	11.29	≤0.01	still	12.10	>0.005	no	2.25	≥0.50
September	25	565	surface	12.87	≤0.005	equal	1.26	<0.50	no	2.00	>0.50
September	18	588	crevice	27.19	<0.005	equal	0.61	<0.50	no	1.81	>0.50
October	45	270	surface	10.10	<0.025	equal	0.78	<0.50	no	3.79	>0.10
October	35	493	surface	28.93	<0.005	still	25.00	>0.005	no	4.28	≥0.10
October	25	568	surface	37.78	<0.005	equal	0.12	<0.97	no	1.95	>0.50
October	18	587	trees	29.78	<0.005	equal	0.81	<0.50	no	0.96	>0.50
November	18	139	surface	9.53	<0.05	equal	0.58	<0.50	no	2.01	>0.50

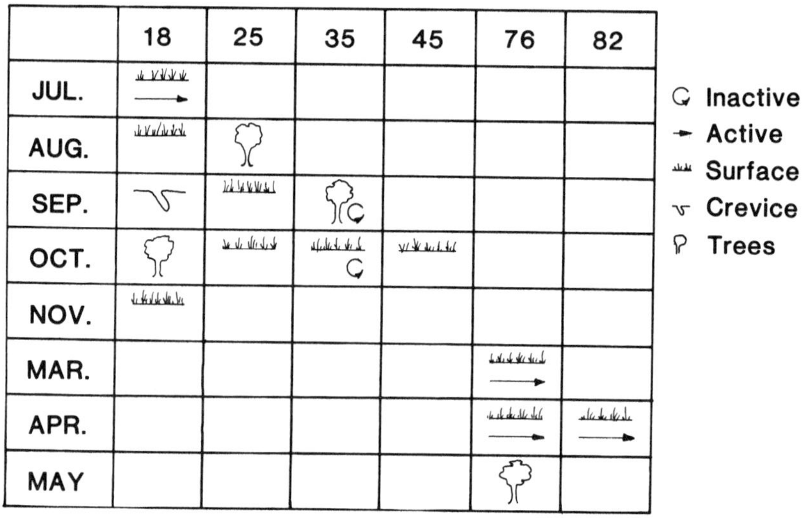

	18	25	35	45	76	82
JUL.	⊥⊥⊥⊥⊥ →					
AUG.	⊥⊥⊥⊥⊥	🌲				
SEP.	⌄	⊥⊥⊥⊥⊥	🌲 C			
OCT.	🌲	⊥⊥⊥⊥⊥	⊥⊥⊥⊥⊥ C	⊥⊥⊥⊥⊥		
NOV.	⊥⊥⊥⊥⊥					
MAR.					⊥⊥⊥⊥⊥ →	
APR.					⊥⊥⊥⊥⊥ →	⊥⊥⊥⊥⊥ →
MAY					🌲	

C Inactive
→ Active
⊥⊥ Surface
⌄ Crevice
🌲 Trees

Fig. 5-14. Graphic representation of table 5-6.

The figure and table show the level of movement intensity per month per individual. For most months the butaan spends an equal amount of time moving and remaining still. The single female represented in these data (No. 35) was the only one of the three showing a statistically significant predominance of nonmovement (inactive); males were either predominantly moving or were equally moving and inactive each month. The male records are even more significant if we compare when their activity patterns were "equal" (movement = nonmovement) and when movement was predominant. Significantly high male movement occurred *only* from March to July. Most of this time males were engaged in courtship (chap. 8). Furthermore, these greater movements tended to be largely on the *surface* of the forest floor—not in trees or crevices. The only exception was in May when No. 76 spent much time in the trees, both moving about and inactive.

Rain and wind.—Most Caramoan hunters believe that butaans are not abroad during rainy weather. We were, in fact, told that none would be available for study during the entire rainy season. Time showed this belief to be incorrect: not only did we obtain all the specimens needed for study during the rainy months, but the telemetric data obtained during that time showed that during storms they moved about regularly—sometimes considerably. Some of this movement is a direct result of the rain, and some is indirect, resulting from greater fruit fall.

Sleeping individuals often move soon after rain starts to fall—presumably to shift to a drier location (figs. 5-1D, 5-15). While such movements seem contrived to avoid rain, other patterns make it clear that rain is not really

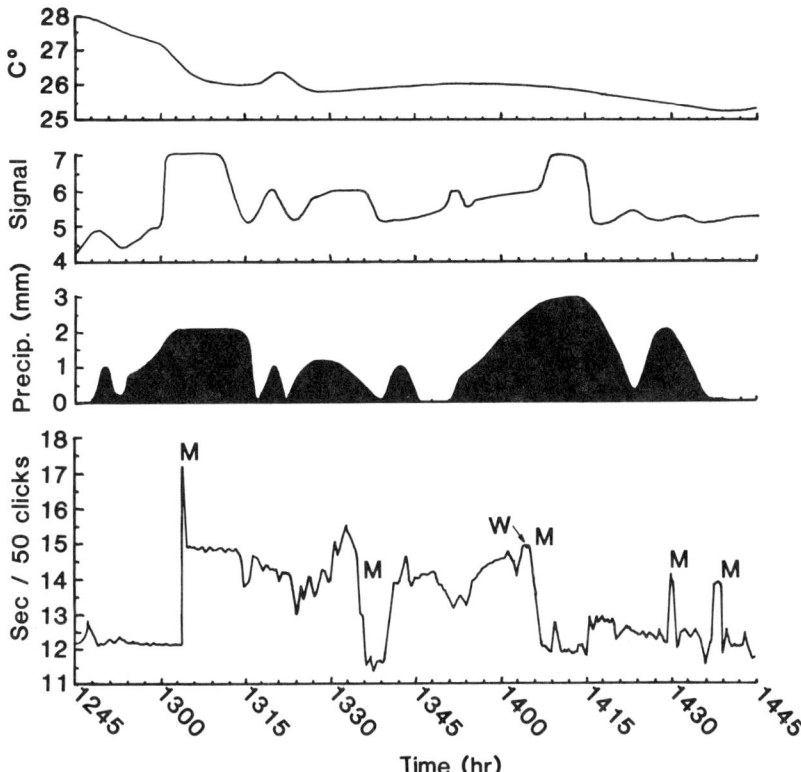

Fig. 5-15. Environmental changes accompanying a rainstorm (25 November 1982) and their effect on butaan 5: *top,* forest ambient temperature cooling curve during period represented (1245–1445 hr); *middle unshaded,* signal strength reflecting movement into and out of rock crevices; *middle shaded,* rain intensity (0, no rain; 1–3, light, medium, and heavy rainfall levels); *bottom,* continuous record of radio telemetric signals from ambient temperature–sensitive transmitter on butaan's hip (see table 5-1 for other data on butaan 5). *M,* major movements of butaan; *W,* wind gusts. At 1245 hr, ambient temperature in forest dropped with approaching storm front. 1249 hr, positional shift of butaan resulted in slight shift in signal strength. 1250 hr, light rain began, transmitter temperature remained steady, suggesting butaan at or near surface but protected from cool rain. 1307 hr, harder rain; forest ambient temperature fell rapidly, and monitor made a significant move into more sheltered location, resulting in immediately cooler ambient transmitter temperatures. 1314 hr, rain stopped, monitor moved to more exposed place, transmitter ambient temperature dropping. 1320 hr, light rain again, followed by monitor movement for about 20 min; about 1335 hr, monitor finally stopped in a sheltered place, where its transmitter temperature immediately became warmer. A movement to a less sheltered place at 1340 hr resulted in cooler ambient temperatures. 1347 hr, rain stopped, started again at 1352 hr, turning into heavy rain by 1405 hr. Wind gusts at 1407 hr resulted in a major positional shift, causing a higher monitor ambient temperature. The temperature remained fairly stable, though the monitor moved about regularly until end of record and throughout rest of rain shower. During this time it left its most sheltered site and remained at or near the surface.

avoided. Individuals have been noted to eat, court, and move about on the surface and in the trees during rainstorms. A comparison of movement five minutes before a rainstorm and five minutes after rain begins to fall shows that activity level is definitely increased, regardless of substrate or shelter type ($\bar{X}$ CV before rain = 1.5%, $\bar{X}$ CV after rain begins = 3.4%, $t = 51.55$, df = 81, $p < 0.001$). If the rain stops within 15 minutes, activity is reduced ($\bar{X}$ CV = 2.7%, $t = 3.47$, df = 69, $p < 0.001$) but will remain at a sustained high level if it is still raining at that time ($\bar{X}$ CV = 4.3%, t-test not significant at the 0.05 level). Figure 5-15 shows the interaction of butaan behavior and important environmental factors during a rainstorm on 25 November. The overall picture presented is a rest period broken by a rainstorm, resulting in considerable movement during the storm. Before the storm is completely over, the individual has left the deepest rock shelters and is walking over the surface during some of the heaviest precipitation.

Captives behave similarly, though particularly strong storms usually result in individuals seeking shelter. However, the percent of available time spent in rock crevices is not significantly correlated with rainy weather in free individuals—rather, crevice use is more clearly linked to the high winds that accompany severe storms.

On the other hand, intensity of movement is related to rainfall in a direct way. Figure 5-16 shows that butaans tend to be more inactive (but not necessarily in a shelter) when rain per week exceeds 8 cm, and that "inactive" periods (= CV 0.0–1.4%, see chap. 1 for description of methods) are least during times when rainfall is from 5 to 8 cm per week. However, movements of "low" intensity (CV = 1.5–2.9%) remain about the same throughout the year (5.5–9.5% of available total daylight), except for an *increase* of low intensity movements during typhoon "Ruping" in the first week of September. "Medium" (CV = 4.0–6.0%) and "high" (CV $\geq$ 6.1%) intensity levels are not significantly different during the year, suggesting that this differentiation is of little or no importance. Intense activity level readings are produced when butaans walk, run, or climb vigorously. Feeding, thermoregulation, and other movements involving relatively few body movements are registered as inactive. Intense activity levels (high and medium) never span a significant part of any day, constituting only 0.35% to 5.9% of the total available daylight time. However, there is a significant increase in activity level during the driest part of the year (April), as represented by the telemetric data. This increase suggests that while low intensity movements may be generally increased during violent rainstorms, high intensity movements are increased during the driest times. In sum, it is not sufficient to say that movement does or does not increase during certain weather conditions; the patterns of inactivity and low and high intensities are different, and each may be affected differently by the same climatic factors.

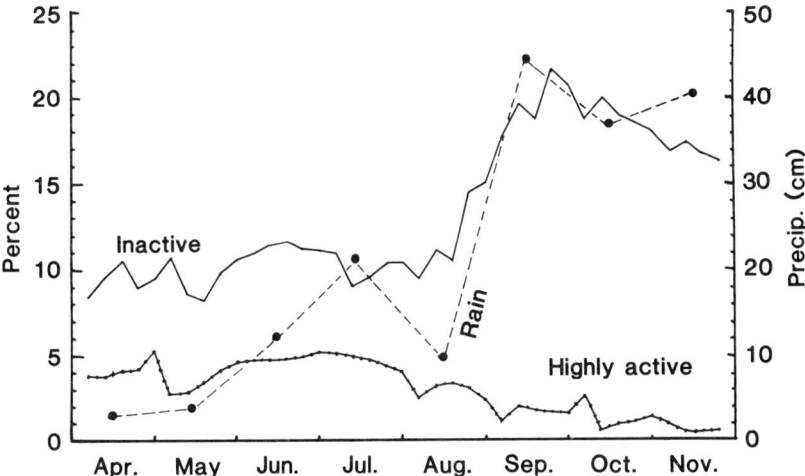

Fig. 5-16. Comparison of the percentage of total time spent per week in significantly inactive and highly active movement by adult *V. olivaceus* ($N = 516$ observations; April–November). Average monthly rainfall is also shown. Note the correspondence of high rainfall and inactivity in September–November and the high activity level in the pre- and early monsoon period (April–July).

While rainfall may affect general movement patterns over longer periods of time, and while rain may cause immediate changes in position of inactive butaans, there is also a significant correlation between the amount of rain per day and the average movement intensity per day (rank correlation $r = 0.72, 0.63, 0.89$, and 0.47 in four individuals for which this analysis was conducted).

Lack of rain is responsible for elevational movement of at least limited extent. During unusual drought conditions in April and May 1983, one butaan was captured while it was lying in a pool at the edge of a mangrove swamp near Lipata. During the same time another had evidently come off the mountain adjacent to Bulang-bugang Spring and was passing through an area of abundant water-filled sinkholes. Another was caught while it was lying in the shallow water of the Manipot River. These are not normal habitats, and all represent movement downward from ridges and mountains into coastal or streamside situations.

Cloudsley-Thompson (1967) has shown that *V. niloticus* (in mesic to hydric environments in Africa) loses water rapidly (6%/24 hr) compared to tropical skinks and geckos. King (1980) has suggested that rate of water loss in *V. gouldii* (xeric environments, Australia), rather than ambient temperature differences, may be responsible for behavioral differences demonstrated in different parts of its range. Captive butaans will soak themselves for days if deprived of water for a week, in spite of the fact that ambient conditions are

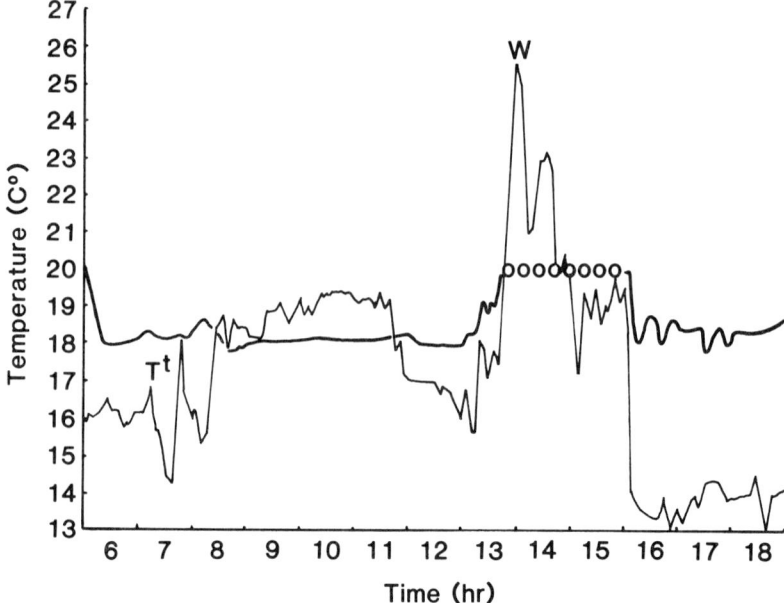

Fig. 5-17. *Heavy line* represents signal strength; *light line* represents immediate lizard ambient temperature. Both curves show dropping transmitter temperature and decreasing signal strength (to 0's) when the butaan enters a water pool deep in rock crevice (*W*). The remainder of the record is as follows: out of rock crevice and climbs tree at 0600 hr, resulting in greatly varying transmitter temperature (T^t) as butaan moves about until 0830 hr; this is followed by slowly cooling T^t as butaan remains in a rather exposed but undoubtedly shady place until 1200 hr. At this time it again climbs tree and T^t rises. At 1300 hr, it descends to ground, with T^t cooling somewhat, enters rock crevice (0's) at 1350 hr and then water pool (*W*), resulting in rapid heat loss. It leaves the pool at 1440 hr but remains in the crevice until 1615 hr; then it goes to the surface, where it remains until the end of the record (1900 hr).

humid. And telemetric recordings show that free individuals of this species regularly go deep into rock crevices to lie in pools and streams (5–30 m below the surface in Ahop A and B Mountains; fig. 5-17).

Sex.—The sex ratio of 118 dissected individuals (examined in the field and in various museum collections) is slightly in favor of males (50 F, 68 M), though the difference is not significant at the 0.05% level. Thus the sex ratio is most likely 1:1.

Sample sizes are too small to evaluate sex ratios on a monthly basis, and the pattern produced is somewhat illogical in terms of known behavioral patterns of the sexes. However, when monthly samples are combined in the following categories, a meaningful pattern is produced: heavy SW monsoon rainy season 16 September to 15 December, post–SW monsoon light rainy period to

15 February, and the dry season to 15 May are all represented by a 1:1 male-female ratio ($N = 70$). However, males predominate in a significant ratio of 1.8:1.0 ($N = 36$) during the early, weak annual monsoon from 16 May to 15 August. This period is when almost all courtship takes place and during which territorial behavior is expected (but not documented in this species) in males. From 16 August to 15 September, males are clearly less active than females, with a ratio of 1:3 ($N = 12$). It is the period when females lay their eggs and apparently search for nest sites in which to do so. Of the females collected during this time, 66% had just laid their eggs. Thus the skewed sexual ratios during a short part of the year are due to reproductive factors—first the males are active during courtship and territorial defense, then the females during nest site selection.

These conclusions are corroborated by records of movement intensity in both captive and wild individuals. Males held captive at Caramoan were less active during a greater part of the day during March–April (28%) than during May–June (72%; X^2 7.1, df = 1, $p < 0.01$). Telemetric field data also show that females are more active than males during July and August (female movement intensity level expressed as $\bar{X}$ CV = 6.0%, males $\bar{X}$ CV = 0.7%, X^2 8.3, df = 1, $p < 0.01$), and considerably less active during September and October (females $\bar{X}$ CV = 2.9%, males $\bar{X}$ CV = 7.0%, X^2 9.2, df = 1, $p < 0.01$; see fig. 5-4 and daylight movement patterns above).

6

Spatial Relations

Horizontal Distribution

PROBABLY no other species of *Varanus* has had so much confusion surrounding its provenance as has the butaan. Only the general locality of "Philippines" accompanied the type, followed by a second specimen labeled "Luzon" and a third labeled "Laguna." Not until 1976 were any precise locality data available, and then for only another single individual (Auffenberg, 1978a) (see chap. 1 for details). Therefore, one of the most important projects during this study was to attempt to establish at least the approximate parameters of the geographic range. To this end short visits were made to eastern Mindanao, southern Samar, northern Leyte, Catanduanes, and southern to central Luzon during the period 1976–83. *V. olivaceus* was documented with specimen materials on only Luzon and Catanduanes islands; no evidence was found on any other site. In the following section I outline the distribution in each of the provinces in which I (sometimes others) found reliable information or specimen materials of this species. Figures 6-1 and 6-2 provide locations of provinces and major physiographic regions discussed below.

Catanduanes.—In 1978 the late Judge Gumersindo Arcilla of Catanduanes told me that butaans occurred in the mountains due north of Virac, but it was not until 1982 that this information was verified when Laurie Wilkins (Florida State Museum) obtained a reference specimen from the forested uplands due west of Hituma. I had been told by hunters from Gigmoto on the east coast that the species is common in the nearby hills. Thus it is now possible to say that the butaan occurs in some numbers probably throughout all the remaining forested areas of Catanduanes, comprising approximately 550 km². It is, perhaps, one of the largest forest blocks remaining in the Bicol area.

Albay (Luzon).—Much of Albay province has been converted to agriculture. Remaining forested areas in which the butaan are found tend to be re-

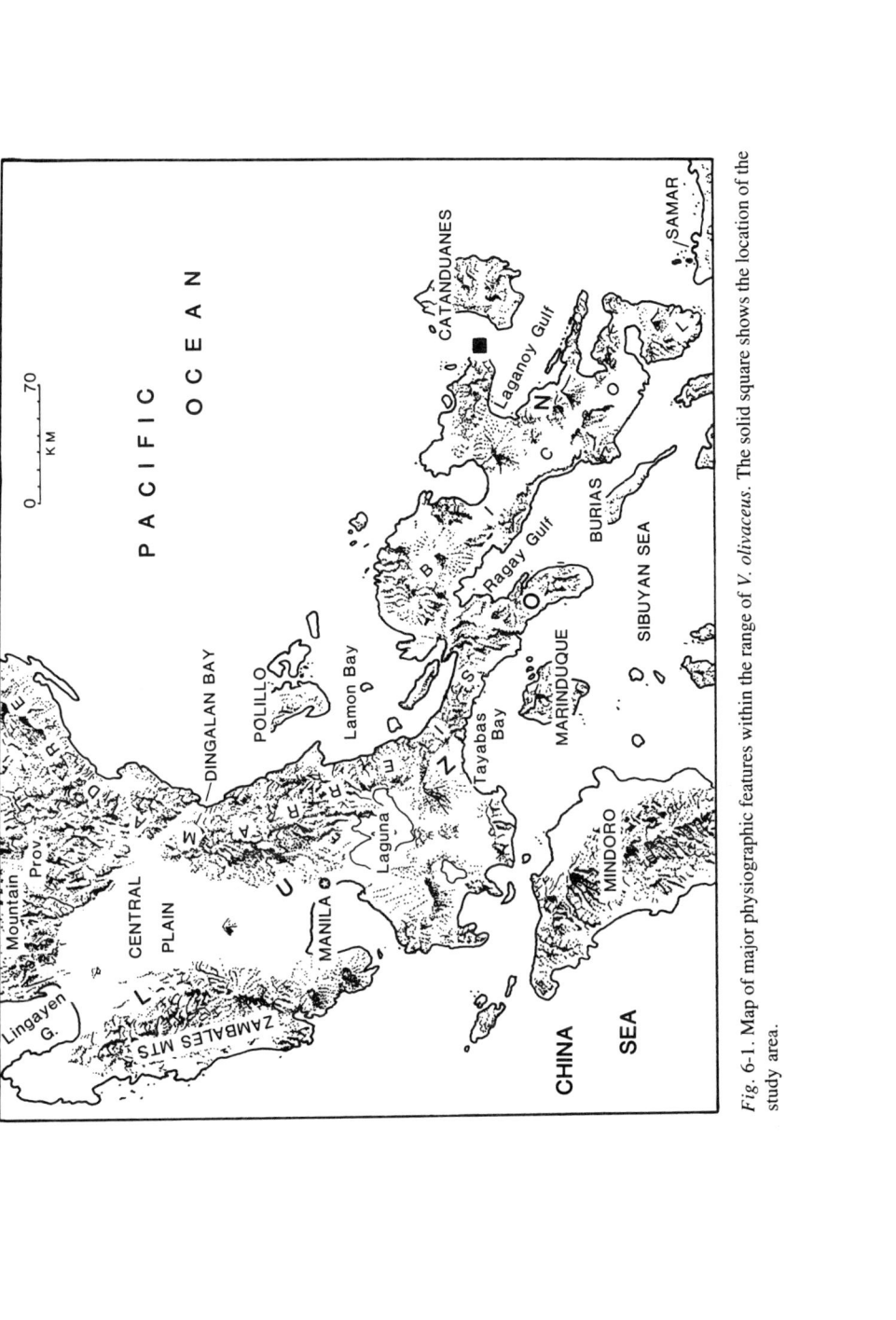

Fig. 6-1. Map of major physiographic features within the range of *V. olivaceus*. The solid square shows the location of the study area.

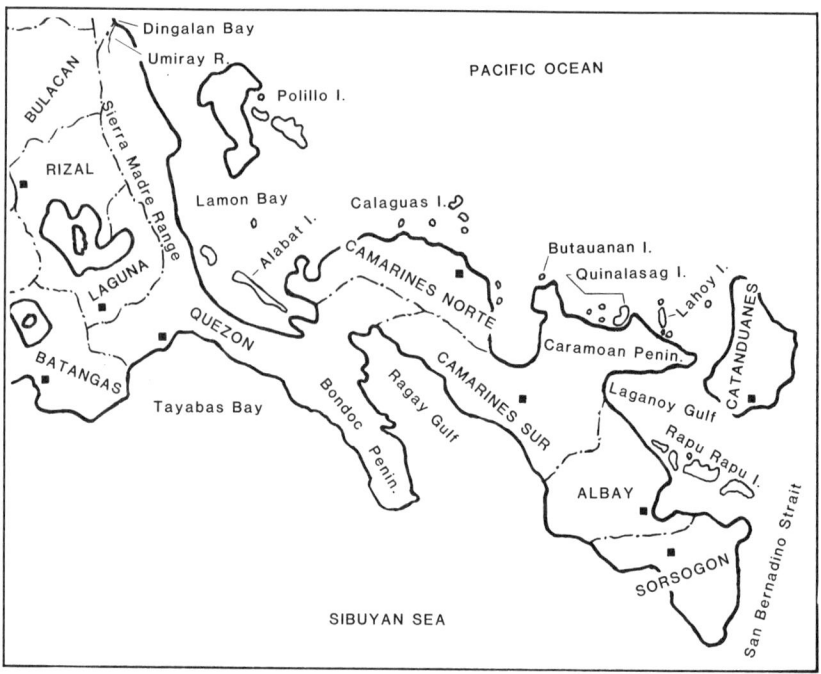

Fig. 6-2. Provincial and other important geographic names mentioned in the text. Squares show location of major provincial towns.

stricted to the northeast part of the province. While Mayon volcano dominates the landscape and has some relatively undisturbed forested areas, most such areas are located higher than the preferred altitudinal limits of the butaan. Most of the lower slopes in which the butaan was once undoubtedly common have long ago been converted to abaca plantations. However, some small populations still remain (ca 25 km^2). The area north-northwest of Mayon, from and including Mt. Masaraga to the northern limit of the province, is well forested and contains good populations of butaans at appropriate lower elevations (ca 300 km^2). In the extreme western part of Albay there is a large mass of highly karstified hills, most of it still fairly well forested. Though no butaan records from this area are available, it is extremely likely that it occurs there, for the habitat is essentially identical to that at Caramoan. There are records from peripheral areas (near Guinobatan). The area is large (ca 300 km^2) and is potentially one of the most important from the standpoint of conservation of the butaan in this province. It is likely that butaans will eventually also be found on the mountains of Rapu Rapu Island, but the forests of the remaining islands of the province were destroyed long ago. There may be a small population in the extreme southeast, several km southeast of Manico (ca 30 km^2), coextensive with a larger area inhabited by them in Sorsogon Province.

Sorsogon (Luzon).—The small province of Sorsogon has been largely converted to agriculture. However, the butaan does occur there, having been found in at least three localities. The least disturbed at present is probably that NW of Sorsogon City (ca 40 km^2), where low mountains of appropriate habitat are not heavily stressed by human populations. The two remaining known populations occur on the slopes of Bulusan volcano (45 km^2) and—separated by a narrow agricultural zone bisected by the National Highway No. 1—Mt. Juba (35 km^2). The Bulusan National Park, for all practical purposes, is above the preferred altitudinal limit of the butaan.

Camarines Sur (Luzon).—Camarines Sur is one of the largest provinces on Luzon in which the butaan occurs. However, much of it has been converted to agriculture, dividing the once extensively distributed *V. olivaceus* into at least four populations. The largest of these is found along the northern province boundary (ca 260 km^2), part of which is included in the Bicol National Forest. This forest may be one of the most important areas from the standpoint of butaan conservation, for the entire area is covered by optimum habitat (chap. 15) at moderately low elevation (to ca 400 m). A smaller mountainous, karstic, forested area occurs along the western coast, from about 4 km NW of Pasacao to near Bantuin Point. It covers approximately 120 km^2, and butaan reports and specimen materials are available from the Pasacao end. There is no protected forest in the area, and current development suggests that much of the forest will be destroyed in the next decade or so. However, cliff areas, crags, and deep ravines are common, and it is likely that the butaan will be found here in small numbers for many years. These two populations are now completely isolated from those in the eastern part of the province by a combination of farms and the low marshy terrain (now largely drained) of the Bicol River Valley. The latter has been an ecological barrier since the last part of the Tertiary (chap. 14).

Two volcanos dominate the central part of the province. The forests of the southern one, Mt. Iriga, have been almost completely destroyed, except on the southern slope, and I found no evidence of butaans living there during my visits in 1976. However, the flanks of Mt. Isarog definitely harbor a small population, for records and specimens are known from several localities. Mt. Isarog National Park comprises the top two-thirds of the mountain. While it is an important protected area for much of the flora and fauna of the higher mountains of southern Luzon, most of the park is located above the altitude limit of the butaan. The area of the slopes inhabited by *V. olivaceus* is estimated at 50 km^2, but deforestation is continuing at a fairly high rate. However, a small population of butaans will probably always be located in the lowest reaches of the park, provided continuing protection is afforded.

The Caramoan Peninsula is the major landform feature of the northwest part of the province, comprising one of the largest tracts of forest habitats in which the butaan is found in all of Bicol. However, the construction of a highway along its length and logging operations in the northwest have reduced the remaining forest. Most of that on the southern slope of the Calinigan Mountains (making up the spine of the peninsula) has already been harvested, though the upper ridges (not generally available for butaans, because many are too high) and the northern slope are still heavily forested (ca 380 km^2). The northwest corner of the peninsula is flat, partly forested land, and may represent the only level mixed dipterocarp forest left in the entire range of the butaan. However, plywood and lumbering industries at both Siruma and Garchitorana, in addition to the spread of agriculture, will rapidly destroy this remaining forest (ca 20 km^2). This butaan population is coextensive with that on the north flanks of the Calinigan Mountains and thus represents one of the largest populations left. It will probably, however, be one of the most rapidly degraded in the years ahead.

The Caramoan National Park is located near the eastern tip of the peninsula. It is an area of forested, highly karstified limestone hills and low mountains. The protection afforded by its park status and the topography make lumbering on a commercial scale impractical (but see chap. 15 for discussion of fuelwood use). Here the butaan remains common. In spite of the area's small size (ca 40 km^2), the area's population will undoubtedly continue to be an important one.

Numerous islands occur on the shallow bank extending northward along the entire peninsula. Many are too small to support viable butaan populations, though the largest can and in some cases do. All of the Lucsuhin Islands, including Lahoy Island (all generally north of Tabgon), are completely deforested and have no butaans, but Quinalasag Island (north of Sipaco) is still partly forested and may harbor some. Butauanan Island (north of Siruma) has been extensively modified and has lost all the butaans for which it evidently was named. It is, incidentally, probably the smallest island that supported a viable population at any time (4 km^2).

Camarines Norte (Luzon).—Though smaller than Camarines Sur, Camarines Norte has several large populations of *V. olivaceus*. Butaans are known to occur in a forested mountain area of about 150 km^2 in the extreme southeast. This habitat is more or less coextensive with the Bicol National Park immediately to the west. The remaining populations have only recently been separated into a northern and southern portion by the agricultural development that accompanied the construction of National Highway No. 1, which bisects the province. Butaans occur in both the northern and southern sections of the mountainous country in this area. However, the northern section (ca 80 km^2) is being rapidly denuded. The southern section is being cleared at the same rate but represents the largest remaining block of forest left in the province (ca 150 km^2). As far as is known, no Camarines Norte islands have butaans. The Calaguas Islands, some distance from shore and due north of Daet, are the largest, but the habitat is open and grassy, with scattered thickets.

Quezon (Luzon).—Unfortunately little is known regarding the distribution of the butaan in the large province of Quezon; it is as long as all of Bicolandia and extends northward to central Luzon along the eastern seaboard. Its northern half is comprised largely of the western slope of the Sierra Madre Mountains, from which the butaan has been reported. There is no reason to believe that it is missing from any part of this vast area wherever the forest has not been completely cleared. It is important to know if it occurs in the lowland dipterocarp forest still present in the valley of the Umiray River in the extreme north. It has probably been extirpated at the coast near General Naker, where likely forest areas have been commercially logged for many years. However, a specimen taken near Real, not too far south, shows that they are still to be found in the general area. The total suitable habitat on the eastern slopes of the Sierra Madre Mountains, including the small portion of the western slopes in this province at the southern tip of the mountain range, is 1590 km^2. It harbors the largest remaining population within any province in which the species is currently known to occur. Land use in the west, toward Batangas from about Lucena City, is highly agricultural, and searches in that area have failed to disclose any butaans. The same is true of the area southeast of the southern tip of the Sierra Madre Range, from near Altimoan to the boundary with Camarines Norte. However, in the extreme eastern part of the province, at its common boundary with Camarines Norte and Camarines Sur, the butaan's range is coextensive with those provinces, particularly along the boundary of Camarines Sur (ca 120 km^2). There is a small forested hilly area (ca 15 km^2) on the Dap Dap Peninsula near the eastern end of the province that may contain a small isolated population (the area at its base, near San Miguel, was converted to agriculture many years ago).

One of the most important projects for the future is to determine whether *V. olivaceus* occurs in the Bondoc Peninsula. Most of this large area is hilly and covered with a scrubby, perhaps secondary forest. Various agricultural practices have recently made great inroads in this area, and this occurrence, plus the fact that the dry season may be a little too long, suggests that the butaan is probably absent. However, the area is remote and so large that it is difficult to survey for the butaan. Probably the best place to check in the future would be near the tip of the peninsula west of San Andres, where primary forest of a somewhat more mesic type still occurs (ca 580 km^2; remaining potentially important habitat in the peninsula: ca 300 km^2). Alibijabon Island off the southeastern peninsular tip is completely cut over, as is Alabat Island north of the base of the peninsula on the Pacific side. At present there is no evidence that the butaan occurs on any of the Polillo Island group. Edward Taylor collected there many years ago but failed to get any information or specimens. Furthermore, almost all the forests have been cut for a long time.

Laguna (Luzon).—Most of Laguna province has been converted to agriculture. This fact, plus the generally high human population density, has apparently been responsible for a major reduction of the original butaan populations. Even so, present evidence suggests that *V. olivaceus* probably never extended far west of the base of the western slopes of the Sierra Madre Mountains, where it is known still to occur, though perhaps in small numbers. The total area included is about 220 km^2.

Rizal (Luzon).—While human population density is high in the western half of Rizal Province (Manila, etc.), much of the eastern third is heavily forested on the western slope of the Sierra Madre, where *V. olivaceus* is known to occur near the southeastern corner. However, it is expected to be found all along this slope to the northern boundary of the province. The area in which the butaan is expected to occur is about 530 km^2.

Bulacan (Luzon).—At this writing there are no butaan specimens available from Bulacan, though they certainly must occur in the eastern third of the province, which is the forested western slopes of the Sierra Madre Mountains. The western half of the province (in the Central Plain; fig. 6-1) is agricultural and has a high density of human population. The area in which the butaan is to be expected is about 645 km^2.

Nueva Ecija (Luzon).—No specimens have been found to date in Nueva Ecija Province, though the extreme eastern edge, which is rather remote (ca 250 km^2), includes the western slopes of the Sierra Madre Mountains and probably contains fairly good populations. However, it is likely that this area does not extend north of the east-west Umiray River Valley in which the highway runs from Dingalan (Aurora Province) on the coast to Bongabon in the completely deforested and highly agricultural Central Plain (fig. 6-1). This valley lies in a deep and geologically significant fault. Several kilometers north the Sierra Madres' mountainous landscape begins again but with an apparently different geological history (chap. 14).

Aurora (Luzon).—No butaans are yet known from Aurora Province, though the species should be expected in at least the extreme southern tip along the eastern flank of the Sierra Madre Mountains from the southern wall of the river valley at Dingalan south to the Quezon boundary (110 km^2).

Other.—I do not expect that the butaan will be found to occur beyond the areas outlined above. If, however, the postulated east-west geomorphological barrier near Dingalan is shown to be unimportant to the distribution of this species, then it should be found throughout much of the hilly to mountainous forested parts of central and northern Luzon. The fact that no specimens have ever come to light, in spite of frequent biological expeditions into this area, suggests that they are not found there. The only remaining possibility is that they may be found to be restricted to appropriate habitats along only the Sierra Madre Mountains on the eastern seaboard all the way to the extreme northeastern tip of Luzon. If so, this northeastern portion would represent about an additional 900 km^2. The determination of the northern limits of the range of the butaan along the Sierra Madre Range is one of the most important studies for the future, because if it is found to occur there the range would be greatly enlarged over that now known.

On previous trips to Marinduque and Mindoro islands, neither I nor others interested in herpetology have found any evidence of the butaan. Nor does it seem to occur on the islands of the Masbate Group (Masbate, Burias, and Ticao), where agriculture has destroyed all of the forests. The little exploration I was able to conduct on Samar, Leyte, Mindanao, and Palawan in 1976 and 1978 makes it reasonably clear that the butaan is not found there either. However, Samar might repay further investigation, particularly at its northern end. *V. olivaceus* is easily confused with juvenile *V. rudicollis* (see chap. 14), and Mindanao populations of *V. salvator* are sometimes similarly patterned (see fig. 14-1).

In summary, the butaan is apparently restricted to Catanduanes and Luzon islands. It is found over about 15% of Catanduanes but a smaller part of Luzon (table 6-1). On the latter, as currently understood, the major part of the range is encompassed in the district unofficially known as Bicol (Camarines Norte southward). Its only known extension beyond this area is along the Sierra Madre range in extreme eastern to about central Luzon. It is not presently known (but may in time be found) on Bondoc Peninsula, which is geologically a southeastern extension of the Sierra Madre. The historical geology of this area and its probable relation to the range of the butaan is discussed in chapter 14.

TABLE 6-1. Approximate area per province in which *V. olivaceus* is found.

Province	Total area (km^2)	*V. olivaceus* habitat (km^2)	Percent of total area
Catanduanes	3,785	550	15
Albay	4,001	475	12
Sorsogon	2,054	140	7
Camarines Sur	5,336	890	17
Camarines Norte	21,147	380	2
Quezon	11,957	1,795[a]	15[b]
Laguna	1,204	215	18
Rizal	2,049	530	26
Bulacan[c]	2,516	645	5
Nueva Ecija[c]	5,492	250	5
Aurora[c]	4,565	110	2
Total	64,106	5,980	9

a. Plus 880 km^2? No records yet.
b. Or 22 percent? No records yet.
c. No records yet.

Assuming that the range as outlined is correct, the total area occupied by the butaan is about 5980 km² (table 6-1). This area is approximately five times larger than that inhabited by *V. komodoensis* (Auffenberg, 1981a), which has the smallest range of any reasonably well known monitor lizard.

Natural ecological factors are of major importance in the distributional patterns, but they are not the only ones. Human population density is high within parts of central and southern Luzon, and throughout most of the range butaans are regularly killed and eaten (see chap. 15). Furthermore, many thousands of square kilometers have been cleared of native forest, destroying the butaan's habitat. Unlike the situation for *V. komodoensis,* in which some agricultural practices actually encourage range/habitat extension (Auffenberg, 1981a), agriculture (with the exception of shade crops, such as coffee) tends to destroy the butaan's habitat.

The following list includes all the localities at which butaans have been found or reliably reported (reference specimens in parentheses). In general the localities are arranged from south to north; elevations are approximate. Figure 6-1 shows the major physiographic features of the area involved, 6-2 the political provinces within the area, and 6-3 the localities from which *V. olivaceus* is known.

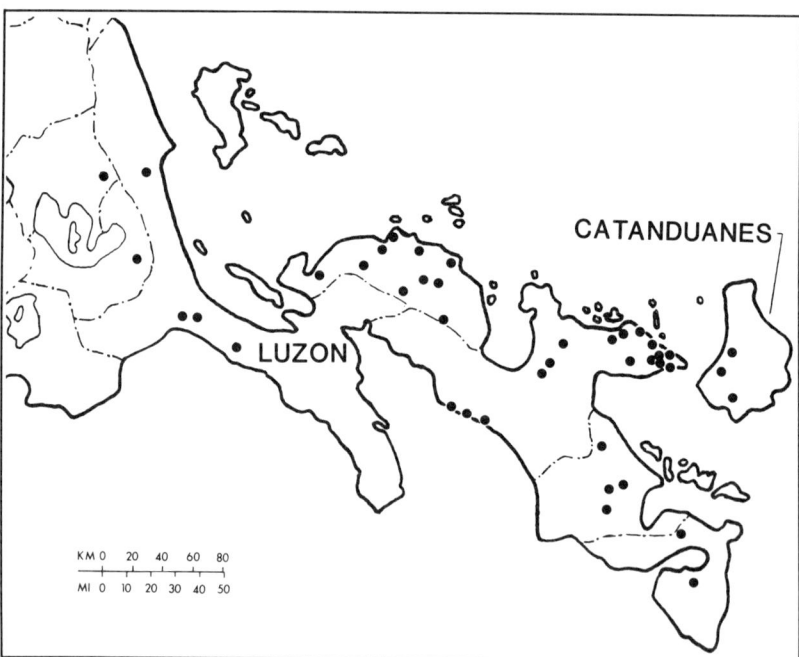

Fig. 6-3. Solid circles show localities in which *V. olivaceus* is currently known. See fig. 6-2 for geographic names.

Catanduanes Island

Talahid District

    8 km W Gigmoto, el 320 m (UF 55116)
    10 km NE Inalmasinan, Hituma area, 380 m
    18 km NNW Virac, el 400 m

Luzon Island

Albay Province

    0.3 km W Amtic, at Barrio Buang, between Mt. Masarauag and Mayon volcano, 3 km E Batang, el 200 m
    20 km NW Legaspi, along Benao River, near Ligao, el 200 m
    SW slope Mayon volcano, 3 km E Masarauag.

Camarines Norte Province

    Barrio Igang, 8 km NW Batbaloni, ca 24 km W Daet, el 100 m
    Barrio Macababale, el 100 m
    Barrio Macagangke, 25 km W Labo, el 150 m
    Barrio Malaya, 16 km S Labo, el 250 m (UF 37969)
    Bicol National Forest, where Highway No. 1 passes through it, near boundary with Camarines Sur, el 300 m
    near Bulalon, el 250 m
    3.5 m W Labo, near Bulhao, el 80 m
    Panganiban Peninsula, 5 km NE Panganiban, near May Cruz, el 75 m
    San Rafiel, 7 km W Panganiban, el 50 m
    mountains near Talisay, ca 20 km Daet, el ?
    Tinaga, 5 km S iron mines, 5 km SW Jose Panganiban, el 100 m

Camarines Sur Province

    Ambuyog, 1.5 km S Gibgas, ca 24 km NW Caramoan, el 50 m
    near Barrio Mayburi, 8 km S Minalabak, 16 km S Naga City, el ca 50 m
    Barrio Minalaba, 5 km NW Caramoan, el 60 m
    Batungan Mt., 4 mi E Caramoan, el 150 m (UF 54777)
    Caputatan, 4 km SW Caramoan, el 50 m (UF 55296-7, 55169, 55058, 55026)
    near Caramoan, ca 16 km NNW of tip of Caramoan Peninsula, el 20 m (UF 56286, 55218, 53917-8)
    2 km W Carolinas at Rayhan River, ca 10 km E Naga City, at base of Mt. Isarog, el 100 m
    near Cayohoson, 3 km E Buhi, near Lake Buhi, E Mt. Iriga, el 300 m

5 km N Dalupaon, 10 km NW Pasacao, el 350 m
   Ili, 2 km SW Caramoan, el 50 m (UF 55193, 53607)
   near Ilawod, 2 km NE Caramoan, Caramoan Peninsula, el 100 m
      (UF 37968, 55025, 43616)
   Kabangan Mt, Caramoan National Park, el 400 m
   6 km NW Lagonoy, ca 20 km E Mt. Isarog, near base of Caramoan
      Peninsula, el 250 m
   Mandiclom, 3 km S Gibgas, ca 24 km NW Caramoan, el 100 m
   Mayruhin, 23 km E Naga City, Mt. Isarog, el 300 m
   near Pacol, 8 km ENE Naga City, el ca 50 m
   near Pagsagnaan, 4 km N Parubcan, S slope Calinigan Mts, Caramoan
      Peninsula, ca 40 km ENE Mt. Isarog, el 350 m
   near Panaquasin, 4 km E Carolinas, along Rayhan River, on slopes of
      Mt. Isarog, el 400 m
   near Pandanan, 4 km NNW of tip of Caramoan Peninsula, el 20 m
   Pasacao, 20 km WSW Naga City, el 250 m (USNM 27776)
   Patag, 7 km SE Caramoan, el 50 m (UF 55055, 55298)
   2 km W Sominabung, 8 km NW Pasacao, el 320 m
   Toytoy, 10 km W Gibgas, el 50 m

Laguna Province

   "Laguna" (de Elera, 1895)
   Siniloan, on highway to Infanta, near Quezon border, el 350 m

Quezon Province

   near Mauban, el 200 m (FMNH 220262-3)
   2.5 km N Menahahan, 6 km NE Pagbilao, el 200 m
   Quezon National Park, el 300 m
   near Real (PNM 891), el 200 m
   near Santa Elena, el 250 m

Rizal Province

   Mayaga, near Tanay, el 250 m

Sorsogon Province

   Mt. Juban, near Barrio Mt. Irosin, 16 km S Sorsogon City, el 350 m
      (UF 37967)
   near Pilar, el 100 m
   8 km NW Sorsogon City, Palge River, near Barrio Capoy, el 250 m

Vertical Distribution

The butaan occurs from sea level to a maximum known elevation of about 400 m (recorded in several different parts of the range). This range is slightly higher than the upper limit of the Lowland Mixed Dipterocarp forest associations in the Philippines (see fig. 4-7). The average elevation for 48 localities is 175 m, though this number simply reflects the average elevation at which forests that contain butaans still remain in this area. At present there is no evidence that densities are significantly different at elevations within this total range, though populations obviously become less dense at the upper limit. While the factors responsible for restricting butaans to the Lowland Dipterocarp level are not known, it is clear that there are important temperature and moisture changes between 300 and 500 m (chap. 4), which are probably responsible for limiting many of the plants important to the butaan to lower elevations. However, on Mt. Isarog (Panaquasin), the mollusc genus *Rysotta* and the plants *Spondias pinnata, Grewia stylocarpa, Canarium hirsutum,* and *Canarium ovatum* (all important foods of *V. olivaceus*) are found at 500 m, though *Pandanus radicans* is absent, as is the mollusc genus *Lepidotrichia*. Many of the food species of the butaan thus extend to higher elevations than any current record for the lizard, suggesting that factors other than food alone may be responsible for the butaan's upper vertical limit. It may be significant that hornbills are apparently also restricted to the same upper limit as the butaan in the Isarog area.

Barriers to dispersal.—Open, grassy habitats, marshy tracts, seacoasts, and elevations above 700 m are all partial to complete barriers to dispersal. Except for a few rare instances during the dry season, butaans are seldom found in water and then only in small, shallow bodies. They undoubtedly swim well, as all monitors do, including desert species (Mertens, 1942), but they have never been seen voluntarily to enter the sea by any of the local people particularly well acquainted with them. As stated above, the present spotty distribution of *V. olivaceus* is the result of human habitat modification (mainly on Luzon) and of geologic factors affecting the area from the Miocene through the Pleistocene, whereby some land masses were broken into smaller units and others fused through submergence, plate tectonics, and volcanism (chap. 14).

Home Range

Considerable effort has been devoted to the analysis of location data from live trapping of reptiles (see especially Jorgensen and Tanner, 1963; Tinkle, 1967; Jennich and Turner, 1969; Turner et al., 1969) and radio tracking (Green and

King, 1978). Sanderson (1966) and others have pointed out the inadequacy of the interpretive techniques in this field, especially in the concept of home range. A major difficulty is the uncertainty that arises about the relationship between an animal's real use of space and fragmentary records of this usage provided by sporadic radio fixes. For scientists studying social systems, this problem is not trivial, for a slight difference in the position of home range boundaries of neighboring animals could have significant implications for their social organization.

In a thorough study of analysis techniques, MacDonald et al. (1980, foxes) and Mohr and Stumpf (1966, monkeys, rodents, and birds) compared home range size estimates based on different approaches (circular normal model: Calhoun and Cosby, 1958; Mazurkiewicz, 1969; bivariate diffusion models: Jennich and Turner, 1969; Dunn, 1977; grid methods: Adams and Davis, 1967; Clutton-Brock, 1974; "field workers' estimates" of hand-drawn ranges based on a combination of direct observations, interpretation of spoor, and radio tracking: MacDonald, 1980). Both concluded that the major failure of all qualitative estimates of spatial organization of populations is that they do not necessarily distinguish between the animal's behavior at different locations within the range. From movement data alone there is no hope of interpreting the behavior of the animal within the home range (see Siniff and Tester, 1965).

In this study I was primarily interested in the adaptive behavior of the butaan to its environment—not only in home range sizes and locations per se. Thus I relied completely on the "field workers' estimate," using radio tracking to verify periodically location in the field estimated by antenna directions from the base camp. These data were coupled with as much observation as was possible from relatively short distances and from various field signs, such as spoor. Only in this way could the interpretations of spatial relations in this species be as finely tuned as its real behavior. The undesirable consequences of preconceived biases creeping into the conclusions were, I hope, countered by increasing field observation time and the number of radio contacts with each individual per day.

Activity range.—Being relatively large animals, varanids are capable of moving many kilometers/day (Auffenberg, 1981a), making home or activity range definitions, particularly for adults, troublesome. In addition, workers are not in complete accord as to what part of the total observed movements should be included (see Milstead, 1961; Rand, 1967a; and, particularly, Wilson, 1975).

The interspecific and seasonal variation in behavioral patterns has led to at least some kind of emendation for the term in almost every species studied. By the butaan's activity range, I mean the total area in which the individual is likely to be found, including all of its retreats. For *V. komodoensis* this activity range is comprised of two usually zonal regions, the scavenging and the foraging areas (Auffenberg, 1981a), of which the former is much larger. The latter includes a core area, where most of the activity takes place. There is no scavenging area within the total activity range of the butaan, though core and daily foraging areas can be discerned. Both parts of the activity range are approximations. Though they are necessarily graded in size by definition, their functions are totally different. Their shape and size are dependent upon topography and food location and its density, as well as on size and sex (and social status, as in the Komodo monitor) of the individuals whose activity range is being considered.

There are relatively few published data on the activity ranges of lizards and practically none on varanids (but see Green and King, 1978; Auffenberg, 1981a, for radio-telemetric studies, and St. Giron and St. Giron, 1959; Pianka, 1968, 1970, for following tracks to determine home ranges). The large number of location fixes established for the butaan ($N = 1638$) and the length of time over which these studies were conducted (July 1982–June 1983) make this one of the most detailed attempts with radiotelemetry to estimate home range in lizards.

The locations of nine butaans were plotted 1638 times using only radio signal direction, from 26 to 432 times each (see chap. 1), over a total of 455 radio-days. (The large number of location plots [minimum 101/individual] significantly reduces the effect of sample size bias on all density estimates, regardless of technique [Jennich and Turner, 1969].) These plots were verified in the forest a total of 142 times (6–23 times for each individual). These data were used as the basis for the calculations of ranges that follow. Table 6-2 provides pertinent data on individuals telemetered, number of plots per individual, etc. There was no apparent relationship between the number of plots or period between the first and last plot and the calculated activity range area.

Calculated annual activity range areas of adult *V. olivaceus* varied from

TABLE 6-2. Activity range area of individuals of *V. olivaceus*.

Butaan no.	Sex	Weight (kg)	No. radio plots	No. verified plots	Days between first and last plot	Home range area (ha)[a]
18	M	4.2	432	23	108	1.82
45	M	1.2	303	18	78	1.60
25	M	2.0	208	16	55	1.08
35	F	2.3	209	21	87	0.32
76	M	3.5	128	23	49	2.40
82	M	8.3	100	10	20	2.67
9	M	8.3	109	13	28	2.71
10	F	2.7	123	12	22	0.51
11	F	2.8	26	6	8	0.22

a. For "field workers' estimate" method, see text in section "Home Range."

0.22 to 2.71 ha, mean 1.48, SE 0.21. This activity area is exceedingly small for a large reptile. It is a mere 7.6% of the mean activity range area of adult *V. gouldii* (Green and King, 1978), even though the mean weight of the experimental individuals of *V. olivaceus* is 300% greater than the mean weight of the experimental *V. gouldii*. The butaan activity range is only 6% of that calculated for equal-sized *V. komodoensis* (Auffenberg, 1981a). Finally, *V. olivaceus* activity range is a mere 0.04% of that of *V. griseus* (200–500 ha; St. Giron and St. Giron, 1959), though the latter is even smaller than *V. gouldii*. The differences in annual activity range size between *V. olivaceus* and *V. gouldii–V. komodoensis* probably are due largely to the frugivory of the former (chap. 11) and the carnivory of the latter species, since carnivorous reptiles tend to have larger activity ranges than herbivorous congenerics (St. Giron and St. Giron, 1959). The large activity range of *V. griseus* is undoubtedly also influenced by the low prey density in the Algerian desert where the study of this species was undertaken. That food availability influences the size of home range has previously been demonstrated in some iguanid lizards (Simon, 1975; Ferguson et al., 1983).

Though *V. olivaceus* searches for most of its food on the surface, it also spends considerable time in both trees and deep rock crevices. Thus its activity range has considerable vertical dimension, much more so than do the almost completely terrestrial species discussed above (at least as adults). When this vertical component is included in the calculation (ca 10 m, utilized in roughly 1-m layers), the average activity range of adult butaan is 14.8/ha. This figure is similar to that calculated for the almost completely terrestrial *V. gouldii* (Green and King, 1978).

The activity ranges demonstrated for *V. gouldii* (Green and King, 1978) and *V. komodoensis* (Auffenberg, 1981a) are significantly larger than their sizes suggested by the mathematical model of lizard home range size proposed by Turner et al. (1969, $A = 171.4\ W^{0.95}$, where W = weight in g, based on a large number of small lizards with a variety of diets). Using the same formula, the actual activity range of *V. olivaceus* ($\bar{X}$ adult wt = 3.13 kg) is smaller than expected.

Unfortunately, there are no data available on the size of the activity range of juvenile *V. olivaceus*, though adults are expected to have larger ranges than juveniles (as demonstrated for *V. komodoensis*, Auffenberg, 1981a).

However, a marked difference in size of activity range can be demonstrated in similar-sized adult male and female butaans. Thus the large activity range of adult males is always over 1 ha and may be as large as 3 ha ($\bar{X}$ = 2.05 ha, ± 0.29, SD 0.65, N = 6), and that of adult females less than 0.5 ha ($\bar{X}$ = 0.35 ha, ± 0.12, SD 0.14, N = 3). The difference between them is statistically significant, in spite of the small sample size (t = 6.30, df = 7, $p <$ 0.001).

Other studies have shown that the size of activity range of male and female lizards of several species is also significantly different (Tinkle, 1967; Ferner, 1974; Simon, 1975). Except for *V. olivaceus*, no data exist demonstrating such sexual differences in size of home range for any varanid. However, the fact that males have generally higher activity level and food intake than females (Auffenberg, 1979b) suggests that all male varanids probably have larger activity ranges.

Figure 6-4 shows typical activity ranges of several butaans during 1982–83. It also shows that most daily movements are centered around particular trees, crevices, or other specific surface sites. Sometimes the monitors remain at these sites for long periods of time (up to 74 days), though the average time spent at any area is 8.2 days. These sites are almost always on hilltops (only 3.1% of the time is spent at the lower third, or base, of the hills), where shelters are numerous and food trees most common (chap. 11). The thermoregulatory sites of *V. komodoensis* are usually located on hills (Auffenberg, 1981a), though most hunting is done in the valleys, and burrow shelters are

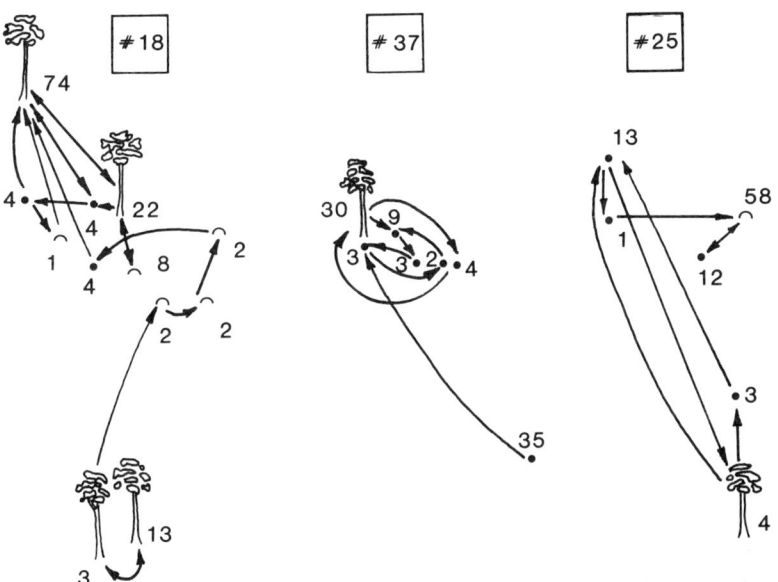

Fig. 6-4. Seasonal home range of three adult *V. olivaceus*, showing movement between and around certain features and their relationship to shape and size of home ranges: *left*, no. 18, male, 22 July–1 December; *middle*, no. 37, female, 12 September–3 November; *right*, no. 25, male, 12 August–10 November. Features shown are solid dots, crevices; half circles, rock outcrops; and trees. Arrows show directions of movements; numbers show days spent at each topographic feature or tree.

located along dry streambeds or on hillsides at the junction of savanna and monsoon forest. Thus from an ecological standpoint the movement pattern of *V. olivaceus* is much more restricted than that of the carnivorous and wide-ranging *V. komodoensis*.

Though butaans occasionally move from hill to hill, they spend most time near the crests and least time in the valleys between. Local topography thus affects the size and shape of the butaan's activity ranges more than it does those of the Komodo monitor's. *V. komodoensis* uses the valleys almost daily for specific activities (burrow use, hunting, and general moving through the large activity ranges), whereas the butaan uses the valleys only for irregular and infrequent movements between hills, restricting its daily activities to the rockier middle slopes and extreme crests, where food and shelter are most abundant.

Green and King (1978) have shown marked seasonal differences in size of the daily activity ranges of *V. gouldii* adults. No seasonal differences have been found in *V. olivaceus*, except for shifts to wetter situations during the dry season (see above), and these shifts may result in increasing the size of the annual activity range. In fact, for No. 76 the daily activity range was increased by 23% when the individual regularly left its arboreal retreat during each of six days and lay near or in a nearby creek during the drought in April. However, for most individuals there is no significant seasonal change in daily activity ranges. Also the ranges are remarkably small compared to those for other species for which such data are available (Green and King, 1978; Auffenberg, 1981a).

Of four individuals for which daily activity range data were available (verified extent of daily movements, minimum 36 plots/individual), the mean is only 0.05 ha ($\pm$ 0.004, SD = 0.012). This low number is presumably due to the restricted sites in which fruits are found in the forest and to the fact that once a food tree is located, there is little reason to leave it for several days (depending on species). This small daily activity range (0.03–0.07 ha) is only 6% of that calculated for *V. gouldii* by Green and King (1978) ($\bar{X}$ = 0.82 ha) and only 5% of the core area of *V. komodoensis* (Auffenberg, 1981a) (9.3 ha for equivalent-sized individuals).

Spatial overlap.— There is considerable evidence suggesting that the annual activity ranges of *V. olivaceus* adults are sometimes broadly overlapping. These data were obtained by capturing individuals with hunting dogs and by radiotelemetric location.

Of 131 captures of butaans in which dogs were used, two individuals were found in the same place (always in a hollow tree) at only three sites, and these individuals were always of opposite sexes. Such occurrences are recorded for August, November, and February, suggesting that spatial overlap between individuals of different sexes is not restricted to the reproductive period but may occur in any season. Thus pair bonding may occur in *V. olivaceus* (as seems more clearly the case in *V. komodoensis;* Auffenberg, 1981a). However, the situation of *komodoensis* is different from that of the butaan in respect to the extent of the overlap, which is great in the former. As I have pointed out elsewhere (Auffenberg, 1978b), this spatial overlap results in frequent meetings between potential breeding partners and may be important in allowing courtship to proceed as smoothly as possible in an otherwise aggressive, rather intolerant species.

The best data on spatial overlap in *V. olivaceus* were obtained through telemetry. They show that individuals sometimes inhabit the same specific area, that the overlap may exist for a number of months, and that such overlap may occur in individuals of the same or opposite sex.

Three individuals were telemetered for the same period on Mt. A (Nos. 18, 25, 35; see table 5-1 for morphological differences). Each lizard's activity range overlapped that of the others to some extent, so that no activity range was used exclusively by one individual. The overlap varied from 300 to 1400 m^2. However, of 112 days in which Nos. 18 and 25 (both males) shared a maximum of 700 m^2 of their adjacent activity ranges, there were no days in which both were in the shared area. Out of 66 days when Nos. 25 (male) and 35 (female) shared a maximum of 300 m^2 of their home ranges, there were no days in which both were in the shared area. Of 47 days during which Nos. 18 (male) and 35 (female) shared a maximum of 1400 m^2 of activity range area, there were 9 days in which they were in the same block, and then for only a short period in June (when the courtship activity of this species is most intense in the Caramoan area).

Though their activity ranges overlap, there appears to be little significant use of an area on the same day by individuals sharing it, suggesting the possibility of mutual avoidance. Though the sample is small, it is suggested that avoidance occurs whether the individuals sharing a part of their activity range areas are of the same or different sexes.

Movement patterns within and outside the activity range.—Extensive displacements of some individuals were registered that included areas distant from the region in which they normally concentrated most of their activity. These displacements were not included in the estimates of average home range. Telemetric data suggest that 11% of the individuals monitored over long periods made such extensive, sudden movements, and all were large males. (Such individuals are probably the "transients" described in Auffenberg 1981a for *V. komodoensis.*) They were not encountered or recaptured in the study area during the succeeding months of work. In all cases the displacements are characterized by almost straight movements, generally following

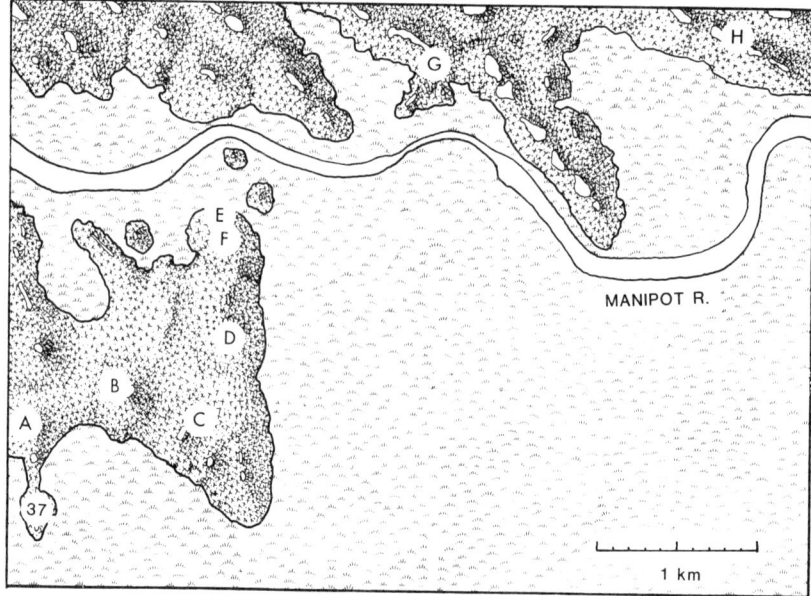

Fig. 6-5. Rapid changes in site use that place individuals in a totally different area are rare in *V. olivaceus*. Number in lower left shows weeks of prior recorded residence at that site. Letters show successive evening resting locations during an 8-day period.

the same direction (fig. 6-5). During the more normal "sedentary" period these individuals moved an average of 6.8 m between successive location plots on a twice/day schedule, but during the "active" phase average distance covered was 121.2 m. Further study may show that displacements are more common and involve greater distances during periods of lower food availability.

Males exhibit a greater variability in annual home range size, their average variation (2.3 ha) being almost three times that of the females (0.8 ha; $t = 5.91$, $p < 0.02$). Why the males' home range size is more variable than the females' is not clear.

Upon applying the test of circularity to the distribution of random points within the activity range of each individual, it was found that none of the individuals exhibited a clear circular distribution in their movements (the density probability function of Calhoun and Cosby, 1958, which assumes circularity, is not applicable to the data). Rather the movement patterns tend to be linear, most movement occurring between certain fruiting trees and some rock crevices (fig. 6-4). The average number of sites regularly visited within the activity range is eight, and the average distance between them is 51.6 m. There is no significant correlation between the distance traveled per day and SVL, because average distances are a direct function of distance between fruiting trees and not entirely at the whim of a wide-ranging, foraging individual. Nor is food density per hectare significantly correlated with home range size, for fruiting phenology and spatial dispersion of food plants tend to produce an even spread of fruits per hectare, regardless of the number of fruiting species that are found in any given area. In general, home ranges tend to be located on ridges and peaks of mountains rather than on the lower slopes.

None of the home ranges studied was located in any part of either the base of the mountains or the valley floors, though these lower locations might be expected in areas where the forest of relatively flat areas has not been completely destroyed.

Shelter Characteristics and Use

The retreats used by varanid lizards are expectedly varied, though burrows in the earth and hollow trees are apparently the most common (Smith, 1932; Mertens, 1942); many species use both. Finding a retreat for the night that is safe from predators and that provides a proper thermal environment is probably not as important for butaans as for most other varanids, because of the availability of shelter almost anywhere on the fissured forest floor and because of the rather uniform temperatures both day and night and seasonally. The retreats of this species do not, to my knowledge, include burrows that they dig into the earth. Hollow trees are used and, even more commonly, existing cavities and crevices in the bedrock, of which there are many—particularly in highly eroded karst areas, such as in the Caramoan area. Here cavities and crevices sometimes extend several tens of meters below the general surface level, and butaans frequently remain there during the windiest days and periods of continuous rain, though often only during the night (chap. 5). Based on current evidence, no particular size or shape of cavity is sought out, except that at night the one chosen is often a narrow slot (horizontal or vertical) into which the butaans wedge themselves rather tightly, usually with the tail coiled in a spiral.

A retreat commonly used in daytime is the tangle of vines, scandant shrubs, and epiphytes found on the trunks and lower limbs of some of the larger forest trees (fig. 6-6). In height these vary from 3 to 40 m above the forest floor, but those most frequently used are 5–20 m ($\bar{X}$ = 6.2 m; 87 local shrubs and vines have an average canopy height of 5.4 m, so most retreats are at the canopy level of most arboreal plants; see fig. 6-6 for additional details). These arboreal thickets in which the butaans often sleep during the night or rest during the day are composed of up to 12 plant species (mode = 9). (In central and southern Luzon there are 26 families comprising 62 species of vines, 17 families of 86 species of scandant shrubs. There are 18 species of purely epiphytic plants: 8 trees [all *Ficus*], 4 ferns, 4 orchids, 2 vines.) The dominant plants

Fig. 6-6. *Left:* arboreal retreats of tangled vegetation are often used by *V. olivaceus* both day and night. *Right:* bar shows where no. 3 spent 95 percent of its time in a *Sterculia* tree; arrow shows its usual resting site in the tree.

usually comprising these tangles are scandant bamboos (genera *Schizostachyum* and *Dinochloa*), climbing araoids (particularly *Rhaphidophora merrillii*), Orchidaceae, and ferns (especially species of *Asplenium, Lygodium,* and *Drynaria*).

Tree hollows are not as commonly used as retreats as are arboreal thickets and rock crevices, probably because they are much less common. Hollows large enough to accommodate the butaan occur only within trunks or major limbs. Normally they become available only after trunks or limbs have been snapped off during storms, exposing the hollow cores. All retreats, of whatever kind, are found on the slopes and tops of mountains; all are rare along highly forested stream valleys. Though sometimes present, they are not usually found in habitats near villages or other habitats highly modified by man.

Radiotelemetric data show 204 observations of where butaans spent the night throughout the year; in 23% of the observations, they slept in deep rock crevices (signal strength > 7); in 21% at or near the surface (signal strength 5–7); and in 56% in arboreal thickets (strength 1–2). In many cases the data were confirmed by concurrent specific site location of the telemetered individuals. However, rock shelters are used as sleep sites more commonly during the rainy season than during the dry part of the year (July–November, $N = 112$, crevices 57%, trees 12%, and at or near the surface 31%; January–May are not significantly different from the annual totals).

All shelter types are also used in daylight, sometimes for long periods of time without any clear correlation with environmental conditions: No. 76 remained in the same tree day and night for 139 hours (April); No. 82 remained in a tree shelter for 163 consecutive hours (May); and so on. It may be of importance that no butaan ever remained in a crevice shelter for longer than 15.5 hours, even during the violent, drenching typhoons of October and November. However, crevices tend to be used more commonly in the morning and early afternoon during the wet season and less often and later in the day during the dry season (fig. 6-7). In a few cases it was clear that the butaan remained in a particular tree (or a nearby arboreal shelter) because it was a food source (chap. 10). In fact, many individuals are caught in food source trees in which they have sought refuge after having been located on the ground by hunting dogs. *Canarium hirsutum* is the modal species in which they are most frequently found (8.7%), though many species are utilized.

Besides food, escape, or shelter, trees have a number of other characteristics that apparently make them attractive to the butaans. These are height of crown, cross-sectional area of the crown, the presence of vines or arboreal thickets, and exposure of the tree crown to the sun (table 6-3).

In general, shelter trees are most often also food species. Twenty-four species were recorded as being used for shelter. Of these, only seven (29%) are

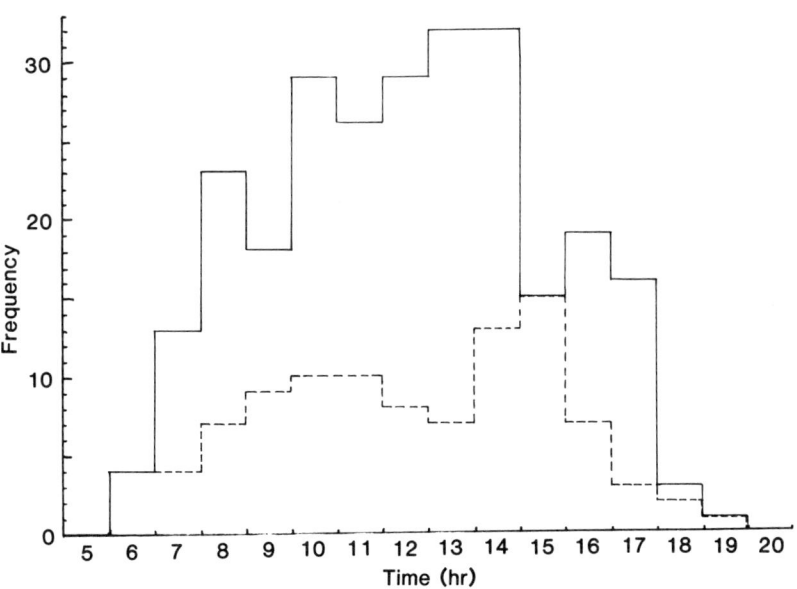

Fig. 6-7. Times of day when *V. olivaceus* individuals enter rock crevice shelters: *solid line*, wet season (July–November, $N = 260$); *dashed line*, dry season (January–March, $N = 100$).

TABLE 6-3. Partial Kendal-τ nonparametric rank correlation coefficients for the attributes of trees with which butaans' use of individual trees is significantly correlated ($p < 0.01$; nonsignificant correlation = n.s.).

	Exposure of crown to sun	Vines, thickets present	Cross-sectional area of crown	Height of crown
Used by butaan	0.31	0.42	0.26	0.19
Exposure of crown to sun		0.21	0.40	0.47
Vines, thickets present			0.18	n.s.
Cross-sectional area of crown				0.52

TABLE 6-4. Relative importance of some nontaxonomic attributes of tree crowns used by *V. olivaceus* (stepwise multiple regression on Kendal nonparametric rank correlation coefficient, τ. (Nonsignificant correlation = n.s.)

Tree attribute	β	F
Exposure of crown to sun	0.15	5.3
Vines in crown	0.24	18.8
Cross-sectional area of crown	0.10	n.s.
Height of crown	0.04	n.s.

TABLE 6-5. Frequency of trees with vines on Mt. Ahop study plot and number of times butaans were located in them.

	Height of trees (in m)		
	0–10	11–20	21–50
Number of trees greater than 10 cm bhd having vines	808	203	41
Number of radio locations of butaans in viney trees	8	72	40

NOTE: Chi^2 = 802338.37; $p < 0.0005$.

food species. However, these make up 69% of all the records of butaans found in trees ($N = 87$). Thus it is clear that basking and nighttime shelter often take place in tree species under which they probably also look for fruits.

The major nontaxonomic factor seems to be the presence of arboreal thickets. This factor is more important than exposure to the sun or the cross-sectional area of the crown (tables 6-3, 6-4), though if vines are found in trees, butaans tend to select the larger trees in which to shelter (table 6-5).

In general, deep rock crevices are more protective and insulative than any other shelter type. Thus it is not surprising that there is more seasonal variation in crevice use than in arboreal thicket use. Even after leaving the evening sleeping shelter (if one is used), butaans commonly enter (or reenter) rock crevices throughout the day (even as early as 0600); most entries are made from 1000 to 1400 hr (fig. 6-7). There is no significant difference in the time of day that rock crevices are (re)entered during daylight hours in the dry or rainy seasons.

Figure 6-8 illustrates the great individual variation in the degree and periodicity of crevice use. In general, use of rock crevices is greater during the wet seasons than the dry ones. As much as 96% of the available hours of daylight may be spent in crevices during severe storms, but it is not clear that rainfall is the only factor determining the extent of daylight crevice use (see chap. 5). Wind is also a major factor, for rank correlation analysis of its strength and the butaan's crevice use shows a correlation of $r = 0.82$, rainfall $r = 0.86$. Temperature variation seems to have no relationship to the extent of crevice use.

Two butaans (of one or both sexes) were found fairly commonly in the same retreat (crevices, arboreal thickets, or hollow trees). Of these types, shared use of arboreal thickets was most common. On the basis of 562 radio loca-

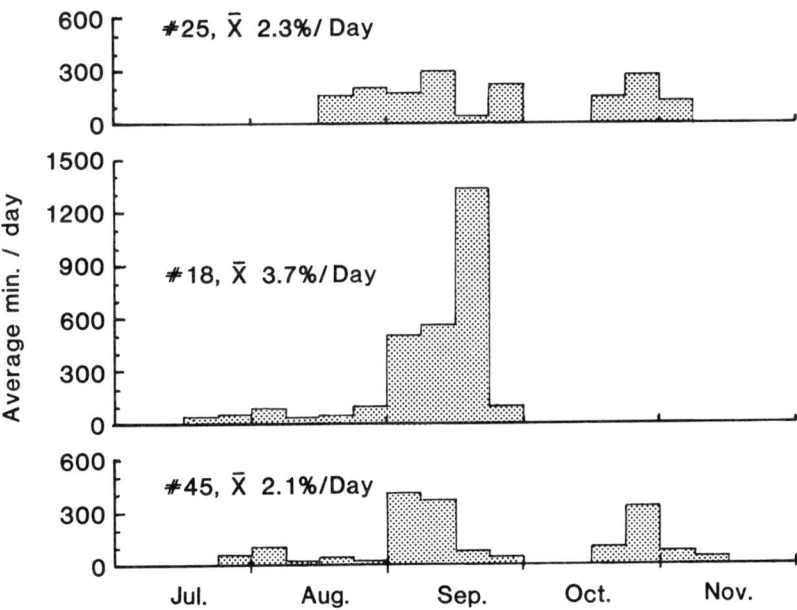

Fig. 6-8. Total minutes per day spent in rock crevices by three adult *V. olivaceus* (22 July–3 November), showing general noncorrespondence in patterns between individual retreat use, except for period during typhoon "Ruping" during the second week of September. Another typhoon passed the area during the last week of October.

tions of butaans from close distances (< 30 m), 1.6% were of individuals using the same retreat. This figure is higher than multiple use of retreats noted for *V. komodoensis*—a more aggressive species than *V. olivaceus*. As shown in chapter 8, long-term associations of male and female butaan adults may be an important factor in reproductive success. Mutual tolerance of two individuals in a single retreat may thus be the rule of this species, at least when the two interacting individuals are adults of the opposite sex.

7

Individuals and Populations

Individual Characteristics

IN addition to ontogenetic and sexual variation in color and morphology treated in chapters 2 and 3, some individuals have distinctive morphological and behavioral traits (usually most noticeable in large adults).

Injuries.—Of 118 butaans examined for signs of external wounds, 32 had indications of previous injuries. None of these could be classed as particularly severe, such as limb or tail loss; most were old wounds representing minor injuries that had long since healed. The smallest individual with signs of injury was 1.2 m long, suggesting that most injuries are sustained after sexual maturity. Some wounds are clearly related to differences in sexual behavior, some to interspecific coactions, and some to injuries received from the coarse environment.

The scars of the butaan reflect injury patterns significantly different in degree and kind (figs. 7-1, 7-2) from those of *V. komodoensis* and *V. salvator.* Though living in the same general environment, *V. salvator* has more injuries (63% of the total investigated) than *V. olivaceus* (52%). Only 8% of the butaans lack a tail tip, whereas 16% of all *V. salvator* do. In studies of other snakes and lizards, tail amputations are believed to result from predation attempts by other animals (see Spears, 1977, for review). While this may be true for more severe tail amputations, I have shown elsewhere (Auffenberg, 1981a) that, at least in *V. komodoensis,* tail lacerations and tip amputations are commonly found in females and are caused during intraspecific aggression in which the females are the main recipients. This may or not be the case for *V. olivaceus,* for the teeth are not shearing but crushing types. Still, it is noteworthy that 75% of instances of amputated tail tips occur in females.

Most butaan wounds (and most wounds of all other varanid species studied) are on the dorsal surface (69.3%), and most of these are thin, longitudinally

148 Gray's Monitor Lizard

oriented scars found primarily along the dorsal midline of both the body and tail. As has been demonstrated in captives, this type of wound is caused by constant rubbing, almost always in a tight nighttime retreat. In fact, such wounds can be caused within a period of two weeks or less. Wild individuals with such wounds undoubtedly received them because of frequent use of narrow crevices. Such median rubbing wounds account for 16% of all wounds on *V. olivaceus* and 11% of all wounds on *V. salvator* from the Caramoan area but

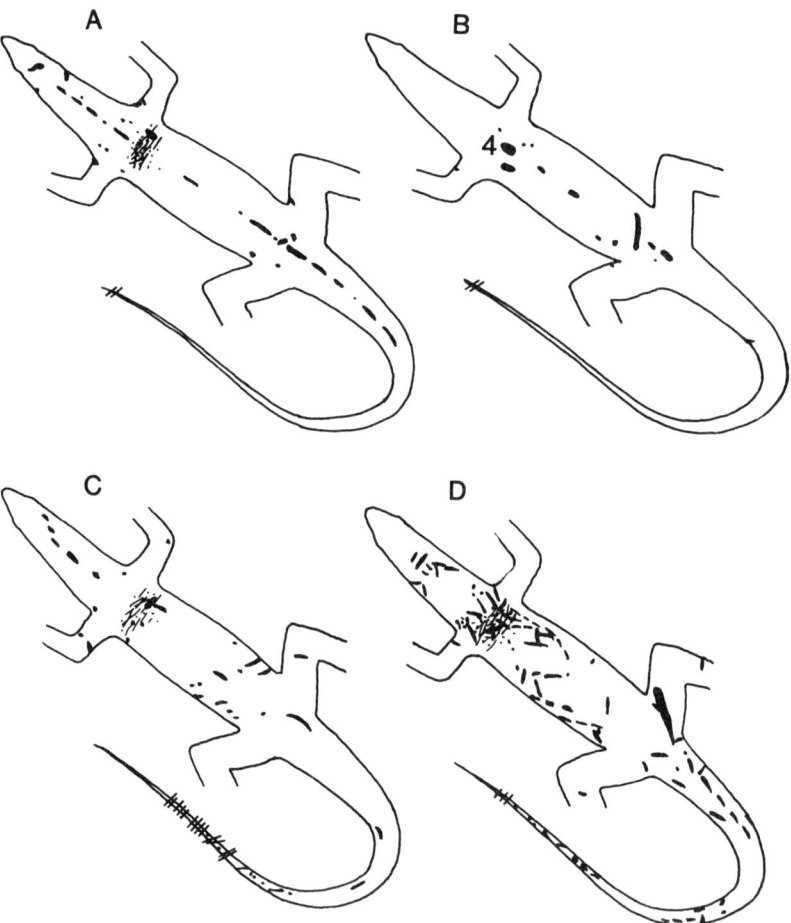

Fig. 7-1. Composite scar patterns of *Varanus* species. *A*, dorsal surface, *B*, ventral surface, *V. olivaceus* ($N = 106$) (the number 4 shows where scars occurred in the same place on four individuals); *C*, dorsal surface, *V. salvator* ($N = 126$, Caramoan area, W. Auffenberg notes); *D*, dorsal surface, *V. komodoensis* ($N = 50$, after Auffenberg, 1981a).

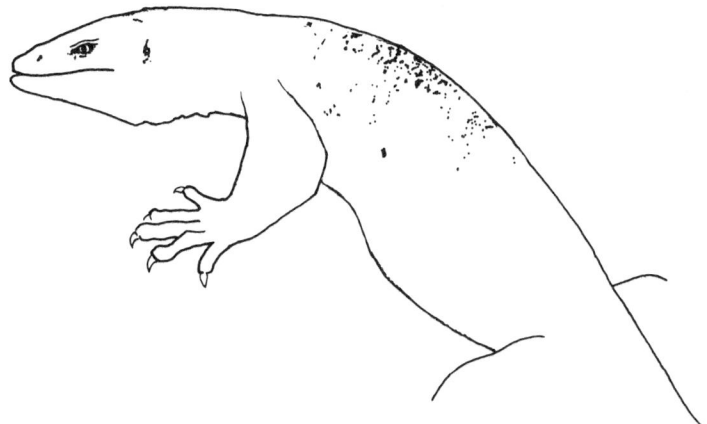

Fig. 7-2. Typical scar pattern on anterior dorsal area of old adult male *V. olivaceus*.

none of the wounds from the same species in Palawan, and none in *V. komodoensis*. Thus it seems that the butaan regularly jams itself into tight places. Observations on captives held under semiwild conditions in Caramoan substantiate this conclusion.

A few large, scattered, laceration-type wounds occur over the dorsum (35% of dorsal wounds). These are similar in shape to many found in *V. komodoensis* and *V. salvator*. Most of these are apparently caused by bites from conspecifics or predators. Puncture wounds are never seen on the butaan (they are common in adult *V. komodoensis*—apparently caused by the defensive tactics of large wild boar which are attacked as prey). In one case fire burns were found on the tail base of an adult *V. salvator*, but no burn scars were found on any butaan.

Wounds are also common on butaan ventral surfaces—more common than in any of the other species studied. These include not only rubbing wound types caused by crevice shelters but some sustained when the butaans fall or leap out of trees onto sharp rocks below. Wounds also occur on claws, most disfigured members resulting from falling or running. No disfigured claws were found on any *V. salvator* examined.

Almost all the injuries tabulated for butaans seem to have resulted from accident. Few wounds were made by other large vertebrates or conspecifics (as in *V. komodoensis*). Probable conspecific-induced wounds include a few on the neck and shoulders, one angled wound on the tail of a large male, a few nipped tail tips, and one on the belly of a female. There are, however, a characteristic series of wounds found in almost all adult male butaans, caused by the claws of other males during bipedal male-male combat (chap. 13). In this type of combat the males clasp each other vigorously and wrestle back and forth in an attempt to throw the opponent to the ground. The resultant wounds tend to be shallow, parallel, and concentrated in the area behind and between the shoulders (figs. 7-1, 7-2). Such wounds are most common in the oldest males, where the entire postscapular area may be covered with scar tissue. The color of the healed skin is often a grizzled grayish yellow, lacking any trace of the original color pattern. The normal scale pattern is often completely destroyed, and reformed new scales are both oddly shaped and irregularly placed. During the combat season (June–July) adult males may show much evidence of recent fighting in the form of bleeding from multiple small wounds in this area. The same type of wound has been described for adult males of *V. bengalensis* (Auffenberg, 1981b).

Certain aspects of butaan postcombat behavior may also result in minor wounds, particularly on the face, brachial, and sacral areas. These are found on individuals that have been at least momentarily subordinate to conspecific adults (usually males). Most are produced when a subordinate individual refuses to lie still when straddled by a dominant one. The latter may then bite and then jerk its head vigorously, occasionally tearing the victim's skin. Biting of this type occurs most commonly on the dorsal surface of the neck, but the enlarged scales in that area provide protection. The enlarged neck scales of the local water monitor subspecies (*V. s. marmoratus*) probably have the same function (see Vogel, 1979, for description of neck scales in breeding *V. salvator;* also see chap. 8).

In summary, butaans have relatively few scars, and most are caused by the physical environment. Severe aggression of the type known to occur between adult *V. komodoensis* is missing in *V. olivaceus*. However, minor wounds resulting from intraspecific coactions are common, though these are mostly accumulated after sexual maturity.

Diseases.—Little is known regarding disease in any monitor species. Though no definitive experiments were conducted, butaans are in my opinion much more susceptible to serious infections than either *V. komodoensis* or *V. salvator*—both carrion-feeders (which may help to develop immune systems of the type that seem less effective in butaans). A number of the *V. olivaceus* and *V. salvator* brought to our Caramoan camp had been bitten by the hunting dogs used to capture them. In practically every instance the *V. salvator* individuals recovered from internal bacterial infections resulting from these bites, whereas such infections were the cause of death of all such wounded *V. olivaceus*. This difference in susceptibility may be one reason why evidence of previous extensive injuries are often seen on *V. salvator* and never on *V. olivaceus*.

One old, battle-scarred male butaan that died in our camp of peritoneal infection resulting from dog bites was later found to have severe arthritis in the proximal end of the right femur—probably trauma-induced and resulting from an old injury to the femoropelvic joint.

Another male had a necrotic right testis. The condition of the left testis suggests how this may have happened. During the spermatogenic cycle the testes sometimes become so swollen that the tunica albuginea and spermatic arteries and veins are greatly stretched. In this individual the edema of the left testis was so great that the network of arteries and veins associated with the organ showed signs of atrophy. Continuation of the process would undoubtedly lead to necrosis, as seen on the right side. The individual was very old, judging by the extent of male-male combat scarring and general appearance.

Growth.—Shedding frequency is generally accepted as an indication of growth in reptiles, and Mertens (1942) shows that shedding frequency in varanids is related to both growth rate and amount of food consumed. Thus a study of the shedding pattern in *V. olivaceus* provides an indication of when growth occurs in this species.

There is no peak shedding period during the year ($N = 85$). There is, in fact, no significant deviation from expected number of sheddings for any month. Growth of *V. olivaceus* seems continuous throughout the year. Scherer (1903) noted a similar aseasonal shedding pattern in *V. niloticus*. However, the varanids living in seasonally variable habitats have seasonal shedding patterns (*V. griseus*, Grijs, 1899; *V. komodoensis*, Auffenberg, 1981a).

Six immature individuals (probably in second year of life) were marked, released in the forest, and later recaptured (after 1–13 months, $\bar{X} = 4.5$ mon). Their records show that individuals of this age with an SVL of 30–40 cm increase SVL at an average rate of 3.5 cm/mon (0–7 cm) and tail length at 2.5 cm/mon (0–4.7 cm). For two adults (40–60 cm SVL) the average is 0.3 cm/mon. Five captive adults (FSM and Dallas Zoo) fed almost exclusively on mice grew at a similar rate ($\bar{X}$ SVL increase 0.4 cm, OR 0–0.6 cm). The few data available do not substantiate a significant difference in growth rate of adult males and adult females, though it is certain to occur (perhaps at smaller SVL), a conclusion based on differential sizes in the sexes and on data on *V. bengalensis* in captivity (Auffenberg, 1979b).

Longevity.—Gibbons (1976) has shown that, with few exceptions, physiological and anatomical phenomena attributable to age in reptiles have not been convincingly distinguished from those attributable to body size. These phenomena include the bands laid down in some body osteoderms of *V. komodoensis* (Auffenberg, 1981a). There is at present a significant lack of adequate techniques for aging most natural populations of reptiles. One of the basic problems is the lack of known-age individuals, a problem clearly applying to *V. olivaceus* as well, so that we have no data bearing on the longevity of this species.

Temperament.—Two of the most striking differences between *V. komo-*

doensis and *V. olivaceus* individuals are their behavior and their temperament. The former is predictably aggressive and excitable, with hissing, body thrashing, and tail lashing all common. But butaans can be described as timid, shy, and frightened. They often remain still when discovered in the forest. In general, biting is rare, but when they do bite, they do it hard, clamping down and holding on for a long time, instead of using a "bite-slash" technique with rapid release, as is frequently used by *V. salvator* and *V. komodoensis*. Under extreme stress and when cornered, they may leap at annoying humans or dogs with their mouths wide open. Hissing is rare, and the roach is rarely raised. To my knowledge, *V. olivaceus* does not "play dead" (= letistimulation), as *V. exanthematicus* has been reported to do (Barbour, 1926). Its tail is often coiled but so tightly that a full swing would be difficult. However, when provoked, it sometimes furiously lashes its tail at an offending object. In captivity butaans are initially shy and retiring but eventually become accustomed to cages and visitors and feed readily.

The Senses

Visual.—Much of the butaan's behavior suggests it is largely scent-oriented. Sight is used mainly in situations needing no scent confirmation. My observations on captives suggest that, like *V. komodoensis* (Piazzini, 1960; Auffenberg, 1981a), this species can discern even slight movements of animals as small as sparrows and house mice from distances of at least 4 m. This ability parallels the efficiency of sight-oriented hunting behavior in captive *V. bengalensis* (Auffenberg, 1984) and *V. komodoensis* (Lederer, 1942). When in the trees in its forest habitat, the butaan will react to the movements of an approaching human at distances of about 100 m by slowly moving to the opposite trunk surface (fig. 7-3) or into an arboreal thicket of vines and epiphytes.

Auditory.—In spite of the fact that some authors claim that varanids are deaf, or nearly so (Waite, 1927; Burden, 1928; Broughton, 1936; Collins, 1956; Piazzini, 1960), the anatomy of the internal ear suggests that they have moderate acoustical ability (Miller, 1969). In addition, some data on living animals imply that they hear certain sounds well (Procter, 1929; de Jong, 1937, Lederer, 1942; Auffenberg, 1981a).

Recent experiments have shown that some reptiles previously considered deaf are capable of hearing sounds produced at the frequency and level of common forest noises. Though no definitive data are available, observations show that the butaan does hear certain noises. Captives will open their eyes or close them, turn their heads or lift them into an attentive position, if sudden noises are made, such as boards being clapped together.

In general, varanids make few sounds themselves. Of these, hissing is probably the most frequent noise produced, but even it is infrequent in the

Fig. 7-3. *Left:* adult *V. olivaceus* "hiding" behind tree trunk; *middle,* tail used as strut in climbing; *right,* tail coiled after entering shelter. Taken from photos.

butaan. In fact, when placed in stressful situations, in which other monitor species would hiss, the butaan rarely makes any noise. Open-mouthed lunges are often made silently, unlike the pattern followed by *V. bengalensis, V. salvator,* and *V. komodoensis.*

Olfactory.—Burghardt (1970) has published a review of chemical perception in reptiles. He concluded that chemical signals are important in lizards and that many of their responses seem almost completely controlled by such cues. Such signaling is probably true for all monitor lizards, especially in regard to prey detection and sexual discrimination (Auffenberg, 1981a, b, 1983b, 1984). As in other varanid species, visual inspection alone seldom convinces a butaan that an unfamiliar object is edible. In almost every instance in which feeding was observed in captivity, both animal and plant foods were eaten only after they had been touched with the tongue.

Locomotion

Spoor.—Because of the nature of the forest in which the butaan is found, and particularly in view of the rocky surface in most of the most highly utilized communities, few spoor are left behind as evidence of its passage. Fecal and gastric pellets (see chap. 10) are rarely encountered because of the overgrown and commonly fissured and eroded substrate. The most obvious signs are claw marks left on the bark of the trunks and larger branches of trees regularly climbed for shelter or basking. Varanid species inhabiting xeric situations leave behind many signs of their recent movements (Green and King, 1978; Auffenberg, 1981a) that could not be used in the study of the butaan.

Escape.—From distances as great as about 100 m and heights of 10 m or more the butaan is able to survey a fairly large area (ca 7800 m^2), considering the density of the forest. Though the area is small compared to that observable by large savanna species (i.e., *V. komodoensis*), it constitutes about two-thirds of the entire seasonal home range (1.48 ha) of large adults, considerably more than is managed by species of more open areas since their home ranges are so much larger.

While most varanids dash off quickly when alarmed, the butaan does not always react this way. Flight distances might be as much as 100 m, but usually much shorter, and may be as little as 2–3 m when the individual is on exposed trunks or branches and even less when it is hidden in epiphytic vegetation. The escape mechanism of butaans in the trees is generally to move slowly to the other side of the trunk (fig. 7-3) and hide there without moving for 15–20 min. Eventually the head slowly moves from behind the trunk, and the entire area is cautiously scanned before any major movements are made. (This general reticence to move, even when approached closely, has apparently given rise to a sometimes-used name that translated means "blind lizard.")

When flushed from the trees, butaans generally do not ascend higher but suddenly throw themselves off the branches, sometimes from heights of as much as 30 m. If descending through the canopy, they usually turn their hind feet in such a way that the claws are dragged through the leaves and terminal branches—evidently to break the fall.

When disturbed on the ground, Gray's monitor will plunge into a nearby hole, if convenient, but otherwise will run up the nearest tree larger than about 10 cm bhd. When in a hole (usually in the rocks), it remains inside with its head toward the end and the tail toward the opening (unlike some varanids; see Waite, 1929; Mahendra, 1931; Auffenberg, 1981a). When the butaan enters a retreat, its tail is always curiously coiled in a simple single loop along the posterior one-third, or in such a loop combined with a bend at the tail base, to form more or less a figure 8 (fig. 7-3). The purpose of this tail coiling remains unknown. When touched in the retreat, individuals simply shrink away.

When *V. bengalensis* (Mahendra, 1931) and *V. gouldii* (Waite, 1929) are in burrows, most of the tail length is pressed tightly against the side of the body, and, at least in *V. bengalensis,* the thick tail base is used to fill the burrow lumen (Auffenberg, field notes). When one pushes against the tail base of a *V. bengalensis* under such conditions, the tail is usually pushed back quickly. In this maneuver even this species, with relatively short spines on the tail scales, can prick a finger so hard that blood is drawn. The effectiveness of a tail used in this fashion by the spiny-tailed varanids of Australia may even be greater.

Swimming.—Nothing is known regarding the swimming ability of the butaan, though even desert species (*V. griseus*) swim well and it is expected that the butaan can do the same. Local hunters in the Caramoan Peninsula claim that some butaans move to creekside habitats in the drier parts of the year and enter and remain in the water for long periods of time. Some of our data support this contention.

Digging.—Though excavation of shelter burrows is apparently common in many varanid species, the butaan is not known to do it. The main reason is that the highly eroded limestone substrate on which it lives provides an abundance of ready-made shelters which cannot generally be modified, even with extensive digging. Nests are apparently most often located in large, hollow, standing trees, which require little excavation. However, observation on both captives and the nature of many foods eaten in the wild shows that scraping and digging in and under surface debris are common during foraging. Leaves and twigs on the forest floor are constantly scratched aside, loose bark torn from standing and fallen trees, and small stones and fallen limbs turned over by butaans in their search for fallen fruits and for most of the slugs and land snails upon which they feed. Though digging is not involved, adult butaans in the Caramoan area spend considerable time walking about under the general ground surface in caves and fissures (chaps. 6, 9).

Locomotion.—Locomotion is always quadrupedal. The only time I have even seen a butaan assume a bipedal stance was during male-male combat; no females have ever been observed to move bipedally under any conditions. However, a bipedal stance is used by adults of both sexes when foraging in rocky areas where fallen fruit might occur in such places as on ledges. In captivity they have no difficulty raising themselves into a bipedal position but always use the tail as a prop.

The butaan's pattern of walking is identical to that described for varanids by Zander (1895), Smith (1931), and Auffenberg (1981a). Timed sequences show that average normal walking speed on nearly level ground is 12.9 m/min, but on highly irregular surfaces (on which most foraging is done) the average drops to 8.1 m/min. Average climbing speed for adults on large vertical trunks is 13.5 m/min, which is higher than normal foraging speed. This rate is much slower on smaller diameter branches, where they appear to have difficulty in moving about.

Climbing.—The young of many varanid species spend much of their lives in trees, even though adults of the same species may be largely or entirely terrestrial (*V. komodoensis, V. bengalensis,* and so on). Butaan hatchlings similarly spend almost all of their time in trees, though most (perhaps all) of their foraging is on the ground. Like the young of *V. komodoensis* (Auffenberg, 1981a), a single clutch of hatchlings tends to remain together for a length of time as yet undetermined (probably less than a month).

Most varanids are good climbers and ascend to heights of 30 m or more (Hagen, 1890; Greene, 1909; Symons, 1912). The desert monitor (*V. griseus*) is the only documented nonclimber (Grijs, 1899). Among the species that climb well, some change their habits ontogenetically, becoming less arboreal with increased size (*V. exanthematicus*, Grote, 1911; *V. komodoensis*, Auffenberg, 1981a). Some species of varanids seem to spend about as much time in the trees as on the ground (*V. dumerilii*, Muller, 1923; *V. rudicollis*, Barbour, 1926). *V. olivaceus* belongs to this group; individuals of all sizes are found in trees, mainly for purposes of basking and sleeping. Still other species are highly adapted to life in the trees and even have prehensile tails (*V. prasinus*, Mertens, 1942). The tail of *V. olivaceus* is not prehensile, but it is commonly used in climbing—usually as a strut in which the anterior three-fourths of the tail is held stiff with the tip placed so that the wider lateral surface is pushed against the substrate (fig. 7-3). The same technique is recorded in Auffenberg (1981a, fig. 10-4) in a 1.2-m *V. komodoensis* individual climbing a tree. The variability of the degree of arboreality in *Varanus* species is shown in table 7-1.

The butaan is clearly more adapted to climbing in trees than adult *V. komodoensis* or *V. salvator*. This ability is proven by its rate of climbing, the apparent ease with which it can hang by even a single toe, and the consistent use of the tail as a supporting strut in difficult situations (fig. 7-3). Butaans are also able to descend small-diameter trees easily—a feat that many varanids find difficult. Adult butaans can move down trunks as slender as 3 cm bhd by using a "fire-pole" descent; they hug the trunk with their front and hind legs, controlling the rate of their "fall" by pressure on the trunk. During this maneuver the tail is frequently twisted around the trunk as well, and descent is headfirst.

TABLE 7-1. Degree of arboreality in *Varanus* species.

Class	Representative species
Climbs little or not at all	*V. griseus*
Known to obtain food in trees but no specific structural specializations for climbing	*V. komodoensis, V. salvator, V. bengalensis*
Scansorial, spending much time in trees, with arboreal specializations	*V. olivaceus, V. rudicollis, V. gilleni*
Arboreal, strongly modified for life in trees, little or no terrestrial foraging	*V. prasinus*
Extreme adaptation for arboreal life, moves on ground only to climb another tree	None

Fig. 7-4. Adult *V. olivaceus* drape themselves in the branches near the canopy when basking.

In general they do not climb or remain at the terminal ends of branches for long periods. Nor do they have special aptitudes that allow them to move about easily in such situations. In fact, their progress through and over twiggy growth seems labored and unsure. Most of their time in the trees is spent basking, usually draped onto the small branches immediately below the canopy (fig. 7-4).

Population Characteristics

Sex ratio.—The sex ratios of varanid lizards in preserved collections often show a marked surplus of males (Pianka, 1968, 1969c, 1971; Auffenberg, 1979b; King and Green, 1979; Horn, 1980). It has been suggested that this bias may be an artifact of behavioral differences between the sexes while the lizards are active, which makes males more liable to be seen and captured (Auffenberg, 1979b; King and Green, 1979), though skewed ratios in favor of males may occur in some species (Auffenberg, 1981a). However, it is likely that a balanced ratio between males and females may be the norm for monitor lizards (*V. acanthurus*, King and Rhodes, 1982). This certainly seems to be the case in *V. olivaceus,* in which the sex ratio is almost exactly 1:1 ($X^2 = 0.26$, df. $= 1$, $N = 106$). In spite of the fact that males seem to be more active and move over greater distances (chap. 5), a balanced sex ratio was probably obtained because all lizards captured were first located on the basis of their scent (using hunting dogs) rather than on the basis of visual contact.

Density and biomass.—With the exception of data for *V. komodoensis* (Auffenberg, 1981a), little information is available on the density and biomass of any varanid lizard, including the butaan.

Capture-recapture methods have been used widely to estimate population densities of animals. Though methods have been continuously refined, even the most modern techniques demand either specialized attributes of the population sampled or high sampling densities and are thus not strictly applicable to many species. Parmentier (1976) recently reviewed the literature in capture-recapture techniques and suggested that even the most refined ones are often grossly inaccurate when applied to the estimation of densities of most natural populations. He warns against wholesale acceptance of the basic assumptions that the conditions demanded by the procedure are always met by the populations under study. The inherent biases are reduced to reasonable levels only by sampling intensities higher than could be obtained in the study of the butaan in the Caramoan area.

Extensive modification throughout much of Luzon's local forests has resulted in major range reductions for the butaan (fig. 15-1). Even where the forest remains reasonably extensive, the core area of their home ranges tends to be centered on ridges and peaks (fig. 6-4). Only a small part of the potential available area on Luzon is regularly inhabited by *V. olivaceus*. On the other hand, considering the butaan's total size, its home range is small ($\bar{X} = 1.48$ ha/individual), and broad spatial overlap is common in some areas, depending on local food supply. Still, even in prime habitat, the number of individuals per hectare is considerably less than one (0.61), due to the usual scatter of fruit trees on which they regularly feed. Given an average weight of 3.17 kg per adult individual, these data suggest that each hectare of prime habitat supports about 1.9 kg, an amount significantly greater than that supported by the habitat of *V. komodoensis* (0.3 kg/ha; Auffenberg, 1981a). This relationship reflects the principle that the biomass of primary consumers is almost always higher than that of secondary consumers within closely related animal groups, which is to be expected when comparing the partly herbivorous butaan to the totally carnivorous *V. komodoensis* (see Harestad and Bunnell, 1979, for recent review). Though the biomass per hectare of the butaan is high, it is still within the upper limits reported for carnivorous species of other groups (5.5 kg/ha for a scincid, Fitch, 1954; 5.0 kg/ha for an agamid, Tinkle, 1967).

8

Reproduction

COMPARED to that for many other reptile groups, information on varanid reproduction in the wild is scanty. While none of the reproductive data for butaans are considered complete, the combination of field observations, discussions with local people, and an intensive series of studies on captives (held in large outdoor enclosures at the base camp; see chap. 1) has provided a rather ample informational base from which at least broad outlines of this species' reproductive biology can be deduced.

Sexual Maturity

Thirty-six females were dissected and examined to determine their reproductive status. Enlarged ova, eventually yolked, and corpora lutea or ovarian follicles greater than 5 mm occur only in females with SVL over 40 cm, or a total length of about 108 cm and a weight of 1 kg. Females in this reproductive state represent 85% of the total sample of females during the breeding season. The smallest female in captivity with which captive males bred was 42 cm SVL. The smallest captive that laid eggs was 44.5 cm. No comparable data are available for any other species of large varanid lizard.

The minimum SVL of females at maturity is 71.4% of the greatest size recorded in the female sample. For *V. komodoensis* the estimated minimum size at sexual maturity (70 cm) is 30% of the maximum size attained by females of that species (data drawn from Auffenberg, 1981a), showing that growth is maintained for a longer period after sexual maturity in the giant *V. komodoensis* than in *V. olivaceus* (growth slopes approximately the same; see chap. 7).

Unfortunately, there is no way of estimating the age of sexually mature butaans, as seems possible in *V. komodoensis* (Auffenberg, 1981a). However, extrapolation from data on butaan annual growth suggests that females become mature in the beginning of their third (perhaps sometimes at the end of the second) year of life. This estimate compares favorably with data for *V. bengalensis* (in captivity, a somewhat similar-sized species; Auffenberg, 1983b).

The testes of 37 males were dissected and examined. Sperm occurred in the seminiferous tubules or epididymis only in individuals with a snout-vent length greater than 45 cm, a total length of 120 cm, a weight of about 1.5 kg, and an estimated age of about 3 years.

Butaan males and females seem to become sexually mature at about the same size, in spite of the fact that males of both this species and others studied obtain greater snout-vent lengths throughout their lives (Auffenberg, 1981a). As in *V. bengalensis* (Auffenberg, 1979b), males have a somewhat faster growth rate than females. Females continue to retain the capability of producing viable gametes to the greatest lengths, but at least some large males show evidence of senility in testicular tissue (chap. 7). Sexual maturity in this species is intermediate between the smallest varanids, which become mature in one year (King and Rhodes, 1982), and the largest species (*V. komodoensis*), which becomes mature in 5–6 years.

The Female Reproductive Cycle

Development of ovaries and oviducts.—To date, the only complete studies on the female reproductive cycle for any species of varanid lizard is that on *V. bengalensis* in northern India (Jacob and Ramaswami, 1976). That study, conducted in an area of highly seasonal climate, showed that the cycle is annual, with one egg clutch deposited in June or July. The current study of *V. olivaceus* shows a similar pattern of a single clutch laid in the monsoon season.

Variation in mean monthly weight of female reproductive tissues in *V. olivaceus* (one ovary only, table 8-1) reflects the progress of vitellogenesis and the development of oviductal eggs. Ovarian tissue weights are least from January through April (regressed phase), followed by a slight increase in weight in May and June, then a dramatic rise in July through August, due largely to yolk deposition. The earliest and latest dates recorded for oviductal and oviposited eggs are 15 July–28 September in the field (ovulation and postovulatory phase) and until 7 November in captivity outside the Philippines (Dallas Zoo). The weight of the ovarian tissue remains high during September but drops in October, when eggs begin to deteriorate (regressed phase). This weight reduction continues through November and December, and the cycle is completed

TABLE 8-1. Size and weight of ovarian follicles and ovary in *V. olivaceus*.

	Avg. diameter of follicles (mm)	Range of diameter of follicles (mm)	Avg. weight of one ovary (g)	Range of weight of one ovary (g)	N
January	6.8	6.3– 7.4	1.4	1.4	1
February	8.5	7.7– 9.5	1.6	1.4– 1.9	3
March	7.7	6.2–10.1	1.8	1.1– 2.6	5
April	8.5	6.8– 9.5	1.6	0.6– 2.1	3
May	9.0	9.0– 9.0	2.3	1.6– 3.0	3
June	8.8	7.1–10.2	3.3	1.0– 6.8	3
July	17.1	10.0–34.3	22.5	18.1–25.3	3
August	25.0	25.0	26.8	26.8	1
September	12.6	5.3–25.0	14.1	8.3–20.0	3
October	6.3	5.2– 7.5	7.6	6.1– 9.1	2
November	5.8	5.8	3.4	3.4	1
December	7.5	7.0– 8.2	2.8	1.8– 3.6	3

as weights again become least from January through April. The ovarian weight of females examined after completion of vitellogenesis (June), but before oviposition, was about 2–3 times as great as in those that had oviposited but not yet initiated the growth of the next crop of follicles. Follicular reabsorption occurs in May, when from 50% to 62% of the follicles are reabsorbed, and this process is completed by at least the time of oviposition. During reabsorption the follicles are orange-yellow in color and become flattened, with a longitudinal groove formed on one side, due partly to the thecal overgrowth. The follicles that are to remain and develop further remain turgid, lack the groove, and turn from a translucent milky white to a pale yellow.

During the resting stage (regressed phase) the oviductal chamber is straight and parallel-walled. At the height of the reproductive cycle several important changes are noted: the wall becomes noticeably thicker, the chamber more convoluted, and the infundibulum significantly enlarged. It is at this stage that the oviductal chamber receives the ovulated eggs. Most mature females retain evidence of egg laying (stretched areas where each egg has been shelled, etc.) until February, some as late as April.

The left ovary is usually larger than the right, with one or two follicles more than are contained in the right ovary. Sometimes the left is much greater, containing up to 70% of the total number of maturing follicles.

Vitellogenic cycle.—The ovarian follicles of each female dissected throughout the study were measured. Table 8-1 shows the monthly ranges and averages of the largest set of follicles. The trend obviously follows that of the ovarian weights, for yolk deposition in the ova is (during much of the year) the greatest contributor to this weight. Though yolking begins as early as De-

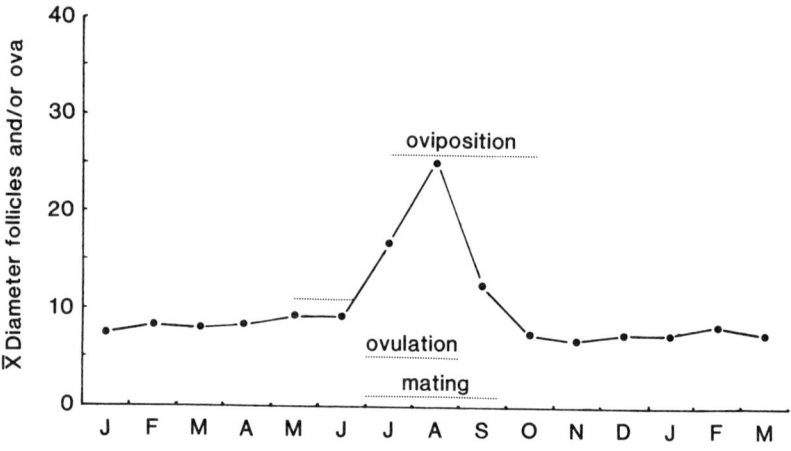

Fig. 8-1. Annual reproductive cycle of female *V. olivaceus*.

cember in some females (undoubtedly contributing to the slowly increasing weight and diameter of the follicles), the most significant increase in both number of follicles involved and percent of weight gain occurs just before ovulation (usually July–August), so that there is no clear separation between the vitellogenic and ovulatory phases of the reproductive cycle.

Figure 8-1 graphically depicts the progress of the vitellogenic cycle in the Caramoan butaan population. Because the climate is so uniform throughout southern Luzon, I believe this summary reflects the reproductive cycle throughout the entire range of the species (though there is some evidence that females on Catanduanes may lay their eggs as much as one month earlier).

Corpora lutea.—Corpora lutea were found in the ovaries of 13 mature butaan females (SVL range 43.0–56.0 cm); none was found in any immature individuals. They are yellowish-orange in color and reniform in shape. Their number equals the number of eggs in each oviduct when the latter are found there (Panigel, 1956), though I could not determine whether the movement of eggs was contralateral (Christiansen, 1973). Jacob and Ramaswami (1976) show that in *V. bengalensis* ovulation is simultaneously from each ovary.

Atretic bodies (never shed into the oviduct) are commonly seen. They are more flaccid and have a paler fluid yolk and a thinner follicular wall than do the developing follicles. Atretic follicles are never seen at the end of the gestation period, for in this species they are already completely degenerated.

Corpora lutea were found in females from September through June. A plot of seasonal changes in luteal diameter (fig. 8-2) shows that there is a rapid decrease immediately after ovulation. By about November the decrease is much reduced, though it continues. The identification of small corpora lutea

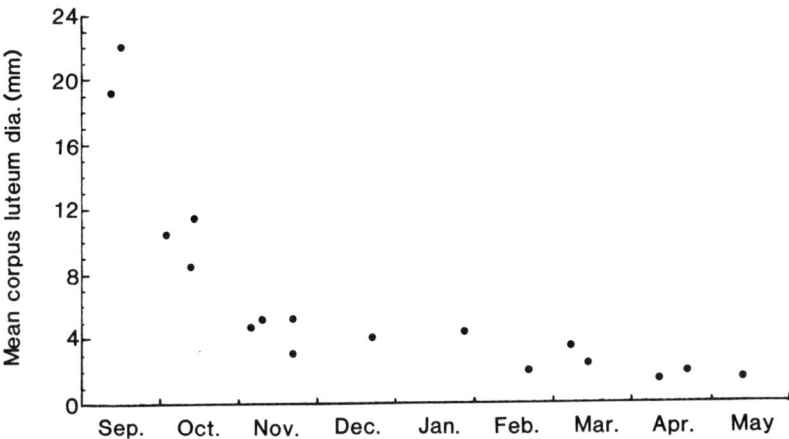

Fig. 8-2. Corpus luteum size reduction after breeding season in mature *V. olivaceus* females.

is difficult, because they tend to become embedded in the surrounding ovarian tissue and there is a lack of color differentiation at late stages. Still, the number of luteal bodies found in the ovaries clearly represents more than a single year's reproductive output. Vestigial corpora lutea appear, therefore, to persist throughout life, as they do in many other lizards (Telford, 1969). Jacob and Ramaswami (1976) show that corpora are transient in *V. bengalensis*. In general, larger butaan females have greater numbers of luteal bodies (fig. 8-3). Dividing the number of bodies in each female by the average number of eggs deposited per year (see below) suggests that females with an SVL of 48 cm or more may have been reproducing for about nine years. Of 34 mature females dissected, 44% either had eggs that year or contained oviductal eggs, and 10% had corpora lutea only, suggesting they had laid eggs in earlier years. The number of corpora lutea varies from 6 to at least 49. These are minimal estimates; some are undoubtedly missed because of their small size. The smallest noted was 1.8 mm in diameter. It is possible that the corpora lutea may have different degrees of importance in different species of reptiles, and the independence of embryos in the oviduct (and their survival) may improve the longer a corpus luteum is retained (Fox, 1977).

Clutches per year.—Table 8-2 shows that usually all eggs comprising a clutch are laid in a single day, but occasionally clutches are laid over several days. Clutches are laid during a relatively short period in the yearly cycle. However, in at least one instance an adult female in the Dallas Zoo (Mitchell, pers. comm.) laid two clutches of seven eggs each in a single year (21 July and 5 November), separated by 107 days. The eggs of the second clutch were significantly smaller (20% shorter and 10% lighter) than those of the first clutch.

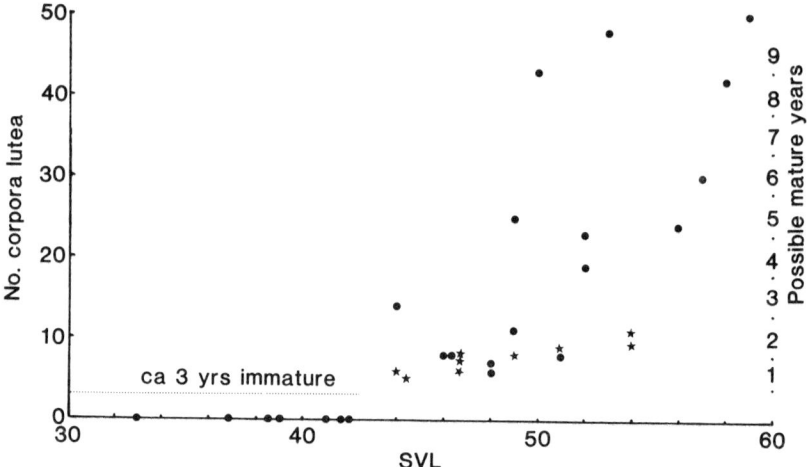

Fig. 8-3. Number of corpora lutea per female *V. olivaceus* compared with SVL (in cm). Immaturity seems to last three years (see text), and females become mature at about 44 cm SVL. *Solid dots* are nonyolking follicles; *stars* are those that are forming yolk. The probable number of reproductive years for females of different SVLs is shown on the right.

TABLE 8-2. Characteristics of eggs ($N = 47$) of *V. olivaceus*.

Date laid	Eggs per clutch	Length (in mm)		Diameter (in mm)		Weight (in g)	
		OR	$\bar{X}$	OR	$\bar{X}$	OR	$\bar{X}$
15 July	8	63.0–70.0	66.7	32.5–37.0	34.6	34.0–44.0	40.0
21 July[a]	7	68.0–73.1	71.0	31.0–39.0	34.8	35.3–64.1	48.9
12 Sept.	5	60.0–89.0	62.9	36.0–37.0	36.5	33.0–43.0	40.0
22 Sept.	6	64.0–70.0	66.8	32.3–36.0	34.5	34.0–44.0	41.1
28 Sept.	7	65.0–70.0	67.4	34.0–40.0	36.7	43.8–55.0	48.5
15 Oct.	7	68.0–70.1	68.5	35.8–36.6	36.1	35.0–56.6	46.6
5–7 Nov.[a]	7	53.0–63.0	56.8	33.0–37.0	34.8	33.1–40.0	37.7
Overall		53.0–73.1	67.2	31.0–40.0	35.5	33.0–64.1	44.2

a. Observations made in Dallas Zoo; all other observations made in Caramoan enclosure.

All courtship occurs from June through September (chap. 13). In the wild most eggs are laid from the middle of June to early November, data confirmed by local hunters. Vitellogenesis occurs once a year, but ovulation may occur twice within a few months; analysis of fat bodies suggests a single annual cycle (chap. 13). Thus most mature females lay one clutch per year, but a few lay two clutches.

Reproductive female cohort.—Practically no data are available on the number of mature female varanid lizards that lay eggs each year; authors simply stated "most" if they refer to the matter at all (i.e., King and Rhodes, 1982). In *V. olivaceus,* on the basis of oviductal or ovarian evidence, 90% of all mature females examined from the Caramoan area had apparently laid eggs the previous year. The significance of this number in terms of recruitment is discussed below.

The Male Reproductive Cycle

Few data are available regarding seasonal testicular changes in varanid lizards (see Fox, 1977, for review). Testes enlargement during the breeding season has been reported in *V. griseus* (Kehl and Combescot, 1955) and *V. bengalensis* (Upadhyay and Gukaya, 1972). Herberer (1930) noted highest testicular activity in *V. komodoensis* during June and July, the months when courtship is most intense in the wild (Auffenberg, 1981a). The same general pattern is apparently typical of *V. olivaceus* (figs. 8-4, 8-5).

Average testes weight is lowest from December through February (fig. 8-4). It increases steadily from January to the annual peak in May through July, when the testes may be so swollen that they bulge around superficial blood vessels. Thereafter testes diameter drops rapidly to the December level. Mean testicular diameter follows the same seasonal pattern, so that testes size and weight are maximum during the courtship period (end of April to middle of October). Testes color also changes seasonally: the usually greyish white testes become more yellow during the height of the reproductive period.

Changes in the tissue types within each male's testes were documented by

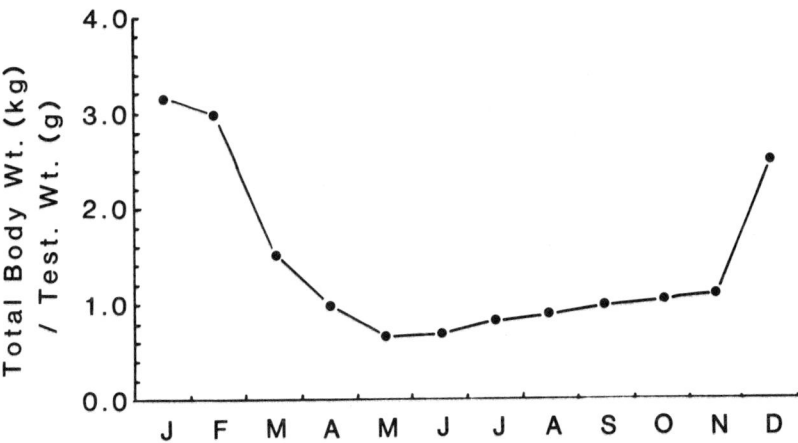

Fig. 8-4. The annual cycle of testes weight, corrected for total body weight of mature *V. olivaceus* males.

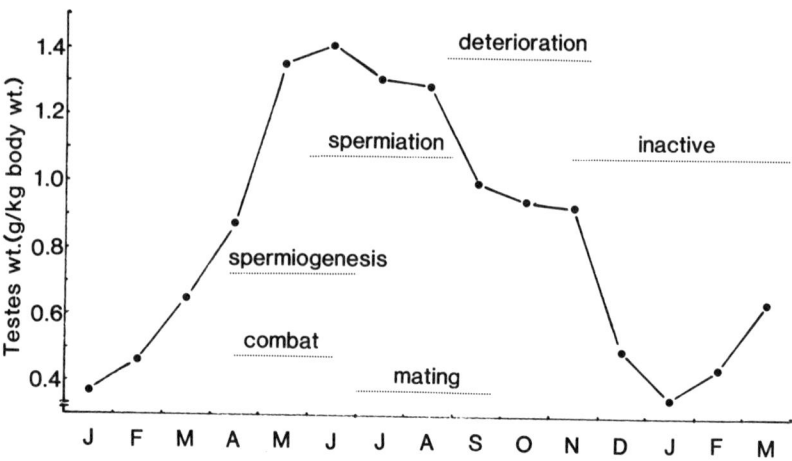

Fig. 8-5. The annual reproductive cycle of *V. olivaceus* males, showing changes in proportional testes weight and major behavioral and physical stages during the year.

two measures: relative proportion of cell types and proportional diameter of the seminiferous tubule lumens. Seminiferous tubule diameter (corrected for SVL differences) is smallest in March through May, then shows a significant increase in June and reaches its maximum size in August. Testes cellular material comprises from 50% to 82% of the tubule cross-sectional area. Greatest tissue mass is present in July. It begins to taper off in November, becoming smallest from January through May. A dramatic increase in cellular intratubular mass occurs in the month of June, the period during which most male-male combat takes place (table 8-3).

The intratubular tissue mass is comprised of spermatogonia, spermatocytes, spermatids, and spermatozoa, though these components vary seasonally in representation. Spermatogonial cells remain more or less uniform in number throughout the year but vary seasonally in respect to mitotic activity, which is greatest from May to July. These mitotic divisions produce spermatocytes, which reach a numerical peak in July. Their number is gradually reduced through August and September, then begins to drop rapidly in November to the lowest level in January, when even the few spermatocytes left have deteriorated. Spermatids can be found throughout the year but reach a decidedly high and distinct peak in July and begin to deteriorate rapidly in September; lowest annual values are found in November–January. Mature spermatozoa, with well-developed tails, etc., can only be found from July through September. There is no marked sharp peak in number; the data suggest only a generally high, flat plateau during this period. The first signs of deterioration (tail dissolution, etc.) are noted in September, and essentially no spermatozoa in any condition are present from November through May. Fibroblasts are present throughout the year in almost equal numbers. Spermiation (free sperm in the tubule lumens) can be demonstrated only from July through mid-November, with a decided peak in July. By the end of August, many sperm have deteriorated, represented by only small fragments. There is no evidence of any sperm in the lumens from December through May. The Sartoli cells remain unchanged during the year— apparently typical of lizards (fide Kehl and Combescot, 1955). Figure 8-5 illustrates the general trends in seasonal cellular activity in butaan testes.

Spermatogenic cycle.—A series of five spermatogenic stages is described to evaluate the cellular changes taking place within the seminiferous tubules of each male. The number of cell layers making up the wall of the tubule, the relative numbers of the cell types contained in the wall, and the presence of sperm in the tubule lumen were considered prior to categorizing each male.

Stage 1 (inactive) represents the period of minimal spermatogenic activity. The tubule wall is only 1–2 cell layers thick and is primarily composed only of spermatogonia. Individuals classified in stage 2 (maturing) exhibit proliferation of spermatogonia, differentiation of spermatogonia into primary spermatocytes, and production of many secondary spermatocytes and some spermatids. The period of maximal sperm production (spermiogenesis) characterizes stage 3. Large numbers of both spermatids and spermatozoa may be found in the tubules of males in this stage, but only one or the other may be abundant, depending upon the exact time of the cycle. During stage 4, sperm leave the seminiferous tubules and pass out to the epididymis (spermiation); few new spermatocytes or spermatids are forming. Stage 5 represents a deterioration of the spermatozoa produced prior to the inactive stages.

The number of males exhibiting each of the spermatogenic stages was recorded by month. The resulting trend in the time of each stage during the annual cycle correlates closely with changes in overall testes size and detailed changes in seminiferous tubule diameter and relative tissue composition (January–March 100% stage 1, April–May 100% stage 2, June 100% stage 3, July 100% stage 4, August 50% stage 4 and 50% stage 5, September 100% stage 5, November 25% stage 5 and 75% stage 1). Individuals exhibiting stage 1 are found from November to March, when testes size, tubule diameter, and relative tubule proportion are all minimal. During stage 2 (April–May) tubules are enlarging and becoming a greater proportion of the testicular tissue. Stage 3 lasts only through June and corresponds to the period when gonadal weight, tubule diameter, and relative tubule abundance are reaching maximum levels and are at their peak throughout stage 4 (July). During stage 5 all these variables are in a state of decline (August-early November).

Changes in the epididymis reflect what is occurring within the testes. During the breeding season the mean diameters of the ductus epididymis decrease

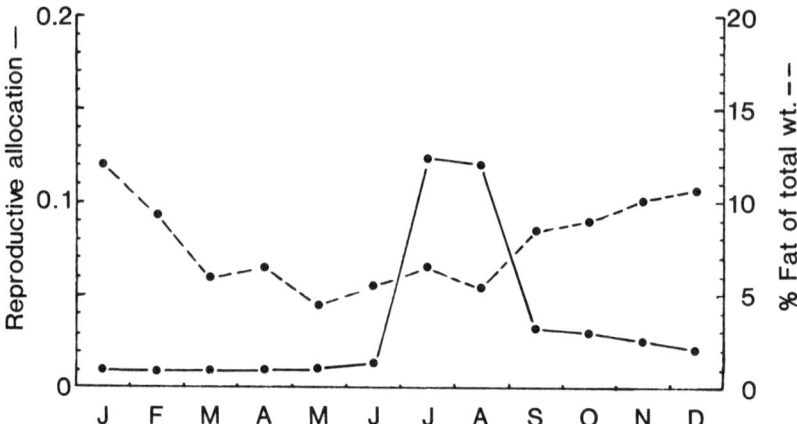

Fig. 8-6. Annual cycles of reproductive allocation (*sensu* Tinkle, 1969) and body fat production in adult female *V. olivaceus*.

slowly and continue to decrease throughout the inactive period. However, in June there is a dramatic increase in size.

Body fat and reproductive cycles.—The annual cycle of body fat in mature butaans was discussed in chapter 3. Figure 3-11 shows that the fat cycles for males and females are different in degree, though similar in seasonality, with the lowest values found from about May or June through August and the highest from about October through December. Thus mature individuals have the least fat during the height of the breeding season and the most fat essentially after the breeding and major fruiting season (fig. 8-6). As shown in chapter 3, the best available explanation is that seasonal changes in body fat of this species are more closely related to seasonal changes in food abundance and quality than to the reproductive cycle. Thus one can only assume that the high fat levels from October through December are necessary to carry these animals over the period of reduced food availability from February through April, when fruits are low in number and diversity and snails are generally difficult to find due to low rainfall (there is less snail movement). Actual yolking of follicles is noted in May at the earliest, when fat bodies had already reached their lowest annual levels in females. Thus one must conclude that the fat still available to males and females after the somewhat lean dry season of February–April must be sufficient for spermato- and oogenesis. However, a lean fruit crop in June–September of one year (resulting in less fat production) may result in total use of body fat during the lean spring period, resulting in low reproductive potential in the following summer.

Behavior and Reproduction

Besides the many physiological changes that accompany the reproductive season, there are many changes in behavior as well. The most striking is combat and courtship, details of which are presented in chapter 13. There are aspects of both, however, that are better discussed here, such as their timing. Table 8-3 gives dates during which male-male combat and courtship have been witnessed in captivity in Caramoan. Combat between males starts before

TABLE 8-3. Recorded instances of courtship with 2 females and male-male combat observed in captivity at Caramoan. Numbers refer to times per day.

Date[a]	Courtship		Combat	Other
	Female A	Female B		
24 April			1	
1 May			1	
19 May			1	
7 June			2	
8 June	1		1	
10 June	1		5	
11 June	1		7	
12 June	3		11	
13 June	2		8	
14 June	4		2	
15 June	6		3	
16 June	10		4	
17 June	11		2	
18 June	3			
19 June	1		1	
22 July				Eggs laid by female A
13 August		1		
16 August		2		
21 August		1		
24 August		3		
25 August		3		
26 August		5		
28 August		8		
29 August		10		
30 August		7		
1 September		3		
3 September		1		
17 September		1		
21 September		1		
15 October				Eggs laid by female B

NOTE: No encounters were observed in July.
a. Captives held together in same pen from 27 February 1982 through November 1983.

any courtship begins and ends long before courtship stops among the resident males and females. The peak of aggressive interactions between the two adult resident males was on 12 June—during sperm maturation and approximately one week before the most intense courtship between the males and female A (16-17 June). This date was the beginning of both spermiation and ovulation and approximately four weeks before that female laid its eggs. On 13 August the males started to court female B; this behavior became most intense two weeks later (on 29 August). This female laid her eggs about six weeks later. Though these individuals were penned together for several months preceding and succeeding these dates, no other combat or courtship behaviors were seen. Thus I conclude that male-male combat is restricted to a short period at the end of sperm maturation and preceding spermiation, courtship, and mating. It reaches a peak at exactly the same time that testes weight is the greatest (fig. 8-5). Though intense courtship is restricted to a few months of the year, it apparently lasted over a slightly longer span of two and a half months in the case of the two females involved.

Both females ceased all feeding about two weeks before laying eggs (8 July for female A, 1 October for female B), and neither resumed feeding for two to three weeks after oviposition. Norris (1953) and Rand (1967b) reported the same in the largely herbivorous iguanid lizards *Dipsosaurus* and *Iguana*. Feeding may cease because there is little space for both food and eggs at the same time (Rand, 1978). There is no evidence in *V. olivaceus* of tail raising behavior prior to egg laying, as reported for *V. bengalensis* by P. Deraniyagala (1957) and for *V. salvator* by Vogel (1979).

Egg Morphology and Complement

The eggs of all varanids have a soft, relatively smooth, leathery shell, without surface ornamentation or crystalline materials. Under magnification the surface is slightly grainy, without the elongate anastomosing wrinkles of *V. komodoensis* (Auffenberg, 1981a). While the eggs of some varanid species adhere to one another (*V. exanthematicus* regularly, *V. salvator* rarely), in one or more masses (Lederer, 1942), those of *V. olivaceus* are always separate. Though Oesman (1967) reported that varanid eggs harden in the sun, all other workers, including me, have found them to remain flexible but turgid throughout incubation.

Forty-seven butaan eggs were measured and weighed (table 8-2). Some misshapen eggs were laid (Mitchell, in litt., and pers. obs.) but were excluded from the data presented. Such eggs are common, particularly among captive varanids (Brongersma, 1932). Mean and overall range in length, diameter, and weight are 67.2 mm (53.0–73.1), 35.5 mm (31.0–40.0), and 44.2 g (33.0–64.1). The average total mass of the entire egg complement is 320 g, or 18.6% of the female's average total weight (1.72 kg, $N = 17$). In general the eggs are much smaller than most local people think they are. Even the negritos in Camarines Norte told me that the eggs were larger than those of the local water monitor, though this is obviously not the case. Of 12 clutches of water monitor eggs from the Caramoan area the average weight is 31.8 ± 0.8 g, average length 66.1 ± 0.3 mm, and average diameter 31.6 ± 0.5 mm. Though the average length is slightly greater in *V. salvator*, the difference is not significant. However the eggs of *V. salvator* are more slender and the average diameter significantly less, which results in the average egg being significantly lighter as well. However, the differences are slight and the eggs are essentially the same size. The total mass of the eggs in *V. salvator* (average) is 254.4 g, or 23.1% of the average weight of females with eggs (1.7–3.1 kg). This difference in percentage of total weight is significant at the 5% level, so that *V. salvator* produces a proportionally larger relative clutch mass.

Based on the number of corpora lutea (for 1983 only), oviductal evidence, and clutches actually laid in captivity (total $N = 17$), the number of eggs per clutch varies from 4 to 11 ($\bar{X} = 7.1 \pm 0.38$). Though sample sizes are small, it is apparent that clutch size is directly related to female size (female SVL 43.0 cm, $\bar{X}$ eggs 5.5; 47.0 cm, $\bar{X}$ eggs 7.0; 49–54 cm, $\bar{X}$ eggs 8.3). The same general relationship exists for most lizards studied (Tinkle, 1969; Tinkle et al., 1970; Andrews and Rand, 1974), including some varanids (Horn, 1980; King and Rhodes, 1982). Even within *V. salvator*, populations in which adults attain larger size on the Asian mainland lay more eggs than populations of smaller adult size on islands (Auffenberg, field notes). Variation in clutch size due to individual or local variation in female size may be greater than representatives of several lizard families analyzed by Vitt and Congdon (1978), who concluded that such differences are usually minimal when compared to other factors, such as foraging type (see below).

Nest Location

No data are available on where butaan females lay their eggs, though Aeta negritos of Camarines Norte told me that it normally takes place in hollow trunks and large limbs of both fallen and, particularly, standing trees. They told me they know of no eggs laid in holes in the ground. On the other hand they said that *V. salvator* regularly lays its eggs in holes in the earth, which is verified by my observations of both island and mainland populations. Therefore it seems that these two species use different nesting sites.

Incubation Time

There are no direct observations of incubation time. However, newly hatched *V. olivaceus* (2 in 1981, 4 in 1982, and 1 in 1983) were always seen during the southwest monsoon (19 May – 17 July). This is almost exactly the time during which eggs are laid by females in the same area. The only conclusion to be drawn at this time is that eggs of this species take from six months to nearly one year to hatch in the Caramoan area. This duration is not unusual; many varanid species have long incubation periods (*V. niloticus* up to 10 months, Cowles, 1930; *V. griseus* up to 10 months, Thilenius, 1897; *V. komodoensis* up to 8.5 months, Auffenberg, 1981a). However, some have shorter periods of 2.5–5.5 months, such as *V. bengalensis* (Smith, 1931; Deraniyagala, 1957; Jacob and Ramaswami, 1976), *V. spenceri* (Peters, 1969), *V. acanthurus* (King and Rhodes, 1982). Even some of the smallest varanid species known have an incubation period as long as 4 months (King and Rhodes, 1982). However, it is possible that not all populations of any given *Varanus* species have the same incubation time. *V. salvator* eggs take from 2.5 to 10 months, depending on local conditions (Kratzer, 1973 and pers. obs.).

In almost every instance in which I could obtain data, it is clear that hatching among varanids occurs during a rainy period and incubation during a preceding dry season of varying length. There is, in fact, some evidence that incubation of *V. komodoensis* eggs takes only 2 months under environmentally controlled conditions (Galstaun, pers. obs.) but 8 months under natural conditions. This difference suggests that, like some land tortoises, embryonic development in varanids may take place during a shorter period than suggested by "incubation" periods and that the fully developed embryos may wait in the shell until rainy weather comes (cf. *Geochelone sulcata*, Cloudsley-Thompson, 1970; *G. pardalis*, Wilson, 1968). In *Geochelone radiata* this period may be as short as 2 months under ideal laboratory conditions (Rowe and Janulaw, 1971) and as long as 15 months under some field conditions (Wilson, 1968). In the Caramoan area most eggs of *V. olivaceus* are laid in one rainy season, June and July. These pass through a short dry season in August and early September, after which another clutch may be laid, at least by some larger females. These eggs, plus those laid earlier, pass through a longer, wetter rainy period lasting until January. It is followed by still another dry period. The eggs probably hatch at the beginning of the southwest monsoon season in May or early June. Thus some eggs may take as many as 10 months until emergence, others as few as 6 months.

Control of the Annual Reproductive Cycle

As illustrated by Baker (1938), factors controlling reproduction can be divided into ultimate and proximal ones. Ultimate factors (sensu Baker) are those that somehow enhance the chances of reproductive success of the species (i.e., recruitment) and include effects of weather and food supply, particularly on the young. Among other things, ultimate factors determine the best time of year for breeding to occur. In reptiles the incubation period becomes an important factor in this timing. Ultimate factors result in the evolution of genetically based physiological mechanisms that influence reproductive timing. Proximate factors are (sensu Baker) those in the changing internal or external environment to which the individual organism responds directly, not indirectly, via evolution; they act as timers of the breeding season in a physiological sense. They are often hormonally controlled, involving important social interactions (see Crews, 1975).

Much evidence has been gathered to show that reproductive behavior in animals (including lizards; see Crews, 1973, 1974; Cox and LeBoeuf, 1977; Crews and Williams, 1977) is not guided by either ultimate or proximal factors but by interactions between them. We are just beginning to understand the ecological and evolutionary importance of physiological-behavioral interrelationships, though the diversity of organisms in which this has been studied is still small.

Proximate factors affecting the timing of butaan reproduction probably include rainfall, day length, temperature, and internal (hormonal) factors. Rainfall is clearly a strong factor; breeding in this species occurs during the early part of the first monsoon of the year (as it does for many other reptiles inhabiting tropical areas of seasonal rainfall). For the butaan it is clearly only one of several factors, for there is no indication that any of the captive monitors bred more during rainy weather than in dry spells—once the rainy season had started. Furthermore, breeding often continues through the drier period between the southwest and northeast monsoons. Thus it is reasonable to conclude that rainfall is only one of several factors inducing reproductive behavior in this species, though it may be more important in the earlier than the later part of the first combat-breeding attempts and the early rains of the first monsoon season.

There is still no general agreement on the role of day length as a proximate factor controlling breeding seasons in the tropics (Snow and Snow, 1964). There are, at present, no definitive data bearing on this subject in regard to tropical lizard breeding seasons. In the Caramoan area, day length varies from a maximum of 12 hrs 52 min in June, to a minimum of 11 hrs 24 min in December. Since breeding occurs at least as late as October, it takes place over a wide range of slowly decreasing day length. However, most breeding seems to coincide with longest day lengths in June and July.

Snow and Snow (1964) are of the opinion that temperature does not play a major role as a proximate factor in the breeding of tropical vertebrates. Nor does it seem to be important in the Caramoan area. It is true, however, that temperatures drop significantly during the monsoon season and remain generally cooler throughout much of the butaan's breeding season.

Reproductive Strategy

Compared with other lizard species (Tinkle et al., 1970), the butaan is a late-maturing form (3 years), with maturity occurring at an average SVL of about 45 cm. It produces one, rarely two broods a year, with an average of about six eggs per clutch, deposited in a nest chamber (usually a hollow tree) over a period of usually less than one day. The reproductive characteristics of *V. olivaceus* are consistent with Tinkle's hypothesis (1969) that lizard species of large size and delayed maturity should also be single-brooded, for my data suggest that most female butaans indeed lay only one clutch per year.

In the wild, most butaan eggs hatch during the first monsoon when insect prey for the young are near the highest level for the entire year (fig. 8-1). Hatching occurs after a long incubation period of 6 to 10 months. Most breeding takes place during an early monsoon season, during a period of high fruit production and diversity (fig. 9-2), when fat deposits are at a minimum. Breeding is resumed in an apparently less intense fashion during the beginning of the next rainy period. Eggs are laid 1–1.5 months after breeding, apparently in standing hollow trees, where they are probably more protected from predation than if laid in burrows in the ground.

Butaan hatchlings are significantly larger than *V. salvator* hatchlings (chap. 2) because *V. olivaceus* eggs are larger in diameter. But the selective advantage of larger butaan hatchlings is offset by a smaller average clutch size. However, a few females apparently lay multiple annual clutches. As a result, a few butaans out of the entire population may produce over a dozen large young each year. Because all babies seem to hatch at about the same time (regardless of when the eggs are laid), emergence is apparently more or less synchronized with the first part of the first monsoon of the next year. The extended period of time over which nesting occurs may increase probability of survival by reducing egg destruction by predators utilizing seasonality to forage especially for eggs. The synchronized emergence of both clutches probably also increases survival, for it occurs during the part of the year when insects are most plentiful. Insects remain abundant during the earliest part of the life of the young and are apparently eaten for the first several months. (Harrison [1955] and Medway [1972] have shown that pregnancy rates are increased in Malaysian forest rats during periods of maximum fruit abundance, while passerine hatchlings are highest when insect abundance is highest.) During the second and heavier rainy season, when insect populations decline, the young feed on land snails (as do adults). Fruit is not eaten until the beginning of the heavy fruiting season of the following year. This general pattern is similar to that of young *V. komodoensis,* which feed on large living invertebrates and small vertebrates during their first year of life. Their hatching is also timed to occur during a relatively wet period when animal food diversity and abundance are high. Many months after emergence, the young switch to carrion (in part) when live prey become less common at the end of the long and severe dry season.

In both species, emergence during the dry seasons would clearly be a high-cost, low-benefit situation, and the young profit by delaying emergence until densities of small animal prey are higher. Immediate emergence after embryonic development (probably ca 85 days) would force butaan hatchlings to face the period during the year at which insect food is the least plentiful, though snails would be abundant during the wet periods. Reviewing similar patterns in turtle reproductive strategies, Gibbons and Nelson (1978) stated, "The primary benefits of delayed emergence from the nest is the proven sanctuary offered during a period when growth benefits are likely to be outweighed by predation or by mortality resulting from harsh environmental conditions." The cost of remaining in the nest longer is a loss in feeding and early growth, though this loss is offset by the large egg and hatchling size of *V. olivaceus,* even if rare double-clutching allows a greater number of these eggs to be produced each year. If delayed emergence occurs in *V. olivaceus* (as it seems to), it is probably because environmental suitability cannot be assured in southern Luzon until the arrival of the June rains. These rains may provide an appropriate environmental cue. As in turtles (Gibbons and Nelson, 1978), delayed emergence is probably more widespread among varanids than current published data indicate. If so, it may allow flexibility in harmonizing the feeding niche with reproductive strategy (especially opportunistic emergence). This flexibility could also explain how single species can be found over such extremely broad geographic and ecologic ranges (i.e., *V. bengalensis, V. niloticus,* etc.).

Life history patterns are often explained on the basis of the concept of reproductive allocation (= total reproductive tissues, divided by total body weight; Hickman, 1975). (Semlitsch and Gibbons [1978] have shown that wet weight of reproductive tissue provides an accurate measure of reproductive allocation and is less variable than lipid ratios and dry weight methods.) Though some investigators consider this as only one of several components of reproductive effort, many believe it to be a suitable measure of the major energy expenditure in reproduction (Tinkle and Hadley, 1973). Patterns of reproductive allocation in lizards have been studied in some iguanids (Tinkle, 1969; Tinkle et al., 1970), but never in varanids. Information pertaining to this relationship is now available for *V. olivaceus*, and its pattern is discussed below.

The absolute mass of body tissue devoted to reproduction increases with SVL in *V. olivaceus* ($r = 0.76$; $p < 0.05$). This relationship is consistent with results reported for other lizards (Fitch, 1970; Tinkle et al., 1970).

Because 90% of mature females collected during the southwest monsoon had enlarged follicles, reproduction is assumed to occur annually. Oviposition occurs from July through mid-October. Most resources should be shunted into developing eggs during this season. Reproductive allocation (sensu Tinkle, 1969) does vary seasonally (fig. 8-6), the means increasing from 0.02 in June to 0.123 in July. Two values in July are 0.20.

This increase in reproductive allocation is not positively related to the weight of abdominal fat bodies (fig. 8-6). The percentage of body weight made up of the fat bodies decreased in females from early January ($\bar{X} = 12.2\%$) to a minimum value in May ($\bar{X} = 4.7\%$), 60 days before peak reproductive allocation. Body fat increased dramatically in September at the same time that reproductive allocation dropped equally rapidly. These data are not consistent with other reports (Hahn and Tinkle, 1965; Gibbons, 1972) which indicate that fat bodies may be an important source of lipids during follicle development. As stated previously, the body fat cycles in this species seem most closely related to food availability.

Comparison of the behavior of *V. olivaceus*, *V. salvator*, and *V. komodoensis* shows that the first is the only one of the three that exhibits a limited foraging technique, combined with a cryptic, semi-arboreal predation escape strategy. Both the other species are widely foraging and exhibit strong flight escape strategies. When reproduction allocation is considered, *V. olivaceus* exhibits the lowest value of the three ($\bar{X} = 0.15 \pm 0.02$, $N = 37$), unless two annual clutches (rare) are added ($\bar{X} = 0.30$). That of *V. komodoensis* is similar ($\bar{X} = 0.19 \pm 0.05$, $N = 12$), and that of *V. salvator* is higher ($\bar{X} = 0.43 \pm 0.04$, $N = 46$). All these values for reproductive allocation are within the range provided by Vitt and Price (1982) in their analysis of reproductive allocation for a large number of lizard species. However, it is not at all clear why *V. salvator* can "afford" a much higher reproductive allocation than the behaviorally similar *V. komodoensis*, unless *V. salvator* has more food resources available because of its generalist feeding behavior. Nor is it clear why the reproductive allocation of the rather sedentary butaan is so low when only a single clutch is considered—unless predation is much greater on adults of this species than I believe it to be. Annual double clutches at least for some (perhaps all) females increase reproductive allocation and recruitment potential to a level closer to that of *V. salvator*. However, in spite of the model divergence suggested, the reproductive system in *V. olivaceus* undoubtedly represents an optimal compromise between reproductive investment and adult and hatchling mortality costs.

9

Food and Feeding

FEW EXTANT reptiles eat plants, though several species of turtles and lizards are known to feed on leaves and fruits, some fairly regularly. Among reptiles, only lizards are arboreal herbivores, though they show great diversity in diet, taxonomic relationships, and geographic distribution. Many different plant parts are eaten (flowers, buds, fruits, and leaves; see Hallinan, 1920; Mayhew, 1963; Harris, 1964; Iverson, 1977; Auffenberg, 1982), but most herbivorous lizards are folivores, or some combination of folivores-frugivores, often with varying degrees of faunivory. (Herbivory is a general term referring to eating any parts of plants. It is often used imprecisely in naming the habits of animals that feed only on specific parts of plants: i.e., folivory, leaf-eating; frugivory, fruit-eating; granivory, seed-eating; algivory, algae-eating; etc.)

There are surprisingly few data available on the specific diets of even the best known "herbivorous" lizards. While some authors provide extensive, apparently reasonably complete lists of plant species eaten (Minnich and Shoemaker, 1970; Dubuis et al., 1971; Nagy, 1973), most rarely categorize them by plant part, and, just as important, the life stage of the part (young leaves and fruits are chemically very different from old ones; see McKay, 1974; Moss, 1977; Hladik, 1978). The studies of Iverson (1979), Auffenberg (1982a), and Wiewandt (1977) are exceptions, but even these are woefully primitive in view of recent discoveries and hypotheses.

Herbivory in lizards is most common in the families Iguanidae, Agamidae, and Scincidae (see Iverson, 1982, for review), though exclusive herbivory is found only in the first two. Among these families there are varying dietary mixtures of fruits, leaves, and animals (Auffenberg, 1982a; Rand, 1978). Granivory is apparently rare among lizards but does occur in species found in xeric areas (*Anglosaurus,* Hamilton and Coetzee, 1969; *Uromastix* is partly folivorous and granivorous, Dubuis et al., 1971). The Galapagos iguana *Amblyrhinchus* is entirely algivorous, feeding under the sea to depths of 10 m (Darwin, 1845); some geckos are even nectivorous (Whitaker, 1968). Though the problem of herbivory in modern reptiles has attracted much attention lately (see Rand, 1978; Iverson, 1982, and VanDevender, 1982, for reviews), much remains to be learned.

Frugivory is never exclusive in any vertebrates (for reasons to be discussed below). Lizards that eat fruits also always eat animals (Auffenberg, 1982a). It is the only type of herbivory reported in the family Varanidae. Mertens (1971) reported that captive *V. prasinus* sometimes eat bananas. (Roux [1936] also reported that captive varanids sometimes eat bananas but was not specific as to species.) Parry (1932) mentioned varanids in northern India feeding on melons, cucumbers, and rice but gave no supporting evidence, and the report is universally doubted. However, *V. komodoensis* has been observed to eat cooked rice from a pot in which kitchen scraps, including fish entrails, were sometimes kept (Auffenberg, 1981a). Finally, the partial frugivory of *V. olivaceus* has been reported earlier (Auffenberg, 1978a, 1979a). This study extends our knowledge of herbivory (frugivory) in varanids, and in lizards in general, in a significant way. It also corrects a few minor errors in my earlier report on feeding of *V. olivaceus* (Auffenberg, 1979a). Although food habits of herbivorous lizards have been studied on the basis of stomach contents and fecal pellet examination, only my earlier paper on the feeding of *Cyclura carinata* (Iguanidae) explores herbivory (largely folivory) among lizards from the standpoint of available resources. It was to obtain better insight into the relationships between resource availability and frugivory in a lizard that the current study was undertaken. The morphological specializations of *V. olivaceus* in respect to its partially frugivorous diet have been reported in chapter 3.

The data and discussions that follow relate to the foods taken by the butaan. They are based on 116 dissections of wild-caught specimens to examine gut contents and 102 fecal pellets, almost all of which were dropped by individuals a few days after being made captive; a few were picked up in the wild. Gut contents were divided into stomach and intestinal categories, and all items in each identified, counted, examined for tooth marks, etc., and, when possible, weighed and measured. However, most comparisons involving the sizes and masses of food taken were based on data obtained from identical food species captured or picked up separately in the forest during other studies on food availability (see below).

General Utilization Patterns

Table 9-1 lists all the food species found in the intestines and feces of *V. olivaceus*, as well as those eaten by captive animals. Of 46 food species recorded,

TABLE 9-1. Food taxa taken by *V. olivaceus* in both the wild and captivity (*N* = 218: 116 gut contents and 102 fecal pellets).

Taxon	Number of samples in which found	Percent of samples in which found
ANIMALS		
Mollusca		
Veronicellidae (slugs, 2 sp.)	1	0.5
Snails		
Cyclophorus ceratodes	7	3.3
Geophorus bothropoma	2	1.0
G. trochiformis	1	0.5
Hemiglypta microglypta	1	0.5
Lepidotrichia abraea	3	1.4
L. semisculpta	49	23.3
Obba columbaria	1	0.5
O. listeri	1	0.5
Ryssota angulata	6	2.9
R. ovum[a]	1	0.5
Trochomorpha boholensis	2	1.0
Crustacea		
Decapoda		
Coenobita rugosa (hermit crabs)	10	4.8
Crab (? species)	1	0.5
Isopoda (pill bugs)	2	1.0
Araneida		
Lycosidae (wolf spiders and relatives)	2	1.0
Insecta		
Coleoptera		
Dynastinae (rhinoceros beetles)	2	1.0
Prioninae (long-horned beetles)	1	0.5
Orthoptera		
Phaneropterinae (bush katydids, sp. A)	1	0.5
Acrididae (locusts, etc.)	1	0.5
Vertebrata		
Aves		
Fringillidae (finches, ?sp.)	1	0.5
Bird egg	1	0.5
Chickens[b]	n.a.	n.a.
Owlets[b] (*Athene* sp.)	n.a.	n.a.
Mammalia		
Mus musculus[b]	n.a.	n.a.
Rattus exultans[b]	n.a.	n.a.
PLANTS		
Anacardiaceae		
Dracontomelum dao (dao)[c]	2	1.0
D. edulis (dao)[c]	n.a.	n.a.

(*continued*)

TABLE 9-1 (*continued*)

Annonaceae		
Uvaria sorsogonensis (ulagak)	2	1.0
Burseraceae		
Canarium hirsutum (kiramo)[a, c]	23	11.0
C. ovatum (pili)	5	2.4
C. vrieseanum (mili-pili)[a, c]	13	6.2
Guttiferae		
Garcinia sp. A	2	1.0
Meliaceae		
Aglaia harmsiana (matobato)[c]	8	3.8
Sandoricum koetjape (santol)[c]	2	1.0
Moraceae		
Ficus (sev. sp., balete)[c]	4	1.9
Malaisia scandans	5	2.4
Palmae		
Caryota cumingii (bagsang)[a, c]	8	3.8
C. Rhumphiana (bagsang)	1	0.5
Corypha elata (buri)[a]	1	0.5
Livistona rotundifolia (anahao)[c]	1	0.5
Pandanaceae		
Pandanus sp. A[d]	1	0.5
P. radicans (ollano)[c]	17	8.1
P. simplex[b]	n.a.	n.a.
Tiliaceae		
Grewia stylocarpa (parong)[c]	14	6.7

n.a. = not applicable: eaten only in captivity.
a. In feces of an individual from Camarines Norte, Luzon, and (except for *R. ovum*) another from Catenduanes.
b. In captivity only.
c. In captivity and wild.
d. In gut of individual from Catenduanes.

27 are animals and 19 are plants. Many species of both are recorded only a few times because they are used as food sporadically; some foods have been recorded only in captivity. In some instances this rarity in the gut is inversely correlated with the abundance of the same food species in the wild, suggesting that some food species are not eaten even when encountered (see below).

Among animal food species eaten by wild butaans, molluscs (11 species) are the most commonly utilized by adults, though only a few species are regularly taken.

Molluscs are also regularly eaten by other varanids (*V. niloticus* and *V. exanthematicus;* Werner, 1908; Lang, 1919; Pilsbry and Bequaert, 1927; Mertens, 1942; Fitzsimons, 1943; Mead, 1961), though there is evidently much geographic variation in this habit (see Cisse, 1972). These species, like *V. olivaceus*, have skulls, head musculature, and teeth adapted to crush the shells of the molluscs (chap. 3). *V. komodoensis* (Auffenberg, 1982a) and *V. indicus*

(Dryden, 1965) feed on snails sporadically. These species lack the shell-crushing specializations of the species that regularly feed on molluscs. Molluscivory is also reflected in the highly specialized teeth and feeding habits of the South American teiid, *Dracaena guianensis* (Edmund, 1969). Scincid species in the Caramoan area sometimes eat small snails of the family Achitinidae (minute locally). Slugs are eaten by *V. salvator* in the Caramoan area but by only one individual out of over 100 dissected. Though centipedes are locally large and quite common and are regularly eaten by other local lizards (Auffenberg et al., MS) and by many species of monitor lizards, none has been recorded as food of *V. olivaceus*. There is no evidence of cannibalism, though it has been reported in other varanids (Auffenberg, 1981a) and even in iguanids that feed largely on plant materials (Iverson, 1977; Auffenberg, 1982a).

In addition to animal foods, butaans also feed on fruits. No leaves, buds, or flowers of any plants are eaten by the butaan (even though villagers sometimes claim the leaves of several species of local plants are eaten). With the exception of *V. prasinus,* no other varanid species is known to feed regularly on any plant material, and certainly none feeds on it as extensively.

The general utilization pattern for plants differs from that for animals. While somewhat fewer total plant species have been recorded, more plant species are regularly eaten. Of the eight families in which species are eaten, the only one that has all local species of the same genus taken regularly is *Canarium* (pili-nuts and relatives) in the family Bursereacae. To a certain extent, the representation of each *Canarium* species in the diet reflects its local abundance (see below). Within the palm family, the fruit of both local (Caramoan area) species of *Caryota* are eaten, though one only rarely. There are three local *Pandanus* species, though only one is regularly eaten, due partly to their ecological distribution; a second species has been eaten in captivity (see below). *Grewia* (Tiliaceae) is another commonly utilized food species, though apparently only one of several locally available species is eaten.

Several crustacean species are eaten, but hermit crabs (*Coenobita rugosa*) most commonly. However, these crabs are rarely eaten entire but rather only the legs—evidently after being broken off the living prey. The largely herbivorous iguanid *Cyclura carinata* also regularly eats only the legs of hermit crabs (Auffenberg, 1982a). Spiders and insects are rarely eaten by adults, in spite of their abundance in the wild. Juvenile butaans, however, do feed on these regularly. The only vertebrates recorded eaten in the wild are birds and their eggs (but see feeding in captivity, below).

Table 9-2 shows the frequency of different food species arranged by lizards' sex and weight. It is clear that males and females eat the same food items, the only significant differences being in the numbers of *Grewia stylocarpa* and

TABLE 9-2. Frequency of food taxa in *V. olivaceus* of different sex and weight.

Taxon	Male	Female	Butaan's weight (kg)		
			<0.50	0.50–1.00	>1.00
ANIMALS					
Mollusca					
Veronicellidae (slugs, 2 sp.)	1		1	1	2
Snails					
Cyclophorus ceratodes	1	6			7
Geophorus bothropoma	1	1			2
G. trochiformis		1		1	
Hemiglypta microglypta	1				1
Lepidotrichia abraea	3			1	2
L. semisculpta	28	21	3	6	40
Obba columbaria		1			1
O. listeri		1			1
Ryssota angulata	3	3			6
Trochomorpha boholensis		2	1		1
Crustacea					
Decapoda					
Coenobita rugosa	5	5	2		8
Crab (sp. ?)	1			1	
Isopoda (pill bugs)	1	1	1	1	
Araneida					
Lycosidae (wolf spiders and relatives)	1	1	1	1	
Insecta					
Coleoptera					
Dynastinae		2	2		
Prioninae	1			1	
Orthoptera					
Oedipodinae	1	1	1		1
Phaneropterinae		1	1		
Vertebrata					
Bird, adult (finch ?)	1			1	
Bird egg		1		1	
Total animal food	49	48	13	15	72
PLANTS					
Anacardiaceae					
Dracontomelum dao	1	1		1	1
Spondias pinnata	2	1			3
Burseraceae					
Canarium hirsutum	13	10		4	19
C. ovatum	1	4			
C. vrieseanum					13
Guttiferae					
Garcinia sp.		2	1		1

(*continued*)

TABLE 9-2 (*continued*)

Taxon	Male	Female	Butaan's weight (kg)		
			<0.50	0.50–1.00	>1.00
Meliaceae					
Aglaia harmsiana	4	4		1	7
Sandoricum koetjape	1	1		1	1
Moraceae					
Ficus (spp.)	3	1			4
Malaisia scandans	3	2	1	2	2
Pandanaceae					
P. radicans	14	3	1	2	14
Palmae					
Caryota cumingii and *C. Rhumphiana*	4	5	1	3	5
Corypha elata	1				1
Livistona rotundifolia	1				1
Tiliaceae					
Grewia stylocarpa	10	4		1	13
Total plant food	58	38	4	15	85

Pandanus radicans fruits eaten. The drupes of the latter are among the largest fruits eaten by butaans, so that their high representation in males may be a reflection of the fact that males are larger than females (chap. 2). On the other hand, *Grewia stylocarpa* fruits are among the smallest eaten by *V. olivaceus*, and the disparity in their representation in the sexes is difficult to explain.

The data also show that small butaans eat proportionally more animal foods than plant foods and that the largest individuals eat equal proportions of both. The same trend has been noted in other herbivorous lizards (see, i.e., Rand, 1978; Iverson, 1982). However, no animal food species can be said to be restricted to juvenile butaans and ignored by adults, though the reverse is true for some plant species eaten.

The number of food *species* in a single gut varies from 0 to 5; the frequency distribution represents not a Poisson curve but one highly skewed to the lower numbers, with 2 food species/gut being modal and a mean of 2.1. The number of food species per gut is not significantly different between males and females or among the three size classes.

The number of food *items* per gut varies from 1 to 116 (tables 9-3, 9-4); males contain a maximum total (116) higher than that of females (50), a factor undoubtedly related to the larger maximum size of the males. The average number of food items/gut does not, however, differ significantly in the two sexes. In general, smaller specimens contain fewer food *items* (but not food *species*) than larger individuals: 2.5 items per gut in the smallest size class, 16.6 in the largest class, and 8.4 in the intermediate ones.

TABLE 9-3. Number of items per food species per stomach ($N = 116$).

Food species	Mean number of items	OR	Total number of items	Percent of total items
Lepidotrichia semisculpta	1.2	1–10	58	27.2
All other animals	1.3	1–3	38	17.8
Aglaia harmsiana	6.5	1–22	11	5.2
Canarium hirsutum	8.6	1–31	22	10.3
C. ovatum	1.6	1–3	5	2.3
C. vrieseanum	14.9	1–112	13	6.1
Caryota cumingii	13.4	2–65	10	4.7
Dracontomelum dao	7.5	2–13	2	0.9
Ficus (strangler types only)	17.6	1–26	5	2.3
Grewia stylocarpa	12.0	1–42	13	6.1
Pandanus radicans	4.9	1–23	24	11.3
Spondias pinnata	7.0	1–6	6	2.8
All other plants	2.5	1–6	6	2.8

TABLE 9-4. Food species and food items found in *V. olivaceus*, arranged by butaan's sex and length class.

Category	Empty	Food species per gut		Food items per gut	
		OR	X̄	OR	X̄
Males ($N = 48$)	2	1–5	2.0 ± 0.14	1–116	16.7 ± 5.20
Females ($N = 37$)	2	1–4	1.3 ± 0.19	1–50	11.8 ± 2.20
<0.50 m SVL ($N = 4$)	0	1–4	2.2 ± 0.11	1–4	2.5 ± 0.26
0.50–1.00 m SVL ($N = 9$)	1	1–3	1.7 ± 0.29	1–18	8.4 ± 2.11
>1.0 m SVL ($N = 70$)	3	1–4	2.1 ± 0.13	1–116	16.6 ± 2.44

Empty guts are found equally among males and females and (when sample sizes are considered) among different weight individuals (table 9-4). Fourteen individuals had no food in either section of the gut; 43 had food in the intestines but nothing in the stomach. With a known clearance rate of 2–6 days ($\bar{X} = 4$), depending on food type, these data suggest that 57 (37.2%) had not eaten 2 to 6 days prior to capture. Thus it is clear that at least some individuals do not eat every day. Still (as shown below), when compared to other varanid species, a large proportion of the individuals had food in their stomachs. Nine had food in the stomach but not in any part of the intestine. However, the differences among those with food only in the stomach (9) are significantly lower than those with food only in the intestine (43), showing that the latter contains the remains of several feedings, or at least that food moves out of the stomach and into the intestines more slowly than the food moves through the intestines. Eighty-seven (55.5%) contained food in both segments of the gut, suggesting that fasting and food shortages are probably rare.

Items in the stomach and intestine were converted to their original average weights, and the totals for each section of the digestive tract were calculated. The average weight of food in the stomach was 34.8 g, SE 9.8, for the intestines 65.2, SE 7.2, and both combined 100.7 SE 12.9. There is no significant correlation between food weight and predator size (corr. coeff. $r = 0.27$), though larger individuals can obviously eat larger amounts of food. The average proportion of food in the stomach in comparison to the predator's weight is 1.1%, in the intestine 2.1%, and stomach and intestine combined 3.2%.

Almost all fruits eaten have a relatively thin pericarp in comparison to the total fruit weight. In general, flesh weight is greater than seed weight in the sugary fruits eaten and less in the waxy or oily fruits eaten (75% and 6.7%, respectively).

Though other local frugivorous vertebrates (palm civets, monkeys, etc.) feed on at least some introduced plants (such as corn and coffee), the butaan is known to feed on only one naturalized plant species (*Sandoricum koetjape*) and that uncommonly. This fact points up its generally highly selective food habits. In addition, the butaan feeds on a small number of the fruit-producing trees of the local forests; many of the others are regularly eaten by local bats, palm civets, and monkeys (chaps. 11, 12). Though the figure cannot be determined accurately because of the great plant diversity in the local forests, conservative estimates place this utilization by butaans at about 4% of the local fruit-producing plants. This number is much less than the utilization percentage calculated for *Cyclura carinata*, which may feed on as much as 70.8% of the macroflora on some West Indian islands on which it lives (Auffenberg, 1982a). Studies by others (i.e., Carey, 1975; Wiewandt, 1977) show that high utilization ratios may be usual for "herbivorous" iguanids. However, preliminary studies I have conducted on *Cyclura cychlura* on larger islands in the Bahamas suggest that when floral diversity increases, food diversity becomes smaller in proportion to the total available flora. Therefore, given the great diversity of fruit-producing plants in Caramoan, the high selectivity of *V. olivaceus* may not be so unusual, though it is still a reflection of its high level of food specialization.

Food Types Taken in Captivity

The pattern of food acceptance in captivity differs from that noted in the digestive tracts of wild-caught individuals. First, though a variety of plant foods might be offered, it is abundantly clear that *V. olivaceus* is extremely selective in what fruits are actually eaten, not only in respect to species but to fruit condition as well. Thus only fruits of the type eaten in the wild are eaten in captivity (table 9-1), and these fruits are accepted only in a state of perfect ripeness—no immature or overripe fruits are taken. Bartels (1964) describes the same selectivity by the palm civet (*Paradoxurus hermaphroditus*) under natural conditions in Indonesia; immature fruits of palms are rejected by marsupial frugivores in Guyana forests (Charles-Dominique et al., 1981); and the literature on primates is full of similar observations (Gittings and Raemaekers, 1978). Not one unripe fruit was found in hundreds taken from the guts of butaans. Ripeness is apparently determined by both scent and touch; in captivity, fruits, whether accepted or not, are regularly licked for long periods of time before being taken or rejected. In some fruits there are, in addition, other cues besides scent to determine maturity, such as color or erect v. recumbent stinging hairs on the surface of unripe *Canarium hirsutum* (see below). *Paradoxurus hermaphroditus* selects ripe fruits while they are still attached to the tree, moving from cluster to cluster and leaving those not quite ripe for a later evening (Bartels, 1964). Other civets and marsupials (Charles-Dominique et al., 1981) and frugivorous monkeys do the same. However, the butaan apparently does not climb trees to get ripe fruits but selects from those that have fallen from the tree (see below).

In captivity, butaans regularly take rats and mice. Yet, no rodents were ever found in the intestines of wild-caught individuals. *Python reticulatus* from the same habitats commonly contained these rodents. The secretiveness and nocturnal habits of local rodents are perhaps the primary reasons why they are not regularly taken by *V. olivaceus*. (Carnivory in captivity is common for some iguanines but not necessarily in the wild [Auffenberg, 1982a].) *V. salvator* eats rodents often, but most are the smaller species found near the base of the hills and along stream courses or young of larger forest species (*Bullimus*) found in nests dug under rocks and fallen trees. In terms of general preference, captive butaans seem to prefer live vertebrates (birds and rodents) to dead ones, though they will eat the latter. (Earlier [1979a] I repeated information that this species refused carrion, though this work shows that this information is not entirely correct.) Unlike the pronounced carrion-feeding of *V. salvator*, however, *V. olivaceus* will not feed on dead animals after they have begun to attract flies (i.e., start to produce a strong odor). The feast or famine feeding strategy of both *V. salvator* and *V. komodoensis* (Auffenberg, 1981a) is reflected in the carrion-feeding of both these species. The butaan does not exhibit a feast or famine strategy (see above), as we can see by its conditional selectivity of food (ripe fruit and live animals).

Even though one bird egg was found in the gut of *V. olivaceus*, eggs placed in the cages of captives are often ignored, though birds' eggs must be frequently encountered, particularly in the arboreal part of their habitat. I cannot explain the butaans's apparent disinterest in birds' eggs, unless, like fruits and rodents, it is highly selective in terms of condition and species (the eggs tried were fresh chicken eggs taken from the nest an hour or so before).

Food piracy among captives is rare—odd behavior when compared with captive *V. bengalensis* and *V. salvator,* which regularly engage in active piracy; some *V. bengalensis* individuals even seem to prefer taking food from others rather than searching for it themselves in the pens (Auffenberg, 1984). This behavior may be related to the fact that *V. olivaceus* lives in a dense habitat, and individuals rarely see one another eating.

Seasonality

Figure 9-1 and table 9-5 show the seasonal utilization of the major groups of food species. There were some individuals with completely empty guts, though they were rare in *V. olivaceus* (only 5.2% of all guts examined). However, in *V. komodoensis,* 34% were empty (Auffenberg, 1981a), as were 35% of 85 *V. salvator* examined from the Caramoan area. Hunger is probably more common in both carnivorous varanid species for which such data are available than it is in the partially frugivorous and specialized *V. olivaceus,* which, for reasons discussed below, is rarely without abundant food throughout the year.

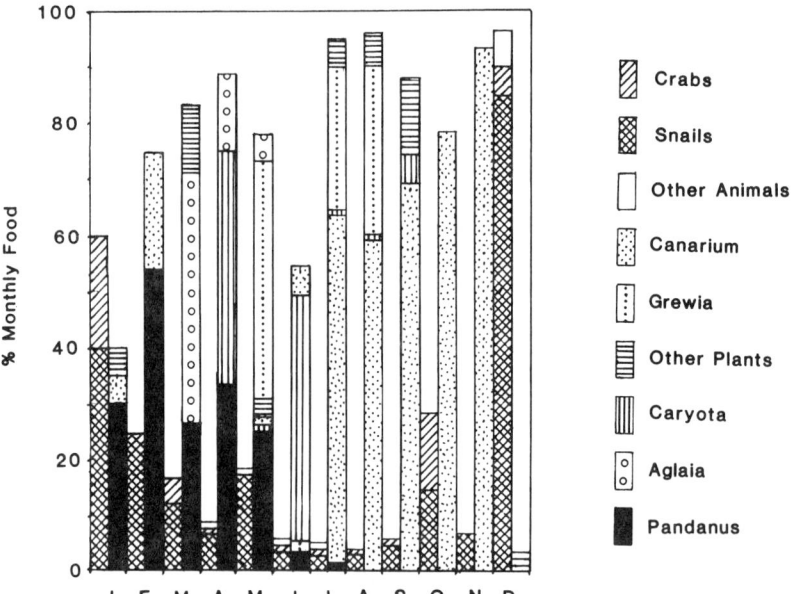

Fig. 9-1. Seasonal utilization of the major food groups eaten by *V. olivaceus* adults.

TABLE 9-5. Seasonal occurrence of food items in the digestive tract of Caramoan *V. olivaceus*.

Food taxa (number of species)	Jan	Feb	Mar	Apr	May	Jun	Jul	Aug	Sep	Oct	Nov	Dec	Total
Cyclophorus (1)			2		1			1	2				6
Lepidotrichia (2)	4	9	4	8	7	3	11	6	2	1	4	17	76
Obba (2)	1				1								2
Ryssota (1)	2	2			1								6
Trochomorpha (2)	1					1	3		1				6
Slugs (1)											1		1
Isopods (1)											1		1
Spiders (1)							1						1
Crabs (2)	4		1	1		1	1	1		1		1	12
Beetles (2)						2	2					1	5
Grasshoppers (2)				1			1						2
Birds (2 ?)					2								2
Aglaia (1)	1		22	4	4	32							62
Canarium (3)		7			1	5	165	136	73	8	86		482
Caryota (1)				86	1	40	4	1	4				136
Dracontomelum (1)									15	2			17
Ficus (3 ?)	1						2	2					6
Garcinia (1)			7										7
Grewia (1)					27	2	67	70					166
Malaisia (1)									3	2			5
Pandanus (1)	6	24	14	49	16	4	1		1				114
Sandoricum (1)								10					11
Spondias (1)							7	2					9
Uvaria (1)					3								3
Total number of items	20	42	50	149	64	90	265	229	102	14	92	21	1138
Number of *V. olivaceus* examined	5	7	6	8	7	10	18	8	11	2	6	4	
Number of taxa per month	8	4	6	6	11	9	12	9	9	5	4	5	
Mean no. of taxa per individual	1.6	.6	1.0	.8	1.6	.9	.7	1.1	.8	2.5	.7	1.3	
Mean no. of items per individual	4.0	6.0	8.3	18.6	9.1	9.0	14.7	28.6	9.3	7.0	15.3	5.3	

Low level feeding (see food amounts below) occurs regularly, without major pauses, through all seasons, so that the feast and famine pattern typical of *V. komodoensis* and *V. salvator* is not a major factor in the life of the butaan. Like most other varanids studied in tropical areas, *V. olivaceus* is an active forager throughout the year. However, females ready to oviposit may stop feeding several weeks before laying eggs and for several weeks after.

The second important feature of figures 9-1 and 9-2 is that some animal food species are eaten throughout the year (i.e., molluscs and crabs); the rest are eaten in a somewhat scattered pattern, judging partly from their rarity in the gut and partly from seasonal availability (i.e., insects). In general, plants show up in a more seasonal pattern. Figure 9-2 shows the general precipitation pattern for the Caramoan area, and it can be correlated with the seasonal utilization of some fruit species. *Pandanus* fruits are eaten through the early drier part of the year and the southwest monsoon but not the mid-monsoon

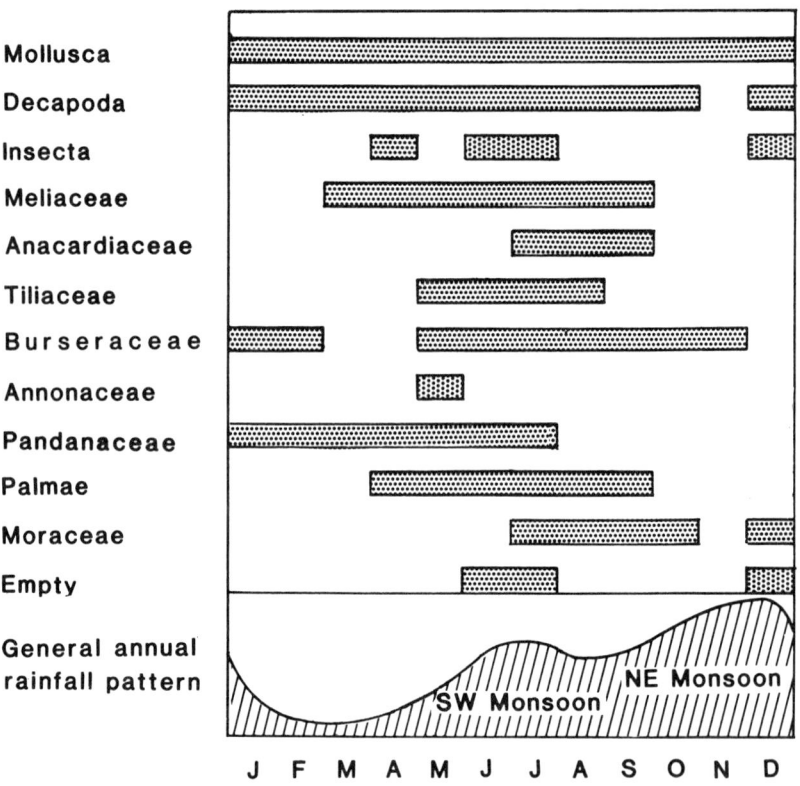

Fig. 9-2. Seasonal utilization of major foods of *V. olivaceus* compared with general rainfall pattern in the Caramoan study area.

drier period and the rainy northeast monsoon. Species of the Burseraceae show a completely different utilization pattern; they are eaten all year except during the wettest and driest parts. Palmae are eaten only during the southwest monsoon and short periods preceding and following it. Tiliaceae are similarly utilized, though over an even shorter period of time, Meliaceae similarly but over a slightly longer period, and Anacardiaceae only during the drier period between the two monsoons. Figs (Moraceae) and some *Canarium* species (Burseraceae) represent the only multiple-pulsed pattern for plant foods in the year; they are eaten for 2–3 months, then not again for 14 months. Further analysis suggests that *V. olivaceus* is highly selective regarding the *Ficus* species on which it feeds (as are several large local frugivorous birds; see chaps. 11, 12).

In general, availability of plant foods is more seasonal than availability of animal foods, and most plant species are eaten between May and September, from just before to just after the southwest monsoon. The majority of plant species eaten do not fruit during the heavier rains and generally higher winds of the northeast monsoon or during the early part of the dry period following it, in February and March. Crab prey are also represented more frequently during this time (fig. 9-1). The diet then becomes highly proteinaceous from December through February, when the fewest plant foods are available.

Table 9-5 provides numerical data for the occurrences of different food species taken by the butaan during the year, from which several important conclusions can be drawn. While more different *types* of animals are eaten than plants, some plant genera are represented by as many as 482 fruit items (*Canarium*), while the highest animal frequencies total only 76 (*Lepidotrichia*). Furthermore, though many molluscs are eaten, it is only this genus that is frequently represented in the gut. Next to *Canarium*, plants most commonly eaten are *Caryota, Pandanus,* and *Grewia,* and several seasonal use patterns are again obvious. *Pandanus* is eaten from January through July (7 months); its frequency distribution builds to a maximum in April and gradually reduces thereafter until July. *Grewia* is eaten only during 4 months, with the maximum frequencies during the last two. *Canarium* has a utilization pattern similar to that of *Pandanus* but shifted to later in the year; maximum use is during the middle of a 7-month period.

Caryota seems to have still another pattern, with high utilization levels at the beginning of a 6-month period. In view of its short use period (4 months) and high representation in the sample (165), *Grewia* is one of the most important of the food plants of *V. olivaceus,* partly because of the heavy fruiting by individual trees of this species (chap. 10). Thus only a few fruiting trees are necessary in an area to increase significantly its utilization rate. The species of *Canarium* tend to be somewhat the opposite: a relatively small biomass of fruit per tree (chap. 10) but high overall usage rate. This pattern suggests that many trees of these species are visited by a foraging butaan to account for their high representation, a notion buttressed by the fact that members of this genus are monoeceous, so that only the female trees in any area bear fruit. The most seasonally restricted fruit eaten is probably *Malaisia scandans,* available only from late August to early October, each plant fruiting only once each year.

Considered on a monthly basis, the variety of foods utilized in terms of the number of species represented in the gut is highest during the southwest monsoon period (May–July) and lowest during the wet and often stormy northeast monsoon (October–December). This pattern is not the same as that for the average frequency of items per gut per month. The lowest averages are from December through May and are generally higher from April through August (when food is apparently abundant). Since general health depends more on quantity, assuming a sufficient variety exists in the diet, the leanest months are December through May (mean number items/gut/month 4.0–8.3), and the best food availability is from April through August (mean number items/gut/month 8.9–28.2). The fruiting phenology of the mixed dipterocarp forest in the Caramoan area is described in chapter 10, and its implications regarding butaan feeding strategy are fully discussed there.

As discussed below, the butaan's diet is comprised of a group of fruits that are high in sugar (*Dracontomelum, Grewia, Spondias*) and another group high in oils (*Canarium, Caryota, Pandanus, Aglaia*) and animals rich in protein and calcium (molluscs and crabs). The frequency distribution of these three food types shows an interesting annual pattern (fig. 9-3): each type is dominant at different times of the year. Proteinaceous foods are frequently eaten in December and January and least eaten from June through September. Rainfall by itself is not the major factor in determining when protein foods are eaten (though both crabs and snails are most active during monsoon periods), because many are eaten during the northeast monsoon and few during the southwest monsoon. Sugary fruits are eaten primarily in the months immediately preceding and during the southwest monsoon (May through July). Oily fruits not only represent the greatest percentage of the diet throughout the year on the basis of number of items eaten but show two major low points in utilization corresponding to the beginning of the southwest monsoon and the end of the northeast monsoon periods. This fruit type dominates the diet from August through November and nearly so in February and March.

As shown above, the butaan eats no leaves, though it is obvious that these would be much easier to harvest than fruits, which are available only periodically. On the other hand, fruits generally contain more available energy than leaves, with nutrients moving into the fruits preferentially (Bollard, 1970). Dependence largely on fruit as a food requires that individuals must be able to move easily throughout the environment; it probably also means that they

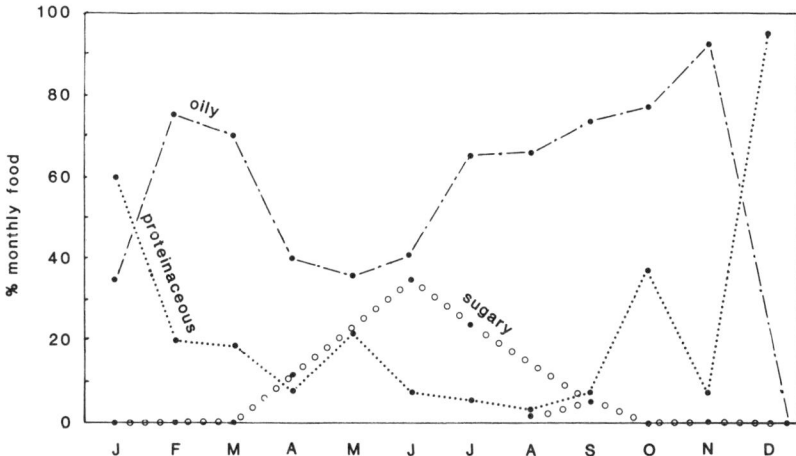

Fig. 9-3. Seasonal use of the food classes oily, sugary, and proteinaceous.

must be enough aware of its geography and the location of fruiting trees that these details would be remembered from one part of the year to another (see chap. 11). The high fiber content of leaves may be enough alone to render them unattractive to *V. olivaceus* (Troyer, 1983), but it is also likely that leaves contain a much greater variety of stronger toxins—all of which require energy to degrade (Freeland and Janzen, 1974). Toxins are also present in unripe fruit to protect the embryo from predation (Janzen, 1971a, b, 1975b), and one may surmise that these monitors pass over unripe fruits for the same reason. Ripe fruits, in contrast, are rich in either oils or sugars, the latter most easily and quickly assimilated (Hladik and Hladik, 1972). There is no current evidence that either the sugary or oily fruits eaten by the butaan are produced successively within each type, as Charles-Dominique et al. (1981) report in some South American forests.

Characteristics of Important Foods

According to Schoener's theory of optimal foraging (1969a, b), the butaan should attempt to maximize the amount of food it harvests per unit time and select the largest food particles it can efficiently find and handle. In general this pattern does not seem to be the one followed.

Table 9-6 shows that the greatest diameters of animal prey (sometimes more appropriately called length in the more cylindrical species) vary from about 15 to 107 mm. However, the latter measurement is represented by hermit crabs. The butaan rarely eats the entire crab; normally only legs are pulled off and swallowed when the crab is violently slung from side to side. The more

TABLE 9-6. Physical characteristics of animal prey of V. olivaceus.

Taxon Habitat	Greatest length X̄ ± SD (ml)	N	Greatest diameter X̄ ± SD (mm)	N	Weight X̄ ± SD (g)	N	Volume X̄ ± SD (ml)	N
Mollusca								
Cyclophorus ceratodes Moist leaf litter in dense forests			3.2± 4.5	39	6.4± 2.4	35		
Geophorus bothropoma Limestone cliffs and large rocks			17.5± 0.6	26	1.0± 0.6	26		
Hemiglypta microglypta Rocky forested situations, under leaf litter			28.8± 2.2	28	3.0± 1.6	22		
Lepidotrichia abraea Under logs and litter in rocky forests			24.9± 3.7	25	5.2± 1.2	23		
Lepidotrichia semisculpta Rocky forested places, usually in litter-filled holes in rock			27.7± 4.8	22	5.6± 1.8	23		
Obba columbaria Rocky forests, generally patchy distribution, on rocks			28.3± 2.4	27	6.0± 1.7	24	6.2± 0.8	24
Obba listeri Scattered, large rocks, cliffs, and strangler fig trunks			41.4± 1.7	27	4.0± 1.7	27		
Ryssota angulata[a] Rocky forested places, in and near floor			26.2± 5.7	84	5.7± 3.2	84		
Trochomorpha boholensis On and under dead vegetation in rocky forests			14.9± 0.1	22	0.9 ± 0.8	22		

Taxon / Habitat							
Slugs (Veronicellidae)[b] Moist, sheltered, forested places under ground litter	51.5± 10.2	24	22.3± 6.5	25	5.2± 2.7	26	3.6± 1.9	25
Arthropoda								
Araneida (Lycosid spider) On forest floor; diurnal			45.6± 5.1	5	4.9± 1.1	11		
Crustacea (hermit crab, *Coenobita rugosa*[c]) Usually rocky forests, but widely distributed; active mainly at night			107.0± 24.0	23	33.0± 14.9	28	27.2± 14.1	29
Insecta								
Coleoptera (rhinoceros beetles[b,d]) Sheltering in rotten wood; nocturnal			40.7± 9.0	24	4.1± 1.0	24		
Orthoptera (locust[b]) Open places with much herbaceous cover; diurnal	61.6± 14.5	27			4.9± 2.4	27	6.3± 2.6	27

a. Only those <45 mm dia.; greatest diameter of all individuals = 56.4 ± 6.5, $N = 117$.
b. Identification pending.
c. Least diameter = 21.1 ± 4.1, $N = 23$.
d. Least diameter = 19.1 ± 3.1, $N = 24$.

commonly ingested food species are more uniform in their diameters, ranging from 27.7 (*Lepidotrichia semisculpta*) to 40.7 mm (beetles). *Ryssota angulata* is the largest land snail of the Philippines; the average diameter for all individuals is 56.4 mm. However, the largest individuals are never eaten, and the largest found in a digestive tract had a diameter of 45 mm. When the average diameter of all those smaller than 45 mm is calculated, it is only 26.2 mm; it is in keeping with the diameters of other common animal food species. Average diameter of all animal food species taken is 30.4 ± 1.2 mm ($N = 13$ species averages, hermit crabs not included).

The average weights of individual particles of animal food vary from 3.0 to 6.4 g (excluding hermit crabs), with the overall average 5.0 ± 0.7. Unfortunately, volume was not calculated on all food species. Data available are also provided in table 9-6.

Pressure generated by adult butaans to crush the shells of the molluscs they regularly eat ranges from 1 to 71 kg (table 9-7) over a maximum area of about 6 mm^2, or up to about 1180 kg/cm^2. This remarkable pressure is possible because of the robust jaw musculature of this species combined with its strong jaws and teeth (chap. 3). There is no clear relationship between the shell strength as determined by this measure and the microecology of each mollusc species or even size (except the smallest, *Trochomorpha* and *Geophorus* sp.; the strongest, *Obba columbaria*, is not nearly as large as *Ryssota angulata*, which is the largest but is relatively weak-shelled). The most commonly predated mollusc (*Lepidotrichia semisculpta*) requires an average pressure to break the shell ($\bar{X} = 23$ kg for all tests combined).

Table 9-8 provides physical characteristics for the plant food of *V. olivaceus*. With the exception of *Pandanus radicans*, the greatest diameters are similar to those for the animal foods, varying from 13.3 to 69.7 mm. However, for elongate fruits (*Pandanus* sp. and *Canarium ovatum*), it is the least diameter that is important, so that the range is from 13.3 to 48.6 mm (overall average 26.6 ± 3.1). Throat diameter of adult butaans (SVL = 31.0–70.0 cm, $N = 22$) varies from 21.2 to 50.0 mm, $\bar{X} = 31.8 \pm 1.7$. Since all fruits are swallowed entire, the upper size limit is determined by throat diameters.

TABLE 9-7. Pressure (in kg) required to crush the shells of mollusc species commonly eaten by *V. olivaceus* (see methods, chap. 1).

Species	kg	Species	kg
Hemiglypta microglypta	13–17	*Cyclophorus ceratodes*	32–43
Lepidotrichia semisculpta	19–26	*Obba columbaria*	60–71
Lepidotrichia abraea	9–14	*Trochomorpha sp.*	1–2
Obba listeri	38–45	*Sulfurina sp.*	3–5
Ryssota angulata[a]	5–12		

a. Individuals eaten by *V. olivaceus*, less than 45 mm in diameter.

TABLE 9-8. Morphometrics of fruits of plant foods of *V. olivaceus*.

Family Species	N	X̄ ± SE	SD
Anacardiaceae			
Dracontomelum dao (dao)			
Greatest diameter (in mm)	32	19.0 ± 0.2	1.1
Weight (in g)	31	1.2 ± 0.1	0.1
Volume (in ml)	31	3.7 ± 0.1	0.7
Spondias pinnata (lubas)			
Greatest diameter (in mm)	32	25.8 ± 1.0	5.8
Weight (in g)	32	4.1 ± 0.3	1.7
Volume (in ml)	24	4.1 ± 0.2	1.0
Flesh weight (in g)	20	3.0 ± 0.2	1.0
Seed weight (in g)	20	1.0 ± 0.1	0.3
Annonaceae			
Uvaria sorsogonensis (ulagak)			
Greatest diameter (in mm)	23	32.4 ± 1.8	8.5
Least diameter[a] (in mm)	23	18.0 ± 0.2	1.0
Weight (in g)	23	3.5 ± 0.4	1.7
Volume (in ml)	23	3.0 ± 0.3	1.5
Burseraceae			
Canarium hirsutum (kiramo)			
Greatest diameter (in mm)	93	27.6 ± 0.2	2.2
Weight (in g)	128	7.1 ± 0.1	n.a.
Volume (in ml)	86	6.5 ± 0.1	1.6
Flesh weight (in g)	20	1.3 ± 0.4	2.0
Seed weight (in g)	20	27.2 ± 0.4	2.0
Canarium ovatum (pili)			
Greatest diameter (in mm)	28	60.3 ± 0.5	2.6
Least diameter[a] (in mm)	28	31.2 ± 0.6	2.6
Weight (in g)	28	32.8 ± 0.7	3.9
Volume (in ml)	30	27.3 ± 0.4	2.4
Flesh weight (in g)	18	27.4 ± 0.5	2.1
Seed weight (in g)	18	2.8 ± 0.2	1.0
Canarium vrieseanum (mili-pili)			
Greatest diameter (in mm)	32	27.6 ± 0.0	1.7
Weight (in g)	48	3.9 ± 0.1	0.5
Volume (in ml)	31	6.8 ± 2.0	1.8
Flesh weight (in g)	21	1.4 ± 0.4	1.7
Seed weight (in g)	21	26.2 ± 0.4	1.8
Meliaceae			
Aglaia harmsiana (matobato)			
Greatest diameter (in mm)	33	17.9 ± 2.6	1.8
Weight (in g)	33	2.7 ± 0.1	0.8
Volume (in ml)	17	3.7 ± 0.1	0.3
Sandoricum koetjape (santol)			
Greatest diameter (in mm)	40	48.6 ± 1.8	6.9
Weight (in g)	16	71.0 ± 0.6	2.8

a. Probably the more significant diameter.

(*continued*)

TABLE 9-8 (*continued*)

Family Species	N	X̄ ± SE	SD
Moraceae			
Ficus (6 species, strangler types only)			
Greatest diameter (in mm)	182	21.1 ± 0.6	8.7
Weight (in g)	182	5.5 ± 0.1	1.2
Palmae			
Caryota cumingii (bagsang)			
Greatest diameter (in mm)	37	13.3 ± 0.1	0.9
Weight (in g)	38	1.7 ± 0.3	1.7
Volume (in ml)	37	2.1 ± 0.2	0.9
Pandanaceae			
Pandanus radicans (ollano)			
Greatest diameter (in mm)	35	43.6 ± 0.5	3.1
Least diameter (in mm)	34	31.0 ± 0.5	2.7
Weight (in g)	35	16.4 ± 0.5	3.1
Volume (in ml)	15	10.1 ± 0.1	0.2
Flesh weight (in g)	35	2.3 ± 0.3	n.a.
Seed weight (in g)	35	14.1 ± 1.0	n.a.
Pandanus simplex (karagumoi)			
Greatest diameter (in mm)	37	69.7 ± 0.3	1.8
Least diameter (in mm)	36	18.3 ± 0.3	1.8
Weight (in g)	36	6.2 ± 0.1	0.9
Pandanus tectorius (pandan)			
Greatest diameter (in mm)	32	42.8 ± 1.3	7.2
Least diameter[a] (in mm)	32	30.1 ± 0.7	4.2
Weight (in g)	32	13.1 ± 0.3	1.7
Tiliaceae			
Grewia stylocarpa (parong)			
Greatest diameter (in mm)	30	20.7 ± 0.3	1.6
Weight (in g)	34	3.2 ± 0.1	0.8
Volume (in ml)	12	3.1 ± 0.3	1.0
Flesh weight (in g)	23	2.7 ± 0.3	1.5
Seed weight (in g)	25	0.5 ± 0.0	0.0

Table 3-1 provides additional data on morphometrics of the head that are potentially important in feeding.

Fruit weights of these species eaten show considerable variation, since some have denser pericarps and endocarps than others or contain more or less water. The overall range is 1.23 to 71.01 g, average 14.6 ± 2.03 g. Volumes of fruits eaten vary from 2.99 to 65.90 ml, average 8.6 ml. For animal prey the weights and volumes are (excluding hermit crabs) OR = 3.0–6.4, X̄ = 5.0 ± 0.34 g; OR = 3.6–6.3, X̄ = 5.4 ± 0.5 ml. Thus the animal prey of the butaan tends to be one-third the weight of the plant food per item, adding to the importance of the latter in the diet; though animals are eaten all months of the year, plant foods are taken in greater numbers of items. The data presented show that plant items tend to weigh more individually. There is, of course, a difference in the nutritive value/g between animal and plant foods (discussed below). The volume of plant food items is approximately twice that of the animal foods taken.

Although a relationship between predator size and its prey has been shown in other species of lizards (Pianka, 1969b; Auffenberg, 1978b, 1981a), no such relationship is demonstrable for *V. olivaceus* (or *V. acanthurus,* King, in press). Table 9-9 provides data on the approximate sizes of common wild fruits of the Caramoan area and indicates which are eaten by the butaan. However, most fruits eaten are medium-sized. The fact that the largest are not eaten is due to the throat width limitation mentioned. It is not clear why the smallest local fruits are not eaten, except that few of these smaller fruiting species are found in primary forest or on ridges; most are ecotonal in habitat, especially along cleared field edges and borders of secondary forest. Foraging efficiency may be part or all of the explanation of why the small fruits are bypassed. Within the four median size classes of fruits eaten, there are 11–12 representative species per class. In the size class from 10 to 19 mm, 33% are eaten; in the 20–29-mm class, 66% are eaten; in the 30–39-mm, 27% are eaten; and in the 40–49-mm, only 8% are eaten. The 20–29-mm size class thus seems to be the most important in terms of number of species utilized. A maximum of 20% of the trees producing *edible* fruits usually dispersed by animals (see chap. 11) are utilized by *V. olivaceus.*

The butaan is very specific in regard to the fig species eaten. To my knowledge, only fruits of the "balete" strangler types are regularly eaten. This pattern is apparently not due to any included pollination insects in the fruits (as found in many figs; see Redford, 1912; Janzen, 1979), for fruits of this type rarely have many insects in the central cavity. In general, cauliferous fruiting fig species tend to produce large numbers of fruits over long periods of time; fruit rarely accumulates on the ground, for rodents regularly clear the surface of any fallen fruits (see T. Auffenberg, MS, for further discussion). The only terminal fruiting type in which the fruits are located on the tip of a long slender, rarely branched trunk (*Ficus pseudopalmae*) is never eaten by the butaan, perhaps because the fruit tends to be overripe before it falls from the tree. Shrub types with fruit produced in small clusters in the axils of leaves are also ignored by *V. olivaceus,* perhaps because they are the preferred food of many species of birds and several bats and few fruits fall to the ground. Only the balete types produce huge amounts of fruit in short periods of time and in abundance; while the fruits tend to be eaten by a large number of mammal, bird, and bat species (see chap. 12), they are produced in such abundance that many of appropriate ripeness fall to the ground, where they are eaten by at least two species of rodents, hermit crabs, and the butaan.

TABLE 9-9. Size of edible fruits of genera common in southern Luzon.

Family	Genus	Eaten (+) or not eaten (−)	<1	1–1.9	2–2.9	3–3.9	4–4.9
Anacardiaceae	Dracontomelum	+			●		
	Semecarpus	−				●	
	Spondias	+			●		
Annonaceae	Polyaltha	−		●			
	Uvaria	+		●	●		
Araceae	Rhaphidophora	−		●			
Burseraceae	Canarium	+			●	●	●
Combretaceae	Eugenia	−			●	●	●
	Psidium	−					●
	Terminalia	−				●	●
Dilleniaceae	Dillenia	−					●
Ebenaceae	Diospyros	−					●
Elaeocarpaceae	Elaeocarpus	−			●		
	Muntingia	−		●			
Euphorbiaceae	Antidesma	−		●			
	Phyllanthus	−		●			
Flacourtiaceae	Flacourtia	−			●		
	Pangium	−			●		
Guttiferae	Garcinia	−					●
Lauraceae	Litsea	−		●			
Melastomaceae	Memecylon	−		●			
Meliaceae	Aglaia	+			●		
	Sandoricum	+					●
Menispermaceae	Cissampelos	−	●				
Moraceae	Artocarpus	−					●
	Ficus (many species)	±[a]	●	●	●	●	●
	Malaisia	−		●			
	Streblus	−		●			
Oleaceae	Olax	−		●			
Palmae	Caryota	+		●			
	Corypha	−		●			
	Livistonia	−		●			
Pandanaceae	Pandanus	+[b]			●	●	
Rutaceae	Citrus	−			●	●	
	Clausena	−			●		
	Glycosmis	−			●		
	Murraya	−			●		
	Triphasia	−			●		
Sapindaceae	Cubilia	−					●
	Nephelium	−					●
Sapotaceae	Mimusops	−				●	
	Palaquium	−				●	
Solanaceae	Solanum	−			●		
Tiliaceae	Grewia	+			●		
Verbenaceae	Gmelina	−				●	
Vitaceae	Cissus	−			●		
	Tetrastigma	−			●		

a. Not all species eaten.
b. Individual drupes, not entire syncarp.

Table 9-10 shows still other characteristics of the fruits eaten (or not) by the butaan in the Caramoan area. It shows that fruit color is not important in the selectivity of the butaan: some eaten are yellow (most), red, purple, green, or brownish, but these colors are also found among fruits not eaten. The one local white fruit is not eaten by *V. olivaceus,* nor are any of the two purple-red ones or the three greenish white ones.

Fruits eaten by the butaan are borne on 11 species of trees, 2 species of shrubs, and 1 species of vine. These proportions are reflected somewhat in the distributions of habit types among the different fruiting species known in the study area, so that there seems to be no selection for plant habit. The plants producing fruits eaten by *V. olivaceus* range from 5 to 10 m; most fruit is found on trees about 15–20 m high. None is found in the lowest forest levels, then, though a few occur in the subcanopy. Most, however, are canopy species. There are no emergent or sentinel trees represented, though none of these produces edible fruits anyway. In general, most of the fruits of the Caramoan dipterocarp forests are produced below the major canopy, thus lower than the level in which most butaan food fruits occur.

Fruit type is also variable, being represented by berries (4), stony drupes (4), normal drupes (3), fibrous drupes (1), receptacles (1), and drupaceous syncarps (1). There is no major fruit type that is not eaten.

Since the butaan feeds largely on fallen fruit, the lasting quality of the fruit on the ground is exceedingly important. Unfortunately, I have no data on this factor, though in captivity individuals will feed only on recently picked ripe fruit. Once it begins to appear just a little shriveled it is ignored. Janzen (1978) has made the interesting suggestion that some secondary compounds are powerful antibiotics and arrest decomposition of fallen fruit. This process, of course, has the advantage of making them available for a longer time to seed dispersers. However, the same antibiotic would also be harmful to the herbivores that also depend on a gut microflora to help digest leaves. One would then expect that primarily folivorous lizards would avoid fallen fruits of many species and would probably avoid those of some species entirely. Unfortunately, the available literature sheds little light, and this study does not help in this instance.

In summary, there is almost as much variation in the fruits eaten by the butaan as in those uneaten—at least in respect to their physical properties. There is no clear selection for any physical parameter other than size (diameter), for the largest and smallest fruits of the forest are apparently avoided. The trees that produce the fruits eaten by the butaan tend to be rather large; none of the fruit is borne on low-growing shrubs or herbs. One naturalized species (*Sandoricum koetjape*) is rarely eaten. Chemical characteristics of the fruits eaten are discussed below.

TABLE 9-10. Some characteristics of common southern Luzon plants bearing edible fruits.

Genus	Plant type	Height (m)	Fruiting body	Fruit type	Fruit color	Remarks
Aglaia[a]	tree	15	panicle	berry	yellow	naturalized
Annona	tree	8	axillary	berry	green	naturalized
Artocarpus	tree	15	cauliferous	receptacle	green	very large
Calophyllum	tree	20	terminal panicle	drupe	green-yellow	thin rind
Canarium[a]	tree	20	axillary panicle	stony drupe	purple	
Caryota[a]	tree	5	axillary spathe	drupe	purple-brown	
Chisocheton	vine	15	axillary, few–many	drupe	red	
Cissampelos	vine	4	axillary racime	drupe	red	
Cissus	vine	10	umbel-cyme	berry	purple	
Citrus	tree	10	axillary racime	berry	yellow-orange	acrid
Clausena	shrub	6	axillary panicle	berry	white	anise smell
Cocos	tree	30	axillary panicle	drupe	green	very large
Corypha[a]	tree	20	terminal spathe	drupe	green	hard pericarp
Dillenia	tree	15	terminal	follicle	greenish white	
Diospyros	tree	15	axillary, solitary	berry	brown	
Diplodiscus	shrub	5	axillary, few–many	drupe	red	
Dracontomelum[a]	tree	15	panicle	berry	yellow	
Dysoxylum	tree	20	axillary panicle	drupe	yellow-white	
Elaeocarpus	tree	10	axillary panicle	berry	red	
Eugenia	tree	30	terminal panicle	drupe	pink–greenish white	
Ficus[a]	shrub-tree	20	cauliferous, axillary or terminal	receptacle	red-yellow to green-purple	
Flacourtia	shrub	3	axillary, few–many	drupe	purple	
Garcinia[a]	shrub-tree	8	axillary, few–many	drupe	reddish	
Garuga	tree	15	axillary panicle	berry	green	
Glycosmis	shrub	5	axillary panicle	berry	red	
Gmelina	tree	8	terminal racime	drupe	yellow	

Genus	Habit	#	Inflorescence	Fruit	Color	Notes
Grewia[a]	tree	8	umbel	drupe	yellow	fibrous seed
Litsea	tree	15	umbel	drupe	yellow	
Livistonia[a]	tree	20	axillary spathe	drupe	green	sweet-smelling
Malaisia	scandant shrub	10	axillary spadix	receptacle	red	
Mangifera	tree	25	axillary, solitary	drupe	green-yellow	
Memecylon	tree	8	cyme	berry	purple	
Mimusops	tree	15	axillary, few–many	berry	red	
Muntingia	tree	10	axillary, solitary	berry	purple	
Murraya	tree	8	axillary cyme	berry	red	
Musa	herb (large)	8	axillary panicle	berry	green-yellow	
Olax	scandant shrub	10	axillary, solitary	berry	yellow	
Palaquium	tree	8	axillary, solitary	berry	green	naturalized
Pandanus[a]	shrub	5	terminal spadix	drupaceous syncarp	yellow-red	
Polyaltha	shrub	4	axillary, solitary	berry	purple	
Psidium	tree	8	axillary peduncle	berry	yellow	naturalized
Rhaphidophora	vine	6	terminal spathe	berry	orange-red	stinging crystals
Sandoricum[a]	tree	15	panicle	berry	yellow	naturalized
Semecarpus	tree	15	panicle	receptacle	purple	poison sap
Spondias[a]	tree	8	fascicle	drupe (stony)	yellow	fibrous seed
Streblus	tree	15	bract, solitary	receptacle	yellow	
Terminalia	tree	25	axillary spike	drupe	purple	
Tetrastigma	vine	10	umbel	berry	purple	
Triphasia	shrub	3	axillary, solitary	berry	red	naturalized
Uvaria[a]	vine	15	axillary cyme	berry	yellow	
Zizyphus	shrub-tree	10	axillary peduncle	drupe	blue-yellow	

a. Eaten by *V. olivaceus*.

Toxicity and Other Chemical Properties

Many of the fruits eaten by the butaan have high oil content, particularly members of the genus *Canarium*. The commercially valuable *C. ovatum* has been analyzed by several workers in respect to the chemical contents of both pulp and seed; the latter is of no importance to the butaan, which does not digest the endosperm. In these studies average lipid content of pulp was 75%, average protein 12%, sugars 1.2%, starch 5.5%, fiber 2.3%, ash 3%, and moisture 3% (Bacon, 1909; Brill and Agcaeoili, 1915). The fruits of *C. villosum* have so much oil of good quality that they are sometimes used as cooking fuel (Bacon, 1910), and other species are used as torches (West and Brown, 1921). It is this oil that is the basis of the resinous sap characteristic of many *Canarium* species. The same oil gives the pulp a bitter "spicy" taste due largely to the oily terpenes, which constitute up to 16% of the total resinous weight. Several terpene oils are present, of which dextrolimonene, dextrophillandrene, terpinene, and terpinolene are most important (Bacon, 1909; Leenhouts, 1955). Other organics contributing to the unique smell and taste of *Canarium* pulp are limonenetetrabromide, nitrosylchloride, dipentene dehydrochloride, and often a large amount of formic acid (Bacon, 1909, 1910). Bacon (1910) has shown that the amount and kind of terpenes vary greatly from one tree to another, suggesting that some trees may be preferred as food over others in the immediate vicinity. This variation has been shown to occur in plant food selection by *Cyclura carinata* (Auffenberg, 1982a), where even specific parts of specific trees are apparently preferred to others.

Two genera within the Anacardiaceae are also regularly eaten by the butaan—*Dracontomelum* and *Spondias*. Resinous saps smelling of terpentine are common in the family, and many species are poisonous and powerful skin irritants, particularly the immature fruits. In spite of those drawbacks, many frugivores regularly feed on the mature fruits of members of the family in both the old world (Corner, 1940) and the new (Auffenberg, 1982a).

Analysis of mature fruits eaten by *V. olivaceus* shows that the caloric content varies from 3934 to 5276 cal/g (corrected for each species), snails from 5336 to 5628 cal/g (not food species; data taken from Cummins, 1967) (table 9-11). The overall average of species averages for fruits is 4581 cal/g, which is not significantly different from the mean value of fruits eaten by *Cyclura carinata* (4593; Auffenberg, 1982a). There is no significant correlation between the percent of fruit species eaten and their caloric content. Thus caloric content, by itself, is not an important guiding factor in short-term dietary structure, because it fails to separate available from nonavailable calories or sugars from oils. However, the analysis does show that in the single species in which data were available (*Canarium ovatum*), ripe fruit has significantly more calories than unripe fruit of the same species ($t = 5.26$, $df = 2.0$, $p < 0.05$).

TABLE 9-11. Chemical characteristics of the major food types consumed by
V. olivaceus in the Caramoan area. ($N = 2$ for each category, except for *Ficus nota*, for which $N = 3$.)

Food type	Crude protein (%)	Average calories per gram			% ash
		Uncorrected	OR	Corrected for H_2O and ash	
Aglaia harmsiana	5.59	4051	3986–4116	4640	5.2
Canarium ovatum (ripe)	7.09	4204	4137–4343	5276	11.2
Canarium ovatum (unripe)	5.57	3679	2691–3702	4542	10.8
Caryota cumingii	3.78	3998	3994–4002	4571	6.5
Ficus nota	9.09	4052	4004–4077	4853	8.0
Grewia stylocarpa	7.24	3378	3376–3381	3934	8.1
Spondias pinnata	5.86	3540	3495–3584	4248	7.0
Land snails[a]	n.a.	5532	5336–5628	n.a.	n.a.

a. After Cummins, 1967.

The average percent of nitrogen in the fruits eaten by butaans is 1.14, ranging from 0.60 to 1.45. The lowest percentage of protein (table 9-11) is found in the fruits of the palm *Caryota cumingii* (3.78) and the highest in *Ficus nota* (9.09). (Vellayan [1981] has shown that most figs, in contrast to their protein content, are very low in nitrogen. The great variety of fig species suggests that the group may include many chemical patterns.) All remaining fruits eaten have protein percentages from 5.57 to 7.24.

Food Location

Studies on captives held in the Caramoan pen show that average foraging speed is 10.5 m/min (details in chap. 11). During this time the head and darting tongue are swung from side to side, covering an arc with a cord dimension of about 0.7 m; thus about 7.35 m^2 are normally covered per minute during normal foraging on a rough surface. During such periods, as well as while they are still, the behavior of the captives shows clearly that moving prey the size of a mouse or rat can be seen up to 4 m, though most charges for live prey are made from distances of less than 2 m ($\bar{X} = 1.5$ m).

Previous studies of feeding behavior have shown that varanids are particularly adept at locating food with their tongue and associated Jacobson organ, both in captivity (Lederer, 1942; Auffenberg, 1984) and in the wild (Auffenberg, 1981a). There is, in fact, a possibility that varanids are able to trail at least certain types of prey for considerable distances (see discussion in Auffenberg, 1981a), as many snakes clearly can (Burghardt, 1970). So the butaan may be capable of following the mucous trails laid down by land snails, and location of such prey may be largely dependent on these trails. Mucous trail following appears to be widespread among gastropods (see Cook, 1977, for review). Indeed, Wells and Buckley (1972) postulate that such following is a general feature of gastropod behavior (though some species clearly do not, or cannot, follow any trails). Among snails, mucous trailing is used in homing, interspecific predation, and courtship (Gonor, 1967). Mucous trails are thus potentially useful to varanids in finding their snail prey as well. However, Townsend (1974) has reported that at least among snails the marker within the mucous trail may start to decay and, in some cases, be completely gone within 30 minutes, so that the usefulness of such a trail to the butaan may be short, if indeed it is used at all.

All snails eaten by the butaan are species regularly found in rocky forested situations, particularly in limestone hills (table 9-6). *Lepidotrichia semisculpta*—the species most commonly eaten—is most often found in forests with limestone showing at the surface. Though nocturnal, individuals of this species generally move about only during rainy periods. During the day they usually hide in small litter-filled holes in the rock—frequently tightly wedged. *Ryssota angulata* lives in similar places, though often resting under litter in fissures and at the base of cliffs. *Obba columbaria* has a microhabitat similar to that of *L. semisculpta* but is normally found in rocky areas, frequently at the tops of ridges. It too hides under litter, usually in holes in the rock. Thus the butaan's technique in finding these three species is to investigate crevices and holes in the rock and flick leaves and other litter out of the way with the snout. (This habit probably explains the slitlike and posteriorly located narial opening in this species. *V. rudicollis, V. bengalensis,* and *V. dumerilii* use their snouts in the same way and also have posteriorly located slitlike nostrils.) A few snail species live on vertical faces—either on rocks or more rarely on tree trunks (*Obba listeri* and *Geophorus bothropoma*)—and sleep exposed on the rock faces though sometimes tucked into small crevices. Most are perched a meter or more above the general forest floor, which may explain why they are not more commonly eaten by the butaan, for at least *Geophorus* is locally common (K. Auffenberg et al., in press). Still other molluscs eaten tend to be restricted to the forest floor, resting under leaf litter (slugs, *Cyclophorus, Hemiglypta*), and are found apparently by the butaan's using its snout to move the litter or scratching the leaves around with its front feet. *Trochomorpha* is normally found on decaying wood but is not common anywhere; specimens are apparently simply picked off the surface whenever encountered. *Lepidotrichia abraea* is usually found in the rotten wood of fallen branches and trunks. The butaan undoubtedly locates them by scratching under such places with the front feet. *Lepidotrichia semisculpta* and *Ryssota angulata* are the only species regularly found in fissures below the ground surface, where they are sometimes even moving about in the daytime if light intensities are low.

The insects eaten by *V. olivaceus* are most often beetles of types sheltering in rotting wood (rhinoceros beetles) on the forest floor (table 9-6). They are undoubtedly located by the butaan's scratching with its front feet. Hermit crabs (a more common food species) are found most frequently on the forest floor in rocky areas. During the day they shelter in crevices and holes in the rocks, where they must be found by foraging monitors. Wolf spiders are restricted to the forest floor. These facts suggest that the young butaans also do most of their foraging on the ground.

Almost all the fruits upon which the butaan feeds are found on the ground as well. In only two instances was there any evidence that fruits might have been eaten in the trees (the stomach contents contained fruits with their peduncles attached). However, in windy weather slightly unripe fruit often breaks off and falls to the ground, so that the presence of attached peduncles is no proof of arboreal feeding. In spite of the fact that *V. olivaceus* claw marks are often found on the trunks of trees whose fruits they eat, the scratches are also found on many other species whose fruits they do not eat. I have never seen any individuals feed in the trees, nor have any hunters stated so with any degree of conviction. Though adults can and do climb out onto terminal branches of some tree species, other species have slender terminal branches that are not capable of supporting them (table 9-12). The butaan is not morphologically or behaviorally adapted to feed effectively in terminal branches, in spite of the fact that some of its food species have remarkably robust terminal twigs. (Robust terminal twigs of many of these species may be related to providing perches for hornbills—heavy birds that feed on the same fruits.) Almost all food selection takes place on the ground. The only regular exception is *Pandanus radicans;* its drupes are evidently picked individually from the syncarp (see chap. 11). All fruits growing in trees are selected after the fruits have fallen.

The fissured substrate surface means that most fallen fruits tumble into holes and crevices. The butaan regularly drops down into the larger of these— often several meters below the level of the forest floor. The foraging technique

TABLE 9-12. Terminal twig diameters (in mm) of trees whose fruits are regularly eaten by *V. olivaceus.*

Family	Species	OR	$\bar{X}$	SE	N
Anacardiaceae	*Spondias pinnata*	8.1– 9.2	8.4	0.23	13
Burseraceae	*Canarium hirsutum*	20.1–29.9	23.8	1.01	24
	C. ovatum	9.5–13.0	12.1	0.34	23
	C. vrieseanum	5.6– 6.2	6.0	0.22	18
Meliaceae	*Sandoricum koetjape*	16.3–10.7	8.9	0.66	24
Palmae	*Caryota cumingii*	20.1–27.3	24.3	1.23	14
Tiliaceae	*Grewia stylocarpa*	3.0– 4.9	3.4	0.15	28

is thus somewhat similar to that described by Mautz (1982) for the xantusid lizard *Lepidophyma smithi,* which lives in caves and fissures and feeds on fruits falling into the openings from the surface.

Consumption Techniques

Butaans swallow slugs entire without mastication, but all shelled snails eaten have had their shells crushed. Of all the snails eaten, only one individual of *Geophorous* sp. was found entire in the gut. To crush the shell the butaan manipulates it so that its umbilical depression rests either on or directly below one of the largest teeth of the upper or lower jaw (numbers 9 and 10 of the dentary and 8 and 9 of the maxillary). These large teeth do most of the crushing of the shell and are the most blunted and chipped with wear (fig. 3-5). One of these teeth (or matching maxillary ones) enters the dimplelike umbilicus of the shell, serving to hold it securely while pressure is applied through adjacent teeth. Initial breakage usually occurs on the superior surface. Additional crushing is common, but the shell is rarely completely smashed before it is swallowed. Crushing may be necessary for the digestion of snails, for even when they are to be preserved in formalin or alcohol, the shell is so impermeable that proper preservation of all parts requires that it be cracked or punctured.

The butaan's primary crushing site is at the rear of the tooth row, close to the insertion of the adductor musculature. The relative shortening of the tooth row in this species and the shortened dentary element represent a reduced length in the load-arm of what constitutes a third-order lever system (force between load and fulcrum), thereby increasing the mechanical advantage of the system. The reduction of the load arm, expressed as a ratio (L. load arm/ L. fulcrum to force), means there will be a greater force applied at the point of maximum bite. In *V. olivaceus* this ratio is 1:2; it produces a greater mechanical advantage than that of all other *Varanus* species, except *V. exanthematicus,* which has an identical ratio. These two species are approached only by *V. niloticus* (Rieppel and Labhardt, 1979), whose ratio is 2.5. All other species have a significantly lower mechanical advantage (higher ratio), and these three are the only varanids known to feed regularly on snails. At least two other families of lizards have species that habitually eat snails, and these also have low ratios (Amphisbaenidae, *A. ridleyi,* 1.9–2.3, Pregill, 1984; Teiidae, *Dracaena guianensis,* 0.88, this study). That the increased mechanical advantage accompanies the major dietary switch from insectivorous to mollusc-frugivorous at about 1 m is shown by the higher ratio in juvenile *V. olivaceus* (3.5 v. 1.2 in adults). A similar ontogenetic change in mechanical advantage of the lower jaw has been demonstrated in *V. niloticus* (Rieppel and Labhardt, 1979).

Data show that there is great variation in the resistance of snail prey species to crushing by *V. olivaceus* adults. *Lepidotrichia semisculpta,* the most common species eaten, requires an adductor muscle power of 19–26 kg to crush its shells. This force translated to the small tip of the (usually) single tooth that cracks the snail shell means that a pressure of 70,000 g/ca 4 mm^2 of tooth surface is needed, or about 2.8 kg/cm^2 at the tooth surface.

Smaller fruits with a soft pericarp (*Grewia, Dracontomelum,* etc.) are apparently swallowed without crushing. So are some of the larger ones with a harder pericarp. However, butaans puncture and crush the rinds of most fruits of *Canarium* before swallowing them. This practice may result in more rapid digestion, particularly in *Canarium,* for the rind is waxy and probably resistant to digestion.

Adult butaans have little difficulty killing even the largest individuals of the several local *Rattus* species (though *Bullimus* adults are sometimes difficult to kill because of their large size). They are usually killed by suffocation, which the butaans accomplish effectively with their powerful jaws and associated musculature. They are, in fact, able to kill large rodents faster than can an equal-sized *V. salvator* and *V. bengalensis* (their skulls and musculature are less bulky, and their teeth are saberlike slashing-puncturing rather than holding-crushing types; see Auffenberg, 1984 for details of *V. bengalensis* feeding behavior). In general, the killing technique of *V. olivaceus* depends less on slamming the prey about than does that of *V. salvator.* After the initial grab the prey is usually pressed against the substrate with considerable downward force; equally strong, deliberate bites follow, so powerful that the entire visceral area is often completely collapsed. This action suffocates the prey. This powerful bite, plus the tendency to retain the strong bite for a long time, makes the bite on a human hand or foot a painful, frustrating affair.

The efficiency with which butaans kill even large rodents shows that the absence of these prey in the sample guts obtained during this study is not an indication that they are unacceptable or that butaans do not possess the wherewithal to kill such animals. Rather, it seems to be a reflection of the difficulty with which such prey are obtained in the highly fissured rocky habitats in which the butaan normally occurs. That rats (at least *Rattus everetii* and *Bullimus luzonicus*) are common in the same habitat is shown by the success with which we trapped them at the bases of fruiting trees.

The butaan's tail is never used in maneuvers during feeding; the front feet are occasionally used to hold or change the position of prey; the claws are used, but rarely, in a way that suggests the predator is trying to pull the large prey apart, though I have never observed them to be successful at it. Techniques concerned with foraging by *V. olivaceus* are discussed in chapter 11.

Consumption Rate and Amount

Food consumption rate of wild vertebrates is a simple concept, but it has usually proved difficult to measure. Many indirect methods of estimation have been attempted, with varying degrees of success.

There have been few attempts to measure the food consumption of lizards, or indeed of any reptiles. There is, however, a considerable amount of information on oxygen consumption of these animals, and such data can be used to measure energy requirements if suitable simplifying assumptions are made about temperature relations and activity levels in the field (see McNab, 1963; Alexander and Whitford, 1968). More direct determinations of food consumption have been made for insectivorous lizards by using fecal pellets and stomach contents (see Harris, 1964; Johnson, 1966). The most detailed is by Avery (1971), who estimated food consumption by four different methods in the insectivorous *Lacerta vivipara,* a European lizard. He showed that consumption values per day vary with light levels (thus also with temperature). For sunny days his simplest (and perhaps best) method provided a consumption rate expressed by $F = 21.8 \times W^{0.74}$, where F = food consumption in mg dry weight/g live prey weight/day and W = live prey weight in g.

For average-sized male *V. olivaceus* this model provides a value of 10.3 kg per year, for the 20 largest males 14.8 kg/yr, for average-sized females 6.3 kg/yr, and for the largest females 7.3 kg/yr. On a daily basis these values are calculated as 28.1 g, 40.5 g, 17.3 g, and 26.8 g, respectively. The ratios of food weight/year/predator weight are calculated as 2.5:1, 2.2:1, 3.1:1, and 2.8:1, respectively. That this model is approximately correct for the butaan is suggested by the number and weights of rats eaten in captivity (data from Ardell Mitchell and pers. obs.), which vary from 200 to 280 g/wk for adult *V. olivaceus* (2.3–3.1 kg). This calculation provides an annual food intake of 7.5 kg/yr, or 20.7 g/day, or a ratio of food to body weight of 2.8:1. Thus the actual probable intake per year is well within the estimate provided by the model. The implications of this annual intake rate are discussed elsewhere.

Handling and swallowing time/item varies from 2 to 247 sec, with an average of 8.6 sec ($N = 16$) for fruit and 10.1 sec ($N = 51$) for all animal prey (molluscs, rodents, and birds combined): 72.5 sec ($N = 22$) for mice, 247.3 sec ($N = 16$) for rats, 70.0 sec ($N = 13$) for birds. These times are essentially identical to those reported for *V. bengalensis* (Auffenberg, 1984).

Digestion

V. olivaceus is like all other varanid species studied in regard to digestive efficiency: all the animal prey, except claws, feathers, teeth, and the shells of molluscs are completely digested. In addition, the pulp of all fruits eaten is digested, though tough seeds and their fibers pass through the entire tract without much if any modification (though such passage may actually enhance seed germination; see chap. 11). The crushed eggs of birds are digested, though eggshells of reptiles are usually passed in the feces of *V. salvator*. Lederer (1942) reported that uncrushed eggs sometimes pass through the tract of *V. komodoensis* without digestion. He also reported that ticks and mites are not digested by that species, and I have found the same to be true of *V. olivaceus*. Traumatized specimens apparently stop (or slow down) digestion and frequently defecate undigested foods.

Though no data are available, digestive efficiency of both plant and animal matter is reasonably high in all varanids. Undigested or partially digested fruits are defecated only when *V. olivaceus* individuals are stressed. Normally most pericarp and almost all snail flesh material is completely digested. Unfortunately, no precise data are available for *V. olivaceus*, but digestive efficiency (sweet potato) reported at 85% in one herbivorous lizard (Throckmorton, 1973) is equivalent to that in insectivorous types (Rand, 1978). However, some folivores may be less efficient when feeding on normal coarse leaves (Nagy, 1973; Auffenberg, 1982a), but not all of them, for dissection of *Hydrosaurus pustulosus* in the Philippines (author's field notes) suggests that digestion of leaves may be complete in this species.

The digestive efficiency of *V. olivaceus* is more typical of carnivorous than herbivorous lizards. Calculated efficiencies of the former vary from 70% to 90% and significantly exceed those of herbivorous species, which are 30–70% (see Iverson, 1979, for review). However, the values for "herbivorous" lizards are based on the efficiencies of largely folivorous species (Troyer, 1983). Throckmorton (1973) and Hansen and Sylber (MS) have reported efficiencies high in the carnivore range, based on cooked sweet potato tubers. While the usefulness of these values has been questioned (Iverson, 1982), since no lizards are known to eat tubers, they are probably nearly equivalent to efficiencies of frugivores, for the tissues involved are more similar to one another than to those of leaves. Studies of digestion efficiencies of lizards after they feed on a variety of foods natural to the species is badly needed (Troyer, 1983).

The digestion pause of the butaan was determined in captivity under semi-wild conditions in the Caramoan area (chap. 1). Various staining crystals (see chap. 1) were placed within the body cavity of dead rats and inside the pulp of some fruits. Some of this stain passed through within 24 hours, but stained, solid feces never appeared until the second day and usually later (48–144 hr, $\bar{X} = 72$ hr), with an average of about three days. However, hair and feathers normally appeared later (120–168 hr, $\bar{X} = 144$ hrs), suggesting a digestion rate similar to that established for the Komodo monitor on the basis of similar fecal data (Auffenberg, 1981a). Mean passage rates for folivores appear to be somewhat slower (4–15 days; see Throckmorton, 1973; Auffenberg, 1982a; Christian et al., 1984), and even *V. olivaceus* may retain fruits in the gut for a longer time than proteinaceous animal foods.

The proximal part of the colon in "herbivorous" iguanids always contains digesta, whether the stomach does or not (Grenot, 1976; Case, 1976; Iverson, 1982). However, the colon of *V. olivaceus* is often completely empty. One can only assume that this difference is due to the need in the largely folivorous reptiles to maintain a microflora in the gut for cellulose digestion (Fenchel et al., 1979; Troyer, 1982), whereas such retention is not necessary in the frugivorous *V. olivaceus*.

While *V. olivaceus* does not apparently carry a specialized gut flora to help process plant foods ingested, other factors may contribute to digestive efficiency. Thus Johnson and Lillywhite (1979) have shown that even herbivorous lizards differ in their ability to digest cellulose, depending on the type of food eaten. Studies with humans have even shown that different individuals have different abilities to digest cellulose in common foods (Slaven and Marlett, 1980). Differences in the degradation of fibrous materials in the large intestine may also account for interspecific differences in digestive efficiencies of herbivorous lizards (Christian et al., 1984). The digestive efficiency of *V. olivaceus* remains unknown for any food type.

Fecal Pellets

Except that they often contain remains of fruits eaten (usually seeds only), butaan fecal pellets are identical to those described for other monitor lizards (Lederer, 1942; Auffenberg, 1981a). As in all reptiles, the nitrogenous wastes are excreted in the form of a white (rarely pink) semisolid (uric acid) at the same time as digestive residues. The uric acid is ordinarily in the form of a pasty nitrogenous cap on the fecal matter. It is normally expelled first, with the darker, odoriferous residue (often containing seeds and bits of snail shell) immediately thereafter as several (1–5) usually slightly elongated pellets. *V. olivaceus*, unlike *V. komodoensis*, apparently never ingests earth.

Individual fecal pellets vary greatly in size; five randomly selected were an average of 3.8 cm long and 1.6 cm in diameter. They do not last long on the ground, for the frequent rains tend to wash them away in about 1–5 weeks, depending on rainfall and position.

Droppings are normally cast in the morning, after the individual has moved about or at least finished morning basking. Most droppings are deposited on the forest floor, some in holes in the ground; there is no evidence that droppings are ever cast in the trees. Since locating fecal pellets on the forest floor is extremely fortuitous because of heavy rainfall, fissures, and plant-covered surface, few were found in the wild during this study. (The fecal samples in tables 9-1 and 9-2 are comprised of both pellets found in the field and a greater number taken from the end of the large intestine of dissected individuals.) However, the 4 field samples were found along small trails and in shallow hollows in overhanging rocks. As in *V. komodoensis*, fecal pellets are apparently important in chemical communication between individuals (Auffenberg, 1981a), for captives regularly investigate fresh fecal matter deposited in the enclosures.

Fecal pellets are passed with much more movement than I have seen in any other varanid species. First the body and tail are raised high off the ground, this hip area wiggled violently from side to side, and the pellet finally passed, often only after repeated contraction of the visceral area. After defecation individuals usually move forward and scrape the cloacal area against the ground. Only later as the individual walks away is the tail dropped back onto the surface.

Gastric Pellets

After feeding on large meals many snakes often disgorge the undigestible portions of their prey in the form of a gastric pellet, containing mainly hair and occasionally teeth, claws, and feathers. The habit is also reported for both captive and wild varanids of several species (Hediger, 1934, for *V. indicus;* Lederer, 1942, and Auffenberg, 1981a, for *V. komodoensis*) and is probably normal in the species that feed on larger mammals, or at least large quantities of flesh with hair or feathers. Pellets seem particularly common in large adults.

Butaan gastric pellets are also occasionally cast by large individuals after big meals. As in other species noted, butaan gastric pellets (all produced under captive conditions) were composed of mammal hair or feathers. A typical pellet is 100–125 mm long and 15–23 mm in diameter at the widest part. Such pellets are much smaller than those cast by *V. komodoensis*, undoubtedly due to the tendency for *V. komodoensis* to gorge itself when large amounts of food are present (Auffenberg, 1981a), which is not a habit of *V. olivaceus*.

The pellet is disgorged in a manner similar to that observed in snakes. The lizard arches the anterior part of the body, neck, and head off the surface. The hyoid apparatus is retracted to enlarge the esophageal passage and the mouth opened widely. The head and neck are then violently thrown from side to side as the belly is pulled in. Such gastric pellets are covered with an odoriferous mucoid substance.

The Feeding Niche

The results of the Caramoan food and feeding analyses show that the butaan lacks specializations of the gastrointestinal tract (Iverson, 1982) for dealing with bulky leaves or the wide array of secondary compounds usually found in them. The species is not sensitive to the toxins of the few fruits upon which it regularly feeds (see chap. 11 for additional discussion). It does, however, possess some behavioral and morphological specializations for dealing with a largely frugivorous diet.

Regarding plant material, the butaan feeds only on the pulp of ripe fallen fruits of a few vine and tree species. It is, in consequence, a selective plant feeder by comparison with certain other more generalized "herbivorous" lizards (i.e., *Cyclura, Iguana;* see VanDevender, 1982; Auffenberg, 1982a). The animal part of its diet is restricted to a few locally abundant land molluscs. Inability to use coarse foliage high in fiber may be the most important factor restricting this species to those relatively aseasonal but floristically diverse evergreen forests within its geographic range. In this respect it differs greatly from nearly all large herbivorous iguanid and agamid lizards, which are able to occupy seasonally drier and floristically simpler forests and which can probably persist longer in badly degraded forest associations.

The most conspicuous elements of butaan diet are land molluscs that provide proteins poor in the remainder of the foods, sugary fruits such as *Grewia stylocarpa* and *Spondias pinnata,* and high oil content pulp such as members of the genus *Canarium.* Land snail shells require considerable crushing power in the jaws, which butaans have (chap. 3); the high sugar fruits are small in diameter and are thus shared with many other forest animals, though the fruit crops tend to be heavy as a predator-satiation device; the *Canarium* fruits tend to be large and are thus not eaten by many potential vertebrate seed dispersers. Furthermore, oily fruit crops tend to be smaller and scattered over a longer part of the year than sugary fruit crops. However, in general, all of these sources are widely scattered and uncommon compared to the foods available for folivorous lizards (Auffenberg, 1982a).

Mollusc feeding is known in many other lizards, including skinks, teiids, and several varanids. Some, like *V. niloticus* and *V. exanthematicus,* feed on live snails found either on land or in the water (Cisse, 1972); others feed on these and dead marine snails thrown onto the beach during storms (*V. komodoensis;* Auffenberg, 1981a). Some species have teeth highly modified to feed on such prey, including the butaan (chap. 3). *V. komodoensis* has, in addition, a powerful skeletomusculature system involving the mandibular area of the skull that enables it to crush some of the strongest of the local snail shells. What is perhaps most important is not that the butaan feeds on so many snails but that it feeds on so few other animals—some of which are common in the same habitat. Second, it is restrictive in what species of snails it eats. Without exception these are all land snails—none are species whose usual habitat is in the trees, in spite of the fact that the butaan frequents the same trees for considerable periods of time (chap. 5). The only explanation is that at least some of these tree snails are thought to possess a fairly powerful toxic substance that, when released, will kill other species of both tree and land snails kept in the same container (G. Auffenberg, pers. comm.). It is possible that such toxins (known for other species of tree snails in other parts of the world as well; see Heatwole and Heatwole, 1978) protect these snails from butaan predation.

10

Food Resources

Animal Foods

ANIMAL density estimates have relied greatly on capture-recapture methods. However, some recent studies suggest that these methods are probably more limited than previously believed (Parmentier, 1976). Parr et al. (1968) and Heatwole and Heatwole (1978) found the Jolly (1965) method unsuitable for studies of snails, and our density studies on some species of limestone-inhabiting species in the Caramoan area could not be brought to satisfactory conclusions using nearest-neighbor methods (Gloyd, 1967), because populations were not always randomly distributed. Our population density estimates of snail species eaten by the butaan resulted in extremely high variances—probably because the assumption that species were equally catchable was violated by species that spend much time in limestone crevices and hollows (i.e., *Lepidotrichia semisculpta*). Consequently, density estimates for the larger snails of Caramoan are based on dead snail shells gathered from fissures and caves in the surrounding limestone hills rather than capture-recapture or nearest-neighbor methods. Since the snail shells collected were about the same size and all were rather large, I expect no significant interspecific differences in deterioration rates. A total of 4339 unbroken shells were collected, identified, and counted (table 10-1).

Lepidotrichia semisculpta is the most common species, representing 51.8% of the shells collected. All remaining taxa are less common (0.1% to 10.8%). In general, rock and leaf litter species are much more abundant than tree snails. Density studies on other species in different parts of the world show similar results (Berry, 1966; Heatwole and Heatwole, 1978; Coney et al., 1982).

Studies on the local density of *Cochlostyla pithogaster* (Auffenberg, 1982–83 field notes), using both the Bailey Triple-Catch and Lincoln Index methods, suggest that in prime habitat there are about 100 individuals per ha (much lower than density estimates for the slightly smaller caeminid tree snail *Caracollus* in similar tropical rain forests of Puerto Rico [Stiven, 1970; Heatwole and Heatwole, 1978]). Assuming (perhaps unjustifiably) proportionally similar death rates in all Caramoan species studied, estimates of the minimum individuals per ha for each of the species are shown in table 10-2.

In general, both tables show that the rock-inhabiting snails are more common ($\bar{X} = 11.3\%$, 325/ha) than the leaf litter types ($\bar{X} = 4.5\%$, 89/ha). The tree snails are least common (2.7%, 71/ha). These results are in accord with the conclusions of others (Coney et al., 1982). Brief studies by our party on Mount Isarog, Camarines Sur, show that all the species except one (*Lepidotrichia semisculpta*) eaten by *V. olivaceus* on limestone substrates are also found on volcanic rocks (at least to 500 m), though less abundantly and of significantly smaller size.

Unfortunately, there are no data on the spatial or temporal availability of the insect species eaten by *V. olivaceus*. In general, these insects are large compared to others of the forest. Local terrestrial insect abundance shows a significant drop from October through December, that is, during the heavy monsoon (fig. 10-1). Most insect species (including food species such as rhinoceros beetles) are generally common through most of the rest of the year but particularly near the end of the dry season and the beginning of the first (southwest) monsoon. Though most movements of rhinoceros beetles occur at

TABLE 10-1. Relative abundance of mollusc species found in the Caramoan area.[a]

Habitat	Species	Abundance (% of total)
Rocky substrate	*Lepidotrichia semisculpta*	51.8
	Ryssota angulata	10.4
	Cyclophorus ceratodes	2.4
	Geophorus (2 species)	2.4
	Obba columbaria	0.9
	Obba listeri	0.1
Leaf litter or rotten logs	*Trochomorpha* sp.	7.4
	Hemiglypta sp.	4.8
	Lepidotrichia abraea	1.2
	Trochomorpha beckiana	0.2
Arboreal	*Calocochlea cailliaudi*	10.8
	Cochlostyla pithogaster	3.8
	Cochlostyla carinata	1.3
	Cochlostyla (new species?)	1.2
	Cochlostyla imperator	1.1
	Cochlostyla virginea	0.1
	Cochlostyla turbinoides	0.1

a. Smallest species (*Geophorus, Cyclophorus,* and *Trochomorpha*) may be underestimated.

TABLE 10-2. Estimates of minimum population densities of common large Caramoan snails based upon data in table 10-1 and extensive studies of *Cochlostyla pithogaster* (see text).

Habitat	Species	Individuals per hectare
Rocky substrate	*Lepidotrichia semisculpta*	1526
	Ryssota angulata	274
	Cyclophorus ceratodes	63
	Geophorus (2 species)	63
	Obba columbaria	24
	Obba listeri	3
Leaf litter or rotten logs	*Trochomorpha* sp.	194
	Hemiglypta sp.	126
	Lepidotrichia abraea	32
	Trochomorpha beckiana	5
Arboreal	*Calocochlea cailliaudi*	284
	Cochlostyla pithogaster	100
	Cochlostyla carinata	34
	Cochlostyla (new species?)	32
	Cochlostyla imperator	29
	Cochlostyla virginea	18
	Cochlostyla turbinoides	3

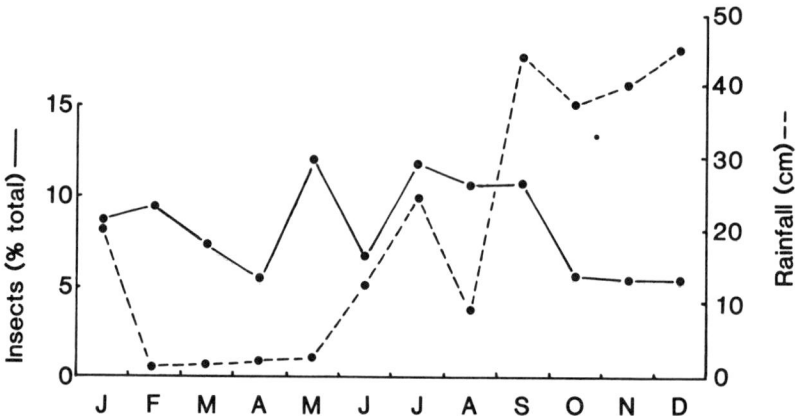

Fig. 10-1. Seasonal variation in insect abundance in Caramoan area ($N = 5259$), compared to local annual rainfall pattern.

night, the monitors undoubtedly locate them on the ground, usually in and under rotten wood.

Hermit crabs are abroad both day and night but usually the latter, especially during rainy weather. No data are available on density, though in general they appear most common in rocky areas, especially close to the sea (though regularly found to at least 300 m el locally). During the day they often hide in hollows and fissures in the rocks or in leaf debris at the base of cliffs or tree buttresses.

The large spiders eaten by the butaan (tarantula relatives without webs) are apparently captured on the ground. We saw them most often in open gaps in the forest or along its edges, where they live under leaves and in rotten wood. As far as I am aware, they are completely diurnal and, unlike snails, rhinoceros beetles, and hermit crabs, are caught while active rather than when resting. There are no data available bearing on density.

Plant Foods

The lowland forests of Malaysia and the Philippines are richer in tree species than any other forests of the world, including those of tropical Africa and America (Whitmore, 1975). Stands of single species are rare, except for a few in swamps, on ridges, along riverbanks, or in disturbed areas. Poore (1964, 1968) reported that of 381 species of trees over 1 m bhd in 23 ha, 157 occurred only once. However, it is not obvious why tropical rain forests should be more diverse than other rain forest types or why those in the Indo-Malaysian area are richer than those elsewhere. One theory holds that tropical forest diversity is promoted by the selective pressure exerted by host-specific insect pests (Janzen, 1970); another is that genetic drift may augment the process of speciation because individuals of already rare species are so isolated from each other that they seldom interbreed (Fedorov, 1966). Rarity may even have distinct evolutionary advantages in predator avoidance (Flenley, 1979). Any viable theory, however, faces the problem that the processes it involves must apply more forcefully in tropical than in temperate regions (Raemaekers et al., 1978).

Density.—Based on a study of 54,000 m^2 immediately surrounding the places where butaans were captured (see chap. 1), it is apparent that the habitat is dense (average number of trees/ha with a bhd < 10 cm 3173.5, OR = 2222–4717). However, most species are represented by very few individuals (fig. 10-2), and the frequency distribution of the number of stems per taxa agrees well with data presented for Malaysian forests by Wong (1967), Whitmore (1975), and Raemaekers et al. (1978). About 70% of tree species are represented by a single stem each, showing the general rarity of each species

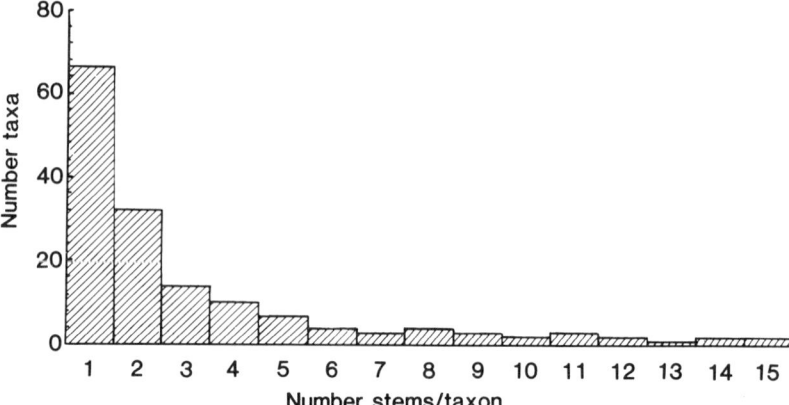

Fig. 10-2. Tree (bhd 10 cm) species diversity in a 1-ha plot in lowland mixed dipterocarp forest, *Vitex* facies; karst mountains near Caramoan.

in these forests. Nearly 43% of the tree species in this block are represented by single individuals with bhd < 2 cm.

It is not surprising, then, that the plants upon which *V. olivaceus* feeds are also highly dispersed in the forests in which it lives and are generally low in density. For the butaan it means that certain food plants are difficult to locate in a forest patch. This sparsity is expected to select for stable and small home ranges for the butaan, accompanied by good spatiotemporal memory to aid in maximizing food location efficiency (attributes of some carnivorous varanids as well; see Auffenberg, 1981a) (see below). A butaan with a daily activity range of ca 1.5 ha (chap. 6) has access to about 47 species of trees whose seeds are animal-dispersed, of which only 9 are utilized for food by *V. olivaceus*. Table 10-3 provides data on the density of individuals of the major food species per ha in the Caramoan forest blocks studied. It also includes the percentages of the total food species per ha for these blocks. The percentage of total food trees/ha in the Caramoan area averages 1.0%; it ranges from 0% (Kabutanan) to 4.3% (Batungan). The table also shows the number of trees of each food species/ha in each block (0–130). *Canarium* and *Ficus* (especially strangling types) are the most widely distributed in the blocks, though *Canarium* is not necessarily the most abundant of the food trees in each block. At Matingapo II there were two butaans taken within a few months (suggesting a fairly high density), and this block also has one of the higher percentages of utilized fruit trees/ha.

Figure 10-3 shows the positions of fruit species within the belt transects of each block and the location where the butaan was caught at each place. At Matingapo II two butaans were caught in the same tree, though several months apart. Belt transect positions were somewhat dictated by local landform. An analysis of the interindividual spacing of trees of the same species in all blocks studied (using Morista's Index of Dispersion, see Lloyd, 1967, for review) shows that *Aglaia harmsiana* has a contiguous distribution (values 1.8–2.2) in all samples studied with more than two trees of this species per transect; all other food species have a completely random dispersion pattern (in spite of the sometimes superficial appearance of clumping in strangling figs, *Ficus*). In general, dense habitats tend to have fewer fruiting trees/ha than open ones (chap. 4).

Almost all fruits eaten by *V. olivaceus* are produced in medium-height trees ($\bar{X} = 12.6$ m; see appendix 1). Thus the fruits characteristically occur in the middle forest levels where there is an optimum combination of vegetation mass, direct sunlight, and a more open stable environment than in the upper canopy. However, as shown earlier, *V. olivaceus* individuals rarely, if ever, forage in the trees for fruits.

Biomass.—Table 10-4 shows fruit production per individual plant (more important species only) per single fruiting season. These data show that a large number of fruits are available at each tree during the fruiting period, averages varying from 342.5 for *Canarium vrieseanum* to 54,747 for *Caryota*

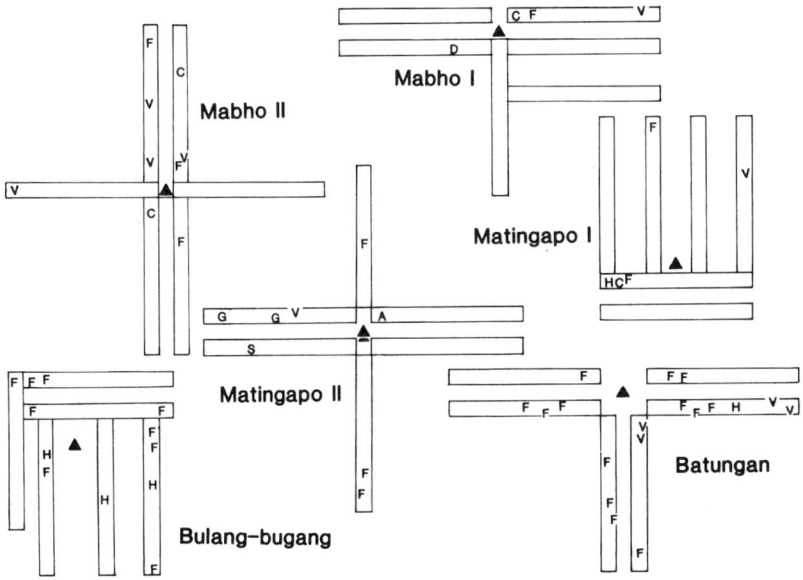

Fig. 10-3. 1000 m² (10 x 100 m) belt transects at specific locations where *V. olivaceus* individuals were caught (triangles), showing distribution of food plants. Abbreviations: A, *Aglaia harmsiana*; H, *Canarium hirsutum*; V, *Canarium vrieseanum*; C, *Caryota cumingii*; D, *Dracontomelum dao*; F, *Ficus* (strangler types); G, *Grewia pinnata*; S, *Spondias pinnata*.

TABLE 10-3. Density of trees used as major food sources in 6000-m² forest blocks studied at nine sites in which *V. olivaceus* were captured.

Study block	Mean number of trees per hectare								% total trees = food trees
	Aglaia	Grewia	Dracontomelum	Caryota	Canarium[a]	Ficus[b]	Spondias		
Mabho I	0	0	1.7	0	1.7	6.0	0		0.3
Mabho II	0	0	0	3.4	7.0	4.5	0		0.5
Matingapo I	0	0	4.0	0	5.0	3.0	0		0.4
Matingapo II	2.5	50.0	0	0	2.5	0	1.7		1.8
Batungan	0	0	0	0	8.0	130.0	0		4.3
Bulang-bugang	0	0	0	0	5.0	16.0	0		0.7
Tiak Lake	10	0	0	5.0	10.0	5.0	0		0.6
Gota	0	0	0	0	0	1.0	0		0.1
Kabutanan	0	0	0	0	0	0	0		0

a. Three species.
b. Strangler type only.

TABLE 10-4. Fruit production per tree per fruiting bout.

	Number of trees studied	Mature fruits per tree		Biomass per tree (g)	Mature fruits on ground		Biomass on ground (g)	Difference in number[a]	Percent difference
		$\bar{X} \pm SD$	OR		$\bar{X}$	OR			
Aglaia harmsiana	8	1,653 ± 82	682–3,667	4,446	926	483–2,156	2,471	727	44.0
Canarium hirsutum	16	5,903 ± 191	1,104–11,660	42,032	97	30–131	689	5,807	98.4
Canarium ovatum	16	1,602 ± 71	88–3,733	52,552	82	51–128	2,690	1,520	94.9
Canarium vrieseanum	12	343 ± 52	104–501	1,333	27	13–93	104	316	92.1
Canarium (tiny species)	5	10,503 ± 261	375–22,386	1,260	43	15–86	5	10,460	99.6
Caryota cumingii	14	21,155 ± 32	15,485–28,913	36,175	101	51–310	173	21,054	99.5
Caryota Rhumphiana	6	54,747 ± 1,621	1,694–107,800	114,969	56	23–617	118	54,691	99.9
Dracontomelum dao	8	29,561 ± 398	687–48,730	36,360	766	708–781	942	28,796	97.4
Grewia stylocarpa	12	5,701 ± 312	1,318–18,689	18,185	1,121	902–1,370	3,576	4,580	80.3
Pandanus radicans	1	180		2,955	156		2,558	24	13.3
Sandoricum koetjape	6	1,681 ± 89	523–5,968	118,644	341	116–1,130	14,921	1,340	79.7
Spondias pinnata	13	4,784 ± 112	1,763–8,514	20,043	965	458–1,787	4,042	3,819	79.8

a. Mean number of fruits on tree less mean number of fruits on ground.

Rhumphiana. This great variation in fruits per tree is not closely associated with food preference ratings (chap. 11); some of the heavily fruiting trees, such as *Caryota Rhumphiana* and *Dracontomelum dao*, are rarely eaten by *V. olivaceus*. Actually, some of the more commonly eaten fruits are produced in smaller numbers per tree, such as *Canarium vrieseanum* (see below).

Table 10-4 also shows that from 79% to nearly 100% of the fruits on all these trees are eaten by various organisms (almost all vertebrates) before they fall to the ground, where the butaan finds and eats most of those upon which it feeds. In October 1983 we observed over 1,000 fruits available on one tree, but an average of only 4.4 mature fruits fell to the ground each week—mainly because the tree was visited almost every night by a palm civet (*Paradoxurus philippinensis*), who fed on the ripe fruits in the tree.

The ground under *Grewia stylocarpa* trees is often littered with fruit ($\bar{X} = 1121$ fruits/wk). *Sandoricum koetjape* has the largest fallen fruit mass per tree, but the species is rather rare in the forest and is seldom eaten by the butaan. Of all the fruits eaten, it is the only one that is introduced (though completely naturalized throughout Luzon; Merrill, 1912).

The importance of these data to foraging strategy is discussed in chapter 11. For the present it is important to realize that fruit trees of the species eaten by *V. olivaceus* do not have a contiguous distribution but are widely scattered in a botanically variable environment. Furthermore, the number of fruits per tree produced by different species varies greatly, and, more important, the number of fruits produced is not correlated with the number that falls to the ground. While a tremendous amount of fruit is produced per average *V. olivaceus* daily activity range, little (20%–0.1%) is actually available as food ($\bar{X} = 0.9\%$), even within the entire home range.

Calories.—The total annual kilocalories available to *V. olivaceus* per ha (fallen mature fruit only) varies from 306 (*Dracontomelum*) to 77,492 (*Grewia*) for an average of 1874. Variability is high (SD 29,650) due primarily to proportion of sugars and oils; per tree they vary from 1016 (*Caryota*) to 16,992 (*Spondias;* average 10,194) but with much less variation (5948); per mean home range area varies from 915 (*Caryota*) to 114,074 (*Grewia*) with normal variation (SD 5919). Highest seasonal caloric values are found during the southwest monsoon and lowest during the heaviest rainfall period of the northeast monsoon. There is no correlation between seasonal or taxonomic caloric availability and food preference in *V. olivaceus*. Nor is there any relation between seasonal use and presumed seasonally different demands of the butaan, such as high lipid fruits in the yolking season, watery fruits in the drier season, etc. However, Herrera (1982) has shown that fruits of species of bird-dispersed plants ripening at different times of the year differ in their nutritional properties in such a way as to match the seasonally changing demands of their bird dispersers. To me this finding suggests that the butaan is not the most important of the animal dispersers of these fruits in Caramoan but rather the hornbill (for which no appropriate data are available; see chap. 12).

Fruit shadows.—Another factor regarding the distribution of food that is of importance in the feeding strategy of *V. olivaceus* is the actual distribution of fallen fruit—the fruit shadow of Janzen (1971a). When the resting places of fallen fruits below a tree are plotted, it is obvious that most fruit shadows are skewed, the highest frequency of fruits on the ground more or less under the outer perimeter of the canopy limits (fig. 10-4). Furthermore, the frequency-distance distributions are different in each tree species: they are most peaked in *Spondias pinnata* (2.5 m $\bar{X}$ canopy dia.) and least in *Caryota cumingii* ($\bar{X} = 4.0$ m) and *Dracontomelum dao* ($\bar{X} = 3.2$ m). These longer-tailed distributions are produced chiefly because of the rounded shapes of *Dracontomelum* and *Caryota* fruits, which make them roll farther from the parent tree. The pear-shaped fruits of *Grewia stylocarpa* tend to roll in a circle, producing short fruit shadows ($\bar{X} = 2.0$ m). The subtriangular cross section of *Canarium* fruits ($\bar{X} = 2.1-2.6$ m) prevents much rolling, even on sloping surfaces, as does the highly angular cross section of *Pandanus* fruits ($\bar{X} = 2.8$ m). The importance of these frequency curves in respect to food location by *V. olivaceus* is discussed in chapter 11.

Figure 10-4 also shows that the fruits of *Grewia stylocarpa* often fall

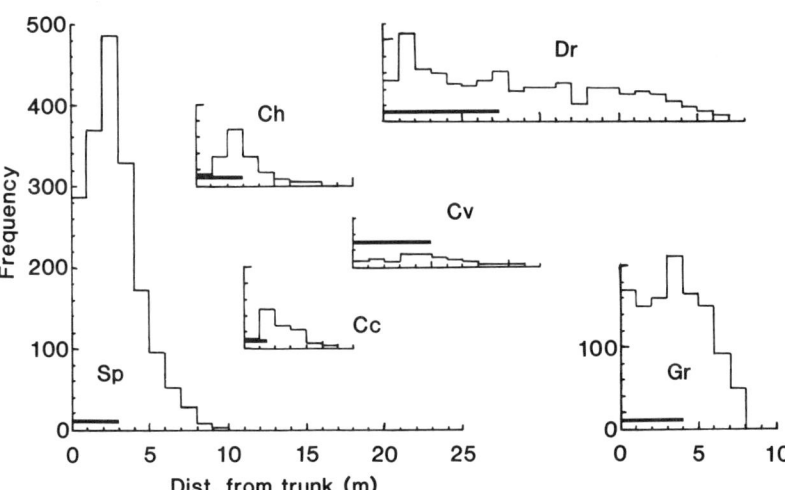

Fig. 10-4. Frequency of fallen fruit of taxa commonly eaten by *V. olivaceus* ($N = 5$ trees/sp) and distance from trunk of producer trees. The heavy line shows the mean canopy limit from the trunk of the tree producing the fruit shadow. Abbreviations: *Cc, Caryota cumingii; Ch, Canarium hirsutum; Cv, Canarium vrieseanum; Dr, Dracontomelum dao; Gr. Grewia stylocarpa; Sp, Spondias pinnata.*

through the canopy, producing a relatively flat frequency curve below the canopy but steeply slanting beyond its limit. *Dracontomelum dao* fruits also fall through the canopy but may roll far from the parent, so that the curve both below and beyond the canopy is flattened. The fruits of *Spondias pinnata* and *Caryota cumingii* tend not to fall through the canopy but to roll off it, resulting in a steep slope under the canopy. Though the fruit of *Spondias* is oval and that of *C. cumingii* a slightly flattened sphere, both roll far beyond the canopy border. The fruits of *Canarium hirsutum* seldom fall through the canopy but, like those of *Spondias,* roll off it; however, their subtriangular cross section keeps them from rolling far. As expected, since the fruits of *Canarium vrieseanum* are more rounded than those of *C. hirsutum,* they roll farther from the canopy edge.

Each species thus has a somewhat distinct fruit distribution pattern that must be of significance to the foraging behavior of vertebrates who are "expected" to feed on fallen fruits. Interestingly, the fruits often eaten in the trees (*Dracontomelum, Spondias, Caryota*) by other vertebrates (palm civets and monkeys; see chap. 12) are rounded. Those fruits (*Canarium* and *Grewia*) often eaten by nonclimbing predators seem specialized in their shape so that they do not roll far from the parent tree. Thus they seem to be specialized to attract mainly density-dependent larger predators, like V. olivaceus, who find their food on the ground. Oily and sugary types of fruit occur in both categories, probably resulting from some form of competitive exclusion. The potential coevolutionary factors involved are discussed further in chapter 11. In summary we can say that the outer shape of the fruit determines, at least partly, the frequency distribution of fallen fruit below the tree, just as seed shape and morphology help to regulate the speed at which the seeds move through the digestive system (chap. 9).

In spite of a great divergence in horizontal cross-sectional area of the canopy of the several different fruit species utilized, there is no relationship between canopy size and the number of fruits on the ground or even the number usually found beneath the canopy itself. While most fallen fruits are beneath the canopy (50%–78.5% of total fallen fruits), and the canopy area varies from 7.1 to 180.1 m^2, the average distance from the trunk shows a close dispersion, from 2.2 m to 8.1 m. The maximum average distance from the trunk is a reflection of how fruit falls through or over the canopy surface, as well as of fruit shape. From the standpoint of the butaan's foraging tactics, the most important figures are total fruits on the ground, average distance from the trunk, and maximum area on the ground over which the fruits might be found. Table 10-4 provides data on the total fruits on the ground/ha; *Grewia* is highest by a significant amount, species of *Canarium* and *Caryota* are lowest, and all others intermediate.

Within an average hectare of Caramoan forest, and throughout the entire year, a total of 56.3% of each hectare can be expected to contain at least some fallen fruit. The area beneath the crown of fruiting trees contains from 50% to 78% of the total fruit fall. The smallest surface areas in which fallen fruit can be expected are found beneath *Spondias, Caryota,* and *Canarium ovatum* trees (0.06%–0.2% of the total area in a forest hectare). The largest areas in which fruits can be expected are under *Dracontomelum, Grewia stylocarpa,* and *Canarium vrieseanum* (1.2%–4.7% of hectare). The maximum areas in which fruits fallen from individual trees can be found are smallest under *Spondias* (0.7%), *Caryota* (0.7%), and *Canarium ovatum* (1.0%), and largest under *Dracontomelum dao* (11.6%), *Grewia stylocarpa* (12.5%), and *Canarium vrieseanum* (22.8%). Thus the subcanopy and maximum areas have the same patterns; the same species are found in the smallest and largest categories of each, in spite of somewhat different patterns of fruit rolling, etc.

Fruit phenology.—In chapter 9 the seasonal pattern of food *utilization* based on gut contents collected throughout the year was discussed. In this section, the seasonal *availability* of food will be described for the Caramoan forest. In spite of the fact that many writers not familiar with tropical forests still refer to them as aseasonal, numerous studies have shown that tropical forests (even those of humid evergreen types) are distinctly pulsed in respect to their phenology (see Richards, 1952; Schnell, 1971; Walter, 1973; Augsberger, 1982; Foster, 1982, for reviews). The most prominent seasonality in the tropics, however, is the change in availability of water through the year rather than the change in temperature experienced in temperate areas.

Still, data on the phenology of fruiting in tropical forests are rare and often incomplete. More work has been conducted in the New than the Old World (see Foster, 1982, for New World review). Long-range studies in Asian forests have been conducted, and some significant data are published for Sri Lanka (Koelmeyer, 1960), the South Pacific (Baker and Baker, 1936), India (Krishnaswamy and Mathauda, 1954), and especially Malaysia (Corner, 1952; Poore, 1968; Medway, 1972; Ng, 1972; Ng and Loh, 1975; Raemaekers et al. 1978). Unfortunately, other than the following data for the Caramoan area, I know of no pertinent data for the Philippines.

The Caramoan mixed lowland dipterocarp forests are distinctly seasonal in regard to leaf flushes, flowering, and fruiting, though not all species follow the same patterns annually, or even all individuals of the same species. Still, there is a rhythmicity in the appearance of fruits throughout the year that is related, in general, to rainfall pattern (fig. 10-5); the major fruiting period tends to precede by two months (and extends through) the earlier southwest monsoon and starts to wane before the heaviest rains of the northeast monsoon. Medway (1972) found a similar pattern in Malaysian forests, though in my study ripe fruits peaked in July–August rather than September–November as in Malaysia. Because *V. olivaceus* is dependent upon the availability of fruit

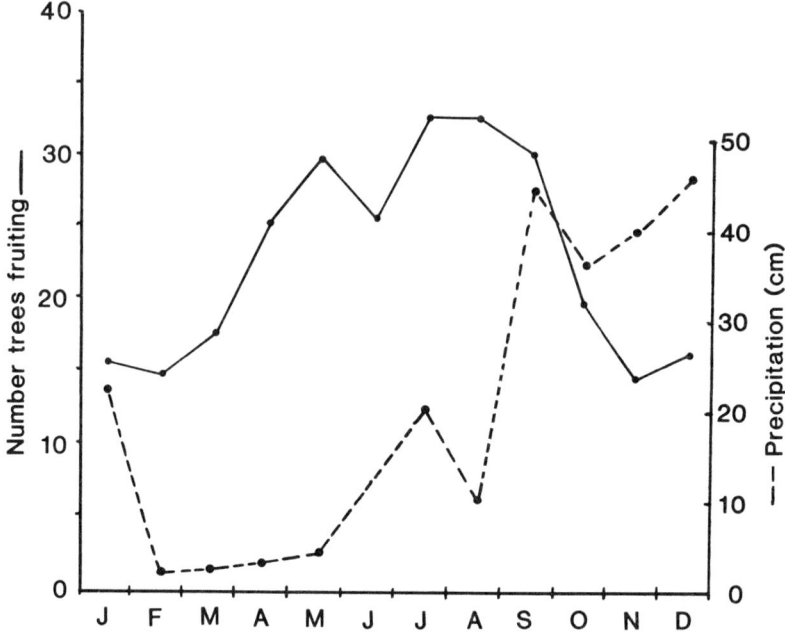

Fig. 10-5. Number of animal-dispersed tree species fruits/month compared with rainfall.

throughout the year, whatever variations or patterns occur in fruiting cycles will be reflected in the kinds and amounts of foods consumed.

Several different patterns of fruiting duration per individual tree are exhibited among the fruit species eaten by *V. olivaceus* (fig. 10-6), ranging from 2 to 25 weeks; all species eaten have only one fruiting bout per year, except *Caryota cumingii,* which has two. *Aglaia harmsiana* has the longest fruiting period, with individual trees fruiting continuously for about half the year. The long steady fall of *A. harmsiana* fruit has important implications for the foraging patterns of the butaan, especially compared to the tree species that fruit for a short period of 1–2 weeks (i.e., *Ficus,* strangler fig types; see appendix 1). The individual trees of most species eaten bear their fruit over a period of about 3 to 7 weeks. The butaan, once having found a fruiting tree within its activity range, could therefore expect to remain in the vicinity for 1 to 2 months, depending on tree species, and be reasonably assured of obtaining fruit from it throughout this time. So, while the density of trees, number of fruits produced/tree/year, and the total weight of food produced (tables 10-3, 10-4) are all important, another significant factor is the length of time over which the food is available, for the butaan must move from one tree to another of the same or another species as some trees complete their fruiting phase (see fig. 10-6 and chap. 11).

This seasonal pattern in food availability at individual trees is further extended by the shape of the fruiting pulse itself (fig. 10-6). Though an individual tree of *Aglaia harmsiana* fruits for almost a half a year, the number of fruits produced per week shows a pulse in which maximum production occurs every two and a half to three weeks, suggesting that some weeks during this entire time are better than others for the butaan's foraging. *Grewia stylocarpa* and *Dracontomelum dao* both fruit only once a year for about 6–7 weeks. *Canarium vrieseanum* produces two crops per year per tree—one long period

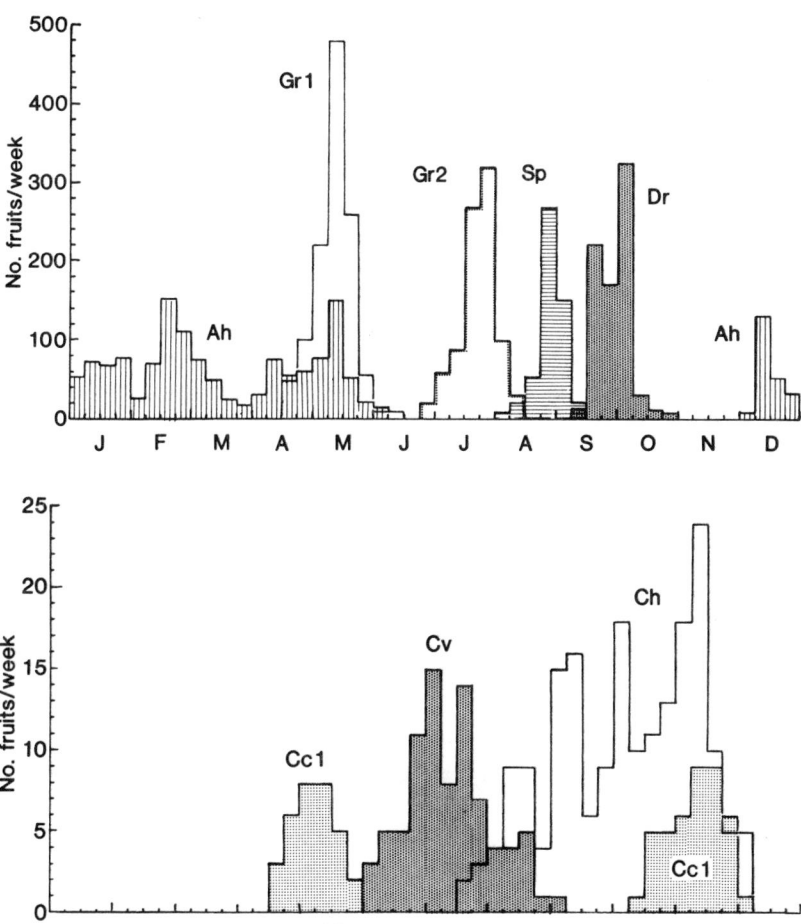

Fig. 10-6. Fruiting pulses of selected single fruiting trees utilized by *V. olivaceus*. Abbreviations: Ah, *Aglaia harmsiana;* Cc1, *Caryota cumingii* (characteristically fruits twice a year); Ch, *Canarium hirsutum;* Cv, *Canarium vrieseanum;* Dr, *Dracontomelum dao;* Gr1 and Gr2, two different individuals of *Grewia stylocarpa;* Sp, *Spondias pinnata.*

of 6–7 weeks and a short one of 1–2 weeks, both in the rainy season. It also produces fewer fruits per week than *Grewia,* and the fruiting pulse lacks the spiked peak of the *Dracontomelum. Grewia*'s high production occurs in the middle of the fruiting flush. Most of the fruit falls to the ground, producing an overabundance from the standpoint of the butaan. In fact, a single tree of *Grewia* characteristically produces far more fruits than could conceivably be eaten by an adult butaan during most of the fruiting pulse, whereas a single tree of *Canarium vrieseanum* produces only about two butaan "meals" per week. The pattern for *Spondias pinnata* is nearly identical to that of *Grewia stylocarpa* (one pulse per year with a high spike). *Canarium hirsutum* has a lower frequency curve similar to that of its congener but has only one pulse per year. Most *Ficus* (strangler types) fruit once, more rarely twice per year, with a large number of fruits produced over a short period. (As in tropical Africa [Alexandre, 1980] and Malaysia [Medway, 1972], species of *Ficus* also often fruit asynchronously [Corner, 1940; Hladik and Hladik, 1969].) *Caryota cumingii* has still another fruiting pattern in which relatively few fruits are produced during two short periods (rarely 1 or 3) each year, separated by about 6 months. No single species of fruit is available to the butaan throughout the year.

All fruits eaten by the butaan in the Caramoan area are either short-pulsed, abundantly fruiting (sugary) species (*Spondias, Dracontomelum, Grewia,* and *Malaisia;* see fig. 10-6) or long-fruiting (oily) species producing smaller fruit loads per week (*Caryota, Canarium*). The oily ones are eaten by relatively few forest vertebrates (presumably due to the inability of most animals to detoxify the high levels of secondary compounds in them; chap. 9). Many invertebrates eat the former group (chap. 12). Since the seeds of all these genera are animal-dispersed, the sugary, short-fruiting types are dispersed by many kinds of vertebrates and the long-fruiting oily types by only a few highly specialized ones. Fruiting pulse length seems directly related to the number of potential seed dispersers available to the plant. This dependence emphasizes the importance of *V. olivaceus* as a seed disperser for, particularly, the species of *Canarium* but makes it nearly as important in the same role for *Spondias,* since dispersal is affected by many animal species. Data suggest that plants depending for seed dispersal on a few highly specialized predators, probably with small home ranges, spread the fruiting pulse over a long time to avoid the waste associated with fruit overproduction. On the other hand, the species depending on many more widely ranging seed dispersers tend to have short, explosive and very productive fruiting pulses.

Figure 10-7 shows still another important temporal factor regarding the fruiting of food plants of *V. olivaceus*. As an example, each *Canarium hirsutum* tree fruits for a long time and only once a year. (*Canarium hirsutum* is one of the largest fruiting species in the forest. Many *Canarium* sp. fruit only

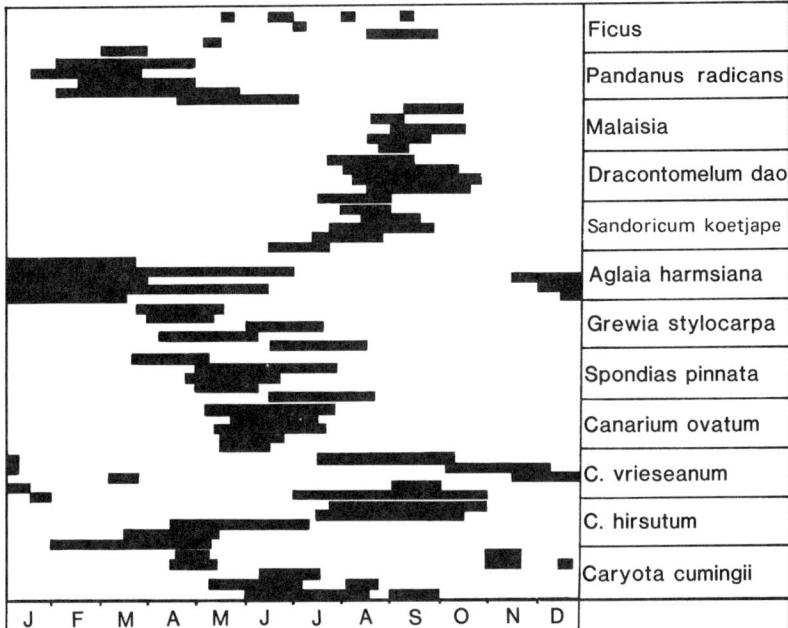

Fig. 10-7. Seasonal fruiting pattern of common plant foods of *V. olivaceus*.

once a year in Africa [Alexandre, 1980], Malaysia [Whitmore, 1972], and Sri Lanka [Koelmeyer, 1960].) However, fruiting is not entirely synchronous between all conspecifics—even adjacent individuals—and the fruiting pulse of each tree tends to be staggered in respect to that of its neighbor (apparently due to a combination of physical conditions acting on the species and an internal physiological rhythm of individuals [Koelmeyer, 1960]). Combined, the fruiting peaks of all mature trees of this species cover all of the dry and earlier southwest monsoon seasons (ca 9 months). *Canarium vrieseanum* tends to fruit after and before its congeners (two fruiting periods/yr), and trees of this species also tend to fruit asynchronously. On the other hand, *Canarium ovatum* fruits synchronously over a total period of only about 8 weeks in the Caramoan area. This pattern is true of almost all of the rest of the species upon which *V. olivaceus* feeds (see fig. 10-7). (Koelmeyer [1960] reported that *Canarium zeylanicum* on Sri Lanka fruits sporadically through the year but that there is a major fruit flush from July through November. His 10-year study showed that not all trees fruited every year but only 9, 7, 6, or 1 time in 10 years.)

The fruits eaten by *V. olivaceus* can be divided into high-oil and high-sugar types (chap. 9). Charles-Dominique et al. (1981) have shown that within both these types in New World tropical forests there is a regular temporal staggering of fruiting from one species to another, so that both types of fruit are available to local frugivores throughout the year. This availability does not occur in the Caramoan forests, at least among the sugary fruits eaten by *V. olivaceus* (but it may, if all sugary fruits of the forest were analyzed). *Grewia* and *Spondias* thus produce fruits from April through August and *Dracontomelum* and *Malaisia* from August through October (fig. 10-6). There are no high-sugar-content fruits produced from November to March. On the other hand, the high-oil-content fruits (*Caryota* and all the *Canarium* species) are staggered in regard to their fruiting pulses. Thus *Canarium hirsutum* tends to fruit for a long period from February through October, with *C. ovatum* squeezed into May-July during a period when *C. hirsutum* is not producing as heavily as at the beginning and end of its fruiting pulse. *Canarium vrieseanum* fruits sporadically from January through March, when *C. hirsutum* is not fruiting, or only slightly, and then again from July through December, with the major fruiting after the last fruiting pulse of *C. hirsutum*. *Caryota cumingii* tends to fruit twice, once during the rainy period at the end of the year when only the fruits of *Canarium vrieseanum* are available and again from April through September, with the peak of this pulse between the two major peaks of *Canarium hirsutum*. The oily fruits of *Pandanus radicans* are produced primarily before and during the earliest fruiting of *Canarium hirsutum*.

Charles-Dominique et al. (1981) have shown that small- and large-fruited species tend to be staggered throughout the year. This was not the case for at least the fruits eaten by *V. olivaceus,* but casual observation suggests that it may indeed be true if all of the fruits of the forest were analyzed.

A typical fruiting sequence for a single tree of *Spondias pinnata* often begins fairly vigorously (23 fruits on ground the first week), and weekly productivity remains more or less at this level until the fifth week; then there is a rapid increase in weekly productivity until the end of the tenth week. This increase is followed by a rapid reduction in fruits produced per week, so that only two weeks of fruiting remain after the major part of the pulse is completed. The early part of the fruiting pulse, when productivity tends to remain low compared to the later peak period, can be considered a priming period, when major fruit seed dispersers are "primed" to the presence of food at this tree. The major fruit pulse that follows seems not only to satiate seed destroyers also attracted to the tree but to produce sufficient fruits for the seed dispersers as well. Such "priming" frequency curves are also typical of the fruiting pulses of *Grewia stylocarpa, Canarium hirsutum,* and, perhaps, also *Canarium vrieseanum*. The food species switching by *V. olivaceus* that must certainly accompany this pattern of fruiting in the Caramoan forests is discussed in chapter 11.

Thus almost every fruit food species has a different reproductive strategy, producing many or few seeds, over one or two seasons, in differently shaped frequency-time curves. Combined with different species densities per hectare of forest, these variables are exceedingly important in the foraging tactics of *V. olivaceus* (discussed in chap. 11).

These various factors are expressed in the actual seasonal abundance of edible fruits available to the butaan at each of four different forest sites in which *V. olivaceus* were caught. One or two trees of certain species may produce great seasonal overabundance in some parts of the forest, while adjacent ones have little or no fruits available (i.e., Matingapo I and II in April and May); seasonal overabundance can rapidly change to underabundance in the same block (i.e., Matingapo II, May–June; Mabho I and II, February–March), due to the fruiting or absence of it in a single tree in that area. Thus the butaan is, under some circumstances and in some areas, forced to move from one part of the forest to another; in other areas it is not so forced, due to the presence of particular individual trees. The quality of an area from the standpoint of the butaan depends on the particular fortuitous local mix of fruit-producing species of a remarkably small number of tree taxa (about 12). (As Koelmeyer [1960] shows, the mere presence of a species in a plot does not necessarily insure an adequate fruit crop in many species; some are dioeceous, some fail to mature their fruit every year, etc. *Sandoricum* is a good local example, for not all trees fruit every year.) In some cases this mix may demand relatively minor movements throughout the year as an accommodation to local fruit abundance and in others more extensive movements. Chance alone is probably not the reason why two or more butaans were found in the areas where fruit production tended to be high throughout the year compared to other, even nearby places (i.e., Matingapo II). Fogden (1972) has illustrated the similar patchy and erratic nature of fruit as a food supply for animals in Borneo's secondary and primary forests.

Thus the critical factor in butaan density is the presence of only a few individuals of a few species of fruiting trees, each of which has a slightly different fruiting pulse in terms of fruit production and phenology. Together, however, these species tend to fill fruiting gaps for other species throughout the year (fig. 10-7). Collectively, with the snails upon which the butaan feeds, these trees provide a reliable annual supply of needed nutrients. The surprise is not this success but rather that entire relationships are based on a remarkably small number of plant species (12 maximum) that the butaan eats, compared to the several hundred species that may be involved in feeding other fauniherbivorous reptiles, such a *Cyclura carinata, C. nubila,* and *C. cornuta* on larger islands in the West Indies (Auffenberg, 1982a).

In dipterocarp forests the patchiness of the fruit supply is largely the result of the fact that few tree species bear fruit that is eaten by *V. olivaceus* and those that do are rare. Even for the frugivorous and omnivorous birds in such forests the proportion of trees with edible fruits may be less than 10% (Fogden, 1972). Of over 3000 trees/ha (bhd < 10 cm) in the Caramoan study blocks, only 288 represented the few species whose fruits are regularly eaten by *V. olivaceus* (10%, as in Fogden's study). Although the proportional numbers of trees that might be utilized by birds as food might be higher (McClure, 1966; Fogden, 1972), even for these animals the number of species is small. As shown by other studies on omnivorous scincid lizards (Auffenberg and Auffenberg, 1987), the fruiting species utilized are a small part of the total fruiting species of the forest that depend on animal dispersal of their seeds. The high degree of species-specific fruit-feeding guilds among bats (Bonaccorso, 1979) suggests a similar pattern of utilizing only a small fraction of the total fruit crop of the forest and reflecting the narrow feeding niches of probably almost all tropical forest frugivores.

11

Foraging Strategy

FOR an individual, the most important life processes are to obtain in sustained, sufficient amounts the nutrients needed for growth and maintenance and to reduce the risks that might threaten its life or its effectiveness in obtaining food. For a species, the most important processes are to maintain itself through reproduction and to improve its reproductive effectiveness through natural selection.

It follows that some of the most important studies of the biology of the individual pertain to the feeding and antipredator strategies that the individual utilizes during its life history to obtain food and remain alive. While nothing is known of the butaan's antipredator strategies, many of the preceding chapters bear on its feeding strategy. (MacArthur and Pianka [1966] presented a model that attempted to predict foraging strategy of a consumer based on the density of food species. Later models emphasized the relative value of the food to the predator rather than food density alone. In general, most authors define feeding strategy as the aggregate morphological, physiological, and behavioral adaptations that maximize some measure of food value per unit time [Emlen, 1966, 1968; Rapport, 1971; Schoener, 1971; Cody, 1974; Pulliam, 1974; Ellis et al., 1976; and Estabrook and Dunham, 1976].) Those chapters have stressed the adaptations by which the butaan is able to find, ingest, and digest the several types of foods upon which it regularly feeds. In this chapter I attempt to draw together important factors bearing on this species' feeding strategy.

The primary task of any theory of feeding strategy is to specify for a given animal that complex of behavior and morphology best suited to gather food energy in a particular environment (Kamil and Sargent, 1981). While such studies may take several approaches, this one concentrates on what foods are maximized and how, from the standpoints of behavior and morphology, they

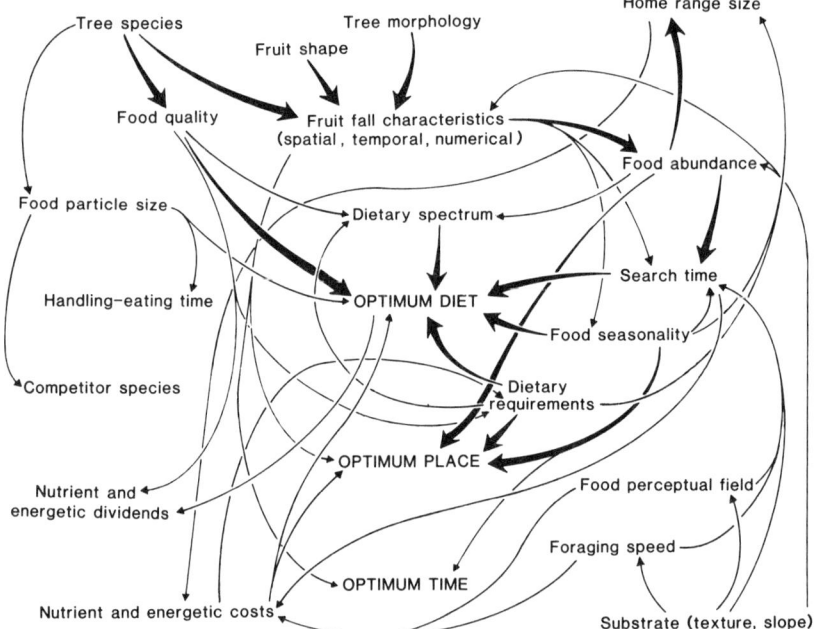

Fig. 11-1. Interrelationships of factors believed to be important in the feeding biology of *V. olivaceus*.

are obtained and processed. Three key aspects to this problem in the butaan are the optimal diet, the optimal foraging space, and the optimal foraging period. A synopsis of what follows is shown in figure 11-1.

Optimum Diet

Morphophysiological adaptations to diet.—With a maximum weight of almost 10 kg, and a length of over 1.5 m, *V. olivaceus* is one of the largest lizards living in the world today (chap. 2). (Pough [1973], reviewing the morphological correlations with herbivory, pointed out that lizard size is one of the most striking. His conclusions were that within lizard families it is the large species that are herbivorous, the moderate ones that are omnivorous, and the small ones that tend to be strictly faunivorous. These categories do not apply well to varanids, because, with the exception of *V. olivaceus,* they are carnivorous regardless of size. Second, *V. olivaceus* is large though omnivorous, and the only other varanid known to eat plant material [*V. prasinus* rarely eats it and then only in captivity] is one of the smallest species in the family.) Despite its size, *V. olivaceus,* compared to other large varanids, feeds on small food particles (Auffenberg, 1981a). The average weight of food particles eaten is about 1/1000th (9.4 g) the average weight of the largest individuals, whereas in *V. komodoensis,* some mammalian prey may be as much as five times more than the weight of the largest individuals. The butaan also consumes less food at each feeding bout than carnivorous monitors (max. stomach food content wt in *V. olivaceus* 2.1% of total average lizard wt; in *V. komodoensis* stomach food weight may be as much as 71% of total weight; in *V. salvator* from the Caramoan area this average is about 19%). This tendency of the partially frugivorous butaan to ingest less food per feeding bout is reflected in its proportionally smaller stomach compared to that of the carnivorous *V. salvator* (chap. 3).

Other anatomical differences between the butaan and the water monitor include the butaan's more massive head, shorter front legs but heavier shoulder girdle, more massive hind legs, and shorter tail (fig. 2-4). From the standpoint of its feeding habits, its proportionally greater head weight is particularly important, for it is due to a well-developed masseter muscle complex. The mass, construction and positioning of these muscles (fig. 3-3) provide the butaan with the capacity to exert great power in its bite. Coupled with the heavy construction of the skull (fig. 3-2), these features make it possible for the butaan to apply as much as 70 kg of pressure through its jaws to the shells of the land snails upon which it frequently feeds (chap. 10). The shells are crushed by a combination of this great muscle power and the highly specialized, blunt, barrel-like teeth of both the upper and lower jaws (fig. 3-5). (The teeth are not an adaptation for frugivory though the butaan eats many fruits. These are swallowed entire, with the skin usually remaining unbruised. Thus the statement of Rand [1978] that herbivorous lizards lack crushing teeth remains correct. Almost all snails are crushed by the teeth, regardless of shell size.) Modification of the bones of the palate allows swallowing of the hard, spherical fruits on which the butaan most commonly subsists (fig. 3-4).

The slitlike external nostril opening, posteriorly located, apparently restricts the amount of debris that would enter the opening if it were rounded and more anteriorly located when the snout is used by juvenile lizards to move forest floor duff in search of insects and by adult lizards seeking land snails (fig. 2-6).

The digestive system is specifically adapted to process efficiently the fruits upon which the butaan feeds most extensively. These specializations include a well-developed caecum, a large intestine of great surface area, and a different organization of the villar surface of the small intestine (fig. 3-7). All are clearly related to frugivory and not to folivory. There are no valvular structures in the gut behind which microbial symbionts reside and multiply in special chambers where they help degrade cellulose (see McBee, 1971, 1977; Iverson, 1982, in press).

It was shown in chapter 9 that at least some fruits upon which the butaan feeds have high levels of tannins and essential oils (i.e., *Canarium* and *Caryota* sp). Some biologists believe that such oils and dozens of secondary chemical compounds (particularly in tropical forests) comprise a major defensive system against insect predation and parasitism. All herbivores must be effective in detoxifying whatever secondary compounds are present in the particular plant foods they eat. Some of these foods have a greater variety of or more powerful toxins than others, even on the same plant (leaves generally possess more toxins than flowers and pollen seldom has any). Since plant predators vary in their ability to detoxify these chemicals, some plants require less energy for processing than others; some require so much energy that the same food may be lethal to one herbivore species but not others (see Janzen, 1978, for general discussion and Auffenberg, 1982a, for examples among lizards). Thus the fruits eaten by the butaan probably represent a wide array of kinds and strengths of defensive compounds, all of which are cost-effective to the butaan in terms of nutrients obtained and energy demands and possible liver damage sustained.

As stated elsewhere (chap. 10), the fruits eaten by *V. olivaceus* fall into two categories, oily and sugary. In general, the sugary fruits (*Grewia, Dracontomelum, Spondias, Uvaria, Malaisia, Sandoricum*) are less preferred (table 11-1), in spite of the fact that they possess generally lower levels of secondary compounds when ripe (explaining the high percentage of insect infestation in wild fruits [see Rudran, 1978]; Redford et al. [1984] suggest that the insects often included in wild fruits are deliberately eaten with the pulp, but there is no evidence that butaans select sugary fruit on this basis). They are eaten by a wider variety of forest vertebrates (*Paradoxurus*, many bats and birds, etc.), so the butaan faces more competitors for these sugary fruits. This

TABLE 11-1. Food preferences of *V. olivaceus* for major fruit species.

Species	Type	Preference Index[a]
Aglaia harmsiana	sugary	0.05
Dracontomelum dao	sugary	0.02
Ficus (all sugar types)	sugary	0.08
Grewia stylocarpa	sugary	0.02
Spondias pinnata	sugary	0.02
Canarium vrieseanum	oily	0.56
Canarium ovatum	oily	0.66
Canarium hirsutum	oily	0.79
Caryota cumingii	oily	1.00
Pandanus radicans	oily	51.10

a. The preference index is calculated as a ratio of a food item in the stomach to its availability in the wild.

pattern has also been demonstrated on Barro Colorado, Panama (Howe, 1982). On the other hand, the endocarps of oily fruits eaten (*Canarium, Caryota,* and *Pandanus*) contain many tannins and phenolic acids. These compounds tend to make small amounts of protein indigestible (large amounts of proteins reduce these effects and may explain the otherwise high insectivory and molluscivory of butaans and hornbills that eat these fruits) and often require rather complicated detoxification processes, presumably accompanied by high energy demands. Few vertebrates, therefore, compete successfully with the butaan for the oily fruits, particularly those of *Canarium* and *Pandanus*.

According to Milton (1979) and Walter (1971) there are two ways for animals to deal with the toxic compounds so prevalent in tropical forests: specialize in one or a few kinds of foods requiring similar detoxification processes or eat small amounts of a number of different plant species. The latter strategy requires a more complex detoxification system, most commonly involving a microbial gut flora important both in degrading cellulose and in parts of the detoxification process itself (Freeland and Janzen, 1974). Among lizards, these are plant generalists, feeding on a wide variety of leaves, fruits, flowers, and buds. Almost all herbivorous lizards of the world (*Cyclura, Iguana, Ctenosaura,* and others; see Burghardt and Rand, 1982) fall into this category, whether the diet is completely restricted to plants or not. Apparently because they are largely folivorous, these genera are all provided with valvular intestines highly modified to house complex microbial floras (Iverson, 1982). These microflora are most often obtained by intraspecific coprophagy in iguanines (Sokol, 1967, 1971; Iverson, 1979, 1982; Troyer, 1982)—hence the absence of coprophagy in *V. olivaceus.*

Most plant specialists are insects (Janzen, 1971a, b, 1973c, 1975b), which restrict their predation either to leaves, seeds, flowers, or fruits of specific plant species or to a small group of related ones. Considering only its plant foods, the butaan is a specialist herbivore. It has a feeding strategy in regard to plant secondary compounds more like that of herbivorous insects than like that of almost all the remaining larger herbivorous lizards of the world (except the algivorous Galapagos marine iguana, *Amblyrhynchus cristatus*), for its plant foods are comprised of only the ripe fruits of a few species of trees. It eats no flowers or leaves. It is therefore not surprising that it lacks specialized structures in its gut to house specialized microbial flora. Most important, its apparent inability to utilize coarse foliage high in fiber may be the chief factor restricting this species to the relatively aseasonal yet floristically diverse evergreen forests within its geographic range. In this respect it contrasts strongly with nearly all other large herbivorous lizards, which are able to occupy seasonally drier and floristically simpler forest associations, including pioneer types whose fruits are adapted for animal dispersal of seeds. That the butaan possesses a secondary compound detoxification system of little effect across a broad spectrum of plant species is suggested by the fact that it feeds exclusively on fruits designed for animal consumption and dispersal.

The dietary spectrum.—The evidence presented in earlier chapters clearly shows that the butaan shifts from a completely molluscivorous-insectivorous diet to a molluscivorous-frugivorous one at about 1 to 1½ years of age. Still, even juveniles are rather specialized, for they are known to feed on only a few species of slugs, land snails, beetles and large spiders—all foraged on the ground. There is no evidence that the butaan feeds on massed social insects, such as termites, as has been demonstrated for *V. rudicollis* (Mertens, 1942) and *V. bengalensis* (Auffenberg, 1983a). The omnivorous *Cyclura carinata* is even known to open active termitaria with its claws in order to feed on the inhabitants (Auffenberg, 1982a), and the related *C. cornuta* feeds on aggregations of caterpillars (Wiewandt, 1977). However, the butaan appears to feed on scattered spiders, beetles, snails, and slugs found more slowly during foraging through fallen leaves on the forest floor.

After analysis of data from lizards of varied diets in other families, Pough (1973) concludes that a weight of 300 g is the upper limit at which insectivory is efficient. This conclusion fits well with data on the ontogenetic shift in *V. olivaceus,* for individuals begin to eat fruit at about this weight and frugivory becomes more or less dominant in individuals heavier than 600 g (although some small lizards eat fruit [i.e., *Lepidophyma smithi,* mean adult total weight 17.5 g; Mautz, 1982]). Molluscs are eaten from near hatchling weight (min. recorded 160 g) to the maximum mass known for both males and females.

Although all individuals will eat animals opportunistically (dead or alive), the diet of all those over one and a half years of age and of both sexes and during all seasons is composed mostly of fruits and molluscs (slugs and land snails). Though birds and rodents are common in the environment, they are rarely eaten. No positive correlation has been demonstrated between butaan size and food particle size, as has been reported for other varanids (Pianka, 1969b; Auffenberg 1978b, 1981a). King (in press) reports no predator-prey size relationship in the largely insectivorous *V. acanthurus* either.

Because half the food of *V. olivaceus* is fruit, much of its food does not come in nutritionally balanced packages. But its total diet is reasonably balanced between easily assimilable carbohydrates (high-sugar-content fruits), required fats (oily fruits), and high-protein and calcium-rich flesh (land snails); the proportion is approximately 1 sugar: 2 protein: 4 oil (by wet weight of all gut contents examined; see details chap. 9).

It has been shown earlier that adult butaans feed primarily upon only 13 food species in a forest remarkable for its great species diversity; of these, one is a snail, and 12 are plants. This selection of a few food species out of many thousands of potential ones and the spatial dispersion of these few food species within the forest both have great importance in determining where and for how long an individual butaan searches for food. As discussed below, the foraging patterns of the adults make it clear that they have learned (1) the food distribution patterns within their activity ranges, and (2) how to gauge the length of time they should spend searching and feeding in a particular place in such a way as to maximize net energy gain.

Of as many as 3500 individual trees/ha in the local forest, the fruits of only about 1–6 individuals/ha are used by the butaan as food. However, these few trees may produce from 160 to over 31,000 food particles (fruits)/ha, though not all at the same time (see below). Of these a large proportion is predated by other animals while the fruit is still in the trees, so that the average total number of fruits falling to the ground and available to the butaan varies from 12 to 6000 per ha (table 10-4). These fruits are scattered in areas under and near the parent tree that vary from 78 m^2 to 1660 m^2, or from 0.7% to 16.6% of a ha. What these facts mean is that most fruit foods will at one time be located on only from 0.4% to 11% of the entire daily home range of adults, depending on the particular tree species bearing fruit at that time. The average distance of these food particles from the parent trunk varies from 2 m to 8 m. Thus plant food resources are localized in space. This fact, in addition to the movement patterns of telemetered individuals, suggests that adults probably know the locations of the trees within their home ranges that are food producers, and they search the areas beneath them regularly for fallen fruits (see below). Land snails are more evenly scattered and available all year. They are found resting in hollows and crevices in rocks or under forest duff, and their location probably requires the greatest energy expenditure for food of all the food eaten by the butaan. Food pursuit is not a factor in the feeding strategy of *V. olivaceus*.

Rank correlation analyses of the number of items of the major food species per stomach of *V. olivaceus* with food diameters, food particle weights, and the number of total weeks fruits are available on the ground suggest that selection is not made on these bases (Rho = 0.02, 0.09, and 0.03, respectively). (More important than the number of fruits on the ground is their lasting quality in a reasonably ripe state, for *V. olivaceus* tends to feed on only perfectly ripe fruits.) Nor is there a statistically valid correlation between items in the stomach and the number of fruits on the ground per average home range area (Rho = 0.22). However, when the fruits are separated into oily and sugary types and these are analyzed separately, utilization of both types is correlated (Rho = 1.0) with home range. This suggests that *within each category* the species are eaten in direct proportion to their availability. So the only factor in fruit selection seems to be fruit type (oily or sugary), not fruit species. Nor is selection based on caloric content only, for within each category there is considerable variation among the several species (see tables 9-11 and 11-5). These results do not support earlier models of optimum diets (MacArthur and Pianka, 1966; Emlen, 1966, 1968; Schoener, 1969a, b, 1971; Estabrook and Dunham, 1976), based on the concept of maximizing *energy yield* per unit foraging time. Rather, *V. olivaceus* feeding seems to fit better the models of optimum diets stressing *nutrient mixtures* (Westoby, 1974; Pulliam, 1975; Milton, 1979), often with dietary constraints related to secondary compounds. The models of Westoby (1974) and Hughes (1979) are particularly interesting, for they are based on foraging patterns mediated by food location learning, as seems to occur in *V. olivaceus* (see below). They show further that the chances of eating the kinds of rare foods consumed by the butaan are not the same for each encounter but that they probably change during each feeding period because the individual is selecting meals rather than a series of nonrelated foods. Food selection by adult butaans is probably constantly adjusted each day on the basis of whether recently eaten foods were high in oils or sugars. Such adjustments control the composition of the next "meal." Carrion feeding (probably rare on the basis of the extremely low frequency with which this species comes to carrion traps set for the sympatric *V. salvator*) and mollusc feeding serve to increase proteins in a diet otherwise low in them (Field, 1967). Westoby's model, based on qualities rather than quantities of food species and reflecting what he calls a "fallible nutritional wisdom," fits the butaan data well when viewed in light of the requirements of both the energetics of the individual and long-term detoxification phenomena (as it does in the feeding strategy of the largely folivorous iguanine *Cyclura carinata;* Auffenberg, 1982a).

Food-specific preference is an important factor to the butaan, for finding and eating snails is clearly different from finding and eating fallen fruits. Switching allows self-selection from a number of possible food choices.

In most studies to date, vertebrate predation on a particular prey species generally increases as a function of prey density in a characteristic way in which the lower part of the curve is a broad sigmoid and the upper, later part a relatively flattened plateau. Such curves suggest that normally the total amount of food eaten remains the same over all food densities, because as the density and consumption of favored food rises, consumption of less favored alternative food declines. What is most important is that even at the highest food densities, the individual is always taking at least some alternative foods. Similar conclusions have been reached for studies on other herbivorous lizards (Auffenberg, 1982a). This eating pattern has been taken to suggest that individual lizards are progammed to maintain nutritionally balanced diets.

In conclusion, the diet of the butaan includes one group of fruits high in sugar, another high in oils, and snails high in protein and calcium. On the basis of the few total species utilized as food, the butaan is highly specialized. Its specialized diet apparently reflects a need for lipids, as well as the fact that competition is low for the fruits from which it obtains them. The lack of competition for these fruits is probably due to this lizard's ability to detoxify a high level of certain secondary compounds in *Canarium* and *Pandanus* that many other forest vertebrates apparently find difficult to do. Second, the specialization is also based on the need for easily processed carbohydrates, which it obtains from sugary fruits of a few species for which there is much competition. Third, its diet is based on the need for proteins and calcium, which it secures almost exclusively through predation of land snails (for which it is morphologically specialized, particularly in its skull). Though some plants have high levels of protein in their leaves (Marshall and Hook, 1960), the protein is available only if a gut microbial flora is developed—which is absent in the butaan. However, the rigidly narrow diet of this species requires other physiological and morphological adaptations that can be demonstrated throughout the entire digestive system. The feeding pattern during the year is based partly on continuous adjustment by the butaan to maintain a nutritional mix of qualitatively different food items. However, within each food category the actual harvest rate of each acceptable species is determined by absolute abundance of fallen fruit on the ground—not because of any preference for one species over another (see below).

Optimum Foraging Space

Spatial and temporal components of the search pattern.—*V. olivaceus* is limited by ecological factors in the techniques of feeding behavior available to it. Unlike *V. komodoensis,* it cannot raise its head above the general level of the vegetation and observe game several hundred meters distant. Nor does wind carry the scent of food to it from great distances, as it does to all carrion-feeding varanids. So from the standpoint of foraging, the butaan's perceptual world is greatly circumscribed, reaching, at the most, 2–3 m in any direction. The only cue that more food may be available at one time than another is that storms often cause many ripe fruits to fall to the ground. Such storms often result in increased movement on the part of local adult butaans (chap. 5).

Though fruit is most abundant in the trees, the butaan does not regularly search for food there, as does the local civet cat or even many folivorous-frugivorous lizards (*Cyclura,* Auffenberg, 1982a; *Iguana,* Rand, 1978; and *Hydrosaurus,* Auffenberg, field notes). Though food might be found from the deep rock crevices and caverns to the high emergent canopy, almost all foraging takes place on the forest floor. To the butaan, the herb layer is a dense, dark environment in which it searches for fallen fruits and snails. In this zone of highest humidity the butaan spends most of its time. Generally cooler than the zones above it, the herb layer is also the most buffered, so its temperature fluctuation is small.

The next higher forest zone is comprised mainly of stems and trunks, through which air passes more freely than through the lower forest layers. Most tree trunks support an important forest canopy layer of dense leaves, which produces most of the fruits upon which the butaan feeds. Though butaans ascend up to and beyond it, they rarely spend much time there and, as far as is known, rarely feed in it. It is, however, the zone in which they often hide in arboreal thickets of epiphytic growth, climbing bamboos, and scandant shrubs.

Extending far above the canopy are emergent trees in which the butaan spends little time, though it sometimes basks there. No fruits eaten by the butaan are produced there (most emergent trees have wind-dispersed seeds).

The feeding patterns of several telemetered adults (chap. 5) show that there are probably only 1–2 daily feeding forays, though most fruits are found and eaten in the morning, and that the mean distance covered in each foray is short compared to the distances covered by other semiherbivorous lizards (Auffenberg, 1982a). Forays are short because the butaan remains in a nearby tree or crevice, rather than returning to a burrow or some other site-specific nocturnal refuge. Observation of captives shows that the mean width of the search path is 0.7 m wide and that the daily mean foraging area of adults is only about 78 m^2. Because the fruit trees upon which they feed are usually scattered, adults move from areas without much fruit to areas with more. They remain in the new area for varying lengths of time, depending on the mix of species and tree characteristics present.

The mean calculated overall annual home range of only 1.48 ha is exceedingly small for a reptile of this size, only 6.8% of that of equal-sized individuals of a carnivorous species (i.e., *V. komodoensis*), only 0.06% of the home range of the deserticulous *V. griseus,* and about 7.5% of adult *V. gouldi* in Australia (see chap. 5 for details). The butaan's small home range is surely associated with its frugivorous habits. Though no data are available for lizards, both Snow (1965) and Charles-Dominique et al. (1981) found that frugivorous mammals also have smaller home ranges than carnivorous ones. Herbivorous lizards are often more sedentary than insectivorous ones (chap. 5).

As in other varanid species, female butaans move shorter distances, cover less area, and are active less often than males (see chap. 5). And, whereas females forage mainly in the morning, males tend to be more uniformly active throughout the day (fig. 5-4). Females, because they are generally smaller, apparently require less food and find it in a shorter period of time after waking from the night's sleep. Adult males, being larger, require more food and must spend more time searching for it. A number of social factors (such as male territorial defense) may demand more time of the males than of the females. And females do not feed at all (or very little) for several weeks preceding and following oviposition (see chap. 8).

As a result of their increased activities, males are exposed to more risks than females. Though fitness increases with increased activity of one behavioral type (such as foraging), the incremental benefits from additional foraging decrease in direct proportion to the increase (Real, 1980). Charnov (1976) argued similarly that increased foraging (in a specific patch) provides diminishing returns after a certain point. Thus one expects butaan males to forage in one fruit-fall area until the energy intake falls to below the expected level and then switch to another patch. The female would, of course, do the same but, because of lower daily food demands, may have fallen fruit available for a longer time. Though both strategies obviously have advantages, the female's reduces the uncertainty of finding an additional food source. The female's strategy tends to reduce the risks associated with moving about—predation and cost-effective energetics. It follows that male life expectancy would be shorter than that of females (as it is in many other vertebrates). But there is at present no way to determine the age of *V. olivaceus* individuals, and size would be a poor measurement for age in this case, because males grow faster than females.

Real (1980) argues that an animal with either little time or little energy available for foraging will always be conservative and avoid risk, foraging only for known food types in known places. Such is the strategy of the butaan: much of its foraging behavior increases the certainty of finding food fruits by concentrating the search to a few species, upon which few other local organisms feed, and by restricting movements once fruits are found. And within the species, females have more restricted total home range than males because of their smaller size and their egg-carrying role.

Another important factor in the feeding strategy of the butaan is learning food location, at least for the most common foods eaten within the home range. Individuals who learn to search in only a few food-rich sites at the few fruit trees in their home range are unquestionably more fit than those searching randomly over a wide area with highly variable food densities. The most interesting search behavior is repetitive foraging at the same tree. The apparent need for food variety and the impingement of seasonal factors (availability of fruits, etc.) suggest that rigid repetitiveness is not always desirable or possible. The location of new food sources depends on some randomness in search patterns, though this is not always energetically cost-effective (see below). In some cases these discoveries may be facilitated by social factors, such as territorial boundary patrolling and the annual search for mates. As pointed out, this behavior is expected more from males than from females.

Long-lived carnivorous monitors would also benefit from learned associations of food concentrations and time of day or season. In fact, much of the hunting behavior of adult Komodo monitors seems related to the location of previously successful hunts (Auffenberg, 1981a). In the largely folivorous iguanine *Cyclura carinata* there is much evidence that certain individuals repeatedly return not only to the same tree to feed but often to the same part of the tree (Auffenberg, 1982a). Like many other vertebrates studied both in the wild and in the laboratory (see Curio, 1976), butaans are apparently able to associate certain parts of their home range with greater and lesser (or no) food density and to search for food only in places where past successes suggest that continued searching is worthwhile. A detailed knowledge of the home range is probably necessary to assure a constant supply of both primary and alternative food species, and in some carnivorous varanids this knowledge is regularly put to use in selecting ambush sites for large vertebrate prey (Auffenberg, 1981a). Both carnivorous and frugivorous monitor lizards regularly miss some parts of the home range during normal searching patterns. Lizards may expend more energy than is justified as they search for these missed food patches on the chance that they may find new food sources (Croze, 1970; Morrison, 1978).

In its search for fallen fruit, the butaan can concentrate its survey in a small area (tables 11-2 to 11-4). This behavior is strengthened by the success of finding one fruit; among all foragers studied so far, a successful search leads either to slowing down forward movement, decreasing the turning radii of subsequent movements, or increasing the width of the search path. In *V. bengalensis*, all three movements have been observed after a successful food search (Auffenberg, 1984). Table 11-2 shows the relationships of several important parameters in the butaan's area-concentrated foraging for fallen fruit. Note that only under *Caryota cumingii* trees can the butaan expect to cover

TABLE 11-2. Area-concentrated search for food by *V. olivaceus* under different tree species. The average search area per foray = 78.2 m^2.

Tree species	Distance between fruits (m)	Maximum area of fallen fruit (m^2)	Avg. number of feeding forays needed to cover entire fruit dispersion area
Dracontomelum dao	2.17	1,661.9	20.00
Canarium ovatum	1.68	980.8	12.50
Canarium vrieseanum	14.09	380.1	5.00
Aglaia harmsiana	0.26	237.8	3.04
Grewia stylocarpa	0.20	227.0	2.94
Canarium hirsutum	1.68	162.9	2.08
Caryota cumingii	0.77	78.5	1.00
Pandanus radicans	0.51	68.4	0.08

TABLE 11-3. Estimated search times for fruits of species commonly eaten by V. olivaceus.

Species	Maximum fruit dispersal		Fruits under canopy	
	Number per m^2	Estimated search time (minutes)	Number per m^2	Estimated search time (minutes)
Dracontomelum dao (s)	0.46	0.29	2.27	0.06
Canarium hirsutum (o)	0.59	0.23	2.08	0.06
Canarium vrieseanum (o)	0.07	1.94	0.26	0.52
Canarium ovatum (o)	0.63	0.22	1.67	0.08
Caryota cumingii (o)	1.29	0.10	7.11	0.02
Grewia stylocarpa (s)	4.90	0.03	12.06	0.01
Aglaia harmsiana (s)	3.89	0.03	25.18	0.005
Spondias pinnata (s)	2.70	0.05	11.78	0.01

NOTE: (s) = sugary fruits; (o) = oily fruits.

the entire area in which fruits occur in one average foraging foray; up to about 20 forays would be necessary to cover the area in which *Dracontomelum dao* fruits occur. If the butaan makes only a few forays each day, many of the fruit species eaten would spoil on the ground before they could be harvested by a single butaan. In the tables the fruit species are ranked in terms of the ease (number of feeding forays, or energetic costs) by which the butaan is able to explore the entire area under the tree. However, for higher fruit densities this information is academic; with an average handling time of 8.6 sec/item, the maximum number of fruits that can be handled per minute is only 7. With an observed average fruit intake of only 4.4 per "meal," at fruit densities of greater than $10/m^2$ a "meal" can be ingested in about half a minute. Food species for which this time is possible are *Spondias pinnata, Aglaia harmsiana,* and *Grewia stylocarpa* (table 11-3), all sugary fruits, for which estimated search time is about 0.01 min/fruit. That is, the fruit is so prevalent under the canopy that walking forward is hardly required, for contact with fruits per minute is much higher than the rate at which they can be manipulated and swallowed. The oily fruits are less dense and require more search time (table 11-3).

Though these data give the impression of superabundance of foods high in sugar, none of these trees fruits all year. Furthermore, within each fruiting pulse, densities of fruits on the ground vary significantly from week to week and from one location under the tree to another. Since foraging success varies directly with fruit density, searching intensity will vary both temporally and spatially under the same tree (table 11-4). On the basis of fruit density/m^2, the areas with the highest, most reliable densities are always near the tree trunk, regardless of species or the point in the entire fruiting cycle. However, fallen fruit dispersion varies from one species to another. Table 11-4 shows that the oily fruits of *Canarium hirsutum* fell over a 7-week period, with the peak drop in the fourth week. The maximum fruit item contacts per minute (at an average foraging speed of 10.5 m/min) in the density zones closest to the trunk varied from about 28 (week 6) to 7 (week 1). Within this central zone little movement is required for the butaan to obtain its entire daily fruit needs. In the next meter-wide zone, fruits contacted per minute drop to 1.5–6.1, and slight forward movement is required. The same general pattern is found in other species upon which the butaan feeds. Thus in *Caryota cumingii* (oily) the number of fruit contacts per minute within the "minimum energy expenditure foraging area" (i.e., that area requiring "no movement" to obtain sufficient daily fruit requirements) varies from 7.5 (3 m from trunk, week 1) to 169.6 (week 2, at trunk). In this species not only is fruit abundant at the trunk throughout the fruiting period of seven weeks, but this "very abundant" zone extends 3 m from the trunk during the early part of the fruiting pulse. The peak production occurs between weeks 2 and 4. In *Dracontomelum dao* (sugary) the minimum energy line extends to 10 meters from the trunk during a sharp fruiting peak in the third week.

The data show mainly that there are both temporal and spatial patterns of fruit fall beneath individual trees bearing fruit eaten by *V. olivaceus*. Most important, they show that once a fruiting tree has been located, the butaans rarely have to move far, if at all, beneath the canopy to obtain a sufficient meal. Therefore, the major time and energy expenditures are moving from one tree to another—either because the fruiting pulse is over, or because the diet requires a nutritional mix of sugary, oily, and proteinaceous foods. Finally, the data also show that not only are the sugary fruits produced in greater absolute abundance than the oily ones but that these high densities are more widely disbursed on the ground (though over a shorter time span). In spite of this wider availability, oily fruits are taken in higher proportion than sugary ones, showing the great preference of this species of frugivore, though caloric content (for equally aged fruits) is not significantly different (chap. 9).

Sugary and oily fruits are found together in the gut about as much as expected by chance. Butaans do not seem to shuttle back and forth between fruiting oily and sugary trees during a day. Rather, they remain near one tree for a long period of time, corresponding more or less with the length of its fruiting pulse (chap. 5). Only oily fruits and animal prey occur more commonly than expected (either separately or in combination) at the 0.05% level (chap. 9).

Table 10-3 expresses the differences in abundance of food species in different parts of the local mixed dipterocarp forest, based on the transects studied during the project. The general rarity of these species among the total flora is not only of considerable importance in determining the distance that must be traversed between potential or real food patches; it has a bearing on territo-

TABLE 11-4. Number of fruits contacted per minute by a foraging *V. olivaceus*, estimated using average foraging speed (10.5 m/min) and path width (0.7 m). Underlined values represent the lowest fruit density at which the forager needs to move only rarely to find fallen fruits.

	Radial distance from trunk (m)	Week of fruiting cycle										
		1	2	3	4	5	6	7	8	9	10	11	12
Canarium hirsutum[o]	1	7.1		9.1	9.1		27.9						
	2	6.1	1.5	2.9	2.9	2.9	2.9	2.9					
	3				3.7	0.8	7.5						
	4				1.5	5.1	1.5	2.9					
	5				0.7								
	6				0.7								
Caryota cumingii[o]	1	56.5	169.6	131.9	103.6	28.3	47.1	9.4					
	2	15.6	37.4	37.4	3.1	9.3		3.1					
	3	7.5	3.7	16.9									
	4	6.7	2.7										
Spondias pinnata[s] (1)[a]	1	56.5	47.1	18.8	47.1	56.5	216.7	405.2	1121.3	716.1	172.0	103.6	
	2	15.6	12.4	6.2	15.6	40.5	57.2	127.7	289.6	258.5	277.2	81.0	
	3	9.4	13.1	9.4	22.5	35.6	97.5	135.0	157.5	238.1	150.0	56.2	1.9
	4	9.3	4.0	4.0	8.0	17.7	54.7	77.4	77.4	81.4	28.0	13.3	6.7
	5	1.0	3.1		4.1	2.1	40.0	28.0	49.8	23.9		9.3	4.1
	6		0.8		1.7	1.7	7.7	17.0	31.5	6.8		1.7	0.8
	7				0.7		1.4	11.5	15.1	2.9			
	8						0.6	4.4	2.5	1.2			
	9						0.5						

NOTE: s = sugary fruits; o = oily fruits.
a. Three observations of *Spondias pinnata*.

(*continued*)

TABLE 11-4 (continued)

	Radial distance from trunk (m)	Week of fruiting cycle											
		1	2	3	4	5	6	7	8	9	10	11	12
Spondias pinnata[s] (2)	1	56.5	376.9	320.4	263.8	103.6							
	2	3.1	99.7	149.5	84.1	59.2	3.1						
	3	5.6	58.1	45.0	30.0	35.6	3.6						
	4	4.0	40.0	30.7	6.7	14.7	1.3						
	5	1.0	30.1	12.4	2.1	4.0							
	6		21.3	11.9	1.7	0.8							
	7		7.9	5.0	2.1								
	8			1.2									
Spondias pinnata[s] (3)	1	56.5	508.8	282.7	131.9								
	2	15.6	196.2	81.0	46.7								
	3	3.6	133.1	48.7	5.6								
	4	4.0	36.0	6.7	4.0								
	5	1.0	33.2	7.3	1.0								
	6			2.5	0.8								
	7			4.3	0.7								
	8			1.7									
	9		4.0	2.2									
	10		2.9	1.0									
	11		6.2	0.4									
	12		8.9	0.8									
	13		4.8	4.1									
	14		5.5	1.7									
	15		1.9	0.3									
	16		1.5										

	Dracontomelum dao[s]						
1		384.1	28.2	103.2	28.3		
2	17.1	260.2	364.3		<u>34.7</u>		
3	13.1	28.1	24.4	5.6	3.8	9.3	9.4
4	5.3	14.7	18.6	2.7	2.7	3.8	<u>3.1</u>
5	6.2	<u>18.7</u>	13.5	5.2	3.1	1.3	3.8
6	<u>16.2</u>	7.7	11.9	5.1	3.4	1.0	1.3
7	5.0	5.0	10.8	5.0	0.7	1.7	
8	5.8	7.1	10.3	3.2		0.7	0.8
9	2.2	4.9	10.4	2.7	1.1		
10	2.4	7.3	<u>10.2</u>	2.9	0.5	0.5	
11		4.0	3.1	1.8			
12	0.8	4.1	8.9	2.4	0.4	0.4	
13	1.1	3.0	8.2	2.2	0.4	0.4	
14	1.0	1.4	6.6	2.4	0.3	0.4	
15	1.0	2.6	6.1	1.3	1.3		
16	1.2	1.8	6.6	1.2	0.3		
17	0.8	2.3	2.8	0.3	0.3		
18	0.8	3.2	0.8		0.3		
19	1.5	1.0	0.5				
20	0.5	0.2					
21	0.4						
22							
23	0.6						
24	0.2						

NOTE: s = sugary fruits; o = oily fruits.

riality and butaan home range size as well. That a relationship exists between home range size and energetics has been shown for many organisms (McNab, 1963 for mammals; Pitelka, 1942 for birds; Simon, 1975 for lizards). The relationship between food abundance and home range size has been demonstrated for at least one herbivorous lizard, *Cyclura carinata*. A comparison of home range data in different environments (Iverson, 1979) with food abundance data in the same study area (Auffenberg, 1982a) shows that lizard activity ranges are clearly larger in areas with less dense food species or less diversity in food species. The length of the feeding foray of the largely grazing land tortoise *Gopherus berlandieri* is also directly related to local food abundance and diversity (Auffenberg and Weaver, 1969). Dugan (1982) estimated that the partially herbivorous *Iguana iguana* spends only 1% of its waking hours feeding, and Moberly (1968a, b) showed that captives of the same species are active only 10% of each day. Minnich and Shoemaker (1970), Nagy (1973), and Grenot (1976) have demonstrated the same pattern in small iguanine herbivorous species. While there are no data available showing that home range size or feeding foray lengths differ from one area to another based on food abundance, one can assume that such would be the case. Furthermore, one would expect seasonal changes in territory boundaries and foray lengths depending on food availability, though data are not available for this species. Green and King (1978) have demonstrated significant seasonal changes in the home range size of *V. gouldii*. Oppenheimer (1982) suggests that in frugivorous-insectivorous primates it is food availability during lean years that determines average home range size.

The butaan is not a true refuging organism (see Hamilton and Watt, 1970), since it does not return to a specific shelter between feeding bouts. Instead, it finds refuge in whatever shelter is available nearby. Nevertheless, the undoubtedly high predation rate characteristic of tropical forests probably results in selection pressure that reduces the distance from the temporary nocturnal refuge to a local food source (= commuting distance). Morrison (1978) showed that the search time component of a foraging foray was more important than the commuting time component. That is, energetic costs were reduced if searching for a closer identical food patch was avoided, since the energy saved in commuting time was more than counter-balanced by the energy lost in increasing search time. The separation of commuting and searching times was not possible for *V. olivaceus*, but in any event, because of shelter use as described above, commuting time is probably unimportant. So it seems that the butaan is able to reduce that part of the foraging foray that is ordinarily the most expensive energetically. However, it is necessary for the butaan to shift regularly the location of its feeding forays to accommodate the need for both oily and sugary fruit types.

Though the butaan seems to search for food in certain places, it also searches for specific kinds of food; i.e., an object-concentrated rather than an area-concentrated foray. Though most optimization models stress food type rather than food patch, it is generally conceded that many mathematical models are too simplistic to provide much guidance in studying behavioral mechanisms (Krebs, 1973). Some students (such as Curio, 1976) are of the opinion that food exploitation stresses food patches rather than food species and that animals rarely rank food patches on the basis of profitability or contained species but rather on the contained food density. Hence a predator under these circumstances would deplete patches of its foraging area in the order of declining food abundance, beginning with those of highest density (Kamil and Sargent, 1981). While density-dependent area concentration is clearly important in the feeding of the varanid lizards I have studied, it is apparent that food quality plays a role equally as important as food quantity. The butaan seems to demonstrate strong choices in regard to food chemistry (oily over sugary) but within each category seems disinterested in any particular species, taking fruits in direct relationship to the number found on the ground (density-dependency).

In *V. bengalensis* the perceptual field of the moving, foraging individual is in the form of an undulating tube, with the amplitude of the movements determined by the individual's size and both the relative and absolute distances between the front and hind legs, as well as the length of the protruded tongue (Auffenberg, 1984). Because the foraging movements are similar in *V. olivaceus* the perceptual field must be similar, though in reality perhaps more limited because of the density of tropical evergreen forests near ground level. This field is increased by the lateral undulations of the body while the individual is moving and is further enhanced by lateral "casting" movements, in which it may remain in the same place while swinging its head, neck, and tongue back and forth in an area-concentrated search pattern.

Probability of feeding success depends upon whether recognition of food is made by contact or by perception from a distance. Although butaans are able to detect movements from as many as about 100 m (chap. 7), food is usually recognized at shorter distances ($\bar{X} = 0.35$ m, based on observations in captivity). For the butaan the probability of contact with fallen fruit can best be expressed by the formula $E^T = 1/2\ rav$ (Morrison, 1978), where E^T = estimated search time (in min) per contact, r = fruit detection radius (0.35 m), a = density of fruit (in number of fruits per m^2), and v = foraging speed ($\bar{X} = 10.5$ m/min). Table 11-3 provides data resulting from analysis of search time for fruits of each of the major species consumed by the butaan. In every instance it is clear that once a fruiting tree is located, search time (E^T) is minimized.

Optimum Time

Within the equable forests in which the butaan lives there are few constraints regarding the periodicity of foraging behavior, though fruits are not equally available throughout the year. However, unlike the situation of many carnivorous varanids, there is no need to correlate daily foraging activity with movement or sleeping patterns of the prey. In species such as *V. komodoensis* the complex behavior of their large mammalian prey necessitates a concomitant complex daily foraging behavior by the monitor. These monitors must be at the right place in the forest at the right time of the day to have any chance of success in obtaining food. And to make matters more complex, these times and places change depending on the activity patterns of each of their major prey species. The land snails on which *V. olivaceus* feeds rest in rock hollows all day, and it is here that they are found by foraging butaans. Similarly, once a fruiting tree has been located within the normal home range, individuals usually remain there, feeding on the fallen fruit below the crown until production ceases.

It is thus no surprise that, unlike most other varanids that have been studied, the butaan exhibits a broad, unimodal foraging pattern during the entire day, without any significant differences in movement intensity from 0800 to 1600 hours (figs. 5-3, 5-4, 5-6); all hours are almost evenly represented (but see below), for there are no apparent pressures to forage here at one specific time and there at another.

There are no significant changes in movement patterns during the seasons—not because of the popularly assumed uniformity of the annual climatic pattern (actually quite seasonal; see chap. 5) but rather because fruits are available (though widely dispersed) in the home range of each individual throughout the year.

Basking is not an important part of the daily time budget, due partly to the generally equable temperatures. From 5% to 55% of the lighted part of each day is spent in dark shelters in rocks or hollow trees (chaps. 4-5). Thus much of the day is not spent in foraging but in safe retreats where the easily obtained foods are digested under thermal conditions that rarely require long exposures to predation dangers encountered by other varanid species.

Critical thermal maximum body temperature (41.6–42.4°C) is the same as for *V. komodoensis* (41.7–42.3°C), a species living in an open habitat of high isolation—in spite of the fact that the ecologies of these two species are different. However, their behavior differs as they approach CTM: *V. komodoensis* acts as though it is in great stress and seems aware of the potential danger of overheating, but the butaan remains passive, seemingly oblivious to the potential danger. This difference clearly reflects the maximum temperature extremes to which both species are subjected during foraging; the ambient temperatures in the butaan's habitat never reach CTM, whereas for *V. komodoensis* they do so regularly. The body temperature of *V. olivaceus* is directly correlated with the vertical movement of individuals from the forest surface high into the canopy (generally higher maximum temperatures) or deep into fissures in the ground (generally lower temperatures). The body temperature varies from 27.8°C to 38.2°C (fig. 5-11) in an environment that ranges from 24.3°C to 38.0°C from below the ground to the top of the canopy. This body temperature is significantly below CTM and well above minimum operative ambient diurnal temperatures. Thermoregulation does seem to be a major part of the activity pattern during days of the year. Water stress is not a critical factor; water is always available, unlike the highly seasonal availability in the environments of many other varanid species.

However, data provided in earlier chapters (as well as in table 11-4) suggest that fruit is not evenly available throughout the entire fruiting pulse. There is always a peak period, when fruit fall is higher than at any other time. Both the height of the peak and its position within the pulse vary intra- and interspecifically, so there is an optimal time to search for fallen fruit below trees of each species. In some species the peak is never very high and often not well marked (i.e., usually the oily types), whereas in others a peak may be pronounced. In most species this peak occurs some time after the beginning of the fruiting pulse, usually during the third week; in others the pulse may start out strongly. In general, the frequency curve is skewed toward the earlier rather than the later part of the fruiting pulse. Thus the optimum time to forage for fallen fruit is several weeks after fruit fall has started. However, the number of new fallen fruit rapidly decreases during weeks 5–7. In some species there seems to be a "priming" period during which the fruit fall is minimal, as though the tree is adjusting its fruit crop maturation to an initial period of advertisement (chap. 10).

Not all species used as food by the butaan produce fruit at the same time (chap. 10). Rather, the fruiting periods of each species are scattered in such a way that, while fruit is available during nearly the entire year, there are rarely periods when more than four species are fruiting at the same time (fig. 10-7). As a result, there is a time of maximum fruit fall of each species on an annual basis, in addition to a period of maximum fruit fall/week for each species. Some species fruit explosively and synchronously. Other species fruit explosively but asynchronously (the local strangler figs, *Ficus indicus, F. benjaminus*, etc., for example); their fruiting pulse is usually short (fig. 10-6). Though many fig species fruit only once a year, others fruit several times (T. Auffenberg, MS; McClure, 1966; Fogden, 1972).

Though there is general uniformity of fruit fall throughout the year, fruit is minimally available during the northeast monsoon (October–December)—the wettest time of the year—when fruit crops are generally much reduced.

This reduction can be seen in such parameters as number of fruits per species per gut and number of alternative (nonfruit) foods in the diet. The largest amount of fallen fruit is found during the earlier southwest monsoon and the period immediately following it (May or June through August). Comparison with other regions suggests that such seasonality in fruit availability and utilization is probably common to most tropical forests (Leck, 1972). Terrestrial habitats with constant food supply throughout the year have not yet been rigorously demonstrated for any part of the world. If such regions do exist they are probably small, for climatic data (Rumney, 1968; Walter, 1971) suggest that only minute parts of the tropics are sufficiently unvarying in rainfall to permit similar levels of fruiting throughout the year (Leck, 1969, 1970). The Caramoan area is such an area, for there is clearly a seasonal pattern in fruiting cycles. The feeding pattern of the butaan is synchronized with the periods of maximum plant food availability, although greater activity of land snails during the rainy seasons may be important in determining the monthly food pattern and seasonal insect flushes are undoubtedly important to juvenile *V. olivaceus*. Other than a greater proportion of land snails eaten during the late and heavy rainy season of the northeast monsoon, there is no real evidence that foods opportunistically exploited at this time are ignored during other parts of the year.

It is likely that in most years there is an abundance of fruit from May through August and a scarcity in November and December. Not only for the butaan, but for most species that include fruit in their diet, the late wet season is a time of low fruit availability and increased competition for the fruit sources that remain (though this cannot be severe, for growth seems continuous throughout the year [chap. 7]). Exactly the same conclusions were reached by Foster (1973, 1982) in his studies of fruiting phenology in Panama. Indirect effects of low fruit production might include a decrease in fruit-eating insects and peak availability for carnivorous vertebrates feeding on frugivorous ones.

Unusual weather may alter normal fruiting patterns and would thus have important effects on frugivorous animals of the forest. A rainy "dry" season in Panama results in few trees setting their fruits (Foster, 1973), apparently because the wet weather puts the trees into a vegetative rather than a reproductive condition. While out-of-phase wet weather was not experienced during this study, it could occur. Presumably due to the effects of El Niño (Canby, 1984) in early 1983, the weather in the Caramoan area was drier than normal. It may have changed the fruiting pattern slightly, though evidence suggests that such changes were probably insignificant. Thus it is likely that too much rain during an ordinarily dry season has a more profound impact on local frugivorous animals than an abnormally dry season has.

The abdominal fat reserves of *V. olivaceus* have been shown to vary seasonally, with the highest values occurring in October and the lowest in June and

July; males display more extreme peaks at both high and low values (fig. 3-11). In general, butaans tend to be least active when body fat deposits are heaviest. They are involved in various reproductive behaviors when fat bodies are lowest. While many explanations have been advanced for the deposition of body fat in lizards (chap. 3), the best one for this species is that the accumulation curves are related to food abundance and quality. The greatest variety of foods is available from July through September (table 9-5) and corresponds with the body fat peak in October. The steepness of the accumulation curve in males is probably due to the higher feeding rate of males than of females during all months of the year. This rate is further suggested by the body fat mass curves of *V. salvator* from the same area, which are clearly correlated with the seasonal availability of their animal prey species (fig. 3-12).

Fat deposits in the butaans are larger than those in *V. salvator* throughout the year, suggesting that food supply is more regular for the former than for the latter. This view is reinforced by the seasonal distribution and numbers of empty stomachs in both species.

Fruiting seasonality is a major factor in the biology of the butaan. Not all food trees are fruiting at the same time, nor is their fruiting spread evenly throughout the year. The same pattern is well documented in plants in many other tropical forests (Snow, 1965; Croat, 1969; Daubenmire, 1970; Foster, 1973; Burger, 1974; Frankie et al., 1974). Differences in seasonal abundance have also been shown for tropical insects (Janzen, 1973a, b; Smythe, 1974) and various vertebrates (Smythe, 1970; Fogden, 1972; Leck, 1972; Karr, 1976), including lizards (Sexton et al., 1971). For most of these organisms the rainfall pattern and associated environmental changes are important. Most insect populations seem to begin to decline shortly before the end of the rains and decline steadily until the end of the dry season (if any) (Robinson and Robinson, 1970; Fogden, 1972; Smythe, 1974). Fruits are normally most abundant following whatever dry season there is (Smythe, 1970; Leck, 1972; Foster, 1973; Frankie et al., 1974), and this study is no exception (figs. 9-2, 9-3, table 9-5). However, the fruits eaten by *V. olivaceus* are available in staggered pulses throughout the year so that some edible fruit is available in every month (fig. 10-7), but resident individuals are required to move around within their home range to take advantage of them. Most tree species utilized as food by butaans bear fruit only once a year, but different species have different fruiting lengths, the duration varying from almost 2 to 25 weeks. Once a fruiting tree is located within the home range, food may be available in fairly large amounts for an extended period of time. However, the need for a mix of nutrients, including different proportions of carbohydrates, fats, and proteins, requires shuttling between two or more trees and feeding on land snails whenever possible during such forays. That adults move back and forth between feeding sources is well documented by radio telemetric data (chap. 5).

Competition and Fruit

As pointed out in chapter 10, there are remarkably few competitors for the fruits upon which the butaan feeds—particularly the *Canarium* species, for which only the hornbills and a few of the larger rodents compete. *Aglaia harmsiana, Malaisia, Uvaria samarensis, Caryota cumingii, C. Rhumphiana, Dracontomelum dao, Sandoricum koetjape,* and *Spondias pinnata* are also eaten by the local palm civet (*Paradoxurus philippinensis*). While there is extensive predation on *Ficus* species, *V. olivaceus* feeds on only a few and then not regularly. There is no competition with *V. salvator* for insects, slugs, and spiders, except among the young of both. Shrews and some birds compete with *V. olivaceus* for land snails.

Some local plant species may have coevolved with local frugivores, some of which are seed predators and others seed dispersers. This evolution is shown partly by the size, shape, and color of the fruit, as well as shape of the seed. Some oily fruits (*Canarium,* etc.) are highly specialized in the sense that they apparently are intended for consumption by only a few specialized seed dispersers. The sugary types listed are more generalized, appealing to a wider array of apparently more opportunistic vertebrate predators (see Howe, 1982 for similar results in Panama). Species of the genus *Pandanus* (also eaten by *V. olivaceus*) also produce highly specialized fruits in the sense that few vertebrates feed on them, the butaan perhaps the major disperser. They are also in the category of oily (or waxy) fruits; the seeds have bristles, which may be a specialization for germination in sandy soils, and they are thus adapted for retention in the gut of the butaan.

Most of the more important plant species in the diet of the butaan are rarely eaten by other forest vertebrates. The feeding niche of the butaan is narrow compared to the niches of other frugivorous vertebrates (table 12-1). Like hornbills, butaans make frequent visits to the few trees that bear for short but staggered periods throughout the year. Here they pick up the few ripe fruits that fall to the ground each day.

Another species of fruit regularly eaten is matobato (*Aglaia harmsiana*). It is eaten by a wide array of forest vertebrates. Though the tree is rare, it fruits more continuously than almost all other species at Caramoan (chap. 10)—all features that promote frequent visits by *V. olivaceus* in its narrow search images.

As shown in chapter 10, the butaan and other fruit-eating animals may exert opposing pressures on the timing of fruits. By destroying seeds (done mainly by rodents) they may cause plants to fruit synchronously so as to swamp predatory activity. On the other hand, by dispersing seeds (butaan, palm civet, and hornbills) they may encourage asynchronous fruiting so that each plant species may increase its chances of being noticed and dispersed. Some plant species may be subjected to both pressures (*Canarium,* for example) by different predators, and the same animals may exert the same pressures on the same plant species; for example, some rodents carry off the seeds of *Canarium* and then fail to find them later. (Because of its small daily home range and tendency to move only between relatively few fruiting trees, the butaan is probably not an important seed disperser.)

If related plant species stagger their fruiting but individuals within each species are synchronized, the butaan can lock onto a search image and benefit from accumulated knowledge of the location of such trees in its home range (chap. 5). Competition between individual plants, however, may lead to asynchrony within species, which will theoretically confound the locational powers of the local butaans. This competition occurs in some plant species in the Caramoan forests, and it places a premium on the butaan's ability to plan economical routes between current food sources, since it increases the ratio of searching to feeding time in a habitat where all fruiting species tend to occur at low densities (chap. 10).

Frugivory as a Feeding Strategy in Lizards

In the evolutionary system existing between the adult butaan and the plants on which it feeds, fruit eating promotes more fruit (through seed dispersal) while insectivory and molluscivory promote adaptations in the prey that protect them from predation. Though varanid predation on animals favors prey inconspicuousness, which in turn promotes many different searching techniques (Auffenberg, 1981a, 1984), competition among plants to attract animals as seed dispersers favors fruit conspicuousness, seed morphology designed to lengthen seed passage rate (fig. 12-2), and such abundance that there is often little competition for the fruit (Morton, 1973). On the other hand, seed or leaf eating often leads to changes designed to slow or stop such feeding by means of chemical or mechanical systems.

One of the primary questions regarding the specialized food of *V. olivaceus* is "Why is nearly total frugivory so rare in lizards when fruits are actually so common in most tropical forests?" It might be expected that totally frugivorous diets in lizards would increasingly evolve and would be more common than now observed, given the mutualistic relationship between fruit and fruit eaters. (Partial frugivory may be more common than presently demonstrated on the basis of published data [many scincids, larger teiids, etc., Auffenberg, notes].) Of course, competition with other organisms for fruit may be a major factor. The abundance of fruit, the short time needed to obtain it, and its year-round availability in the tropics (Snow, 1965; also chap. 10) would provide an adequate food supply for both adults and young. A fruit diet for small lizards (and young) should theoretically be favored by selection, since it would provide more food than the relatively rare insects do. However, fruit is not generally used as food by young lizards (or even the young of frugivorous birds; Lack, 1962; Morton, 1973), even if they are herbivorous in later life.

Total frugivory (adults and young feeding on nothing but fruits) is unknown in any reptile. Many species generally considered herbivorous, such as *Cyclura,* are not totally so, for both young and adults feed on many smaller animals, dead or alive (Auffenberg, 1982).

Nutritional reasons do not render it impossible for reptiles to be totally frugivorous, though fruits are generally low in proteins and calcium (chap. 9). Moreover, fruit could, and doubtless has, improved nutritionally through selection caused by frugivores "choosing" to feed from plant species with more nutritious fruit (Snow, 1962; Gardarsson and Moss, 1970), though nutrition cannot be considered in terms of only caloric content (Auffenberg, 1982a).

More general herbivory is fairly common, however, among larger iguanid and agamid lizards (see Burghardt and Rand, 1982). Van Devender (1982) has shown that adults of some partly herbivorous iguanines are able to sustain the rapid growth rates characteristic of all young lizards of many different families. These partially herbivorous species grow at an essentially constant rate of $0.27-0.32$ mm/day for a longer period of time than any insectivorous types (through the first two years of life). Thus the general pattern for these species is one of extended rapid growth rate normally characteristic of the earlier life history stages.

The almost constant linear growth of these species results in an exponential gain in mass. Such logarithmic growth in mass is particularly interesting in partially herbivorous lizards since plants are generally thought to be relatively poor food (Pough, 1973; Iverson, 1982; McBee and McBee, 1982), but these observations are based largely on lizards that eat leafy foods.

Pough (1973) observed the correlation between large body size and degree of herbivory in several lizard families, leading to his suggestion that herbivory was related to underlying physiological processes and concluding that large body size allowed lizards to take advantage of the lower weight-specific metabolic rate and the lower foraging costs to become herbivorous. His statement that evolution of a large lizard both requires and permits a switch from carnivory to herbivory is clearly not applicable to varanids, since all of the largest species in the family are completely carnivorous. There are, however, certain morphological specializations required for folivory in lizards, regardless of size (Lonnberg, 1902; Iverson, 1982), including great enlargement of the gastrointestinal tract and production of fermentation chambers. However, these modifications are not needed in frugivory (though other, seemingly less complex modifications are required; see chaps. 3, 9).

Of the smaller lizard species that often or regularly eat fruit (such as skinks; see Rand, 1978 for review), none is even partially folivorous (except a few iguanines). These are all species that are primarily insectivorous and secondarily frugivorous, often during only certain seasons (Auffenberg and Auffenberg, 1987). These species form intermediate steps between pure insectivory and the butaan's situation in which half its food is fruit. Only after further modification of the gut as a fermentation chamber can leaves be added to the diet. With the possible exception of adults of *Hydrosaurus pustulosus,* no lizard species is known to be completely folivorous—probably largely because of their limited mobility compared to frugivorous birds, bats, monkeys, etc. I conclude that for lizards the evolutionary sequence is from insectivory to insectivory-frugivory to frugivory-folivory to exclusive folivory.

As pointed out, young lizards, regardless of their adult diet, are almost entirely insectivorous (some iguanines are exceptions), and no young of any species feed on fruits, even partially. That juvenile butaans fail to eat the smaller fruits that are eaten by many local birds and skinks seems rather strange, particularly in view of Van Devender's (1982) discoveries regarding the high sustained growth rate of some partially herbivorous iguanines. I believe that the most plausible hypothesis for why post-hatchling butaans do not eat fruit involves the interaction between predation and diet. Predation pressure is strong on *V. olivaceus,* as suggested by both its shyness and its coloration. Thus high predation during early stages would favor the occurrence of a short juvenile stage. Rapid growth (depending on a high protein diet) during this phase allows the shortest possible time between hatching and the size at which some capacity for actively escaping predators is achieved.

In spite of the faster growth rate of subadults of lizards feeding on a mixed diet (Van Devender, 1982), fruit alone seems to represent a poor diet for rapid growth since it is notoriously low in both proteins and calcium (table 11-5). On the other hand, insects, in spite of the more difficult access to them, represent a proper diet upon which the rapid growth of young can be based. The faster growth achieved through insectivory results in relatively less predation on the butaan at its most vulnerable size. Morton (1973) has shown that frugivorous birds grow more slowly than insectivorous ones and that the species that feed their young both insects and fruits do not feed them fruits during the earliest period of nestling when growth rate is the highest. He also showed that birds that are totally frugivorous have longer nestling stages than those fed at least some insect food. While some of the lack of proteins and calcium in tropical fruits could be overome by higher food intake by juvenile butaans, the high selectivity of fruit and the distance between fruiting trees in both time and space make this alternative less viable for the smallest size classes because of the resulting higher energy expenditure accompanying the propor-

TABLE 11-5. Nutritional data for 100-g samples of fruits and insects.

Food	Kcal	H₂O (%)	Protein (g)	Liquid (ml)	Carbo-hydrate (g)
FRUIT					
Ficus (several species)	54	84.0	1.2	0.1	9.3
Spondias mombin	42	88.0	0.8	0.2	10.1
Average, all fruits[a]	121	75.4	1.4	8.3	12.3
INSECTS					
Termites	n.a.	60.0	10.1	1.3	n.a.
Beetles (several species)	192	56.2	27.1	3.7	4.8
Caterpillars (several species)	86	81.1	10.6	2.7	1.4
Crickets	117	76.0	13.7	5.3	0.0
Grasshoppers	170	62.7	26.8	3.8	3.1
Average, all insects	141	67.2	17.7	3.4	2.3

a. Data from Leung, 1968; Jenkins, 1969; Morton, 1973; and this study.

tionally larger foraging area required. There is, furthermore, the likely possibility of higher predation rates that could be overcome as a population factor only by increased clutch size. If slow growth is the result of a purely frugivorous diet for the young, predation pressure becomes an increasingly strong source of selection.

Competition may also have been an important factor in the evolution of frugivory in the butaan. Thus competition with *V. salvator* (and perhaps many nonreptilian species feeding on vertebrate prey) could force *V. olivaceus* into a totally different food spectrum, at least as adults. This change would have the possible additional benefit of completely obliterating all competition with young of their own species. Competition between young *V. olivaceus* and young *V. salvator* (plus an even greater number of other organisms) for insect food remains as high as ever, but perhaps it is cost-effective because of the benefits accruing to the faster growth rates associated with insectivory and with the benefits associated with the possibility of smaller clutch sizes (chap. 8).

Frugivory and the adult butaan.—While there are no extant completely frugivorous lizards, many tropical lizards (especially Scincidae, Iguanidae, and Agamidae) are partially frugivorous. Partial frugivory may occur to various degrees even in species generally thought to be completely insectivorous (Auffenberg and Auffenberg, 1987), and the same pattern can be found in many tropical birds (Morton, 1973). This common incidence of partial frugivory attests to selective advantages, some of which are obvious, others that are not.

There is much variation in the amount of frugivory exhibited by lizards in the Caramoan area during the course of a single year. Many otherwise insectivorous species eat fruit as it becomes available periodically, especially if insect densities at that time are low but even if they are not (Auffenberg and Auffenberg, 1987). This behavior does not necessarily mean that fruit is of lesser importance to those species than others, for the habitat characteristics and competitive abilities of these species may be determined in part by the probability of certain fruit species occurring within their home ranges. This occurrence of fruit makes the distribution of tropical scincids extremely complex and may explain the patchy distribution found for many of them; it may explain the restricted range of *V. olivaceus* as well—at least in part (see chap. 14).

The emphasis on either fruit, leaf, or insect food has resulted in striking differences in the lizards that exploit each type. These differences relate to adult size, breeding period, microenvironment, morphology, and foraging patterns. In the Caramoan area these lizards can be divided into those that take a mixed diet of fruit and animals (mainly scincids but the largest *V. olivaceus*), those purely insectivorous (local smaller agamids, such as *Draco*, some of the smaller scincids, and those living in "special" environments, such as *Emoia* near the sea, and many others), and the folivorous types (represented locally only by *Hydrosaurus*). Insectivorous types occur in the greatest range of local habitats—some restricted to such diverse environments as mangroves and sea shores (*Emoia*) and others to grassy hillsides (*Brachymeles*). Almost all are terrestrial, though a few are fossorial (*Brachymeles*) or even arboreal (*Draco*). The group feeding on both animals and fruits is either terrestrial or somewhat arboreal (*Lamprolepis*, some *Mabuya* species, and *V. olivaceus*). However, within this group, *V. olivaceus* is the only species that feeds primarily on fruits. *Hydrosaurus* is largely arboreal, which is expected on the basis of its apparently complete folivory.

Of these, the butaan fills a distinctly mesic tropical forest niche; not only is it more dependent on fruits throughout the year, but its alternative foods are land snails of large size found only in this type of forest. It uses a strategy similar to what Morton (1973) calls "fruit-searching" in birds—effective in exploring the spatial-temporal mosaic of fruiting patterns. Moreover, it is intimately tied to the shorter-lived subcanopy trees most characteristic of forest gaps and other openings which "use" the butaan as part of their means of seed dispersal. The fact that so few forest trees "use" the butaan for seed dispersal is a reflection of niche complexity among tropical forest trees (see Ashton, 1969 for discussion). These tree species produce fruits of high nutritive quality but defend themselves with phenolics and tannins against more wasteful mammalian frugivores (palm civets), who tend to deposit dozens of seeds in each dropping. The resulting bouquet of seedlings results in high seedling competition (see Howe, 1982 for discussion of a similar situation in Panamanian forests).

A decided decrease in fruit abundance in the late wet season in Caramoan (October–November) has a great effect upon the largely frugivorous butaan. This effect bears on the evolution of fruit and animal eating generally. Competition for these few fruits increases in the late wet season, and there are decided switches to other foods which might be considered "less preferred," such as snails. Though snails are needed in the diet, the search for them requires more expenditure of energy than the search for fruits, for they are never area-concentrated. In terms of frugivory, one may divide the year into easier and less easy times (high and lower fruit density). The relative proportion of each of these must vary appreciably both temporally and spatially (see Terbough and Diamond, 1970). There may be intense selection pressure favoring morphological and behavioral abilities to find and eat snails, even if the probability of unusually dry conditions is low. Butaan morphology and behavior that contributes to food finding must be adapted to hard times when limiting conditions are approached and competition is most severe.

If the omnivorous butaan competes well for molluscs during hard times, with shrews and other competitors, it is obviously better off, evolutionarily, than if it is morphologically and behaviorally close to being a complete frugivore. The effect of the seasonal aspects of fruiting in the Caramoan area is that selection may not favor wholly frugivorous adult diets, because the frugivore must be adapted for survival during hard times when fruit is less available. It is apparently for this reason that *V. olivaceus* will feed even on carrion (if not too old), in spite of the fact that it is apparently low on its preference scale. Wholly frugivorous diets for the butaan would be possible only under completely aseasonal climatic regimes—a situation probably not available anywhere in the world, even in the most aseasonal climatic belts.

The high fruit diet of the butaan can be viewed only as the evolutionary reaction to a competitively superior species (*V. salvator*), enabling it to coexist in the same area. Morton (1973), discussing the same topic but as it concerns tropical birds, viewed the insectivorous species as better fit—chiefly because the young were fed insects rather than fruit. Insects are, in general, also rarer than fruits, or at least require more energy to obtain, and thus function more as a limiting factor. The same general argument can be applied to the butaan and its molluscivorous-frugivorous diet.

Fruit availability and the frugivorous diet.—The use of plants by tropical lizards contrasts markedly with their use in temperate areas. The common tropical strategy wherein adults of several families eat fruit and the young eat insects is generally not available in temperate areas, because no plants produce edible fruits throughout the year, or even during the entire period that lizards are active. (*Dipsosaurus, Sauromalus,* and *Uromastix* all have partly herbivorous species that range into temperate areas [Norris, 1953; Berry, 1974; Case, 1976; Grenot, 1976].) There is some reason to believe that the availability of fruits in tropical forests accounts for some of the diversity found in agamids, iguanids, and scincids in those habitats. Such fruit is important for it not only allows more species to inhabit an area but it also probably allows larger adult populations to exist there. The butaan, because of its frugivorous diet, is not limited by food in the same way as the water monitor is in the same area; nor are adult butaans limited by the same factors as are the juveniles, for the foods of adults are entirely different from those of *V. salvator* and those of juvenile *V. olivaceus*.

In summary, there are both positive and negative aspects regarding frugivory in the butaan. Pure frugivory has not been demonstrated for any lizard in the world, the reasons undoubtedly related to the generally insufficient protein and calcium levels of almost all fruits. The presumed slower growth rates caused by such a diet (even though energy-efficient in terms of foraging costs) are probably most significant for the youngest size class after hatching, when rapid growth is probably correlated with escape from predation. Young butaans thus remain completely faunivorous until about 500 g, when they are apparently able to contend more easily with potential predators. Total frugivory of the adults is undoubtedly opposed by selection for molluscs (and other animals, though rarely obtained) during the annual period of low fruit abundance associated with the period of heaviest precipitation. Actually, snails represent a more "complete" food than any one species of fruit. However, the advantages of high food density at known sources (under specific trees) reduce the energetic costs of foraging so much that an emphasis on fruits becomes an acceptable life-style.

The combination of high predator pressure and reduced need for long foraging forays has resulted in locomotor behavior at variance with that of carnivorous varanids. In general, movement of butaans tends to be slow, more like that of an arboreal monitor lizard than a largely terrestrial one. This pattern is undoubtedly related to reduced risks from potential predators, since rapid movements, such as prey pursuit, are unnecessary, given the animals upon which it feeds (land snails) and the fallen fruits that make up most of its diet. When disturbed in the forest, its reaction is usually to "freeze" if in the trees or to enter any nearby crevice in the rocks, regardless of depth, and to remain relatively still in either place. This behavior is in marked contrast to the headlong, crashing flight of most varanids (Auffenberg, 1981a), including the sympatric *V. salvator*. When approached while hiding in a tree, the butaan slowly moves around the trunk, keeping the tree between it and the observer. Thus there is a minimum energy expenditure, even in escape, for it more often relies on cryptic color, pattern, and behavior than on flight to escape detection.

I have tried in this section to specify for the butaan that complex of behavior and morphology best suited to gather food energy in the mixed lowland dipterocarp forests of southern Luzon. Using the definitions of Schoener (1971) in regard to feeding strategies, juvenile *V. olivaceus* are probably *energy maximizers*, for growth demands are high at this stage and a diet high in protein is necessary. Individuals of this size are apparently too small to crunch the shells of land snails to avail themselves of this protein source, as the adults do. Adult butaans are *time maximizers;* they search for trees of certain species that are in a fruiting pulse, and around these trees, where food is sometimes exceedingly abundant, they tend to remain for varying lengths of time.

Conclusions

Based on the above discussion, several consequences of feeding on small, scattered sources can be identified in features of the butaan foraging strategy.

1. The need to locate the inconspicuous trees that form the major food sources and to monitor the fruiting periods necessitates a limited foraging area (see chap. 7). The butaan lives in smaller home ranges than equal-sized individuals of the highly predaceous *V. komodoensis*. Both organisms seem to know their home areas well, though the butaan seems to show a clearer affinity with its home range. The frequent transient individuals one meets throughout the range of *V. komodoensis* relative to the low number of residents in any area (Auffenberg, 1981a) are not seen in the butaan's range.

2. Sources may have to be visited daily or even more than once a day to cull the ripe fruit that has fallen to the ground before other predators (particularly nocturnal rodents and civets) locate it. Thus butaans move about during all parts of the day to obtain their food in a local environment that imposes few, if any, thermal constraints on them (unlike *V. komodoensis;* Auffenberg, 1981a). Like most varanids (perhaps all), chemical perception through the tongue and vomeronasal structures of the butaan is particularly efficient at detection of fruit sources, fruit ripeness, and probably commuting "paths" between potential fruiting trees (similar to paths demonstrated for *V. komodoensis,* over which individuals sometimes pass repeatedly with extraordinary precision; Auffenberg, 1981a).

3. A diet of low-density fruit sources would be expected to result in low densities and biomass of the butaan compared to sympatric frugivores. However, being a reptile, the butaan is efficient in metabolizing what it obtains (see Rand, 1978) and actually retains fairly high densities (chap. 7).

The water monitor and butaan resemble each other in most morphological parameters (chaps. 2, 3), though in the Caramoan area the former does not attain the size of the latter. But their overall diets are entirely different. They use dissimilar foraging strategies to exploit their respective food supplies. The home range of the butaan is much smaller than that of the water monitor, in spite of its larger size, and the butaan travels shorter distances each day. Both species travel and forage on the forest floor in generally the same way, though they seek different prey. The foraging range of the water monitor is generally closer to local streams (Auffenberg, 1980a) and that of the butaan more restricted to the tops and ridges of mountains. The ecological distribution of sympatric populations of *V. komodoensis* and *V. salvator* in Indonesia (Auffenberg, 1981a) shows evidence of spatial nonoverlap, which is expected in view of the similar food habits of the two species. *V. salvator* and *V. olivaceus* do not overlap in food sources, suggesting that their ecological separation is related either to food distribution or to interactions between individuals. Individuals of both species placed in the same large outdoor pen in the Philippines seemed to ignore one another completely, though there was much aggressive interaction between individuals of the same species. I conclude that it is the difference in food that allows geographic sympatry and ecological allopatry between *V. olivaceus* and *V. salvator* in the Caramoan area.

The water monitor travels much greater distances for its food each day than the butaan, but its smaller total size (locally) should make its travel costs smaller. However, limb proportions in both species are about the same, so that large butaans use fewer strides to cover a given distance than the largest local water monitors. In general, the fewer strides taken, the cheaper the movement (Dawson, 1977). Thus the generally cheaper travel of the butaan (shorter distances and fewer strides to cover equal distances) enables it to reach high-energy foods (sugary and oily fruits) more easily than can the water monitor. On the other hand, the usually imbalanced meals resulting from frugivory (Auffenberg, 1982), particularly the protein-poor aspects of such a diet, force the butaan to feed on animal flesh. In its habitat the usually abundant land molluscs are easily found during the day in the same place as the fruits on which it feeds. Foraging for the few types of insects on which the juveniles feed may not be cost-effective for large adults.

These patterns contrast with prevailing conditions for most varanids in that larger species of varanids tend to travel farther when foraging than smaller ones (Auffenberg, 1980a, 1981a), chiefly because travel costs (in calories) are less proportionally in larger individuals than in smaller ones (Schmidt-Nielson, 1971). This is because stride varies in direct proportion to body weight in all varanids. However, travel cost as related to stride length is less important in *V. olivaceus* because of the significant difference in its food base. The smaller home range of the butaan (in spite of its larger size) is due chiefly to its herbivorous diet. One would expect the home range of a purely folivorous lizard (i.e., *Hydrosaurus*) to be smaller than that of a frugivorous one on the basis of the greater availability of leafy foods, but no comparative studies have been made (see Clutton-Brock and Harvey, 1977, for pertinent studies on primates).

It has been shown elsewhere that while there is some overlap in the home ranges of adult butaans, they tend to be scattered and individuals do not concentrate under heavily fruiting trees, as bats (Bonaccorso, 1979), birds (McClure, 1966), and even some iguanid lizards (Auffenberg, 1982) regularly do. The apparent reason is that it would be exceedingly wasteful to wander about looking for inconspicuous food sources of a few fallen fruits under widely scattered trees without a thorough knowledge of the local area. It would benefit an adult to stray across any home range boundary (whether defended or not) to feed only when there is a large conspicuous food source easily located from some distance, such as the carrion eaten by large gatherings (up to 17 individuals) of *V. komodoensis* (Auffenberg, 1978b). Food concentrations easily recognized from a distance probably do not exist for the butaan, though they clearly do so for the sympatric carnivorous *V. salvator* at appropriate carrion (Vogel, 1979; Auffenberg, 1981a). With each butaan remaining more or less inside its home range boundary (even without defense), no time is lost searching for food, and an appropriate amount of a dependable source can be eaten in a short period. Because butaans may have an easily defensible food supply, they are minimally engaged in territorial defense and disputes—probably because of the cost-effective strategy of concentrating food search in relatively small areas about which a detailed knowledge has been accumulated over a period of residency rather than wasting time and energy in wandering. The same technique seems operative for frugivorous bats (Morrison, 1978) and to a certain (though lesser) extent for the largely herbivorous *Cyclura carinata* (Iverson, 1977; Auffenberg, 1982a).

The type of food supply the butaans have, then, allows individuals to restrict movements to a limited area. Defense is not necessary, nor is the requisite boundary patrolling at ground level in a densely forested situation, which is expensive in terms of nutrient demand.

Pair bonding has previously been reported for at least some adults of *V. komodoensis* (Auffenberg, 1981a). Monogamy tends to evolve either when more than a single individual is needed to raise the young or when the carrying capacity of the habitat is insufficient to permit more than one female to add to local recruitment in the same home range at the same time (Kleiman, 1977). Since, with the exception of some (perhaps all) crocodilians, no reptiles take an active role in caring for their young, it is carrying capacity that is important in pair bonding in the Komodo monitor and may explain why it appears not to occur in all adult individuals (= facultative monogamy of Kleiman, 1977).

12

Interspecific Interactions

Potential Competition

ALL of the research carried out at Caramoan was conducted from the standpoint of the butaan. The importance of the trees in which they live and on which they feed has been emphasized and the main features of the forest described (chap. 4). The continuing references in subsequent chapters to plant species supplying food for *V. olivaceus* also point out the importance of studying the local plants as well as the animals. The main foods eaten by adults are fruits and molluscs, and juveniles eat primarily large insects. Some birds and mammals and a few invertebrates share these foods. Competition is assumed to occur when the availability of a food resource required by one species is reduced by the activity of another, so any partially or wholly arboreal bird or mammal taking food normally consumed by the butaan may be regarded as a potential competitor.

Of these potential competitors insects are undoubtedly the most important, but no data exist on this competition. (On the importance of insect predation in tropical forests, see Tomlinson and Zimmermann [1978].) Among potential vertebrate food competitors (appendix 2) the most important are the two local species of hornbill birds (*Buceros hydrocorax* and *Penelopides panini*), the dog-faced, fruit, and flying fox bats (*Ptenochirus jagorii, Pteropus vampyrus,* and *Acerodon jubatus,* respectively), the Philippine macaque monkey (*Macaca philippinensis*), the wild pig (*Sus* cf. *S. scrofa*), and the palm civet (*Paradoxurus philippinensis*). Hermit crabs (*Coenobita rugosa*) are minor competitors for fallen fruits. For juvenile butaans the list is longer and includes many of the larger insectivorous birds, some of the skinks, the Malay civet (*Viverra tungalunga*), and the water monitor (*Varanus salvator*).

Molluscs, shown to play a major role in the butaan's diet, are eaten by local birds, reptiles, mammals, and invertebrates but are not important in their diets. Other than predaceous ants and some carnivorous snails (Bullock *in* Berry, 1966), the most important molluscivores are probably shrews (*Suncus* sp.), as shown by the many empty snail shells on the forest floor that have been characteristically opened by these nocturnal animals. The local primate *Macaca philippinensis* occasionally eats the large land snail *Rysotta angulata,* upon which the butaan also feeds, but not commonly. Most snails eaten by butaans are too large to be preyed upon successfully by the majority of molluscivorous vertebrates. Bullock (1963) reported that the bird *Monticola solitaria* feeds on large numbers of *Cyclophorus* (also eaten by the butaan).

Fruits are important food for many arboreal birds, mammals, and a few reptiles (generally small), including the butaan. It is highly likely that the activity patterns of most species are determined mainly by the dispersion of fruit sources (monkeys, Payne, 1978; Wheatley, 1978; Gittings and Raemaekers, 1978; palm civets, Bartels, 1964; hornbills and the butaan, this study).

Birds.—The ability to fly increases access to widely spaced food sources. Since plant species occur at low densities in the local forests, many fruits are widely scattered and many are rare. Perhaps for these reasons, the most specialized frugivores have evolved among flying animals—medium to large birds and bats (Payne, 1978). Local birds known or believed to be partially or wholly frugivorous are listed in appendix 2 and divided into major groups in table 12-1.

Pigeons are believed to be exclusively frugivorous (Medway and Wells, 1976; Rabor, 1977), but they concentrate on "unspecialized" fruits, such as those of species of *Ficus, Eugenia,* and Euphorbiaceae, whose seeds they disperse. Local parrots are fruit specialists, probably unique among families of Philippine forest birds in that they are primarily seed destroyers. All these birds range widely for food. Payne (1978) stated that there are no highly specialized, exclusively frugivorous birds in Southeast Asia (as occur in neotropical forests; Snow, 1971) that have close coevolutionary relationships with certain tree taxa; he may be incorrect in the case of hornbills. The kinds of fruit that have evolved specifically for dispersal by large birds (such as hornbills) bear dry, nutritious pulp with large seeds (see chap. 9).

There are great size differences among the common frugivorous birds of the Caramoan forests, allowing specialization on fruits of different sizes and on feeding in different parts of the tree crowns. This specialization applies particularly to pigeons and doves (appendix 2). To my knowledge, no frugivorous birds feed on fallen fruits on the forest floor.

The two local hornbill species forage by flying over the forest and settling to feed (only) in the taller trees. Hornbills sometimes eat small animals (Payne, 1978). They are the only other vertebrates that regularly eat the nutfruits (*Canarium* sp.) that comprise about 14% of the diet of the butaan. In addition, about 41% of the hornbill's fruit diet overlaps that of the butaan,

TABLE 12-1. Local plant genera the fruits of which are eaten[a] by frugivores other than civet cats (see table 12-2). Based on droppings and observations.

Family / Genus	Frugivore groups								
	Bat	Dov	Man	Mon	Par	Rat	Crb	Hrn	Brd
Anacardiaceae									
Dracontomelum[b]	●		●	●		●	●	●	●
Mangifera	●		(●)[a]	●		●	●		
Gluta				●					
Annonaceae									
Uvaria[b]	(●)		(●)	●		●	●	●	
Burseraceae									
Canarium[b]			●	(●)		●	●	●	
Garuga	●			●		●	●		
Santira				●				●	●	●
Curcurbaceae									
Trichoranthus								●	
Momordica			●					●	
Dilleniaceae									
Dillenia			●	●			●		
Ebenaceae									
Diospyros	●		●	●					
Elaeocarpaceae									
Elaeocarpus				●					
Euphorbiaceae									
Antidesma		●		●					●
Macaranga				●	●				
Fagaceae									
Castinopsis		●		●					●
Guttiferae									
Calophyllum	●			●					
Garcinia	●		●						
Leguminosae									
Cynometra				●					
Intsia				●					
Milletia				●					
Parkia				●	●				
Pithecellobium				●					
Sindora				●					
Malvaceae									
Hibiscus			●		●				●
Melastomaceae									
Memecylon				●					
Meliaceae									
Aglaia[b]				●	●			●	
Amoora				●					
Chisocheton	●			●			●		
Dysoxylum	●	●		●				●	
Lansium	●		●	●	●				
Sandoricum[b]	●		●	●		●			

a. (●) = eaten only rarely. b. Fruits eaten by *V. olivaceus*.

(*continued*)

TABLE 12-1 (*continued*)

Family / Genus	\multicolumn{9}{c}{Frugivore groups}								
	Bat	Dov	Man	Mon	Par	Rat	Crb	Hrn	Brd
Moraceae									
Artocarpus			●	●		●		●	
Ficus[b]	●	●	(●)	●	●	●	●	●	●
Musaceae									
Musa	●	●	●	●	●	●	●	●	
Myristicaceae									
Myristica								●	
Myrtaceae									
Eugenia	●		●	●	●		●	●	●
Oxalidaceae									
Averrhoa	●		●	●				●?	
Palmae									
Caryota[b]							●	●	
Pandanaceae									
Pandanus[b]			●	(●)					
Rosaceae									
Parinarium	●		●						
Rubiaceae									
Morinda			●						
Randia			●						
Coffea		●	●						
Sapindaceae									
Nephelium	●	●	●	●			●	●	
Sterculiaceae									
Pterospermum			●						
Sterculia					●			●	
Tiliaceae									
Diplodiscus	●	●		●	●	●			●
Grewia[b]	●	●	●	●	●			●	
Muntingia			●	●					
Verbenaceae									
Vitex			●						

ABBREVIATIONS USED:

Bat = pteropid fruit bats
Brd = birds other than hornbills
Crb = hermit crabs
Dov = doves and pigeons
Hrn = hornbills
Man = human beings
Mon = macaque monkeys
Par = parrots
Rat = murid rodents

so they may be among the most important food competitors of the butaan (fig. 12-1).

Mammals.—Mammalian frugivores at Caramoan are both nocturnal and diurnal, arboreal and terrestrial. Of the arboreal forms, only the Philippine macaque is diurnal. Several species of large rats inhabit the local forests, all of them nocturnal. Though some species feed on fallen fruits, most—both seed and pulp eaters—search for fruits in the trees. Those that forage on the ground for fallen fruits are important food competitors, but because these rodents are nocturnal and the butaans are diurnal, they are temporally partitioned (table 12-1). No density studies of rodents were conducted during this work, but data reported by Harrison (1969) for similar forests in Malaysia show that there are probably from 3.0 to 6.5 individuals (several species involved) per hectare, which seems much too low for the forests in which the butaan is found. These data can be considered as representing only minimum densities in the Caramoan area. Local rats are seed destroyers, not dispersers, since they eat the endocarp.

Three species of large bats regularly feed on edible forest fruits (pulp only). All are nocturnal, and no local species feed on fallen fruits. Twenty-six percent of their diet consists of genera also eaten by the butaan. Data from Panama (Morrison, 1978) suggest that bats take approximately 7% of the annual fruit production of the trees on which they feed. I conclude that bats are not major food competitors of the butaan (table 12-1), in spite of the fact that bats are often specialists—some species even partitioning different fig species on the basis of fruit size (Bonaccorso, 1979). These specialist bats tend to have large home ranges (Heithaus et al., 1974; Morrison, 1978), while generalist bats have smaller home ranges (Bonaccorso, 1979). While many fruit bats are im-

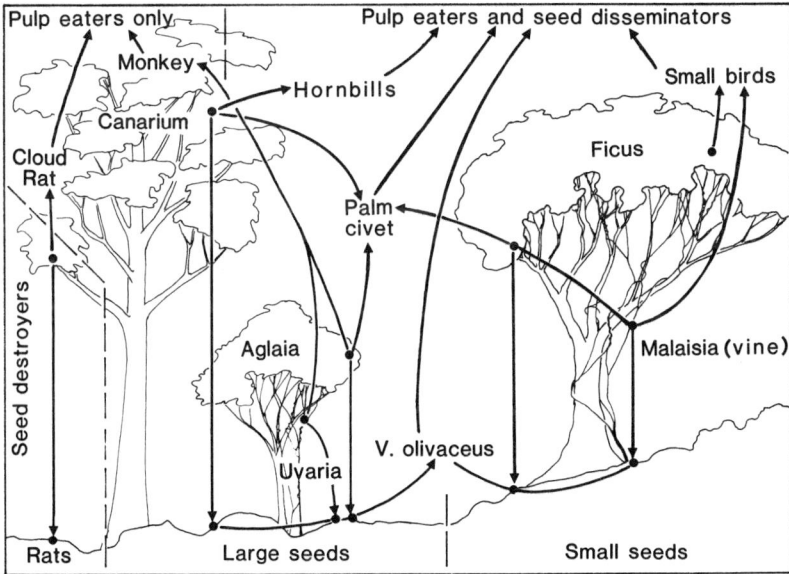

Fig. 12-1. Schematic representation of feeding interactions at trees of major fruit species of *V. olivaceus*. Two major types are shown: those with small and with large seeds. Some fruit predators find their food in the trees, others on the ground. Fruit predators are divided into seed destroyers and seed disseminators.

portant seed dispersers, defecating in flight (Bonaccorso, 1979), most species in the Caramoan area defecate at their roosting sites (as also reported by Fleming and Heithaus, 1981).

Humans obviously feed on a wide variety of domestic and wild fruits, some of which are also eaten by the butaan. They usually pick the fruit from the tree, though they may take others from the forest floor. In general, the fruits selected tend to be in the earliest stages of ripeness, the strategy being for them to ripen later, then be eaten. In general, humans are not important food competitors of the butaan—only 30% of the fruit species are shared, and several of those are not commonly eaten by humans. Only in one area are they an important competitor, and that is in the commercially salable "pili" nut, *Canarium ovatum*. The locations of practically all trees, even deep in the forest, are known to many locals who usually harvest all the nuts. Only a small fraction of the annual pili nut crop is available each year for consumption by the other frugivores of the forest. This nut species is undoubtedly a more important food source for the butaan in larger forested tracts than it is in the Caramoan area.

Of the animal species listed, *Macaca philippinensis* feeds on the greatest variety of fruit species (84% of those listed), of which 19% are also eaten by the butaan (table 12-1). It rarely feeds on the major plant food of the butaan—species of the genus *Canarium*. The same general avoidance of members of this group has been demonstrated by Malayan gibbons (Gittings and Raemaekers, 1978). In one study in Panama, howler monkeys were found to eat about 9% of the annual fruit crop of the species upon which they fed (Morrison, 1978). The diet of *Macaca philippinensis* during the day tends to be more mixed than that of *Paradoxurus philippinensis*. Thus the number of items taken from a single tree by the macaque is lower, but it is offset by the fact that a troop of monkeys feeds from the same tree at the same time. Locally, monkeys are geographically more restricted than the butaan (due to local human predation). In addition, they occur in lower densities than butaans in areas where they occur together (though no solid data are available for local macaque bands). The monkeys are somewhat more restricted ecologically than the butaan, since the former usually remain close to water throughout the year—a pattern similar to that of *Macaca fascicularis* in Borneo (Wheatley, 1976). Finally, though largely frugivorous, most macaque species also feed on leaves and flowers (Payne, 1978; Wheatley, 1978; Chivers and Hladik, 1980).

Of all potential competitors of adult butaans, *Paradoxurus philippinensis* is the most important, for it eats more of the same fruits as the butaan than any other vertebrates in the immediate area. Table 12-2 lists all the fruits known to be eaten by local civet cats. As Bartels (1964) and Prater (1965) reported for *P. hermaphroditus,* closely related, *P. philippinensis* feeds on a variety of fruits, though no nuts, which make up a fair proportion of the diet of *V. oliva-*

TABLE 12-2. Local plants the fruits of which are eaten by civet cats, *Paradoxurus philippinensis*. Based on 212 droppings.

Trees	Herbs
Aglaia harmsiana[a]	*Musa zebrina*
Antidesma bunius	*Pandanus radicans*[a]
Bombicidendron sp.	*Pandanus simplex*[a]
Cicca disticha	*Pandanus tectorius*
Dioxyros sp.	
Dracontomelum dao[a]	Palms
Dracontomelum edule[a]	*Caryota cumingii*
Dysoxylum sp. A	*Caryota Rhumpiana*[a]
Dysoxylum sp. B	
Eugenia sp. A	Shrubs, vines
Eugenia sp. B	*Coffea* sp.
Eugenia calubcob	*Malaisia scandans*[a]
Ficus balete[a]	*Momordica* (cf. *M. cochinchiensis*)
Ficus benjamina[a]	*Psidium guajava*
Ficus indica	*Trichoranthes cucumeriana*
Ficus ulmifolia	*Trichoranthes quinquangulata*
Ficus variagata[a]	*Uvaria rufa*
Garcinia sp. A	*Uvaria sorsogonensis*[a]
Grewia stylocarpa[a]	
"Lumacao"	
Sandoricum koetjape[a]	
Spondias pinnata[a]	

a. Fruits eaten by *V. olivaceus*.

ceus (e.g., *Canarium* sp.). While the Philippine palm civet in the Caramoan area eats fewer kinds of fruit than the Javan species (Bartels, 1964), it still eats many kinds, some of which are also eaten by the butaan. It eats fruits growing on trees, shrubs, palms, and vines, though mostly on trees (fig. 12-1). Like the butaan (and many other forest frugivores, *fide* McClure, 1966), it eats all fruits in a perfectly ripe condition, but fruit selection is in the trees, not on the ground. "*Paradoxurus* [*hermaphroditus*] only takes really ripe fruit. It picks the fruits with painstaking care, apparently leaving the less ripe ones for the following night . . . The organ of smell is extremely acute and most selective, with the emphasis on selectivity. . . . The climbing ability is phenomenal. . . . No tree seems to be too high, no twigs too fragile, almost no vine too weak" (Bartels, 1964).

As does the butaan, the palm civet passes seeds that are unharmed and retain their germinative power. In fact, gut passage may be necessary for some seeds to germinate. The palm civet's wide nocturnal wandering in search of fruits is attested to by Bartels, who found droppings with tufts of pineapple fibers as much as 3 km from his pineapple grove in Java (the only local one).

About 50% of the plant foods recorded for *P. philippinensis* are also eaten by the butaan, though the preferences are different. This overlap varies from 10% to 87% seasonally (table 12-2). One of the palm civet's major fruit foods during the year is *Aglaia harmsiana,* which has a long fruiting season (fig. 10-6). These fruits are also eaten extensively by the butaan. However, *Caryota, Ficus,* and *Spondias*—all eaten by the butaan—sometimes constitute over 10% of the monthly diet of the palm civet. The remaining fruits, eaten by both the local palm civet and the butaan, are rarely represented in the droppings of the palm civet (table 12-3).

Local palm civets feed on a few animals (0.2–3.2% of monthly diet), most of which are insects, though a few land snails are also occasionally eaten. (*Viverra tungalunga,* the remaining local civet, is largely carnivorous, feeding mostly on freshwater crabs and forest rodents in the Caramoan area.) It is reasonably clear that the several species of palm civets are opportunistic, and the diet of even a single species may vary greatly from one place to another; *P. hermaphroditus* is completely frugivorous in Java (Bartels, 1964), completely mixed in India (Prater, 1965), and almost completely insectivorous on Komodo where fruits are rare most of the year (Auffenberg, 1981a). Rabor (1977) claimed that *P. philippinensis* not only eats fruits, but "also feeds on birds and poultry. In fact, it is considered *by settlers* [italics mine] as one of the worst predators of domestic chickens. It also feeds on numerous rats and mice." Thus the current data suggest that *P. philippinensis,* like *P. hermaphroditus,* may be highly opportunistic, feeding regularly on fruits but shifting to an even more frugivorous diet when conditions dictate it. In the Caramoan area, most animal foods are eaten in the wet northeastern monsoon season, when fruits are generally less common and few local species are fruiting (compare fig. 9-2 and table 12-3).

The palm civet is a common animal in the Caramoan area. Individuals were seen at night and during the day in fissures and rock piles; spoor and droppings were common along paths, field edges, and creek beds. It is among the common medium-sized vertebrates of the forest. In Malaysia, *P. hermaphroditus* is about 68 times more common than *Viverra tungalunga* (Harrison, 1969). In Africa, Charles-Dominique (1978) reported a home range of about 15 ha for *Nandinia binota,* a closely related paradoxurine with similar habits. Unfortunately, no data are available on the density of the Philippine palm civet, but Charles-Dominique et al. (1981) have shown that similar-sized nocturnal frugivores of neotropical forests have densities of 20 to 30 individuals/ km^2, which seems about right for the palm civet in the Caramoan area as well.

Appendix 2 and figure 12-1 provide additional information on other species that eat the same food as the butaan.

Some Caramoan fruits, such as figs and bananas, are eaten by many different kinds of forest animals. McClure (1966) reported that in Malaya, 14 of a

TABLE 12-3. Seasonal variation (in %) in foods of *Paradoxurus philippensis* in the Caramoan area based on analysis of droppings (*N* = total items in combined droppings).

Butaan foods N =	Jan 187	Feb 454	Mar 97	Apr 108	May 508	Jun 230	Jul 202	Aug 85	Sep 31	Oct 86	Nov 138	Dec 183
Plants												
Aglaia	33.2	33.2	76.3	11.1	31.7	7.8					6.2	6.5
Caryota		0.5			0.2	15.3		31.7	17.0	11.3	8.7	11.6
Ficus	5.9		10.3		3.7	8.7	14.8		10.0	1.3		6.5
Malaisia								3.5	4.0	1.2		
Pandanus												0.1
Sandoricum						0.4	2.0		27.1			
Spondias							13.4	2.3				
Other plants	59.3	66.3	13.4	88.9	64.2	67.8	69.8	62.5	40.9	83.5	81.9	73.1
Animals	1.6				0.2				1.0	2.7	3.2	2.2

total of 32 fig species are heavily predated by vertebrates. I found the same pattern in Caramoan, for of 84 *Ficus* species found locally, 21 are utilized by many forest vertebrates, including the butaan (though infrequently). *F. variagata, F. balete, F. benjamina,* and *F. ulmifolia* are examples of widely used species in the Caramoan area. Bananas (both domestic and wild varieties) are eaten by almost all forest vertebrates, and *Nephelium* fruits are also commonly eaten by many animals in the forest, but neither by the butaan.

Other fruits eaten by a wide variety of vertebrates and the butaan include the local species of *Grewia, Dracontomelum,* and *Spondias*—all sugary species. However, most local edible fruits eaten by potential competitors are not eaten by the butaan (82%), so that food competition between it and other forest animals must be limited.

There is no proof of competition for space between the butaan and other forest inhabitants. Rock crevices in which *V. olivaceus* sometimes secretes itself during the day are also used by a variety of other animals, including several species of rodents, two local civets, rock gekkos (*Gekko monarchus,* which also lays its eggs there), and land snails, particularly species of the genera *Ryssota* and *Lepidotrichia*. Some of these mollusc crevice coinhabitors are eaten by *V. olivaceus* whenever found (chap. 9).

Interrelations with V. salvator.—The water monitor (*goto* locally, *biawak* in most other Philippine dialects) is common throughout the entire archipelago, as it is in the Caramoan study area. There the water monitor is not restricted to streamside habitats, as it is in drier environments (i.e., Flores Island; Auffenberg, 1980a). Though uncommon away from streams, it occurs in most Caramoan habitats with appropriate cover, including rocky mountainsides far from surface water. *V. olivaceus* is restricted to rocky forested ridges

TABLE 12-4. Animal food of *V. olivaceus* and *V. salvator* in the Caramoan area.

Animal taxa eaten	*V. olivaceus* (N = 186)		*V. salvator* (N = 126)	
	Juveniles	Adults	Juveniles	Adults
Mollusca	−	+++	−	+
Crustacea	−	−	−	+++
Insecta	+++	−	+++	+
Mammalia	−	a	−	+++
Aves	−	+	−	+
Reptilia	−	−	−	++
Amphibia	−	−	−	++
Carrion of any of the above	−	−	−	+++

SYMBOLS USED: − = none; +, ++, +++ = increasing frequency.
a. Only in captivity as far as is known.

and mountainsides. It does not occur in freshwater marshes, mangrove, nipa swamp, and beach scrub, where the water monitor is usually common. Thus whatever interactions occur between them are restricted to certain habitats—chiefly primary and secondary mountain forests.

Locally the water monitor is found almost to the top of the highest mountains (400 m), whether water occurs there or not. Radio telemetry of several individuals shows that this species tends to have larger activity ranges than does the butaan (chap. 6). Furthermore, *V. salvator* seems to occur in higher densities; it is not certain because this species can be drawn to a trap by carrion, whereas *V. olivaceus* cannot.

In mixed dipterocarp habitats both species use the same type of rocky surface and subsurface retreats, though the water monitors do not ascend trees as regularly, nor do they remain there as long, as the butaan. It is not known what, if any, interactions occur between them when individuals meet in the forest. In captivity there was no reaction when individuals of different-sized *V. salvator* were placed with butaan adults.

While the food of *V. olivaceus* is about half fruit and half animal (chap. 9), that of *V. salvator* is entirely animal. There is, therefore, little competition between the two species for food, except by butaan young. Table 12-4 shows the overlap in animal food resources of these two species in the Caramoan study area.

Parasitism

Ectoparasitemia.—Ticks and mites are apparently common on most species of the genus *Varanus,* and several monitor species are strict-total specific, in the terminology of Hoogstraal and Aechlimann (1982). Other ectoparasites are rare on varanids, though a leech (*Glossiphinia* sp.) has been reported to have attacked *V. niloticus* (Mertens, 1942).

The only ectoparasites found on the butaan are the ticks *Amblyomma helvolum* and *Aponomma fimbriatum*. Both genera are common parasites of *Varanus* species and are widely distributed in Southeast Asia and the Indo-Australian archipelago (see Anastos, 1950; Kaufmann, 1972). Both species have been reported from *V. salvator; Amblyomma helvolum* is also known from *V. komodoensis* (Auffenberg, 1981a) and several species of agamid and scincid lizards in the Caramoan area (T. Auffenberg, in press). It has not been reported previously from *V. olivaceus*, nor has *Aponomma fimbriatum*.

Aponomma fimbriatum is apparently rare as a parasite of the butaan, for it was found on only 3 of 116 individuals examined for ectoparasites—one individual of *V. olivaceus* in each of the following areas: Catanduanes Island, near Hituma; Luzon Island, Camarines Sur, near Caramoan; and (identity not certain) Sorsogon, near Bulusan Volcano National Park. *Amblyomma helvolum* was clearly the most common tick species, found on most larger individuals of all populations, but not on the smallest specimens, during all months of the year, showing a fairly even monthly distribution.

Of 116 butaans examined for ticks, 96% had one or more *Amblyomma helvolum*. Infestation was more or less equally distributed between males and females. However, there was a decided difference in occurrence of infestation and *V. olivaceus* size as represented by weight. Only one of 9 individuals weighing 1 kg or less had any ticks (1.8% of total sample studied); all those weighing more than 1 kg had ticks. The heaviest infestation was found in individuals weighing between 1 and 3 kg (52%); it was less in individuals over 3 kg (32% of total). The number of ticks per individual lizard for the entire sample varied from 0 to 62 ($\bar{X} = 11.28 \pm 12.56$); for individuals having ticks the number varied from 1 to 62 ($\bar{X} = 13.17 \pm 12.62$, $N = 48$). These numbers show a great variation in the number of ticks per individual. There is a sexual difference of 3.58 in the mean number of ticks on female and male monitors, which is not statistically significant at the 5% level.

Female ticks are much less common than males; the ratio of females to males is 1:10.96. The only other sex ratio available for this tick species is that on the host *V. salvator* on Flores Island, Indonesia, the ratio is 1:18 (Auffenberg, 1981a). It is not, however, significantly different from that reported here for the butaan.

Male ticks are generally found either on the bases of the claws of the front or back feet (90% of 537 ticks) or scattered over the body (a few); female ticks are rarely found on the claw bases (10%) and never on the posterior parts of the body. Instead, female ticks are most common on the skin between the toes of the front feet (82%) but also occur on the head, neck, and anterior third of the body (including the axillary region). Table 12-5 and the resulting chi

TABLE 12-5. Occurrence of the reptile tick *Amblyomma helvolum* by sex and by frequency in different body areas of *V. olivaceus* ($N = 118$).

Tick sex	Areas of host's body			Total ticks
	Claws	Between toe bases	Other areas	
Females	0	40	9	49
Males	481	0	56	537
Totals	481	40	65	586

square test shows that these distributions reflect a definite relationship with tick sex ($X^2 = 90.33$, df $= 2$, $p < 0.005$). Also, ticks of both sexes often congregate around wounds.

Studies of this tick species on other varanid hosts (T. Auffenberg, in press; Auffenberg, 1981a) show that parasitic distributions are not always identically site-specific. There is, in fact, some reason to believe that *Amblyomma helvolum* has completely different distributional patterns in the several species of reptiles it is known to parasitize (T. Auffenberg, in press).

In a study of ectoparasites of *V. komodoensis*, I found that the abundance of ticks (three species) varied seasonally, with sharp increases (140%) in infestation rate from the wet into the dry season (Auffenberg, 1981a). There is no evidence of significant seasonal differences in tick infestation in the butaan. The Komodo monitor carries approximately the same total number of ticks of three species as *V. olivaceus* carries of only one species throughout the entire year. Another significant difference in tick infestation in these two varanids is that the number of ticks per *V. komodoensis* individual seems directly proportional to the SVL, the largest lizards having the greatest number of ticks. On butaans up to 30.9 cm SVL the average number of ticks is 0.3 ± 0.2; in those with an SVL of 31.0–49.9 cm, the average is 5.5 ± 1.0; in all those larger than 50.0 cm the average is 15.4 ± 2.0 (the last two are not significantly different; data from T. Auffenberg, in press). Thus the medium and largest individuals have the same infestation rate, which is significantly higher than infestation of the smaller size class.

Endoparasitemia.—Adamson (1986) has described a new species of nematode (*Meteterakis vaucheri:* Heterakoidea) from the posterior intestine of a specimen of *V. olivaceus* from Dipapulao Aurora Province, Luzon. The genus typically has direct life cycles, so that the most likely source of infection was an infective ova picked up on the surface or in water. Most species of the genus live in amphibian hosts; a closely related genus occurs in chickens. In general all members of the family tend to be highly host-specific.

Data on the number and anatomical positions of gut parasites were available. All gut parasites found were nematodes. The maximum number found in any individual was 34 ($\bar{X} = 7.13 \pm 0.21$). Only 63% of all adult butaans had more than one parasite in the gut; 13.2% had none. No subadults had any nematodes. Butaans differ from *V. salvator* in the same area, in which many more nematodes are found per host ($\bar{X} = 14.6 \pm 0.62$; Auffenberg, field notes). A more striking finding is that all other herbivorous lizards in which gut parasites have been studied are packed with nematodes (see Iverson, 1982). However, these are all folivores, and the reduced nematode load of the butaan may be related to its frugivorous habits.

A series of blood slides made from 34 specimens of *V. olivaceus* in the field, examined by S. R. Telford, shows that this monitor lizard is parasitized by a small, undescribed malarial plasmodium, which is distinct from but probably closely related to *Plasmodium clelandi* from Sri Lanka. *P. clelandi* is known only from *V. bengalensis* and has not been found in other Sri Lanka lizards, including *V. salvator*. The same pattern is found in southern Luzon, Philippines, for none of 12 specimens of *V. salvator* tested possessed any plasmodia, but all of the *V. olivaceus* individuals examined from Luzon had them. The new plasmodium species found in *V. olivaceus* is being studied by S. R. Telford, and he will publish a separate description.

Enemies

Juvenile butaans are undoubtedly preyed upon by a number of larger animals, though no direct evidence is available. The cryptic coloration of the young particularly (bright greens and disruptive dark bands) suggests that, at least at this stage of life, predation by visual-oriented carnivores, such as raptors, may be high. Mammalian predators probably include the singalong (*Viverra tungalunga*) and several larger reptiles, among which the python (*Python reticulatus*) is probably the most important. It has been reported as a predator of *V. salvator* (Mertens, 1942), and I have considerable evidence that, in India, *V. bengalensis* is frequently preyed upon by *Python molurus*. Smith (1931) reported a 4.25-m king cobra (*Ophiophagus hannah*) eating a 1.29-m *V. salvator* (the snake is reasonably common in the Caramoan area). Mertens (1942) reported that at least some species of monitors are not immune to the bite of poisonous snakes, although Spatz (1931) thought that *V. griseus* was immune to the bite of the viper *Cerastes cerastes* (there is no such experimental evidence, however). That at least some varanids eat poisonous snakes is proven by the feeding habits of *V. komodoensis* (Auffenberg, 1981a).

Man is clearly the most important predator of the butaan, for the species is regularly hunted (chap. 15). Table 12-6 lists its probable predators and the likely life history stage of the butaan affected. Though wild pigs (*Sus* cf. *S. scrofa*) occur in some of the habitats in which *V. olivaceus* is found, the facts

TABLE 12-6. Probable predators of *V. olivaceus* in the Caramoan area.

Predator species	Probable life history stage and location of *V. olivaceus* affected
Reptiles	
Lizards	
Varanus salvator	Eggs and young (notorious reptile-egg predator); on and in trees.
Snakes	
Python reticulatus	Medium to large adults; in burrows.
Ophiophagus hannah	Small to medium; on ground, in holes.
Birds	
Raptors	
Haliaster indus	Young; open country, from trees.
Aviceda jerdoni	Young; closed forests, from trees.
Ichthyophaga ichthyaetus	Young; riverine forests, from trees.
Spilornis cheela	Young; various habitats, from trees.
Mammals	
Carnivores	
Viverra tungalunga	Eggs and young; on and in trees.
Man	
Homo sapiens	Adults (mainly); with trained dogs, throughout habitat.

that this lizard lays most of its eggs high in hollow trees and that the young are largely arboreal suggest that they are not important predators of this monitor species (as they are of some other species, i.e., *V. bengalensis,* Auffenberg, field notes, and *V. komodoensis,* Auffenberg, 1981a).

Plants and the Butaan

Seed dispersal.—The best-studied interaction between plants and vertebrate animals is the effect of the latter on the morphology of fruit and the timing of fruit production. Because this interaction is of central interest to the studies reported here, it will be discussed in some detail.

In evolutionary terms, the most significant reproductive parts of plants are flowers, fruits, and seeds. The dispersal of seeds is especially important in the tropics, since concentrations of seeds are likely to attract and perhaps favor the increase of animals that will destroy them. Plants must continually evolve measures to reduce seed destruction; if seed destroyers learn to overcome them, the plants will cease to produce offspring. Dispersal is a way of reducing seed destruction; in the widest sense it includes the actions of microorganisms and insects as well as higher vertebrates. Seed predation is believed to be a major factor in maintaining plant diversity in tropical rain forests (Janzen, 1971b).

The term frugivore applies equally to animals that eat or otherwise destroy the seed (seed predators) and to those that eat the pulp (pulp predators) and later excrete or spit out the seed, thereby acting as seed dispersers. The butaan is a pulp predator and seed disperser—not a seed predator. Seed dispersal is particularly common in tropical rain forests (90% of species in Costa Rica, according to Frankie et al., 1974; 67–73% in Malaysia, Raemaekers et al., 1978; still unknown in the Philippines). The importance of the distinction between seed dispersers and seed destroyers becomes evident when we consider some examples of the coevolution of the butaan and some of its food plants.

Nutritive pulp around the seed attracts animals (*V. olivaceus,* among others) which disperse seeds that might otherwise be taken by seed destroyers (some local forest rats). There are fruits that evidently have evolved in conjunction with birds. They have a thin rind or one that splits open when the seed is mature, and all are brightly colored to attract attention. In primary forest such fruits are often produced in the upper canopy (Fogden, 1972), but in clearings and secondary forests many are found at lower levels. Fruits that appear to have evolved in response to mammalian predators (palm civets, primates, etc.) tend to have a tougher rind, do not split open, and are brightly colored only if taken by diurnal species. Such fruits eaten by bats are often on pendulous stalks hanging from branches easily reached by flying animals. They are frequently white, yellow, or another light color.

Raemaekers et al. (1978) classified "bird" fruits as unspecialized, nonpulpy, or specialized. The first are small, yielding little nutritive value from each fruit in return for dispersal. Local examples are *Ficus* and *Eugenia* (Vellayan, 1981). The palm civet feeds on many that fall into this class, but the butaan does not, except for a few species of strangler figs eaten quite sparingly. The nonpulpy type has many seeds; they are sometimes digested but also can pass unharmed through the digestive tract. Local examples are members of the Cucurbitaceae. The butaan does not eat any of these fruits either. The third group (specialized) is composed of species that produce a more nutritious pulp, either sweet or oily. They are taken primarily by specialist frugivores, such as the hornbills and pigeons, not by palm civets or monkeys (though they eat them). It is this class upon which the butaan feeds most extensively.

Smythe (1970) and Charles-Dominique et al. (1981) similarly divided seeds and fruits, respectively, into three size classes; the one representing the largest size class is most important in seed dispersal by animals that swallow them and later drop them in their feces as much as several kilometers from the parent tree (Ridley, 1930; Snow, 1962; Hladik and Hladik, 1969). As Snow (1965) pointed out, it may be advantageous for plants producing seeds dispersed in this way to stagger their fruiting seasons, thus reducing competition for dispersal. As shown in chapter 10, this staggering is what takes place in the Caramoan forest when only these fruits are considered. Plants using animal-dispersing mechanisms also tend to fruit over a longer period of the year than Caramoan species dispersed by wind.

The efficiency of seed dispersal by the butaan depends on the proportional number of seeds ingested. If there are more seeds than can be eaten, the surplus is wasted. Conversely, if there are not enough fruits of any kind at any one time of the year, the butaan must switch its food base to something else. The latter behavior occurs during the wettest part of the northeastern monsoon, when fruit loads are low (chap. 10) and it eats more animal prey (chap. 9). The molluscs, however, are not simply filler but represent a rich and important source of protein and calcium in a diet otherwise poor in these nutrients. Any plants fruiting during lean times will increase their chances of dispersal—probably the reason that animal-dispersed plant species tend to have longer fruiting times than wind-dispersed species (Smyth, 1970; data from this study).

The best strategy for the plants bearing fruits attractive to pulp predators-seed dispersers is to produce them at different times than other plants bearing similar fruit. This phenomenon has been documented for the genus *Micronia* in Trinidad (Snow, 1965) and for a number of figs in Malaysia (Medway, 1972) and the Philippines (T. Auffenberg, MS). Such asynchrony occurs more between than within species and is shown among butaan food species by fruits of the genus *Canarium* in the Caramoan area.

Figs are a special case among fruits evolved for and with rain forest pulp feeders. These are not actually true fruits but collections of flowers completely enclosed by a pulpy receptacle. Some are sterile and produced expressly to provide food for the larvae of "fig wasps," which are obligate pollinators of the female flowers (Corner, 1940; Janzen, 1979). They may sometimes be eaten more for the contained pollinating insects than for the fruit pulp (see Redford et al., 1984). It is perhaps the protection afforded by the pulpy container that allows figs to be produced at any time of year, even when the fruit of other genera may fail due to dry weather (as during the 1983 Philippine drought). This constant availability is important to forest animals in the Caramoan area and elsewhere (Medway, 1972); *Ficus* provides food more continuously throughout the year than any other genus of "fruit" tree in the forest (McClure, 1966; Fogden, 1972). It is thus not surprising that the species of this genus are utilized as food by many animal species (Medway, 1972). However, *Ficus* species are not important in the feeding ecology of the butaan (chap. 9), in spite of their frequent fruiting pulses and high fruit yields (T. Auffenberg, MS).

Another strategy of plants producing fruits to be dispersed by vertebrates is for individuals of one or more taxonomic groups to fruit rarely in synchrony. However, this mechanism of satiating potential predators is incompatible with mechanisms of seed dispersal, and most plant species exhibiting it are wind-dispersed (Janzen, 1971a). Most of the emergent trees of the forest fall into this category (*Sterculia* and many genera of local Dipterocarpaceae). The butaan eats the fruits of none of these, though some parrots eat a few.

Another important factor in seed dispersal is mechanical protection—making it difficult for seed predators to extract seeds from the fruits or for invertebrates to penetrate them. Such devices include thick tough rinds (*Xylocarpa*), woody shells around the seeds (coconut), tiny irritant hairs on the fruit surfaces (*Canarium hirsutum*), and sticky resins in the rind or pulp, particularly when immature (*Ficus, Artocarpus*). Mechanical or chemical protection does not entirely protect such seeds; the large cloud rat (*Phloemys cumingi*) can open even the hardest fruits (such as coconuts) with ease, and the butaan regularly swallows some fruits with a hard rind entire. The butaan also feeds on the milipili (*Canarium hirsutum*), in spite of its irritating hairs. Similar stinging hairs are also found on the immature fruits of *Caryota cumingii,* and the ripe ones (regularly eaten by *V. olivaceus*) are protected by calcium oxylate crystals (Sulit, 1918). Various forest rats also feed on both these fruits, as do several local species of hornbills.

While there have been several studies of tropical forest fruits that are eaten by seed destroyers and seed dispersers or are wind-dispersed, few have addressed the problem of seed shape as a factor in gut retention by seed dispersers. As pointed out earlier, all fruits eaten by the butaan tend to be large, and all (except *Malaisia*) are usually single-seeded. In all these fruits, the pulp from around a central seed is digested and the seed later expelled some distance from where it was eaten. It would be advantageous for these fruits (also commonly eaten by other seed-dispersing pulp eaters, such as hornbills) to remain in the gut as long as possible to increase the distance between the parent tree and the place where they are dropped.

Three characteristics of the fruits eaten by the butaan aid in gut retention: all seeds are covered either with long soft fibrous "hair" or stiff bristles, or they have acutely pointed projections at one or both ends of the seed (fig. 12-2) and often have a triangular cross section. Upon dissection of fresh butaan digestive tracts, it can be seen that seeds with pointed ends (*Canarium* species) tend to "catch" in the mucosal lining, particularly at bends. It is not known how long they might remain there, but the fact that the pulp of some fruits in the gut is completely digested from the seeds and the pulp of others of the same species may not be suggests that some fruits pass one another in traversing the gut and that some catch and remain for longer periods than others. (The holding of fruits in the caecum for longer than usual passage time is discussed in chap. 3.)

The seeds of other fruits eaten by seed dispersers are sometimes similarly modified morphologically to slow passage through the gut. The flat seeds of

Momordica cochinchiensis (Cucurbitaceae) have numerous small spines around the periphery that are adapted to lodge in the guts of hornbills, which regularly swallow large mouthfuls of the ripe pulp (containing many seeds, as in a watermelon).

Another shape adaptation of the seeds of fruits swallowed by the butaan is the distinctly subtriangular cross section of many of them (fig.12-2). There is a progressive sequence from the seed of *Aglaia harmsiana* with three low ridges to the almost star-shaped seed of *Spondias pinnata*. The probable advantage of this cross-sectional shape is that as the pulp is digesting, the seed ridges press against the gut wall, tending to slow down the movement of the seed through the gut but still allowing more liquefied food to move past the partly digested fruit. The digested portion does not force the still undigested part remaining on the seed to move to the cloaca. Similar subtriangular seeds seem to be eaten by frugivores in South America (*Richardella*, illustrated in Charles-Dominique et al., 1981), but I have no knowledge that this seed shape is as common there as in Asian forests. However, a study of the seeds of the fruits most commonly eaten by toucans and other large fruit-eating birds may show that adaptations in seed shape that increase passage time are probably more common than now documented.

Another major type of "reluctant" seed is covered by fibrous filaments. Some of these fibers are stiff and concentrated at the seed end; the brushlike ends tend to lodge the seeds in the small intestine particularly, where the inner vilous lining serves as a nonslip surface. Others are long and soft, originating from the entire seed surface; these filamentous fibers tend to become entangled in the villar surfaces of the intestine. The only fruits in the first category eaten by *V. olivaceus* are *Pandanus*, especially *P. radicans*, though seeds of other local fruits that are animal-dispersed have the same fiber structure, such as *Arenga, Cerberus,* etc. Fruits eaten from the second category are *Spondias pinnata, Grewia stylocarpa,* and *Dracontomelum dao*. A palm fruit regularly eaten by the butaan, *Caryota cumingii*, has a hard, waxy-coated seed that is apparently meant to be passed quickly.

The butaan and plant toxins.—Forest plants of the tropics have also developed a number of important protective chemicals (= secondary compounds). There is an immense diversity of phenolics, alkaloids, and other secondary compounds—substances of no obvious metabolic use to the plants (Ehrlich and Raven, 1964; Feeny, 1975; Janzen, 1975a, 1978). The pressures on plants to defend themselves chemically against insects, and on insects to find ways around these defenses, are thought to have contributed to the massive diversification of plants and insects found in rain forests today. Such toxic secondary compounds are also expected to affect mammals, especially since they have more varied diets than host-specific insects; they can evolve narrowly specific means of defeating defenses (Freeland and Janzen, 1974). These sec-

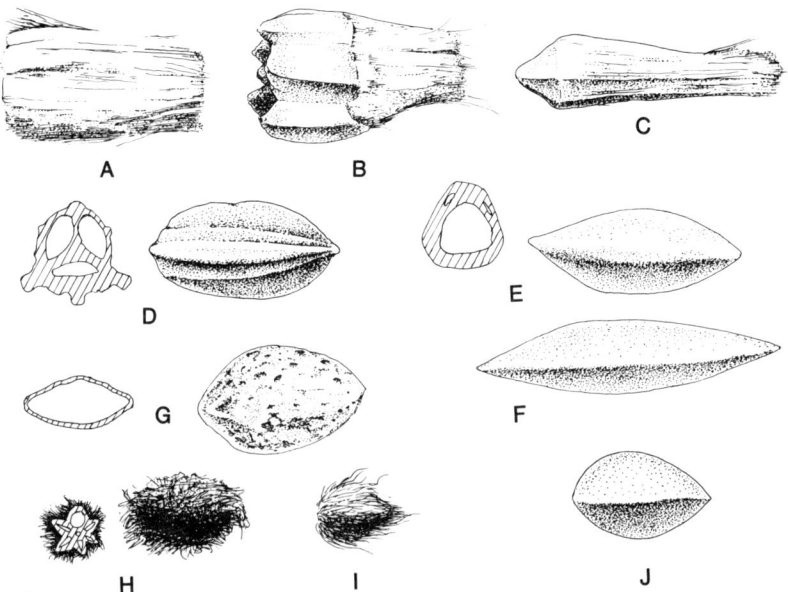

Fig. 12-2. Seeds of fruits regularly eaten by *V. olivaceus: A, Pandanus radicans; B, Pandanus tectorius; C, Pandanus simplex; D, Canarium hirsutum* (showing cross section as well); *E, Canarium ovatum* (with cross section); *F, Canarium* sp.; *G, Uvaria sorsogonensis* (with cross section); *H, Grewia stylocarpa* (with cross section); *I, Dracontomelum dao;* and *J, Canarium vrieseanum.*

ondary compounds affect the maximum quantities that can be consumed, the time that must elapse before the same species can be eaten again, and the differential exploitation of plant parts (see Janzen, 1978, for discussion).

The digestive system of the butaan is apparently immune to many of the secondary compounds that irritate the human digestive tract. One good example is the calcium oxylate crystals found in the hairs covering even the ripe fruits of *Canarium hirsutum* (and several species of palm fruits, including the common butaan food *Caryota cumingii*). These hairs cause severe swelling of human mucous membranes and may even lead to death. Some iguanines also are known to feed on fruits toxic to humans (Wiewandt, 1977; Auffenberg, 1982a), and palm civets will eat the fruits of *Semicarpus* (Bartels, 1964), which is known to cause festering wounds on the skin of most humans.

Many fruits are protected prior to ripening by milder toxins, for the most part chemicals that make the unripe fruits unpalatable. Tannins and other toxins found in unripe fruits are believed to repel most seed dispersers. However, some monkeys and rodents in India regularly eat the fruits of *Terminalia chebula,* which contain up to 53% tannins (Hathaway, 1959) even when ripe.

While exceptionally high tannin content is thought to have been evolved to discourage predation, Howes (1953) made the interesting point that tannins inhibit potent fungal spore germination, thus retarding deterioration of fallen fruit. On this basis, Janzen (1978) suggested that tannins may thus increase the eating of these fruits, at least by the vertebrates that have coevolved a nonsensitivity to high toxic levels of specific types. He made the further suggestion that such antibiotics in fallen fruit may be expected to affect adversely the microflora of certain animals that depend on such flora for efficient digestion and may explain why wild pigs often ignore fallen fruits with a high tannic acid content. It may explain, incidentally, why the partially herbivorous iguanine lizard *Cyclura carinata* feeds on fewer fruits than expected on the basis of availability (Auffenberg, 1982a). It may also explain why so many leaves pass through its gut relatively undisturbed by the digestive process (Auffenberg, 1982a; Iverson, 1982). However, high tannin content is obviously not a major problem for the butaan, for some fruits regularly eaten (*Canarium*, all species) contain high tannin and terpin levels (chap. 9). So few vertebrates regularly eat the fruits of *Canarium* that one can only surmise that the energetic costs of detoxifying them must be great. (Of all the frugivorous vertebrates of the Caramoan area, only the hornbills and the butaan regularly feed on the pulp of *Canarium* species. Hornbills are particularly efficient in detoxifying some fruits poisonous to other animals, such as *Strychnos nux-vomica* [Dixon, 1894].) I presume that *V. olivaceus* is able to withstand regular minor liver damage in exchange for a large amount of nutrients (see Freeland and Janzen, 1974, for discussion of other herbivores). This tradeoff is expected in *V. olivaceus*, where liver damage can be rectified at another time of the year by switching to other less toxic food resources than the *Canarium* species—which is exactly what happens (chap. 9).

Virtually all carnivores are periodically confronted with large numbers of prey at one time of the year and few at another. The pattern is, in fact, characteristic of the life history of some varanids (e.g., *V. komodoensis*, Auffenberg, 1981a). It is also characteristic of many plant eaters (Janzen, 1978), of which the butaan is a good example. During part of the year butaan adults avail themselves of a large amount of high-quality food that is converted to body fat (chap. 3), which is thus available in harder times when even higher energy demands may be placed on the body (gametogenesis, territorial defense, etc.). Vertebrates restricted to the upper canopy usually have fewer opportunities to build fat reserves than those occurring lower in the forest levels, particularly on the ground (Janzen, 1978). Thus one would expect the largely herb-eating land tortoises (genera *Gopherus*, *Goechelone*, etc.) to be the most likely both to build up fat reserves and to use them during periods of food shortage. Largely arboreal, mainly folivorous reptiles (*Iguana*, *Hydrosaurus*, etc.) should show less annual variation in fat reserves and should shift onto foods rich in both nutrients and defensive compounds during times of short food supplies. What is known suggests that this pattern applies to *Cyclura carinata* (Auffenberg, 1982a). However, the entire matter deserves study in a series of genera belonging to several partially to completely different feeding guilds.

Janzen (1978) also pointed out that lack of fat may cause other complications for herbivorous vertebrates; when they are forced to eat more toxic foods, the defensive compounds may circulate in the female's system. While the embryos do not develop inside the mother in egg-laying reptiles, the body fat is apparently metabolized and somehow important in fecundity in most reptiles (chap. 8). This fact may explain why the livers of hatchling *V. olivaceus* and *V. salvator* are so large in proportion to those of adults. Adult *V. salvator* ingest large amounts of carrion. Much toxic material is produced from the respiration/digestion of the bacteria ingested, and it is likely reflected in the chemistry of the fat and liver of the mother. The entire matter of toxic accumulation in the liver and fat and the metabolism of body fat in reptiles needs careful study.

Relationships between the butaan and many of the plants it eats suggest that, along with the hornbill, the butaan is important in seed dispersal, particularly of the oily, tannin-, phenolic-, and terpin-loaded *Canarium* species, and that, along with many other animals (chiefly palm civets, birds, and some of the bats), it is a less important disperser of the seeds of the high-sugar fruits it eats. To this extent, certain adaptations of the butaan's gut, as well as the shape and size of these fruits, ensure that the seeds are moved as far from the parent tree as possible. Several studies (Smythe 1970; Smythe et al., 1982) have shown that in Central American forests the seeds too large to be swallowed entire by any local frugivore tend to be scatterhoarded by rodents (Morris, 1962). Scatterhoarding is unknown in Southeast Asian forests—probably because hornbills *can* swallow large fruits and thus disperse the seeds, as can the butaan, at least in some parts of the Philippines.

While some environmental factors, such as rainfall, are probably most important in determining fruiting pulses (Foster, 1982), competition for seed dispersal is probably also a selective factor. As a result, fruits meant to be eaten by vertebrates such as the butaan tend to fruit synchronously in respect to species but staggered in respect to the entire suite of plants with similar reproductive strategies. Herrera (1982) has shown that fruiting species ripening at different times of the year in Spain differ in their nutritional properties in such a way as to match the seasonally changing demands of their major dispersers (birds in this case). While the principle may be applicable to most tropical ecosystems as well, there is no current evidence that seasonal pulses of fruits are related to the seasonal nutritional needs of the butaan. The butaan needs to have its protein supplied by animal rather than plant food, and high caloric oily foods are available to it throughout the entire year, though species-switching is required to ensure continuous availability.

As shown above, plant protection is another important factor in utilization by certain vertebrates. While many plants protect themselves by chemical means from being eaten (or eaten too soon), the butaan is evidently among the vertebrates that have developed a reasonably effective detoxification system for such chemicals. Other tropical frugivores may protect themselves by reducing intake to minimal amounts or by use of microsomal enzymes (Raemaekers et al., 1978), but the butaan's gut is not greatly modified for a microflora aiding digestion of cellulose-rich plant products (chap. 3). The butaan's lack of gorging during times of plenty suggests that minimal but consistent feeding is a major part of its strategy, unlike the "feast or famine" regime of the Komodo monitor (Auffenberg, 1981a). The summary of butaan feeding in chapters 9–11 suggests that this species probably selects food on the basis of either its protein-calcium content (molluscs) or its carbohydrate content. Carbohydrate foods are either oily or sugary. This utilization of sugary and oily foods is similar to that reported for some tropical forest mammalian frugivores (Hladik, 1978; Milton, 1979). The energy required by *V. olivaceus* to detoxify fruits high in secondary compounds (such as *Canarium* species) is well spent on the basis of the carbohydrate-rich nutrients obtained from them. The sugary fruits, eaten by a much larger number of opportunist forest frugivores, contain far fewer secondary compounds—especially when ripe. They are considered "cheaper" fruits (produced more quickly, in greater abundance, and with less energetic cost) by Howe (1982), who also shows that many such fruits in Panama are wasted because two-thirds fall to the ground uneaten. Higher quality fruits are produced regularly, in small numbers over a longer period of time, for a few specialist frugivores, with the result that proportionally fewer fruits are wasted.

Finally, while seed dispersal is important to many tropical tree species, it may be even more important to secondary growth species, since they rely for their survival on the rapidity with which they can colonize bare ground and gaps in the forest. Almost all of the plant species whose fruits are eaten by the butaan fall into this category (chap. 9). This plant spreading is facilitated by the fact that the butaan often deposits its feces on bare ground. The local palm civet behaves similarly, and Bartels (1964) believes that the Javan palm civet plays an important role in the reforestation of landslides and abandoned agricultural lands.

Germination.—The literature reveals numerous examples of the effects of animal digestion upon the viability of plant seeds. It is not known if germination is facilitated after seeds pass through the digestive tract of *V. olivaceus*, but it is presumed to occur. This passage may be the case especially for *Canarium hirsutum*, in which germination takes place through a trap door-like structure that loosens and falls away when the connective tissue is digested or rots away. It is less clear, but possible, that gut passage enhances germination in other species eaten by the butaan, since germination of some seeds is presumably enhanced when the seeds move through the digestive system of some mammals (Hladik and Hladik, 1969) and tortoises (Rick and Bowman, 1961). *Aglaia harmsiana* and all species of *Canarium* germinate in a few months and do well as saplings in open sunny areas. Germination of these seeds does not require a shady environment. That these species are common pioneer forms is evident from their distribution in the forest, probably originally most common in tree fall and other natural clearings but now widely dispersed because of forest modifications by man. Furthermore, at least *Canarium hirsutum* may bear fruit when the sapling is only a few years old and 2 m high.

13

Intraspecific Interactions

INTERACTIONS between butaans vary greatly, and only a few can be dealt with here. Competition for food and space, agonistic interactions, and courtship behaviors are probably the most important categories and are discussed in this chapter and in other sections of the book. The emphasis here is on combat and courtship.

The excellent review of reptile social behavior by Carpenter and Ferguson (1977) makes it clear that a considerable amount of information has accrued. However, almost all critical observations and experiments have been conducted on the closely related lizard families Iguanidae and Agamidae. There are some important recent contributions on the behavior of varanids (Murphy and Mitchell, 1974; Carpenter et al., 1976; Vogel, 1979; Auffenberg, 1981a, b, 1983b). In general, they seem to show that color is not an important visual releaser, as it is for many agamids and iguanids; rather, shapes, positions, movements, and, particularly, odors associated with special glands in the skin or the cloacal area serve this purpose.

Monitor lizards have few external sexually discriminative characteristics (Mertens, 1942, 1958), though males generally attain larger size and, by this means, are usually dominant to females (Auffenberg, 1979b). Aggressiveness between individuals of all species studied is common and frequently leads to severe injuries and even death (Auffenberg, 1981a). This fact is important in intraspecific coactions, as it accounts both for high stress levels during casual encounters and for a tendency for long-term associations between adult males and females to avoid the potential dangers of new associations (Auffenberg, 1983b).

Behavioral Acts

The behavioral acts listed, defined, and briefly discussed below in regard to combat and courtship in the butaan include releasers, primers, and special tactics that tend to improve success in fighting and courting (behavioral act = particular behavior or display; releaser = stimulus eliciting immediate response [Lorenz, 1937]; primer = stimulus changing the responsiveness of individual [Wilson and Bossert, 1963]). Complete behavioral act inventories for a varanid (*V. komodoensis*) can be found in Auffenberg (1981a) and in Carpenter and Ferguson (1977) for all lizards, so they are not repeated here. Butaan courtship and combat acts are similar (often identical) to those I have described and illustrated elsewhere for *V. bengalensis* (Auffenberg, 1981b, 1983b), except for minor modifications in classification and interpretation. Following some of these earlier papers, I have organized the butaan's behavioral acts into several contextual and functionary classes: warning, weaponry use, investigation, stress reaction, social (dominant and subordinant), reproductive, combat tactics, and miscellaneous. These classes are equivalent to the adaptive behavioral types of Scott (1950).

Warning Class

The aggressive acts comprising this class are exhibited during either offensive or defensive coactions, most frequently by adults but also by juveniles, and are used in a variety of contexts (combat, courtship, etc.).

Tail coil.—Partially or completely coiling the tail in the horizontal plane usually precedes and leads to a tail slap. In varanids it is a common act that has been reported in the literature for several species and observed for more in captivity. It does *not* commonly precede tail slap in *V. olivaceus* as it does in *V. komodoensis* or *V. bengalensis* but is as common in intraspecific encounters as in those two species.

Gape.—A strong cue, common among many varanid species, gaping is always used by individuals of all sizes in a defensive context, most often in intraspecific encounters but also (though rarely) during interspecific coactions. Gaping also occurs at different intensity levels in other varanids (Auffenberg, 1981a), but "full gape" is rarely exhibited by *V. olivaceus* compared to its frequent use by *V. komodoensis, V. salvator,* and *V. bengalensis.*

Hiss.—The common auditory act of hissing sometimes takes the form of repeated huffing, particularly during intraspecific coactions. It is rarely used in defensive interspecific coactions but is frequent in intraspecific activities—particularly by the dominant male during male-male combat. Huffing often accompanies mounting and associated acts by the dorsally positioned male (whether lower individual is male or female). I have never heard juvenile *V. olivaceus* hiss, though baby *V. komodoensis* and *V. salvator* do so.

Lateral compression.—Lateral compression of the body presents the broadest lateral view and is always in a defensive context. *V. olivaceus* rarely use it; when they do, it is weak in intensity compared to the same act in *V. bengalensis* (Auffenberg, 1981b) or some Australian varanids (Stirling, 1912).

Dorsal arch.—A dorsal bending of the midbody region almost always accompanies lateral compression but is rarely of strong intensity in *V. olivaceus*.

Lateral orientation.—The body is placed in lateral orientation so that it presents the least perpendicular view. This movement often is made in conjunction with body high, lateral compression, and dorsal arch, and is used always in a defensive context. Together these serve to increase apparent body size.

Dorsal flattening.—Dorsoventral flattening of the trunk is usually accompanied by tilting the broadened surface toward the rival (fig. 3G in Carpenter et al., 1976). It is always used in a defensive context, usually during interspecific contexts. It is rare in *V. olivaceus*.

Head tilted.—The neck is bowed or bent, placing the head in a sloping position relative to the substrate. Also, the neck is often twisted so that the top of the head is directed toward the adversary. This act is common in *V. olivaceus,* particularly during intraspecific coactions.

Weaponry Class

These acts are used in offensive situations (inter- or intraspecific), usually following one or more acts in the warning class.

Biting.—Clearly the strongest reinforcement of intent, biting is used in courtship and sometimes following male-male combat. Because of the butaan's considerable jaw power, its bite is particularly severe when used defensively (see chap. 10) and may be maintained with bulldog tenacity for up to 25 minutes. Bleeding is uncommon, though tissue damage is often extensive. Biting in a social context (during courtship or after male-male combat) is limited to the neck skin, where scales are largest. Also, it is noticeably gentler than that *preceding* male-male combat (usually by the dominant individual); it may be accompanied by vigorous sideward or backward head jerks, which are more apt to tear the skin. Biting in all contexts seems as common, if not more so, as in either *V. komodoensis* or *V. bengalensis*.

Tail slap.—A rapid lateral swing of the tail often but not always follows tail coiling. It usually terminates intraspecific interactions. The blow is swift and smartly delivered, usually striking the other individual on the side of the body or, less often, the head. Tail slapping in combat has been reported for *V. niloticus* (Cowles, 1930), *V. varius, V. spenceri, V mertensi,* and *V. salvadori* (Murphy and Mitchell, 1974), and *V. bengalensis* (Auffenberg, 1981b).

Investigative Class

Investigative behaviors are associated with learning "who" the other interactant is. They are used at close range. Tonguing is probably the most important, for it occurs during any behavior in which identification of an object is important.

Tonguing.—The act of tonguing includes making contact with the tongue or simply extending the tongue toward another individual (fig. 13-1). It occurs prior to combat or courtship. Concomitant pressing of the snout (= touch of Gillingham, 1979, in snakes) against the body (fig. 13-1) is common in *V. komodoensis* but rare in *V. bengalensis* (Auffenberg, 1981b, 1983a) and *V. olivaceus*. Vogel (1979) stated that tonguing among *V. salvator* usually occurs over the frontal area of coactants, but this sort is not common in *V. olivaceus* (see below).

Following.—Pursuit of one individual by another (= chase of Gillingham, 1979; Vogel, 1979) occurs chiefly during courtship in varanids but also during agonistic coactions in *V. salvator* (Vogel, 1979) and *V. olivaceus*.

Fig. 13-1. Common intraspecific interactions in *V. olivaceus: upper left,* 1, pushing with the snout, 2, head up; *upper right,* 1, chinning; *bottom,* 1, tonguing, 2, head swung away, 3, eye(s) closed.

Stress Reaction Class

Stress reaction acts are common in *V. komodoensis* (Auffenberg, 1981a) and probably in many other species as well. This class is reduced in kind and intensity in *V. olivaceus*; during our studies a raised roach or intense body-high position was never seen in that species.

Gular extension.—In gular extension the throat is enlarged by lowering and pulling the hyoid apparatus posteriorly, sometimes accompanied by a hiss (= inflated throat of Carpenter and Ferguson, 1977; Vogel, 1979). It is common in all varanid species as far as is known. It occurs in *V. olivaceus* but is infrequent and usually low in intensity.

Social Status Class

Social status acts are common behavioral acts, probably used by all varanid species and frequently employed by *V. olivaceus* in several contexts. They are apparently important in conveying information about the social status of one or both of the individuals in the interaction *at that immediate time* but are not an accurate reflection of the overall status of the individual within its social group. Thus large (usually dominant) individuals may, particularly during agonistic intraspecific encounters, show temporary (sometimes protracted) subordinate behaviors to another individual with whom they frequently interact and *might* best in combat.

Head raised.—The head is lifted above the trunk axis and held there (fig. 13-1), indicating social dominance or attention.

Head down.—The head is dropped below the trunk axis and held there, indicating social subordinance.

Head away (= facing away, Auffenberg, 1981b).—The head is moved lateral to the central body axis and held there. It often occurs in response to tonguing of the head or to the placement of an antagonist's head close to that of the respondent's. The act is a common interaction, particularly during the earlier parts of an encounter, when it is used alternately with investigative tonguing (fig. 13-1).

Topping.—Topping occurs when one individual (usually the female in a courtship context) crawls on the back of the male, sometimes remaining there (= riding of Carpenter and Ferguson, 1977). It occurs only during or after courtship or combat and apparently denotes temporary rejection of dominance. However, it may also be used to prevent mounting. (The reference to the figure for topping in Auffenberg [1981b:59] should read figure 2-I only, not also figure 2-E, which is scratching.)

Eye(s) closed.—Usually only the eye facing the opposing individual is closed and held that way—a common act in *V. olivaceus* (fig. 13-1). It is often associated with head away in the butaan and in *Iguana iguana* (Distal and Veazey, 1982).

Body adpression.—The entire body is pressed against the substrate by the subordinate individual, possibly serving to decrease apparent size, to make it difficult for the other to mount, or to straddle or otherwise treat the other as a female (which it might be) (fig. 13-2). When intense, this act may include neck and chin adpression as well. In a courtship context the female may bend the head sharply (fig. 13-2) during adpression, making it difficult for the male to retain an appropriate copulatory position (see courtship below). This act also occurs in males following combative interactions.

Moving away (= walks away, Auffenberg, 1983b).—One individual walks (or runs) away from the more dominant one, but it may be a neutral signal in many situations. It is reported in a subordinate context in *V. komodoensis* (Auffenberg, 1978b, 1981a), *V. bengalensis* (Auffenberg, 1981b, 1983), and *V. salvator* (Vogel, 1979).

Fig. 13-2. Common intraspecific interactions in *V. olivaceus: upper left,* 1, foot over back; upper right, 1, hind foot of mounted male on leg of female, 2, hemipenial insertion; *lower,* 1, front foot of mounted male standing on front foot of female, 2, tail lift by male (arrow indicates stroking movement by male). See text for detailed descriptions.

Reproductive Class

Reproductive acts include those in which *V. olivaceus* engage only during courtship and breeding or immediately following male-male combat (such as mounting).

Mounting.—One individual (always male) assumes a superior position on top of the other (usually female). Either one or both pairs of legs straddle the bottom individual. It is called "straddle" in some publications on reptiles (fig. 13-2).

Scratching.—The male (dominant, following male-male combat) scratches feebly with one foot, usually on the shoulder or middorsal area of the female (or male following male-male combat). It often occurs when the latter is walking away and may function to detain it (fig. 13-2). It is known in *V. komodoensis* and *V. bengalensis* (Auffenberg, 1981a, 1983b), as well as *V. olivaceus*. There is often insufficient time for scratches, and the foot is only placed on the back.

Chin rubbing.—A male in mounted position jerkily rubs its chin back and forth on the dorsal part of the neck and head of the lower individual, usually while pressing downward with the chin (fig. 13-1). It is common in the courtship of snakes (Carpenter and Ferguson, 1977; Gillingham, 1979) and varanid lizards (Auffenberg, 1983b).

Neck bite.—A mounted male bites the neck (occasionally the head) of female (or male after male-male combat). Neck biting rarely occurs in *V. bengalensis* (Auffenberg, 1981b, 1983b) but is common in *V. olivaceus*.

Tail lift.—A mounted male strokes the tail base of the female with one of its hind feet by stroking upward with the claws (fig. 13-2). This act is common in the courtship of most lizards (Carpenter and Ferguson, 1977).

Combat Tactics Class

Combat tactics include the acts characteristic of male-male "ritual" combat. The aggressive acts exchanged between fighting juveniles and those involving females are different (see Auffenberg, 1981b). In every case these acts are meant to enhance the chance of one male pushing or otherwise forcing the other to a prone position.

Bipedal stance.—Rearing smoothly (though sometimes quickly) onto the hind legs, the individual usually uses the tail as a support. In a defensive non-"ritual" context (any sex, any size) the front legs may hang limply at the sides (fig. 2C; Auffenberg, 1981b), but in offensive male-male "ritual" combat the front legs are employed in a brachial embrace. The defensive form often follows a quick unexpected advance or attack by another individual. The offensive "ritual" bipedal stance has been described in several varanids:

V. varius and *V. spenceri* (Fleay, 1958), *V. gouldii* (Johnson, 1976), *V. bengalensis* (Ali, 1944; Deraniyagala, 1958; Auffenberg, 1981b), *V. salvator* (Lederer, 1933; Vogel, 1979), and *V. komodoensis* (Auffenberg, 1981a). It is also characteristic of male-male "ritual" combat in *V. olivaceus*.

Brachial embrace.—In an agonistic context, brachial embrace is used by bipedal males during ritualistic combat. The combatants clasp one another with their front legs in order to hold each other in ventrally addressed bipedal positions during the wrestling phase. Brachial embrace of the female by the mounted male is also a conspicuous part of the later phases of courtship, prior to copulation, and may occur in exactly the same form after male-male combat. As in the agonistic context, the primary function seems to be immobilization of the other individual, in this case by pressing the front legs of the female (male) to her sides and making it difficult for her to move out from under the mounted male. In a sense, it provides the same function as the partially immobilizing xiphiplastral ramming in land tortoises (Auffenberg, 1964, 1978a), the biting of female lizards on the neck by mounted males (Carpenter and Ferguson, 1977), and the laterally draped body of the male over that of the female during snake courtship (Carpenter, 1977).

Wrestle.—In wrestling, bipedally situated individuals twist and turn to try to push one another to the ground. In *V. olivaceus* it is remarkably sudden and vigorous. The splayed hind legs of the opponents help to resist lateral movement when force is applied by either combatant. The main tactic is to push the opponent so that its weight is shifted onto its tail base, thereby raising its hind legs off the ground and making it easier to topple it to the side. The sudden forward force of one individual bends the back of the other in a sharp arc. It is apparently painful, for males so bent are seemingly loathe to fight again immediately and sometimes walk as if injured.

Lunge.—Lunging consists of a rapid movement toward another individual; the distance involved is usually less than the lizard's own length.

Head push.—In a sideward push with the head and neck to force the other off balance during wrestling, necks and heads are sometimes so crossed that they appear entwined (especially in long-necked species, such as *V. salvator*; see Vogel, 1979), reminding one of the more highly twisted necks of snakes (Carpenter, 1977).

Body arch.—The body arch always follows bipedal wrestling, when one individual is nearly pushed to the ground by the other. The tactic is apparently meant to keep the momentarily disadvantaged individual from being pushed to the ground and into a subordinate position. In another paper on combat in *V. bengalensis* (1981b), I discussed this tactic at some length. Body arching of both individuals has been described in ritual combat of male *V. gilleni* (Murphy and Mitchell, 1974; Carpenter et al., 1976), *V. bengalensis* (Deraniyagala, 1957; Auffenberg, 1981b), and *V. salvator* (Vogel, 1979).

Miscellaneous Class

Some miscellaneous acts are used in various contexts. Their purpose may remain obscure, and they do not fit well into the outline presented.

Do nothing.—While the act of "doing nothing" is intuitively neutral in effect, I agree with Hazlett and Bossert (1965) that for a respondent individual to "do nothing" may be a meaningful act in many contexts. It is apparently important in the coactions of *V. olivaceus,* for the signal of a female making no response to a courting male seems to encourage further courtship by the male, whereas "doing nothing" in a male-male combat context usually leads to the termination of the encounter.

Tail spiral.—The last half of the tail is often coiled into a tight spiral (fig. 7-3), especially when the individual is entering a shelter. The function is unknown.

Head jerk.—Head jerking is common, particularly when two individuals encounter one another face to face. It is often accompanied by mutual tonguing and seems related to a high stress level during the encounter, as though one individual expects to be attacked by the other. It is often mutual. In all cases the head is quickly jerked to the side, usually after the tongue is extruded. The eyes are sometimes closed during the jerk itself. It is also engaged in by the male during early phases of courtship, and during later phases it may change into chinning.

Combat

There are few published quantitative data on the combat behavior of varanids. Though important because they outline the major features of varanid combat, most earlier papers were necessarily anecdotal and based on only a few observations (Abdoessoeki, n.d.; Sterling, 1912; Schmidt, 1927; Lederer, 1929, 1933; Vogel, 1929; Smith, 1931; Mertens, 1942; Ali, 1944; Ditmars, 1955; Deraniyagala, 1958; Honegger and Heuser, 1969; Hoogerwerf, 1970; Murphy and Mitchell, 1974; Carpenter et al., 1976). The exceptions are Vogel (1979), who reported on the details of combat among wild *V. salvator,* and my report (1981b) in which the quantitative aspects of male-male combat were analyzed for *V. bengalensis* in captivity.

The following accounts are based on 51 circumstances of agonistic behavior between three adult male *V. olivaceus* held in captivity under natural weather conditions at the Caramoan base camp from 12 February to 18 July 1983. All individuals were acclimated to captivity before observations were made. Data were taken on 38 encounters from 14 April, when agonistic behavior was first observed, until 21 June. During the most active period

(28 April–11 June, 85% of all encounters), observations were continued throughout the day (0700–1700 hr).

Twenty-one behavioral acts were noted during male-male agonistic coactions. Table 13-1 includes the frequencies of all acts tabulated and the probability that they are due to chance alone. Some acts are important because they are so frequent in male-male encounters (walking to, walking away, tonguing, head jerking); others are important because they occur so rarely, though they are expected to be more common (tail coil, tail slap, hyoid enlargement, etc.). The only behavioral acts restricted to male-male agonistic encounters are bipedal position and mutual brachial embrace, neither of which occurs more or less than expected. Some acts are clearly related to dominant-subordinant social status (hides head, closes eyes, lowers head, neck bent into U, turns head away, and all opposites of these acts, such as turns head toward, etc.). Almost all acts in the social status category involve head movements of some sort. Chinning, mounting, and foot on back are all clearly related to courtship (see below). Bipedal position, brachial embrace, and related acts are defensive

TABLE 13-1. Behavioral acts recorded during 51 combat sequences of *V. olivaceus*, showing number of times recorded, *p* values for each (binomial test), and act categories into which they are most likely to be placed.

Act	Number of occurrences	*p*	Category
Walks away	41	0.005	Social status
Walks toward	30	0.005	Social status
Turns head away	2	0.005	Social status
Turns head to	2	0.005	Social status
Raises head	3	0.005	Social status
Lowers head	6	0.005	Social status
Closes eyes	8	0.025	Social status
Hides head	15	0.500[a]	Social status
Tongues	62	0.005	Investigatory
Jerks head	72	0.005	?
Bends neck in U	11	0.100[a]	Neck bite, avoidance tactic?
Follows	14	0.100[a]	Investigatory
Foot on back	3	0.005	Courtship
Mounts	13	0.100[a]	Courtship
Chinning	8	0.025	Courtship
Bipedal stance	24	0.100[a]	Fighting tactic
Brachial embrace	25	0.100[a]	Fighting tactic
Tail coil	2	0.005	Warning
Hisses	17	0.500[a]	Warning
Bites	13	0.100[a]	Weaponry use
Hyoid enlargement	2	0.005	Stress reaction

a. Not significant.

and offensive *tactics* of ritualized fighting, while biting, tail lash, tail coil, gaping, and hissing are warning and actual weaponry use. Tonguing is an investigatory act. The same categories of acts occur in *V. bengalensis* (Auffenberg, 1981b), but some can be considered as specific communicative signals or cues and not all are included in the actual combative phase of agonistic behavior. Some acts may have multiple functions. Tonguing (investigative category) is an act apparently intended to obtain chemical information about the opponent, but it can be also a visual releaser indicating interest. Smelling of the head, shell, and cloacal area in land tortoises has similar dual interpretations (Auffenberg, 1964, 1965; Weaver, 1970). That this interpretation is highly probable in *V. olivaceus* is suggested by the fact that tongue flick rates are significantly higher during precourtship and precombat ($p < 0.05$) than during foraging (precourtship and/or combat mean rate = 10 tongue flicks per 5.6 ± 0.24 sec; foraging rate 10 flicks per 8.7 ± 0.23 sec).

Table 13-2 reorganizes the statistically significant different acts by whether they tend to be initiatory or secondary in the act system (sequentially adjacent). While some acts are more common than expected, as either type, others are less common only when performed as an initiatory and not a following act (e.g. following, turns head away, and chinning). Closes eyes is usually a following act (as expected). With the exception of walking to or away, all social status acts are represented significantly less than expected during agonistic encounters between males. As expected, combative tactics are represented in quantity. On the other hand, figures of use of acts in both the weaponry and warning categories are equal to or less than expected by chance alone. Likewise, the acts associated with courtship are equal to or less than expected. Head jerking (category unclear) occurs more than expected. Movements concerned with the spatial placement of potential combatants in respect to one another, with mutual chemical investigation, and with combat tactics are all major components of ritualized male-male combat. Weaponry use, acts normally associated more with courtship (related to social status when both co-actants are males), and a few of the most extreme subordinate social status releasers (closing eyes and hiding head under leaves, etc.) are not significantly different than expected (i.e., they have median values in accord with their frequencies, if all acts were equally distributed). Occurrence of some of the acts is significantly lower than expected, particularly those that tend to occur *before* actual combat (i.e., before dominance is established *at that meeting*); all are minor though cogent visual releasers in which head position seems to be most important (head up, down, turned to, away, etc.).

Tables 13-2 and 13-3 show the probabilities of certain acts occurring (and whether they are more, less, or about what expected) for all acts recorded as adjacent (i.e., a sequential act pair). Only those for which frequencies exceed a statistically significant level ($p < 0.05$) are shown. These data suggest that

TABLE 13-2. Acts of *V. olivaceus* during combat, arranged from investigative through combative to social status phases (see Auffenberg, 1981b) and indicating the probability of whether the act precedes or follows another in an act pair (no specific pairs considered).

Category	Act	[a]	p if act: Precedes	Follows	[a]
Investigative	Tongues	+	<0.005	<0.005	+
Investigative	Follows	−	=0.050	>0.500[b]	±
Stress reaction	Hyoid enlargement	−	<0.005	<0.005	−
Warning	Tail coil	−	<0.005	<0.005	−
Warning	Hisses	±	>0.050[b]	≈0.900[b]	±
Weaponry use	Bites	±	>0.050[b]	≈0.900[b]	±
Combat tactic	Bipedal stance	+	<0.005	<0.005	+
Combat tactic	Brachial embrace	+	<0.005	<0.005	+
Social status	Walks away	+	<0.025	<0.005	+
Social status	Walks toward	+	<0.005	<0.005	−
Social status	Turns head away	−	<0.010	>0.050[b]	−
Social status	Turns head to	−	<0.005	<0.005	−
Social status	Raises head	−	<0.005	<0.005	−
Social status	Lowers head	−	<0.005	<0.005	−
Social status	Closes eyes	±	>0.100[b]	=0.005[b]	−
Social status	Hides head (under leaves)	±	>0.900[b]	<0.900[b]	±
Avoidance (bite)	Bends neck in U	±	≈0.010[b]	>0.050[b]	±
?	Jerks head	+	<0.005	<0.005	+
Courtship	Foot on back	−	<0.005	<0.005[b]	−
Courtship	Mounts	±	>0.100[b]	≈0.900[b]	±
Courtship	Chinning	−	<0.025	>0.050[b]	±

a. + = more than expected; − = less than expected; ± = about what was expected on the basis of proportion of total acts involved.
b. Not significant at <5 percent level.

walks away can be expected to lead to head jerks, tonguing, or following; walks to will likely lead to walks to on the part of the other potential combatant, or tonguing, or head jerking; tonguing most often will lead to walks away, tonguing, head jerks, or turning the head away by the respondent; initiatory head jerking leads to the greatest array of possible acts on the part of the respondent (walking away, tonguing, head jerks, ritualized bipedal position, mounting, hissing). Turning the neck in a U is always followed by closes eyes in the *same* individual. One must assume, on the basis of the data, that bipedal position of one male leads to the same in a second individual of about the same size and sexual condition and to mutual brachial embrace. After brachial embrace (and wrestling) the loser most often walks away or bites. Biting usually leads to walking away on the part of the recipient. Mounting by one individual tends to result in head jerking, tonguing, or biting by the *same* individual.

TABLE 13-3. Statistically significant act pairs ($p < 0.050$) recorded during combat of adult male *V. olivaceus*.

Following act	Preceding act									
	WA	WT	T	HJ	NU	HH	BS	BE	B	M
Walks away			●	●			●	●		
Walks to		●								
Tongues	●	●	●	●						
Head jerks	●	●	●	●						●[a]
Turns head away			●							
Bends neck in U						●[a]				
Closes eyes					●[a]					
Bipedal stance				●				●		
Brachial embrace								●	●	
Bites									●[b]	●[a]
Mounts				●						
Hisses				●						
Follows	●									

SYMBOLS USED:

B = bites
BE = brachial embrace
BS = bipedal stance
HH = hides head

HJ = head jerks
M = mounts
NU = bends neck in U

T = tongues
WA = walks away
WT = walks to

a. Sequential acts in *same* individual.
b. Biting after combat.

From the data it is clear that *V. olivaceus* males engage in combat of a ritualized type in which bipedal position and brachial embrace are predominant behavioral acts. Cumulative evidence suggests that these two acts are unique behaviors shared by all members of the family Varanidae. They are not seen in any other lizards, though they are common (in modified form) in snakes (Carpenter and Ferguson, 1977). The reported biting between ritually combative male *V. bengalensis* by Deraniyagala (1957) was not noted in my study of that species (though biting is clearly a part of this behavioral pattern in *V. gilleni* [Murphy and Mitchell, 1974] and *V. salvator* [Vogel, 1979]). It occurs in male-male combat of *V. olivaceus* as well but only after one of the combatants has been pushed to the ground and the "winner" has mounted or otherwise treated the "loser" as a female. The role and extent of biting during male-male combat in varanids need further clarification.

No females or juveniles were observed fighting because there was no opportunity to do so. The high aggressiveness of the females of *V. bengalensis* during conspecific fights (Auffenberg, 1981b) could therefore not be demonstrated in *V. olivaceus*. However, biting is more common among males and by males (directed against females) than in *V. bengalensis* (though limited to the postritualistic phase, as pointed out). A significantly higher use of weaponry by females than by males during agonistic encounters also has been reported in *V. komodoensis* (Auffenberg, 1981a) and may be a familial characteristic. Large adult males of *V. olivaceus* show few evidences of high stress levels during the ritualistic part of the combat.

Visual releasers in the form of stereotyped displays are not employed by potential rivals (less than in *V. bengalensis*, in fact). However, the frequent use of tongue flicking during the initial phases of combative interaction suggests that pheromones are probably important chemical cues used to distinguish sex and sexual maturity (see chap. 8), and that they are probably surface releaser types (see chap. 2). Fecal sign-posting in and on the periphery of the activity range of *V. komodoensis* also has been reported (Auffenberg, 1981a), but it has not been demonstrated as important or even present in other varanids, including *V. olivaceus*. Captives were not noted to pay particular attention to feces of individuals newly introduced into the cage. Vogel (1979) reported that tongue flicking is a preliminary act in about half the ritual encounters of wild male *V. salvator*.

Male-male nonritualized combat is commonly preceded by various acts in the social status category, which appear to be important in mediating actual combat. They are also common after ritualistic combat, when the "winner" and "loser" have been designated during wrestling in a bipedal position and one of the combatants has been pushed to the ground. In general, the social acts in this phase of the coaction probably tend to reduce further aggression on the part of both individuals—which may be particularly important at this point, because the winner may severely bite the loser when the latter is forced to the ground.

Though the investigative and sexual acts preceding nonritualized combat in *V. olivaceus* are identical to those preceding ritualized combat, the sequence of acts is variable. Figure 13-3 shows the most likely sequence in behavioral acts from leading up to through completion of ritualistic bipedal combat in males.

Vogel's study (1979) of wild *V. salvator* showed that the winner of one instance of male-male combat did not necessarily remain dominant to the loser (conclusions at variance with those of Honegger and Heusser [1969] on captives of the same species). I have shown (1981b) that dominance among male *V. bengalensis* was determined partly by size, though dominance shifts were common. In this study, dominance was not determined by size, for the winner of most of the bouts was slightly smaller (though also slightly heavier) than the nearly consistent loser. In spite of a heavy nematode infestation, the larger (common loser) generated 70% of the bouts between them during the first 15 agonistic encounters, though it lost nearly every one. However, during the last 28 encounters only 5% were initiated by him, and he lost every one. Near the end of the series of bouts he immediately hid his head in the leaves whenever

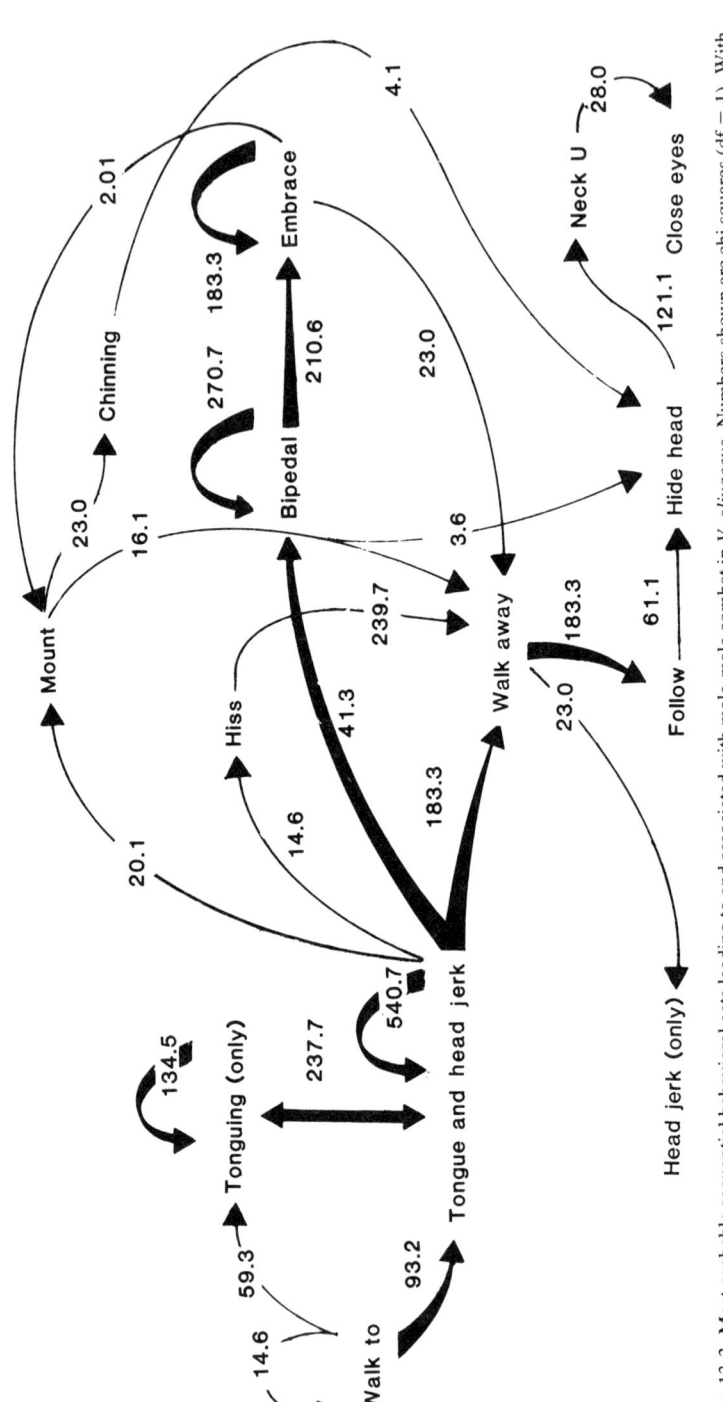

Fig. 13-3. Most probable sequential behavioral acts leading to and associated with male-male combat in *V. olivaceus*. Numbers shown are chi squares (df = 1). With the exception of embrace-mount, chinning–hides head, and mounting–hides head, all act pairs shown are significant ($p < 0.050$). While the sequence may end at any point, the major behavioral dilemma occurs at tongue and head jerk, for there are several common sequences possible at that point. Heavy arrows show the most significant sequences.

the other acted the least bit aggressive. The modal number of acts exchanged between them during the first 15 bouts (in which both showed more or less equal interest) was 13 ($\pm$ 2); during the last 29 bouts it dropped to 9 ($\pm$ 1). So, as the outcome became more stereotyped, dominance was maintained with fewer agonistic acts between them.

There is little evidence that winning led to a significantly higher level of despotism in the winner's interactions with either the previous losers or other cagemates, including a mature female. Nor is there any evidence that consistent losers in male-male combat became less interested in courting female cagemates; on the contrary, interest remained high. However, successful courtship occurred only once with the loser and 29 times with the winner. As in *V. bengalensis,* females sometimes walked in front of the combatants when the ritualistic portion was completed, suggesting a presentation display (Auffenberg, 1981b).

Most male-male agonistic coactions take a surprisingly short time when only the wrestling phase is considered (8–41 sec), the result of the vigorousness of the combative pair once they raise themselves into a bipedal position. Every ounce of energy seems to be thrown into the initial few seconds—particularly in respect to the back-wrenching tactics used by one to bend the other backward sharply and throw its weight against the other's chest and neck. Often one such "backbreaker" is all that is required to determine the outcome of the encounter. The one knocked or otherwise forced down rarely makes an attempt to stand bipedially immediately and usually walks (or runs) away. If held in a brachial embrace that pinions its front legs against its body, the subordinate individual usually attempts to hide its head under leaves and other debris. Neck biting that would place the straddled male's cloaca in proper position for breeding is apparently completely thwarted by the submissive male bending its neck in a sharp U so that his head lies against his own neck and shoulders. A typical male-male agonistic sequence follows:

Males Y and B 0.3 m apart when Y seems to take interest in B.
Y walks to B, tongues it on head when only 10 cm away.
B turns head away.
Y starts head jerk, tongues elbow.
B pulls leg away.
Y tongues shoulder (2X), neck and shoulder (2X), back and shoulder (2X), neck (3X); places both front feet on back of B; tongues neck (2X), shoulder (7X), tail (4X), and back.
B starts head jerk (rate 10 jerks per 5.8 sec).
Y moves closer to head and alternately tongues and head jerks in rapid succession.
B does same.

B   suddenly lunges and raises to bipedal position.
Y   does same and they clasp each other in brachial embrace.
Y   almost immediately topples B in a burst of initial energy.
B   runs away 1 m and stops.
Y   follows quickly and tongues B on hind leg (2X), back and neck (3X).
B   hides head under leaves, closes eyes.
Y   mounts, tongues neck (4X), head (2X).
B   turns neck into U.
Y   remains straddled for few moments, then walks away.
B   remains hidden for 12 min, then resumes walking about enclosure.

Though the interacting males seemed to notice and show interest in one another at distances of 1.7 to 10 m (81% of all encounters), combatants approached one another from less than 1 m; stress was rarely apparent until their heads were about 10 cm apart. At this distance both individuals usually exchanged mutual tonguing and head jerking—the latter probably provoked because of the possibility of being bitten if they remained too close. As a result, each individual tended to tongue the other's head when the latter was turned to the side. Thus they alternated tonguing and head jerking, usually with one tongue touch and one head jerk in each act pair sequence. There was no evidence that males repeatedly tongued one another's head before a wrestling bout (as suggested for *V. salvator* by Vogel, 1979), though the eventually dominant one frequently did so after the bout. However, in *V. olivaceus* other parts of the body were repeatedly tongued at all stages of the encounter. Tonguing the head seems to result in more stress than tonguing other parts of the body, and it frequently leads to a lunge and bipedal position by the one being tongued, resulting in a defensive bipedal position by the tonguer. Figure 13-3 shows the most common behavioral sequences involved in male-male encounters during the courtship period.

In general, larger individuals were always dominant to smaller ones, particularly in the same sex. There were no obvious assertion displays by juvenile and subadult butaans, as are common among these sizes of *V. komodoensis*, particularly when a number of individuals gather around carrion. The fact that higher density results in a greater number of assertion displays by iguanid lizards has been demonstrated by Tubbs (1972) and Stamps (1973). The same pattern probably applies to varanids, so that assertion displays are to be expected only in high density situations, or among species that often gather in groups, even if only temporarily (as in *V. komodoensis*). The generally low density of *V. olivaceus* throughout the habitat and the fact that they seem never to congregate for any purpose suggest that assertion displays are probably rare in this species.

Courtship

A strict definition of courtship that includes display as a principal component is not applicable to varanid lizards (Auffenberg, 1982, 1983b). The butaan is no exception; it lacks the distinctive visual displays that often characterize premating behavior in other lizards. A broader definition of courtship as a "heterosexual communication system leading to copulation" (Morris, 1956) would include the precopulatory informational exchange between males and females, and it is the definition used here.

Many excellent studies have been conducted on the courtship of lizards, indicating that the behavioral patterns generally differ more between families than between genera and species of any one family. These studies also have shown that courtship in most lizards is often stereotyped, with behaviors stressing the importance of visual displays in agamids and iguanids (see, e.g., Carpenter and Ferguson, 1977; Ferguson, 1977, for reviews), though not in varanids (Auffenberg, 1982, 1983b).

While there is considerable information on reproduction of *Varanus* species, there is actually little on courtship itself. Falk (1921) provided data for *V. exanthematicus,* and Werner (1983) for *V. griseus,* though important details are missing. The most complete studies are on *V. bengalensis* (Auffenberg, 1981b) and *V. timorensis* (Moehn, 1984). This study of the courtship in *V. olivaceus* is not as complete, for it is based on fewer individuals and interactions. Nevertheless, the data available suggest that most varanid species have similar patterns, which differ only in detail, and that odor is probably important in sexual and species differentiation.

The data and conclusions to be presented here were all obtained from two pairs of captive animals, maintained in a large enclosure (15 m^2) in the habitat of *V. olivaceus*. Climbing, hiding, etc., were all possible in the large space provided. Observations were made throughout the day when sexual activity was intense but on an irregular basis at other times. Photographs form the bases for the accompanying drawings. One pair was found together in a hollow log in January 1983 and kept together in captivity until October 1983. They may have been associated with one another in the wild for some time before being caught and certainly remained closely associated throughout the entire captive period. The other two individuals were caught separately, with no chance of previous association in the wild. This notation concerning the length of association of potential mates is important in view of the long-term pair bonding that may occur in *V. komodoensis* (Auffenberg, 1981a) and the tendency for captive *V. bengalensis* to breed more readily if the individuals involved are long-term cagemates (Auffenberg, 1981b). All four individuals in the current studies were mature, the smallest female weighing 1.6 kg and the smallest male 6.2 kg.

Of 91 sexual interactions recorded, 12 (13.2%) were successful in the sense that coitus occurred (*V. bengalensis,* 6%; Auffenberg, 1981b).

Seventeen behavioral acts are common components in the courtship of *V. olivaceus*. These represent several different contextual and functional classes. The most common is the reproductive class (N = 156 acts, including scratching, mounting, chinning, tail lift, and brachial embrace), then investigatory (140 acts, including only tonguing and following), then social (129 acts, including head down, head away, neck in U, head up, close eyes, walk away), warning (3 acts, including only hiss), and weaponry use (4 acts, including only neck biting).

In spite of the fact that the original pair were taken together in the forest and held together in captivity from January onward, no courtship occurred until June. The first sign of interest by either individual was 21 April, when the female was noticed lying across the male's back; they remained in that position for much of the afternoon. Identical positions were noted on 27 April and 11, 18, 23, and 29 May. On 5 June several combats occurred between the male of this pair and another placed in the cage several weeks previously (though there had never been any indication of combat or even interest between them prior to this time). (Crews [1975] reported a similar pattern in *Anolis,* in which a constant high level of male-male combat preceded courtship when apparently receptive adult females were exposed to sexually active males.) The same males engaged in combat several more times on 6 and 7 June. On the latter date courtship began suddenly, with 11 courtship attempts (none successful) involving both males and the single female. (Captives in the Dallas Zoo courted rather suddenly on 6, 10, 20, and 21 March, then sporadically throughout the rest of the year [A. Mitchell, pers. comm.]. The noncoincidence of courtship dates for captive and wild animals is, of course, common.) This intense courtship continued for several days intermittently, with repeated combat bouts between the males. In 6% of the 91 courtship bouts initiated by the males, the female had left her place of resting or concealment and situated herself in a conspicuous position near the combatants. This behavior has been noted in the courtship of *V. bengalensis* and is probably important in successful courtship, as the female's location could hardly be missed by the victorious male.

Table 13-4 illustrates the sudden initiation of courtship activity and the fact that the intense period lasted for only a few days, when both males were extremely interested in the same female—in spite of the fact that just before this surge of interest in mating, another female, similar in size, had been placed in the same pen. There was frequent interest in the newer female as she moved about the enclosure, but most investigatory behavior on the part of both males was short lived and they continued to be particularly interested in the longer resident female. (Crews and Williams [1977] showed that female *Anolis* were

TABLE 13-4. Dates and number of male behavioral acts of captive *V. olivaceus* from initiation of courtship in each instance to the first mounting attempts.

Inclusive dates	Number of courtships	Average number of male acts (SE) per courtship bout
January–6 June	0	0
7 June	11	5.7 (1.6)
8 June	14	4.1 (1.7)
9 June	7	4.6 (1.6)
10 June	3	3.3 (0.9)
17 June	1	
13 August	1	
22 August	1	
3 September	1	
6 September	1	3.1 (0.5)
9 September	1	
11 September	1	
12 September–24 October	0	0

sexually receptive only during ovarian activity when preovulatory follicles are enlarged.) Though there was never any sustained interest in the newer female, an even newer one placed in the enclosure in early September elicited some interest from the older male but nothing leading to successful breeding. The female that bred with both males laid eggs on 21 July, about 2 weeks after the earliest successful courtship and 10 days after the last active day.

As in *V. bengalensis*, all courtship sequences were initiated by males. In only 7% of the cases was courtship initiated by the male after the female approached him; 93% of the time the male walked to the female. In most of the courtship bouts (62%), the female was not moving. In spite of the fact that the male often moved to the female after seeing her move, courtship was nevertheless more common with resting females.

The length of time from courtship initiation to the first copulation attempts varied from 0.5 to 2.8 minutes ($\bar{X} = 1.8$ min ± 0.5), significantly longer than in *V. bengalensis*. I have no idea why the investigative and display phase prior to coitus should be longer in *V. olivaceus*, but it may be an artifact of the small number of individuals observed. On the other hand, *V. olivaceus* spent a much longer time in-copula (2.2–12.0 min) than *V. bengalensis* ($\bar{X} = 1.5$ min).

Each succeeding courtship by the same male is accomplished by inserting alternately the right or left hemipenis. There was no exception to this pattern, which is similar to that described by Crews (pers. comm.) for garter snakes. It may be related to sperm maturation in each individual testis, as it seems to be in *Thamnophis*. While the hemipenis is inserted, the male pelvic thrusts varied from 4 to 33 ($\bar{X} = 13 \pm 4.2$). Thrusts are definite but rather feeble and make up only a very small part of the time the couple is in coitus (usually lying very still).

The following brief description is typical of courtship in this species. In this case the entire courtship passed through 14 steps. As will be explained, this number varies from instance to instance, as does the period in which the courtship took place. As in *V. bengalensis,* the number of acts leading to mounting tends to be smaller with repeated bouts (table 13-4).

> Female walks across view of male at distance of 1.5 m; he follows and places one foot on female's shoulder.
> She stops, but does nothing.
> Male then tongues female's neck several times in rapid succession (3X).
> She continues to do nothing.
> He tongues neck (2X) and head (2X).
> She starts to walk away.
> He head jerks.
> She walks.
> He tongues neck (2X).
> She closes eye.
> He mounts and employs brachial embrace.
> She does nothing.
> He tail lifts.
> She does same.

The modal number of steps in successful courtship is 8, but the average is much less (table 13-4), with an overall range of 1–16 acts (table 13–5). This general range is approximately the same as in *V. bengalensis,* so that the complexity of courtship in terms of number of acts is roughly the same for both species. The longer sequences tend to be unsuccessful, for there is a greater tendency for the female (sometimes the male) to break off the relationship. Figure 13-4 illustrates the relationships existing between the most common behavioral acts in courtship.

Table 13-5 shows that while female tail lifting (precedes coitus) can take place anytime during the sequence, it normally occurs from the third through the eighth act. This timing suggests that, as in *V. bengalensis,* the most important discriminatory communication interchanges occur early (steps 1–7 in *V. olivaceus,* 1–5 in *V. bengalensis,* though the difference probably is not significant). The strongest rejection signal by the female appears to be bending the head sharply in a U. It may occur in any part of the sequence but is more common during the first three acts of the female. Its effect on a successful courtship sequence is that it becomes impossible for the male to align his head over that of the female and at the same align the tail bases of both prior to tail

TABLE 13-5. Frequency of acts by *V. olivaceus* occurring in steps 1–16 of 91 initiated courtship bouts.

Step no.	Tp	T	WA	F	HJ	N	S	M	CE	C	HU	U	TL	H	HA	HD	HT	BE	B
1		43			14		15	2		2									
2		34		1	35			6			2	4	1	1	3	3	1		
3		15	11	7	2	16	7		8			5	1					2	1
4		19		2	18			3		1	5	4		1	3	3			
5		14	10	10	3	1	2		4			4						5	2
6		16		1	12			3			4	3							
7	2	6		4	5	1	4	6	2			3						4	1
8		10			7			1			2	3							
9		3		1	2	1	2	2	3			2			1				
10		5			1						2	1							
11		5		2					1										
12		4			1			2		1									
13		4	1				2		1										
14		3			1			1			1	1							
15		3							1										
16												2	1						

NOTE: Even-numbered steps were performed by respondent females, odd-numbered steps by initiator males.

B = bites
BE = brachial embrace
C = chin rubbing
CE = closes eyes
F = follows
H = hisses
HA = head away
HD = head down
HJ = head jerks
HT = head toward
HU = head up
M = mounts
N = does nothing
S = scratches
T = tongues
TL = lifts tail
Tp = tops
U = head in U
WA = walks away

lifting manuevers. Most unsuccessful courtships are terminated when a female walks away and the male does not follow. At no time did I notice any courtship sequences terminated because of female aggressiveness (common in *V. bengalensis*). Biting, when it occurs, is always by the male and is almost always on the neck in the region of the largest scales. Similar behavior has been reported in many other lizards (see Distal and Veazey, 1982). The most common initiatory acts by the male are tonguing and head jerking—none of the others is significant. These acts remain important in male steps 3 and 5 but are greatly reduced in frequency at step 7. The most common respondent female acts are walking away and do nothing. These acts remain significantly high through step 8 (her fourth respondent act). Scratching by the male is common only in his second act (step 3). Mounting occurs without brachial embrace, but the reverse is impossible, hence the greater variation in the steps when mounting occurs. Both sexes hiss, though rarely and only in earlier steps. Social class acts involving the head position (head up, head away, head down) are rare and are utilized only during early stages. On the other hand, head U is engaged in only by females; it may occur throughout the sequence and is often associated with closed eyes. Head jerking is predominantly a male act, though females engage in it. Mounting, chinning, and scratching also are used only by males. Brachial embrace is almost always a male act but is also sometimes used during aggressive combat by females. Unlike *V. bengalensis*, the females of *V. olivaceus* do not head jerk during courtship. All acts in the social class tend to be performed by females (different head positions); almost all investigatory class acts are performed by males (tonguing, following), though in other contexts females will also perform them (such as feeding). The neutral class act (do nothing) is more common to females, and the apparently subordinate act of walking away also is performed largely by females, as is closing the eyes. Though not shown in the table, females (and subordinate males) often avoid amorous males by hiding under rocks, logs, and leaves, though usually only the head is actually covered.

In this study only adjacent lower order interactions were analyzed (e.g., walk away and follow). Higher order interactions of three or more adjacent acts (walk away–follow–bite, etc.) were believed not to be associated on the basis of complete analyses of *V. bengalensis* (Auffenberg, 1983b).

There are 5 acts that might be performed by males in step 1, and in response females may perform 7 acts (step 2). The number of possible acts increases to 10 in the second act of the male (step 3) but remains 6–7 in all the female's steps. In steps 3–5 the number of acts performed by the male is 8–9. Thus variability of acts in the male increases and then decreases, whereas the acts of the female remain the same (table 13–6). This pattern differs from *V. bengalensis*, where the variability in the acts performed from early to late in the sequence first increases, then decreases, for both sexes (Auffenberg, 1983b).

The general pattern in these two species is similar with respect to the number of act pairs performed by the males and females for each step in the sequence. Table 13-7 shows that the total number of act pairs performed during courtship in each step of the sequence increases for *V. olivaceus* in steps 2–3, then decreases steadily through steps 5–6. *V. bengalensis* acts similarly,

TABLE 13-6. Total number of acts by male and female *V. olivaceus* during successive steps in the courtship sequence.

Steps	Male acts	Female acts
1–2	5	7
2–3	10	7
3–4	8	7
4–5	9	6
5–6	8	6

TABLE 13-7. Variation in total number of different behavioral act pairs reported for both *V. olivaceus* and *V. bengalensis* during steps 1–6 of courtship and percent of all acts reported.

Steps	V. olivaceus		V. bengalensis	
	No. of act pairs	%	No. of act pairs	%
1–2	18	17	21	18
2–3	30	10	30	63
3–4	27	11	39	69
4–5	23	22	27	70
5–6	18	11	18	68

though the highest variation in potential act pairs performed occurs in steps 3–4. However, the number of act pairs per step expressed as a percentage of the total number performed during the entire courtship sequence varies from 10% to 21.7% in *V. olivaceus* and from 18% to 70% in *V. bengalensis*, so that the latter exhibits a greater range of acts per step phase (though this may be due to the larger sample size of *V. bengalensis*). Courtship is thus somewhat more structured for *V. olivaceus* than *V. bengalensis*, undoubtedly because aggressive acts, particularly by females, are not as evident in *V. olivaceus* as in *V. bengalensis*.

Steps 1–2.—In the initial phase of courtship the male *V. olivaceus* initiates the interaction in step 1, and the female responds in step 2. About half of the significantly high step 1 male acts of the pair belong in the investigatory class (tonguing), and the remainder include only acts in the reproductive class (head jerking). Other reproductive class acts, such as mounting, chinning, etc., are not higher than expected on the basis of chance ($p > 0.05$). Respondent female acts (step 2) of significant act pairs that are higher than expected are do nothing and socially important acts (walk away, head in U). There are no reproductive class acts that are higher than expected at this stage. One major difference from *V. bengalensis* is that aggressive acts by the female against the male are unknown in *V. olivaceus*. Thus male investigatory and reproductive class acts are most often followed by neutral and social acts by the female. The greatest departure from the expected is in the male-female sequence tonguing–head U. As in *V. bengalensis*, the respondent acts of females show greater variability than the initiatory acts of males. The usual female respondent act involves head movement in the social class, apparently reflecting an emphasis by the female on agonistic rather than reproductive motivation at this early courtship phase.

Steps 2–3.—The second act pair includes the male's response (step 3) to the female's step 2. The most obvious difference from steps 1–2 is that the number of different kinds of act pairs performed is greater (table 13–7), 30 rather than 18. In addition, the levels of significance of steps 2–3 are generally lower than those of steps 1–2, suggesting a less close association between the component act pairs than in steps 1–2, probably because of the greater variability of respondent acts of the male. The pattern for this stage is the opposite of that of *V. bengalensis,* which is believed to be due to the reaction of the male *V. bengalensis* to the often aggressive acts of the females (which can be avoided or modified by the male only in specific ways).

The component acts of the female in the steps 2–3 act pair are largely social (walk away, close eyes, head up, head down) or do nothing. These are also the acts of female *V. bengalensis* at this stage, with the added aggressive acts characteristic of that species. Of the respondent male acts the most significant are investigatory (following, tonguing), reproductive (chinning, head jerking, scratching, brachial embrace), or loss of interest (walking away). Tail lifting appears for the first time. Other than the last, the male's acts are essentially like those of his initiatory act. In general his response to the female's act (2) shows a greater association with it than the female's response (2) to his initial act (1).

The highest departures from expected values are found in act pairs in which the male's acts (reproductive and investigatory) are clearly in response to the largely social acts of the female. However, the highest association (= highest significance) is between walking away by the female and following by the male. Mounting attempts by the male are weakly associated with walking away by the female. Most do nothing acts by the female are followed by male reproduction or investigatory acts.

Included within the significantly associated acts in steps 2–3 are 7 different component female acts and 10 male component responses. Thus male act variation has doubled from step 1.

There is no obvious pattern within act pairs exhibiting less than expected frequencies, except that many are social (biting also occurs here, though rarely). This lack of pattern suggests that these social acts probably reduce the frequency of associative acts and that their expression tends to reduce the possibility of further communicative sequencing toward successful coition.

Steps 3-4.—In steps 3-4, the female is responding (step 4) to the previous male act (step 3). It constitutes her second act in the sequence. Twenty-seven act pairs occur in steps 3–4, so that variability in potentially communicative adjacent acts is about the same as in steps 2–3. I believe that steps 3–4 (and those that follow closely) are the most important of the courtship sequence because there is a reduction in the number of associated acts in later steps, remaining steps are often repetitive, and tail lifts by both interacting individuals are common.

About one-half of the significant associations of the male component acts (step 3) are reproductive and about two-thirds are investigatory. Common associated female acts are walking away and do nothing (step 4). Unlike *V. bengalensis,* there is no aggression by the female. Female reproductive acts (tail lifts) make their first significant appearance at this stage but are not associated with a broad range of male acts.

Unlike the two earlier steps, there is no clear grouping of act pairs into those that are closely associated and others that are less associated. All associations tend to be high, with the highest between the tail lift of the male and the corresponding respondent tail lift of the female prior to coitus.

The female response to tonguing (investigation by the male), though variable, is usually a social act. She may also walk away, but the tonguing–walking away act pair has a low association level in steps 3–4. Of perhaps high communicative importance is the fact that the females often do nothing after being scratched, leading to further reproductive class acts by the males.

There are 7 different component male and 6 different component female acts in the highly significant act pairs. Thus variation in component acts increased from the first to the second communicative act in males (steps 1 and 3) but not in females (steps 2 and 4), where there is a noticeable decrease. The result is that the range of probably important communicative acts is greater at stage 2–3 than at any preceding or succeeding ones. In stage 3–4 the number of different acts drops in males, not in females, which tend to have the same number of acts (though different ones) in either initiator or respondent contexts.

Steps 4–5.—In steps 4–5, the male initiates his third behavioral act (step 5) and is responding to a female act for the second time. There is a reduction in the total number of associative acts from 27 in steps 2–3 to 24. Thus communication potential, as suggested by the breadth of associative acts, is reduced almost to the level of steps 1–2, illustrating a general deterioration of the interaction at this stage if mounting has not yet taken place.

The female acts (step 4) remain about two-thirds social (head down, head U, walk away) or do nothing and about one-third reproductive (tail lift, head jerk). There is no investigative behavior on the part of the female. Female reproductive acts appear in higher frequencies than previously. Yet the male acts (step 5) have not changed greatly, for about two-thirds continue to be reproductive (mounting, scratching, chinning, tail lifting, neck biting, head jerking), while investigative acts (tonguing, following) are still common. However, the socially "neutral" act of do nothing is significantly higher and is often associated with preceding similar or socially subordinate acts by the female. These high associations may reflect increasing loss of interest on the part of the male.

The highest level of act association is found in the precopulatory tail lift by both individuals (as in steps 3–4). Female aggressive acts (rare) always are followed by the neutral acts of do nothing or walk away. In general, the association level between male and female acts remain high. During this phase there are six different female acts included in the significant associations and nine in the following male acts.

Steps 5–6.—In steps 5–6, the male acts for the third time (step 5), followed by the third reciprocal act of the female (step 6). There is a further reduction of total associative acts (18) to the level of steps 1–2, showing decrease in act variability. The decrease probably occurs because many "introductory" communicative statements by either partner already have been dropped from the interchange, because the association already has deteriorated and is in a repetitive phase, or because one member is losing interest.

The female acts (step 6) are significantly higher only in her tail lifts in response to the tail lift maneuvers of the male. All the remaining acts are social (walk away, close eyes, head down, head U, topping) and indicate disinterest in the courtship. The acts of the male are about one-half reproductive (head jerk, tail lift, chinning), more or less neutral (do nothing, walk away), or sometimes investigatory (tonguing), though none but tail lift are part of any act pair in which the frequency is significantly higher than expected.

Figure 13-4 shows the most highly significant initiatory and respondent acts by males and females during courtship. In that part of the diagram covering steps 1–2, it is obvious that the first act of the male is usually investigative or reproductive—usually head jerk or tonguing—and that it is significantly associated with do nothing and social acts on the part of the female. The social acts are walk away and head in U—both believed to represent social subordination. These social acts seem particularly important to successful completion of courtship, for in steps 2–3 these same acts are significantly associated with following and scratching (investigative and reproductive acts, respectively). Do nothing on the part of the female is significantly associated with tonguing, as a duplication of step 1 of the male. Rejection acts by the female are often ignored by the male. During steps 3–4 success is often achieved via mutal tail lifts. Tail lifts in the male, however, also are followed by walk away by the female. In steps 4–5 we see that the female rejections are still largely ignored by the male, for tail lifting (and mounting) are still closely associated with the rejections. In steps 5–6 further tonguing (investigative act) leads to a close association with the female subordinate (and lack-of-interest) act of closing eyes. However, in the courtships that had proceeded appropriately up to this point, male tail lift is closely associated with the same act by the female.

Steps 6–7.—Mutual tail lifts are still closely associated; walking away by the female leads to further reproductive acts (scratching) by the male.

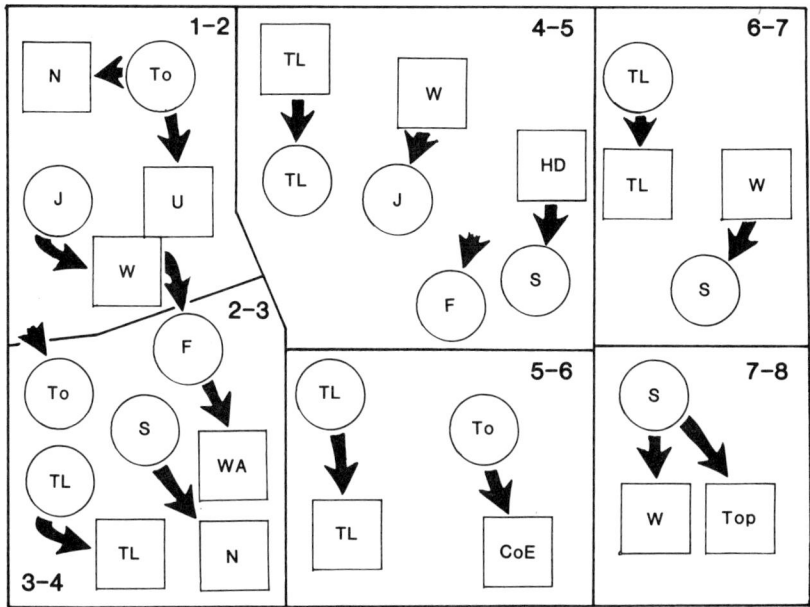

Fig. 13-4. Most likely behavioral act sequences in courtship of *V. olivaceus*. Numbers refer to steps in the sequence. Circles are male acts, squares are female acts. Abbreviations (see text for details): *CoE*, eyes closed; *F*, follows; *HD*, head down; *J*, jerks head; *N*, nothing; *S*, Scratches; *TL*, tail lift; *To*, walks to; *Top*, tops; *U*, head in U; *W*, walks to; *WA*, walks away.

Steps 7–8.—Finally, the reproductive act (scratching) of the male is most closely associated with complete rejection by the female, usually in the form of walking away or topping. In either case, the courtship is ended by this stage if intromission has not yet occurred. Therefore, all the important communicatory cues are exchanged before steps 7–8. The simplest sequence leading to successful breeding is probably tonguing–do nothing–mounting–do nothing–tail lift–tail lift. However, it happens rarely and, if it does occur, it is probably only after the male and female have already courted several times.

Unlike the situation of *V. bengalensis*, most acts by female *V. olivaceus* in unsuccessful sequences are not aggressive. In *V. bengalensis*, this female aggression is overcome by the male's being particularly vigorous and applying a strong brachial embrace and using avoidance behaviors of several kinds (walk away, etc.). These aggression-dominance or aggression-escape behavioral patterns were not noted in the courtship of *V. olivaceus*. Furthermore, long mutual acquaintance between the interacting partners does not seem as important to *V. olivaceus* as to *V. bengalensis*. However, it is still true that females tend to breed with the strongest males—those who win "ritual" combat coactions with other males and are able to manipulate the female harmlessly through brachial embrace, etc. The females themselves often place themselves close to males engaged in combat—presumably for potential breeding purposes when the combat is finished. In the interactions studied it was common for winning males to breed with such available females.

In general, male acts are more closely associated with what the female does immediately preceding them and in no way dependent on what the male has done previously. He repeats parts of the courtship sequence as necessary, depending on the apparent interest of the female. If the female does nothing or remains relatively subordinate, the male presses on to more reproductive-oriented acts; if the female then shows disinterest, aggression, or other less neutral acts, the male tends to revert to earlier steps in the sequence. However, too much repetition suggests the relationship is deteriorating—as seen chiefly through the responses of the female. Therefore, as in *V. bengalensis,* the courtship of the male remains flexible to contend with the more varied reciprocal acts of the female.

Releasers

Important releasers in social communication of reptiles are visual (color and positions and changes in both), chemical (pheromone), and tactile (biting, scratching, etc.).

Unlike the courtship of many agamid and iguanid lizards (see Carpenter and Ferguson, 1977, for review), color is not an important releaser for denoting breeding condition (seasonal changes) or sex in varanids (Auffenberg, 1981a, b, 1983b). The butaan is no exception, for there is no evidence of sexual or seasonal color differences. The important releasers in its courtship are all chemical or tactile, as the previous account of courtship demonstrates.

Table 13-8 shows the relationship of tactile releasers (there are six) in the courtship of *V. olivaceus*. Almost all are performed by the male and seem designed to communicate his interest in breeding with the female. However, chin rubbing also may be an autostimulant in male varanids, as it seems to be in natricine snakes (Noble, 1937). It is also probably important in proper orientation of the male in respect to the female's body for successful intromission. Visual releasers are apparently restricted to head jerking movements, most commonly performed by males but sometimes by females as well, which occurs in both combat and reproductive contexts.

In general, the frequent tonguing by the male (and female, when it occurs) is an indication of the importance of chemical communicative substances in feces (Auffenberg, 1981a), the cloacal area, and the skin (Auffenberg, 1981a, b, 1983b) of varanids. These substances are believed to be important in identifying sex and species, although there is as yet no evidence to prove it. Even iguanid, agamid, teiid, and lacertid lizards, while largely visual, rely to some

TABLE 13-8. Tactile releasing mechanisms used during the courtship of *V. olivaceus*.

Releaser	Act class	User	Releaser group
Neck bite	Reproductive	Male	Tactile
Scratch	Reproductive	Male	Tactile
Tonguing nudge	Reproductive	Both sexes	Tactile
Chin rubbing	Reproductive	Male	Tactile
Tail lift	Reproductive	Male	Tactile
Brachial embrace	Reproductive[a]	Male(s)	Tactile
Head jerk	Reproductive and combat	Both sexes	Visual
Tonguing	All contexts	Both sexes	Chemical releaser?[b]

a. Also in combat between males.
b. Suggests presence of chemical releaser(s) (see text).

extent on olfaction during social encounters (Carpenter and Ferguson, 1977, for review). Hunsaker (after Cole, 1966) demonstrated amino acid differences between the pore secretions on the skin of different species of *Sceloporus* and suggested that these might provide cues for species recognition. Gekkonids, scincids, anguids, and varanids seem to rely heavily on olfactory cues during all stages of courtship and aggression, whereas iguanid, agamid, teiid, and lacertid lizards may use them mainly during the terminal phases of courtship and aggression (Carpenter and Ferguson, 1977), probably as a final proof of identification.

To some extent the region of the body tongued during courtship should be some indicator of where such cues are found in the different species. I have pointed out that in *V. komodoensis* the head, inguinal area, and cloacal regions are regularly tongued by males during courtship bouts (Auffenberg, 1981a) and that courting male *V. bengalensis* behave similarly (Auffenberg, 1983b). Figure 13-5 shows the parts of the bodies of the females touched by the tongues of courting males of *V. olivaceus* (N = 1621 contacts) compared with mutual tongue contacts between male-male aggressive coactions (N = 967 contacts). The illustration clearly shows a similar pattern in both contexts; the heads are more frequently investigated than any other part of the body, followed by the dorsal part of the neck and body, then by the limbs and tails, generally least represented. The cloacal area is not tongued frequently. However, in male-male interactions the axillary and inguinal areas are investigated more often than the same parts in male-female interactions. In general, tonguing occurs over a broader part of the body during courtship than during male-male aggression. A more detailed study of this tonguing pattern in *V. bengalensis* and *V. olivaceus*, coupled with histological studies of the skin on different parts of the body, is being conducted currently. It should provide considerably more information on why these patterns are present and what seasonal and sexual differences exist in microanatomy of the skin of adult varanids.

Table 13-9 shows the distribution of male tonguing acts in three different periods during the courtship sequence (early to step 3, middle to step 7, and late to step 11). In general the number of tonguing acts is reduced from the early to late steps, though the totals are not significantly different. However, it is clear that the same kind of reduction takes place in most categories of areas tongued by the males. Several factors affect the frequencies shown. Tonguing of the head is high in the middle and late phases partly because the male often tongues the top of the head while he is in a mounted position and cannot tongue other parts of the body. Incidents of tail tonguing are higher during the early steps because the male often approaches the female from behind and quickly moves to the anterior part, where he remains during subsequent steps, unless she walks away. Note that the cloaca is relatively uninteresting to the male at any step. The frequencies for other body areas indicate an interest in the anterior part of the female's body throughout the sequence; with additional steps there is a general reduction in tonguing any remaining part of the body, showing the male's shift from investigative to reproductive acts in the later steps. Tonguing in the inguinal area is not frequent in the butaan in courtship and only slightly more common during male-male aggressive interactions.

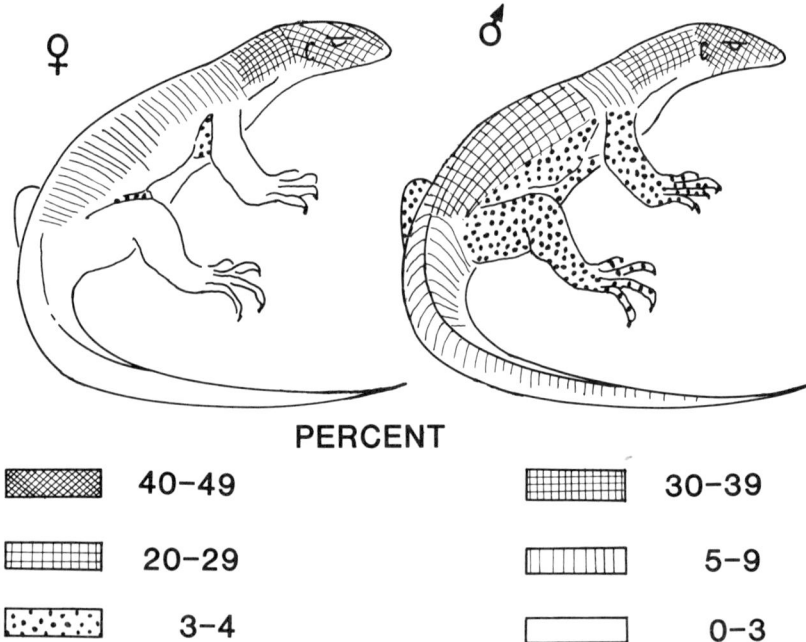

Fig. 13-5. Percentage of times different parts of the body are tongued by males when investigating females and other males.

TABLE 13-9. Temporal-spatial distribution of tonguing acts in *V. olivaceus* males during courtship.

Places tongue on female	Steps 1 and 3		Steps 5 and 7		Steps 9 and 11	
	Freq.	%	Freq.	%	Freq.	%
Head	165	26.1	229	39.1	214	53.0
Neck	254	40.2	184	31.4	112	27.7
Shoulders	62	9.8	37	6.3	35	8.7
Axillary	31	4.9	25	4.3	12	3.0
Front leg	17	2.7	17	2.9	6	1.5
Back	62	9.8	60	10.2	12	3.0
Sides	11	1.7	8	1.4	5	1.2
Inguinal	4	0.6	7	1.2	4	1.0
Hind leg	7	1.2	11	2.0	2	0.5
Cloaca	3	0.5	3	0.5	1	0.2
Tail	16	2.5	4	0.7	1	0.2
Total	632		585		404	

Priming Mechanisms

In addition to releasers or external stimuli that cause an immediate predictable behavioral response, some stimuli change the responsiveness of one individual to another, such as social status (including age), experience, relative size of the two interactants, etc.

Social status is often expressed in dominance hierarchies; in varanids these are determined largely on the basis of size (Auffenberg, 1981a, b, 1983b), with adult males tending to be dominant over all others because of their greater mass. (Size also has been shown to be important in establishing hierarchies in iguanids [Evans, 1936; Greenberg and Noble, 1944; Rand 1967a].) Dominance shifts among adult males are common. Females tend to be bullied by males and during courtship usually are manipulated by them through the use of considerable force via brachial embrace, neck bites, etc. The force used is related to the female's level of aggressiveness; *V. bengalensis* females are often aggressive, and the brachial embrace and other acts by the male are vigorously employed to dominate the female. In that species, females tend to breed more with the males dominant enough to manipulate them, for a high level of dominance is necessary because of the aggressiveness of the females. Thus strength among adult males seems to be an important attribute of success in breeding; it is important first in male-male "ritual" combat encounters, then in managing an often uncooperative female. Although less is known about the courtship of *V. komodoensis,* it is clear that female aggressiveness is manifest in some courtship sequences, particularly when nonresident males are involved (Auffenberg, 1981a). Thus male strength possibly is also important to that species. However, butaan females are definitely less aggressive than females of *V. bengalensis,* and the brachial embrace is less vigorous. Complete manipulation of the female is not nearly as important to the male butaan. The fact that both *V. bengalensis* and *V. olivaceus* females often position themselves near males engaged in (or just having finished) combat is an indication of the importance of social status to successful breeding in these species. In *V. komodoensis* social status among a number of local adult males may be less important because of the presumed pair-bonding found in resident adults of this species (Auffenberg, 1981a).

During the courtship of *V. olivaceus* it is the female that postures in ways suggesting subordinate social status (head away, head tilted, head down or up, head in U, walking (or running) away, hiding head under leaves, closed eyes). In all larger varanid species the male rarely displays his social standing—presumably because of his obviously larger size (Auffenberg, 1981a). Social status among males is partly determined (and largely so among adults) by male-male combat, but in male-female interactions it is determined apparently on the basis of the differential in mass. Female *V. olivaceus* acknowledge this differential by performing many subordinate head postures. For the most part it seems to be these subordinate movements combined with doing nothing that encourage the male to press his courtship. On the other hand, aggressive displays by the male of both this species and *V. bengalensis* lengthen the period during which success may be achieved and often lead to reciprocal social or investigative rather than reproductive behavior on his part. It is not known to what extent size of home site and territory affect social status or whether it functions as a priming mechanism in *V. olivaceus* (though it is of great importance in *V. komodoensis*; Auffenberg, 1981a).

Male courtship undoubtedly primes female receptivity in butaans (and all other species studied) and may establish and maintain bonds of familiarity between them. The latter does not seem to be important to *V. olivaceus*: males apparently court any females placed in the same enclosure with them, though it is not clear whether they would successfully breed with them. In *V. bengalensis* and *V. komodoensis* familiarity definitely enhances the chances of successful breeding. In all these species repeated courtship probably stimulates ovarian development (see Crews, 1975) and, in turn, female receptivity. Female newcomers in the enclosures with "resident" butaan males always elicited courtship behavior, but it rarely led to completed breeding during the four months when the observations were made.

Several pairs of *V. olivaceus* were found in the same hollow tree trunk at various times of the year. This fact suggests that adult males and females frequently cohabit the same shelter for at least part of the year (but probably on an irregular basis), in spite of the fact that shelter seems abundant—at least in the fissured limestone in which they often hide.

In captivity, it was the female that had been captured with a male that regularly bred with that male. Only rarely did she breed with other males in the enclosure and then only late in the season. No newcomer females successfully bred with any males at any time, though they were regularly courted by all males in the enclosure.

Crews (1973, 1975) showed that the iguanine *Anolis* female is not receptive to courtship by males of the same species for some minutes after successful courtship by another male. This same tendency is noted in butaans. Though up to four successful breedings took place in one day, there is a period (about one hour) after coitus during which females are not receptive to courtship attempts. During the nonreceptive period the breeding male often lay next to the female (sometimes for several hours) with one front leg draped over her back and usually her shoulder. As soon as she began to move the male would again begin his courtship. If the female ignored him and settled down, the male again lay next to her with his leg draped across her (fig. 13-2). The same behavior was noted in *V. komodoensis* and *V. bengalensis,* but I failed to realize the importance of this postcopulatory behavior on the part of the male. The importance of courtship persistence and female guarding by nonreptilian male vertebrates has been discussed by Parker (1974).

14

Systematics and Evolution

Systematics

A BRIEF outline of the discovery and nomenclatorial history of *Varanus olivaceus* is provided in chapter 1. What is presented here is an extension of that description. The abbreviated synonymy of *V. olivaceus* follows.

Varanus olivaceus Hallowell

1845   *Uaranus ornatus* Gray (not *Tupinambis ornatus* Daudin 1803), Cat. Liz. Brit. Mus. page 10 (type locality: "Philippines"; H. Cuming, collector).

1857   *Varanus olivaceus* Hallowell. Proc. Acad. Nat. Sci. Phila., page 150 (type locality: Manila, Luzon, Philippines; Dr. Kane, collector), considered a synonym of *V. bengalensis* by Mertens (1942).

1885   *Varanus grayi* Boulenger. Cat. Liz. Brit. Mus., 2, page 312 (new name for the homonym *U. ornatus* Gray 1845, but now a synonym of *V. olivaceus* Hallowell 1857, by priority rule).

Mertens's (1942) diagnosis of *V. grayi* Boulenger reads (translation mine): "A monitor with oblique slit-like narial opening, about the same distance from the snout tip to the eye; supraocular (scales) not widened; smooth (dorsal) neck scales, about as large as those of the posterior superior part of the head, and that blend into the smaller dorsal scales posteriorly; keeled belly scales in 100 transverse rows (from the gular fold to [insertion] of hind leg);

and laterally compressed tail, that has a very low double-toothed keel dorsally. Teeth low, round in cross-section." In addition the holotype is "green, with broad, black bands . . . with lighter flecks. (Parts) of it are olive-brown, with lighter flecks . . . blackish flecks on the dorsal surface of the head."

The salient features of Hallowell's (1857) description of *V. olivaceus* are "large, polygonal scales on the dorsal surface of the head (including the supraoculars) . . . with the largest at the dorsal snout extremity; the nostrils are two oblique openings placed about midway between the extremity of the snout and the anterior canthus of the eye . . . scales upon neck and body small, oval, those of the back more distinctly carinated. . . . tail . . . with a slightly developed crest, tapering gradually toward the end, where it is somewhat compressed. . . . Ground color olive above, lighter below, with obscure dark colored spots and markings."

Hallowell mentioned *V. bengalensis* in his remarks, so he was obviously aware of its description and believed his species represented a distinct one. He stated that *V. olivaceus* had large scales over the orbits (= supraoculars) rather than the small ones found in other species (presumably including *V. bengalensis*).

Earlier in the history of the discovery of varanid species, *V. nebulosus* was described (Gray, 1831) as having, among other distinguishing characters, larger supraoculars and a more banded pattern than *V. bengalensis*. Still later, *V. nebulosus* was regarded as a race of *V. bengalensis* (Mertens, 1937), as it still is today (i.e., *V. bengalensis nebulosus*). Both *V. bengalensis* subspecies have oblique, slitlike nostrils located slightly closer to the snout tip than the eye (like *V. grayi* and *V. olivaceus* types).

Mertens apparently placed Hallowell's *V. olivaceus* in the *V. bengalensis* synonymy on the basis of the following facts and conclusions. (1) At the time of his study, no adult *V. grayi* with an entire preserved head was available. (2) There was thus no evidence on variation of supraocular scale sizes in *V. grayi* (the type has small supraoculars)—these are now known to vary greatly. (3) *V. nebulosus* has a pattern and color much like Hallowell's *V. olivaceus,* as well as a similar nostril position and shape, and with larger supraoculars than *V. b. bengalensis*. (4) No specimen of *V. grayi* then known was provided with detailed locality data—Mertens must have disbelieved the locality data for Hallowell's type, as Manila was frequently visited yet no additional *grayi*-like specimens had come to light in the intervening 100 years (in spite of de Elera's record [1895] for a *V. grayi* from "Laguna"—a province near Manila). Mertens believed the record was worthless because of so many errors in de Elera's Philippines localities. (5) Hallowell's type specimen was evidently purchased in a market by a doctor in the U.S. Navy, and the exact source was thus unknown (specimens are still occasionally available in Manila markets). At the time Mertens's action seemed completely justified. However, the many individuals of this species collected during the course of this study show that there is great variation in both the size of its supraocular scales (fig. 14-1) and its color and pattern. Thus I am led to the conclusion that *V. grayi* and *V. olivaceus* are different names that have been applied to the same population of

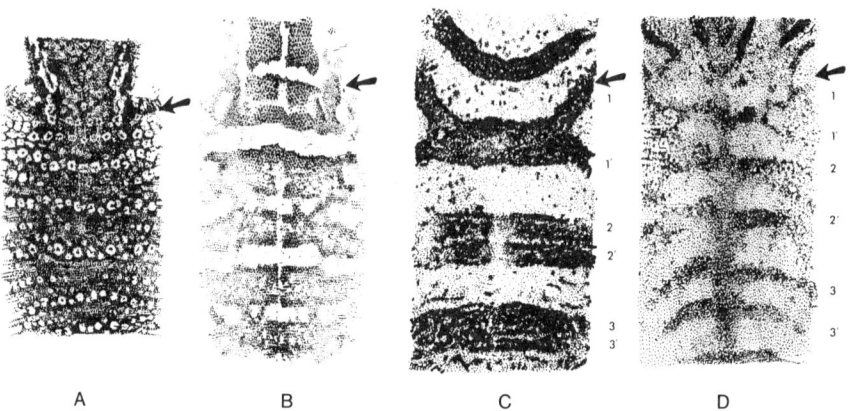

Fig. 14-1. Dorsal color patterns of subadult *Varanus* species. *A*, *V. salvator salvator*, UF 30247, "Sumatra," showing typical pattern of the transverse body bands comprised of ocelli and light lateral neck stripe. *B*, *V. s. cummingi*, UF uncataloged, "Mindanao," showing fusion of dorsal ocelli to form bands superficially similar to those of *V. olivaceus;* light neck stripe still visible. *C*, *V. olivaceus*, UF 55026, Luzon, Camarines Sur, Caramoan area, showing V-shaped dark neck bands; no light lateral neck stripe, wide transverse bands comprised of two compressed bands (see also figs. 2-8, 2-9). *D, V. bengalensis nebulosis,* UF 30223, West Malaysia, 8 mi N Kuala Lumpur, showing dark neck V-stripes, lack of light lateral neck stripes, and presence of well-separated dark transverse bands. *Arrows* indicate insertion level of right front limb. *Numbers* refer to probable band homologs.

monitor lizard. The only major discrepancy other than supraocular size in the diagnoses prepared by Gray and Hallowell is that the color of Hallowell's type (holotype of *V. olivaceus* ANSP 9916, a skin, not located to date in the academy collections) was obviously much darker and more olive than that of the decidedly greenish type of *V. grayi* with its vivid black cross bands. But these differences are exactly the ontogenetic changes that occur in the color and pattern from juvenile to adult stages of this species, particularly in large males (chap. 2). That Hallowell had an adult male in front of him is obvious from the measurements he provided of his type ("total length 5 feet 2 1/2 inches; circumference 1 foot 5 1/2 inches").

Systematics and Evolution

Phylogenetic Relations

General.—One of the most comprehensive studies on varanid lizards and their relationships was published by Mertens in 1942. This work still stands as the most important single publication on the members of the family, and few changes of any significance have been suggested by modern studies. He concluded on the basis of scalation and skull construction that modern family members can be organized into several groups (subgenera). Of these he presented convincing arguments that species within his subgenus *Varanus* are closest to the stem form—particularly in respect to their skulls. Of the several species he placed in this subgenus (type *V. varius*), *V. salvator* (and the closely related *V. komodoensis*) was probably closest to the ancestral type. All other subgenera are thus believed to have evolved from the subgenus *Varanus*. Closely related to this basic stock are all the remaining Southeast Asian monitor species. However, the latter, unlike the species comprising the subgenus *Varanus* (and other subgenera), have oblique, slitlike nostrils. Furthermore, though skull structure is basically similar, the remaining slit-nosed species represent great divergence in cranial anatomy among them.

As a result, Mertens divided these species into several subgenera: *Tectovaranus*, including the two subspecies of *V. dumerilii* (*dumerilii* and *heterophilus*) only; *Dendrovaranus*, including only *V. rudicollis*; and *Indovaranus*, comprised of both subspecies of *V. bengalensis* (*V. b. bengalensis* and *V. b. nebulosus*). Later (1962) he erected the subgenus *Philippinosaurus* to receive the single species *V. olivaceus* (as *grayi*). The latter was obviously closely related to the *V. salvator* basic "stock" but highly specialized in a number of skull features. In general appearance the skull is most similar to that of *V. niloticus*, but, as Mertens showed, these similarities are due to convergence. That is about as much as he was willing to concede regarding the relationships of *V. olivaceus*.

Karyotype and blood protein.—In 1975 two important studies on the phyletic relations of *Varanus* species were published. King and King studied the karyotypes of 16 species (no *V. olivaceus*), confirming many of Mertens's (1942) conclusions, though suggesting possible future rearrangement of some species currently placed in the subgenus *Varanus*. On several grounds they concluded that their Southeast Asian *salvator* group (*V. salvator, V. b. bengalensis, V. b. nebulosus*, and *V. rudicollis*; *V. dumerilii* was not examined) represents the common primordial pattern from which all other karyotypes evolved. The species comprising this group cannot be further separated, so that karyotypic analysis sheds little light on the relationship of *V. olivaceus* to other constituent members of the group to which it so obviously belongs. (Karyotype analysis of *V. olivaceus*, using harvesting and slide methods described in Gorman et al. [1967], was based on short-term leucocyte cultures obtained from blood samples of sacrificed animals. Results are 40 (2n) chromosomes, with arm of acrocentric chromosome pair 5 long, as in *V. salvator*; females without heteromorphic chromosome 9 sex determinant; no acrocentrics in pairs 5–8, only one subacrocentric in some pairs; acrocentric microchromosomals present; but no metacentric members. This formula is found in all members of the Southeast Asian "*salvator* group" [King and King, 1975].)

In the same year, Holmes, King, and King studied the electrophoretic properties of four blood proteins of 18 species of *Varanus* (no *V. olivaceus* included), showing the same general grouping as in the karyotype studies. However, the Southeast Asian *salvator* group could not be further separated by this means either (though the work sheds light on important relationships among species in other varanid groups, particularly some of the Australian forms).

Color.—It is not coincidental that when *V. olivaceus* has been confused with another species it has been with *V. bengalensis*. Of the two subspecies comprising *V. bengalensis*, the one that approaches *V. olivaceus* most closely in color and pattern is *V. b. nebulosus*; it is distinctly banded in all stages of its life history (particularly when young) (fig. 14-1). These bands are black and are more or less evenly spaced along the neck and body. Each band is made up of an anterior and a posterior darker transverse zone as in *V. olivaceus*. Similar double banding can also be seen in *V. b. bengalensis*, though the melanic interband portion is reduced. The neck bands are distinctly V-shaped, the apex directed posteriorly, as they are in *V. olivaceus* (and in *V. b. bengalensis*, though they are often dim), without the longitudinal lateral cervical stripe of the remaining species of the "primordial" group (*V. salvator, V. dumerilii, V. rudicollis*). The body bands in *V. b. nebulosus* are variable in degree of darkness, distinctness, and width but are always evident (except sometimes the oldest males) and brownish black to black in color. Those of *V. olivaceus* appear to be composed of a double band caused by the anterocaudal compression of bands previously more evenly and widely spaced, as in *V. b. nebulosus*. The only species without a distinctive light scapular band are *V. olivaceus* and *V. bengalensis* (see fig. 14-1A, B); instead, they have a dark scapular band that becomes Y-shaped just above the insertion of the front leg (fig. 14-1C).

The background color of *V. dumerilii* and *V. rudicollis* (especially when young) and all life history stages of *V. olivaceus* and *V. bengalensis* are some shade of olive, usually olive-gray (young of *V. olivaceus* are distinctly green). All these species are largely or partially arboreal, and the ground color and pattern are evidently cryptic for tropical forested situations. There are no known seasonal or temperature color or pattern changes as have been reported in *V. flavescens* (d'Abreu, 1932). Young *V. olivaceus* are much less variously flecked and spotted with dark and light dots or ocelli than the young of *V. b. bengalensis*.

As adults *V. bengalensis* (both subspecies), *V. olivaceus,* and *V. flavescens* usually have yellow heads (particularly the males); all other species in the Asian slit-nosed group have blackish heads. The belly ground color is gray in *V. olivaceus,* as it is in *V. bengalensis* (both subspecies) and *V. rudicollis*; it is cream to yellow in all the others.

Nostril.—Of the primordial group (King and King, 1975), only *V. salvator* has an oval or rounded nostril opening near the snout tip. All others have an oblique, slitlike nostril closer to the eye (fig. 14-2), under which the narial tubelike passageway is long and bent into a U (Wegner, 1922; see chap. 2), apparently accounting for associated differences in the bones of the upper snout as well (see below). These osteological characteristics are not found in *V. salvator,* nor is its nasal tube long and bent.

The slit-nosed monitors are evidently more closely related to one another than to any other species, including *V. salvator.* (*V. griseus, V. niloticus,* and *V. exanthematicus* [i.e., the African varanids, each representing a distinctive subgenus] are believed to be derived from the primordial group [King and King, 1975]. *V. griseus* and *V. exanthematicus* have slit-nostrils and are probably derived from *V. flavescens,* thence from *V. bengalensis* [*nebulosus?*].) As shown in chapter 4, the foraging behavior of all these species involves using the snout to move surface debris—much as ducks might use their

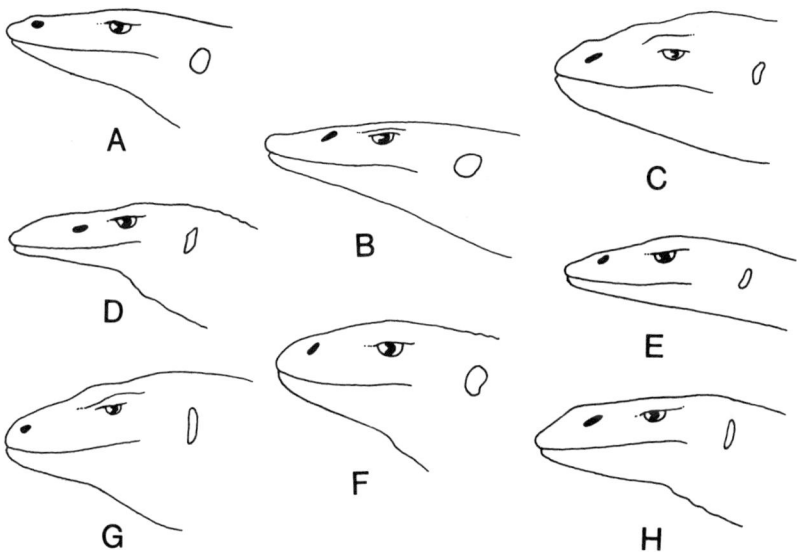

Fig. 14-2. Narial shape in Southeast Asian varanids (all adults unless specified). *A, V. salvator; B, V. dumerilii; C, V. olivaceus; D, V. rudicollis; E,* juvenile *V. olivaceus; F, V. flavescens; G, V. komodoensis; H, V. bengalensis.*

bills. That the slit-nostril was derived from the rounded, more anteriorly located nostril is suggested by the fact that the shape of the opening is more rounded in the juveniles (fig. 14-2; Mertens, 1942). (Gray [1845] even established different genera from juveniles and adults of *V. bengalensis* on the basis of ontogenetic differences in the shape of the nostrilar slit.)

Scalation.—The scale characters of *V.olivaceus* have played a major role in some of the confusion surrounding this species in the past. For many years only the scalation of the type was available for study, and not all its characters were typical for the species (i.e., size of the supraocular scales). The supraocular scales are about as broad as long in the type, and Mertens stressed their nonenlargement in both his diagnosis of the species and his key to the genus (1942). However, with a large series available for study, considerable variation in size and shape of these scales is noted. Most individuals have supraoculars that are distinctly enlarged, to twice the dimension in one direction as the other (fig. 2-7). The pattern of small scales, as found in the type, occurs in only 9% of the individuals in the Caramoan area (122 examined). Transversely enlarged supraoculars are found in *V. bengalensis nebulosus, V. flavescens, V. rudicollis,* and *V. salvator* and less so (sometimes not at all) in *V. b. bengalensis*. Those of *V. olivaceus* and *V. dumerilii* are variable, though most individuals have enlarged supraoculars.

In general, the head scales of *V. olivaceus* are large compared to those of other members of the primordial group. Mertens (1942) is of the opinion that the deserticolous varanid species have the smaller head scales, and this characteristic seems to fit even within the primordial group; of these *V. b. bengalensis* has the smallest, and those of *V. b. nebulosus* are generally larger.

Enlarged neck scales occur in *V. olivaceus* only immediately behind the head. Similar scales (though often covering more of the neck) are found in most other species of the primordial group (especially *V. dumerilii, V. salvator,* and *V. rudicollis*). Those on the neck of *V. bengalensis* are not particularly enlarged anywhere. *V. olivaceus* has neck scalation somewhat intermediate between various species of the primordial group.

In general, all members of the primordial group, with the exception of *V. dumerilii,* have rather small dorsal scales. The midbody scales in *V. olivaceus* are the smallest and are disposed in 186 rows (169–214). *V. dumerilii* is unique in this regard, since it has from 80 to 88 transverse dorsal scales; all other species have closer to the number of *V. olivaceus* but with decidedly fewer than *V. olivaceus* ($\bar{X}$ = 147 to 160 in different *V. salvator* races, 156 in both *V. rudicollis* and *V. b. nebulosus,* 151 in *V. b. bengalensis*).

The number of longitudinal belly scales of the butaan is also fewer than that of any other member of the primordial group ($\bar{X}$ = 109). It is approached more closely by *V. b. bengalensis* ($\bar{X}$ = 96) than by the various *V. salvator* subspecies ($\bar{X}$ = 79–89), by *V. dumerilii* ($\bar{X}$ = 79–82), or by *V. rudicollis*

($\bar{X} = 83$), and the most different is *V. b. nebulosus* ($\bar{X} = 76$). The differences in the dorsal and ventral scales of the tail are distinct in *V. dumerilii*, less so in all other species. *V. olivaceus* and *V. bengalensis* are identical in this regard and are not specialized.

Proportions.—Among the Asiatic monitors with slit-nostrils there is some variation in the placement of the nostril. In *V. olivaceus* it is midway between the snout tip and the anterior edge of the eye opening (1.0, or the same distance to the snout as to the eye). In *V. flavescens* it is closer to the snout (1.3); in all others it is somewhat closer to the eye (*V. bengalensis*, 0.7, *V. dumerilii* and *V. rudicollis*, 0.4). *V. salvator* has rounded nostril openings placed near the snout tip (2.0–2.4).

Body/tail proportions have also been used to suggest relationships among monitor lizards. In *V. olivaceus* and *V. bengalensis*, this proportion is 1.4. Both of these species are partially arboreal, and this habit is generally associated with medium-length tails. *V. dumerilii* and *V. rudicollis* also spend much time in the trees and have tails about the same length ($\bar{X} = 1.5$). More fully arboreal species have longer tails (i.e., *V. prasinus* and *V. salvadorii*, 2.1–2.4); completely terrestrial species usually have shorter tails (i.e., *V. komodoensis*, 1.0 in adults; *V. exanthematicus*, 1.2). The tails of aquatic species are somewhat longer (*V. salvator* subspecies, 1.6, *V. indicus*, 1.8, *V. niloticus*, 1.7).

While the tails of different varanid species are differentially laterally compressed (see Mertens, 1942, for examples), that of *V. olivaceus* is intermediate—not as compressed as *V. salvator* and *V. niloticus* and not as rounded as members of the subgenus *Ondatra* or *V. griseus*.

Neck length also varies among the species of the primordial group. That of *V. rudicollis* is proportionally longest, *V. salvator* next longest, and all remaining ones, including *V. olivaceus*, somewhat shorter.

The powerful, recurved, but stout claws and proportionally long toes have been mentioned (Hallowell, 1857; Mertens, 1942). In general shape and construction, they are most similar to the claws and toes of other arboreal species, including *V. bengalensis*. Though recurved, the claws of *V. salvator* are more delicately constructed. Those of terrestrial species (i.e., *V. komodoensis* and, particularly, *V. griseus*) are generally both less laterally compressed and less recurved.

Scale microanatomy.—There are no preanal pores in either *V. bengalensis* or *V. olivaceus*, though they are present in *V. salvator* (but variable), *V. rudicollis*, and *V. dumerilii*. (The head, neck, and lateral body surfaces associated with sex recognition perhaps function as both tactile receptors and pheromone-producing sites. They are being investigated separately. Auffenberg (1981b) reported on a similar association of behavior and microstructures in other varanid species.)

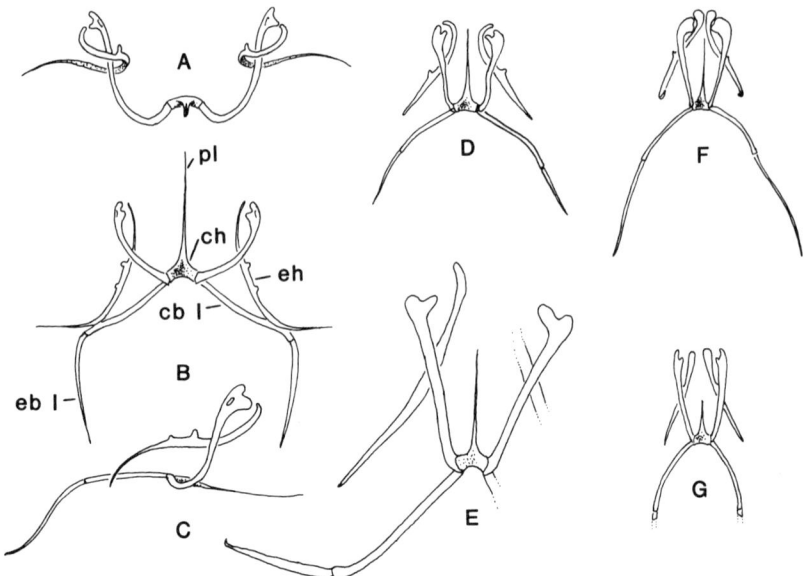

Fig. 14-3. Hyoid apparatus in several *Varanus* species. All ventral views unless noted. *A–C*, anterior, ventral, and lateral views of *V. olivaceus* (UF 50990). *D, V. exanthematicus* (UF 54071). *E, V. salvator* (UF 33161). *F, V. bengalensis* (UF 45593). *G, V. dumerilii* (UF 54209), all ventral views. Abbreviations for hyoid parts: *cb1*, ceratobranchial 1; *ch*, ceratohyal; *eb1*, epibranchial 1; *eh*, epihyal; *pl*, process lingualis.

Osteoderms.—*V. olivaceus* has no small bones in the skin, nor do the other slit-nosed Asiatic varanids, though they are characteristic of many species of the subgenus *Varanus* (*V. salvator, V. komodoensis, V. giganteus*, etc.).

Hyoid apparatus.—In general, this interconnected series is more heavily ossified in varanids compared to other lizards. In *V. olivaceus* it is especially well developed (see chap. 3) and specially shaped to allow the passage of hard, spherical food particles; the remaining monitor species studied have an apparatus designed to bend vertically and to accommodate food particles of a variety of shapes and sizes. At present there is no clear indication from which species *V. olivaceus* evolved on the basis of hyoid construction (fig. 14-3).

Teeth.—The teeth of *V. olivaceus* are short and blunt (fig. 3-5, chap. 3) during the subadult and adult stages and sharp and slightly compressed in juveniles. Other monitors with blunt teeth as adults (*V. niloticus, V. exanthematicus*, though geographically variable) also have compressed teeth when young (fig. 14-4). All other known monitors have sharpened teeth throughout life, except *V. bengalensis* (both subspecies), in which the large adults often have blunted teeth (see Mertens, 1942: figs. 196 and 287). These are identical to the butaan's, though they are produced earlier in life in *V. olivaceus*.

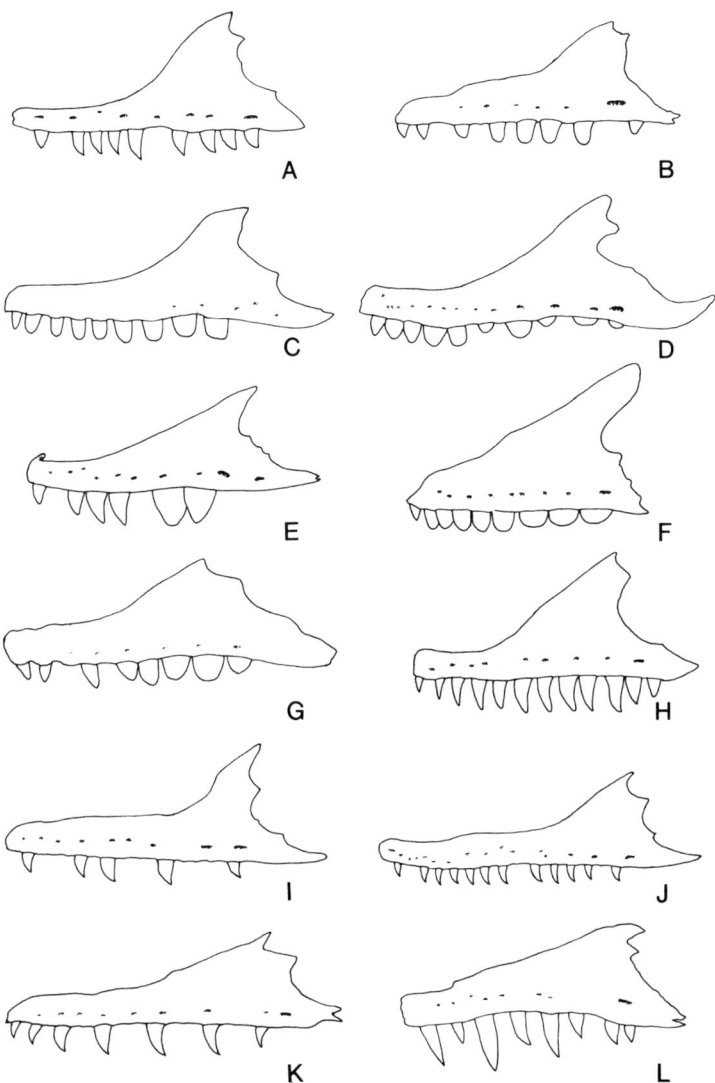

Fig. 14-4. Left maxillary bone of adults of various *Varanus* species. *A, V. b. bengalensis* (ZSM 2639/0). *B,* very old adult *V. b. bengalensis* (SMF 11550), showing blunted posterior teeth. *C, V. niloticus* (SMF 34427). *D,* very old *V. niloticus* (SMF 33251), showing blunted teeth in old age. *E, V. e. exanthematicus* (SMF 90/192). *F, V. e. microstictus* (UF 33261), showing geographic variation in degree of tooth blunting. *G, V. olivaceus* (ZSM 2640/0). *H, V. flavescens* (SMF 11546). *I, V. dumerilii* (SMF 11556). *J, V. rudicollis* (ZSM 18/1919b). *K, V. salvator* (SMF 40176). *L, V. salvadorii* (AMNH 59873). *V. dumerilli* and *V. rudicollis* feed largely on insects (short teeth), and the former also eats many crabs; *V. salvadorii* probably feeds mainly on birds (long stabbing teeth); *V. salvator, V. flavescens,* and *V. bengalensis* are generalist feeders, though the first eats many crabs and the last eats many large beetles; *V. niloticus* and *V. exanthematicus* are also generalists but often feed on molluscs, at least in certain parts of their range.

Blunt teeth are closely related to diet, but it is not certain how. They are not involved in crustacean-feeding, for several species that feed extensively on both freshwater and brackish-water crabs (*V. salvator* and *V. indicus,* for example) have compressed pointed teeth; others that feed on crabs have blunted teeth (*V. niloticus*); still others have blunted teeth but do not eat crustaceans (*V. olivaceus*). However, all varanid species that eat molluscs (terrestrial or aquatic species) have blunted teeth, though not all populations of all species regularly eat them (as far as is known, *V. olivaceus* does throughout its range). To what purpose *V. bengalensis* puts the blunted teeth when it gets them in advanced age remains unknown, for there is no present evidence that it feeds then on more (if any) snails.

The number of teeth also varies among varanids, though not nearly to the extent that it does in most other lizard families. Almost all varanids have 7–9 premaxillary teeth (all species discussed here do), and all have 11–12 dentary teeth. However, butaans have only 9–10 maxillary teeth—a lower number than other varanids (10–12 in *V. dumerilii* and *V. bengalensis,* 10–13 in *V. salvator,* and over 15 in *V. rudicollis*).

Skull.—Many similarities between the skulls of *V. olivaceus* and *V. niloticus* (and *V. exanthematicus*) were shown by Mertens (1942) to be convergent (strongly built skull, tooth shape, etc.). In general the skull of *V. olivaceus* belongs to Mertens's medium-length group, though it is proportionally somewhat higher than that of many other species. The supratemporal fenestrae are large and the surrounding bones strongly constructed. Both of these characters are associated with the great muscle power that this species has in its jaws (see chap. 3). The mandibular length varies among *Varanus* species; that of *V. olivaceus* is of medium size, as it is in *V. salvator, V. bengalensis,* and many others to which *V. olivaceus* might be related. The lower jaws of *V. dumerilii* and *V. rudicollis* are proportionally long.

The premaxillary bone of *V. olivaceus* is similar to that of most other varanid species. The dorsoposterior process is broader in *V. olivaceus* than in all other species of the "primordial group," except *V. bengalensis,* where it is laterally compressed (reflected in the convex dorsal snout profile of *V. bengalensis*).

The maxillary of *V. olivaceus* is mesoprosopic (see Fejervary, 1918, 1935), as it is in both *V. bengalensis* subspecies (and many other varanids), hypsiprosopic in *V. salvator* (and *V. komodoensis,* among others), and platyprosopic in *V. dumerilli* (and others). This terminology reflects the extent to which the maxillary bone is excavated dorsally for the septomaxillae. In general, slit-nosed monitors are platyprosopic, and round-nosed species are hypsoprosopic. However, the fact that both *V. olivaceus* and *V. bengalensis* are slit-nosed and intermediate (mesoprosopic) is a reflection of their less specialized condition compared to *V. dumerilii, V. rudicollis,* and *V. exanthemati-*

cus—slit-nosed species in which the maxillary has become additionally specialized as a result of the posterior extension of the narial opening.

The maxillary has similar proportions in almost all species, including *V. olivaceus*. Its length is about 50% of the basicondylar length. However, the height of this bone is greater in *V. olivaceus* and *V. bengalensis* (39.2 and 39.8% of basicondylar length, respectively) than in other monitor species. Among the primordial group, *V. dumerilii* is closest (38.8%), the remaining species less (*V. salvator*, 28.6%, *V. rudicollis*, 31.8%).

In most species the apertura maxilla-premaxillaris is small to moderate, as it is in *V. salvator*. In a few other species it has become larger. In the *bengalensis-olivaceus-rudicollis-dumerilii* group it is specialized in the opposite way, being tiny or even completely absent.

Though the septomaxillae are of variable size and shape, surface shape and texture dorsally are important taxonomic characters. Among the "primordial" species it is flat and smooth in *V. olivaceus, V. dumerilii,* and *V. salvator,* with longitudinal ridges in *V. bengalensis,* and it is extremely large in *V. rudicollis*.

The shape of the nasals on their dorsal surface is also of value in defining species of varanids. Two anterior processes usually embrace the dorsoposterior process of the premaxillary. Two shorter processes articulate with the frontal. Though Mertens (1942) stated that there are three equal posterior processes in *V. olivaceus*, it is not the usual condition. Almost all gradations can be found between this condition and the presence of a tiny central element (fig. 14-5). In general, the bone is proportionally shorter and wider in *V. olivaceus* than in *V. salvator*. The posterior edge of the external nares is near the middle of the nasal rather than near the posterior end of the nasal or posterior to it. Thus the condition in *V. olivaceus* is like that in *V. bengalen-*

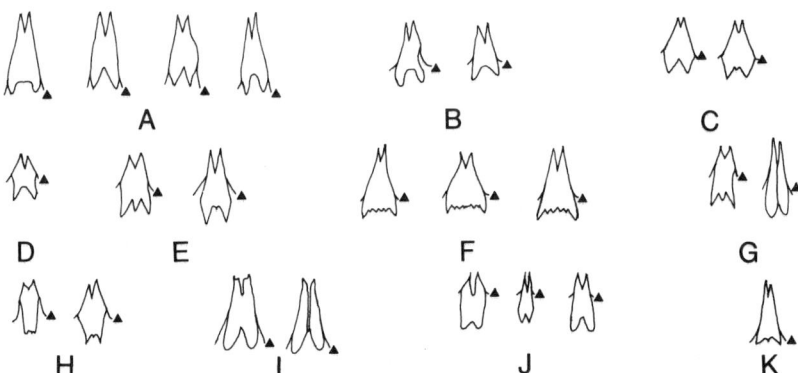

Fig. 14-5. Nasal bones in *Varanus* species. Solid triangles show the most posterior extent of the nasal foramin. A, *V. salvator;* B, *V. niloticus;* C, *V. bengalensis;* D, *V. rudicollis;* E, *V. olivaceus;* F, *V. dumerilii;* G, *V. salvadorii;* H, *V. flavescens;* I, *V. indicus;* J, *V. exanthematicus;* K, *V. varius*.

sis, though the latter lacks a central posterior median process. The nasal of *V. rudicollis* is decidedly wider than *V. olivaceus* (in spite of the proportionally long, narrow skull). The nasal of *V. dumerilii* is the most modified of all; it is widest posteriorly and has a series of small processes medial to the two larger external posterior processes. Relative to the nasal shape of *V. dumerilii* and *V. rudicollis,* that of *V. olivaceus* is not too specialized from the apparent ancestral condition for all varanids.

Most of the characteristic shape of the parietal of *V. olivaceus* (a small anterior plate, due to posterior laterodorsal excavation to form a solid anchorage for the powerful jaw muscles) is clearly related to the ontogenetic development of the peculiar adaptations attending molluscivory. The parietal of *V. dumerilii* is probably the most derived of any other species in the primordial group, since in old age it becomes broad through lateral expansion.

Most of the changes in the bones of the palatal area can be clearly associated with the specialized feeding of this species. Whatever correspondence there is between these bones and those of other molluscivorous species (i.e., *V. niloticus, V. exanthematicus,* etc.) seems due to convergence (see Mertens, 1942; chap. 3).

The only remaining bone of the head that seems to be of taxonomic interest is the dentary. In *V. olivaceus* it is longer than in any other species known (60.3% basicondylar length; 52.0–57.0% in all others). Its length is related to the great mechanical advantage of the jaw mechanism of this species (chap. 9). The longer dentary places the teeth used to crush snail shells more posteriorly—thus shortening the working lever arm in respect to the force lever arm posterior to the coronoid process.

Malarial parasites currently are known only from *V. bengalensis* and *V. olivaceus* (see chap. 12). Furthermore, the two plasmodium species represented seem to be related closely (S.R. Telford, pers. comm.).

Conclusions.—The results of a reconsideration of Mertens's (1942) analysis of skull characters and the more recent karyotyping and electrophoretic studies by several Australians suggest that the recognition of a single Asiatic slit-nosed group of monitor species seems justified. This group is comprised of *V. rudicollis, V. dumerilii, V. bengalensis,* and *V. olivaceus.* It is derived apparently from the considerably less specialized *V. salvator* (and *V. komodoensis*). The slit-nosed monitors represent a major and perhaps early derivation from the most primitive group of living varanids. The most important departure in morphology of this derived group is the posterior movement of the narial opening and its slit-like shape.

Mertens (1942, and later papers) recognized several subgenera within this slit-nosed group. The two *V. bengalensis* subspecies (*V. b. bengalensis* and *V. b. nebulosus*) comprise the subgenus *Indovaranus; V. rudicollis* represents *Dendrovaranus; V. dumerilii* (*V. d. dumerilii* and *d. heterophilus*) represents

Tectovaranus; and *V. olivaceus* represents *Philippinosaurus*. The group from which they are believed to have evolved is the subgenus *Varanus* (though see King and King, 1975, for probable phyletic problems for, at least, the Australian species currently placed there). *Indovaranus* may be more closely related to *V. flavescens* (currently placed in the subgenus *Empagusia*) than Mertens (1942) believed (Holmes, King, and King, 1975; King and King, 1975), though *V. flavescens* probably gave rise to the remaining members of *Empagusia* (*V. exanthematicus*). It is thus not presently clear where *V. flavescens* belongs; in my view it is not a member of the subgenus *Indovaranus* on the basis of a number of morphological and ecological characteristics.

Examination of the variation in 14 characters among the Asiatic slit-nosed monitors shows that *V. rudicollis* is most different from *V. olivaceus*. Important differentiating characters are the former's enlarged keeled neck scales, elongated head, low number of acute teeth, large septomaxillae, and small number of belly scales. *V. dumerilii* is almost equally removed from *V. olivaceus*, particularly in number of dorsal scales and tooth and nasal shape but also in size of the supratemporal fenestrae. *V. bengalensis* is similar or identical to *V. olivaceus* in over two-thirds of all characters compared. Only the color of juvenile *V. b. bengalensis* (not *V. b. nebulosus*), number of belly scales of *V. b. nebulosus* (not *V. b. bengalensis*), and neck scales of *V. b. bengalensis* (less so in *V. b. nebulosus*) are different from *V. olivaceus*. Of the two subspecies, *V. b. nebulosus* shares more characters with *V. olivaceus*. It is also currently the closest geographically and the closest in terms of probable past land connections (see below).

On these bases I conclude that the butaan is a highly specialized species that evolved from a "Malaysian" continental species similar to *V. bengalensis nebulosus*. I also believe that the latter is close to the stem form for all Asian slit-nosed monitors. *V. rudicollis* and *V. dumerilli* are apparently not closely related to one another (Mertens, 1942), based on a comparison of a number of skull and scale characters, and were undoubtedly independently derived from a *bengalensis*-like (rather than *salvator*-like, *fide* Mertens, 1942) stem form.

Ecology and Phylogeny

The "primordial group" (subgenera *Varanus* in part, *Indovaranus*, *Dendrovaranus*, *Tectovaranus*, and *Philippinosaurus*), on the basis of electrophoretic and karyotypic analyses, contains species inhabiting a broad spectrum of environments, from a semiaquatic form (*V. salvator*) to a thorn brush dweller (*V. b. bengalensis*, in part) and three arboreal forms (*V. rudicollis*, *V. dumerilii*, *V. olivaceus*). However, when only the slit-nosed Asiatic species are considered, it is clear that all are partially or almost completely arboreal and that most live in dense tropical forests. The only form that lives outside of tropical forests (*V. b. bengalensis*) is at least partly arboreal wherever trees are common in its range but less so where trees are more widely spaced and lower (northwestern India through Pakistan; pers. observ.). (Mertens [1942] stated that it is not often arboreal, but this is not correct, for, if the habitat contains trees, it will regularly climb to 20 m [Willey, 1909; Deraniyagala, 1931; pers. observ.].)

With the exception of *V. dumerilii* and *V. rudicollis* (both highly specialized morphologically), all members of the "primordial group" tend to be found in the drier sections of tropical forests (when surrounding forest associations are mainly mesic). The butaan tends to be found on the driest ridges and lower mountain slopes. The same is true of *V. bengalensis nebulosus* in Southeast Asia, including the Sunda Islands on which it is found (fig. 14-6). In fact, on

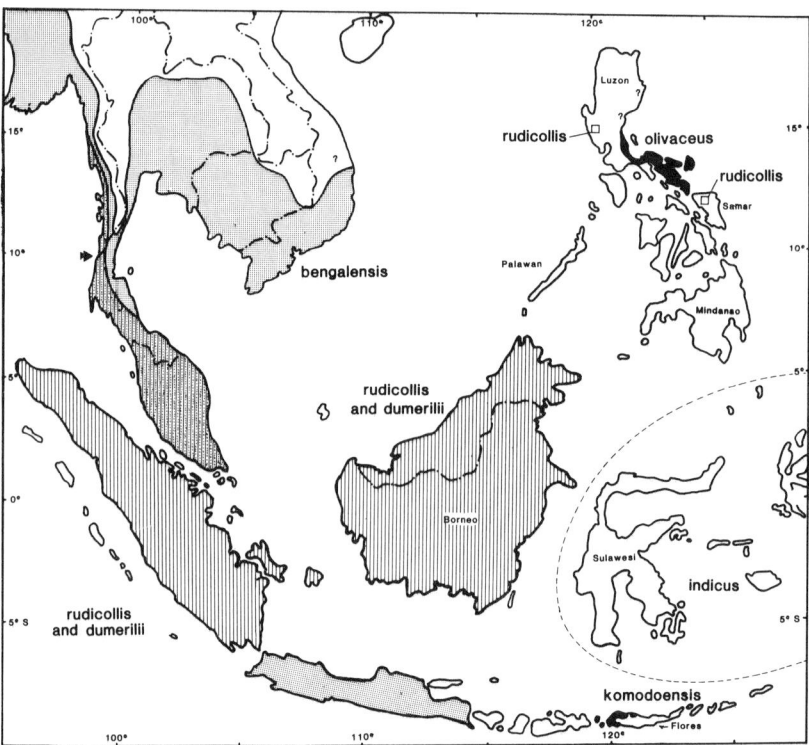

Fig. 14-6. Distribution of slit-nosed monitor lizards in southeastern Asia. The range of *V. komodoensis* is also shown to illustrate its geographic position in respect to the other monitor species of the area. *V. salvator* is found throughout the entire area shown. *V. indicus* is found only on those islands east of the dotted line. The arrow indicates the most northern limit of *V. rudicollis* in the Malay peninsula; the hollow squares represent isolated populations of *V. rudicollis* in the Philippines.

Java it has a relictual distribution, being restricted to small pockets of drier environments within the more widely disbursed mesic tropical forests (pers. observ.). It probably represents part of the relictual monsoon forest reptile and amphibian faunas that were in former times more widely distributed throughout the Sunda Islands (Auffenberg, 1980a). Even in Malaysia and Thailand, it reaches its maximum abundance in sections of drier forest, being absent from large areas of pure mesic evergreen types in central to southern Malaysia and Sumatra (pers. observ.).

It was from this species adapted to somewhat drier conditions that *V. b. bengalensis* probably evolved in the west, becoming specialized to live in even drier environments and lead a less arboreal life-style. The *V. b. nebulosus*-like ancestor also probably gave rise to the marsh-inhabiting *V. flavescens* (it does not live only in dry savanna habitats as often mistakenly stated) and through this species presumably to *V. exanthematicus*. This evolution was accompanied by many changes in both morphology and karyotype. From the basic type, or close to it, there evolved the two highly arboreal mesic tropical forest species, *V. dumerilii* (Barbour, 1921; H.C. Smith, 1931) and *V. rudicollis* (Horn and Petters, 1982; pers. observ.)—the former feeding chiefly on insects on the forest floor but remaining somewhat more generalized in feeding habits when compared to the highly specialized *V. rudicollis* which feeds on termites, tree phasmids, and arboreal centipedes. A third independent line in the Philippines led to the molluscivorous-frugivorous *V. olivaceus*. It is somewhat more arboreal than *V. bengalensis*, frequently using its tail in climbing (fig. 7-3).

The butaan is not primitive but specialized in its gut and head morphology. The derived morphological characters from the stem type (probably near *V. b. nebulosus*) are the large nape scales, large head scales, small dorsal and belly scales, increased mechanical advantage of lower jaw closure due to changes in both muscles and bones, blunted stout teeth, and cryptic color and pattern. Behaviorally it is also unique, moving more slowly through a smaller home range and feeding on a remarkably narrower food spectrum than any other monitor species known.

In general, there are probably no "open" niches in mesic primary forest, except in regard to food specialization. It is this feeding specialization that is so important in the Asian slit-nosed group. The butaan is apparently specialized to compete with other frugivores in an evergreen tropical forest of considerable seasonal stability and in which conspecific tree species tend toward mutual avoidance (Symington, 1943; Ashton, 1969) in a complex ecosystem of high productive efficiency. Here it occupies a biotic niche that is fairly narrow. It feeds on only a few of the thousands of potential local tree and land snail species. No other local reptile shares this niche, even partly. Other lizard species that do take fruit eat only small types untouched by the butaan. The only other more completely herbivorous lizards are largely foliovores; no other local reptiles regularly eat land molluscs of the large species taken by the butaan. Among mammals, only the palm civet is an important food competitor of the butaan, and its major competitors are probably the two local species of hornbill birds.

Phytogeography of the habitat.—The Philippine flora is rich, composed of approximately 10,000 species of flowering plants and ferns. Many of these species are endemic or nearly so (Merrill, 1926). The flora's regional relationships are mainly with the flora of Malaysia, though distinctly continental (Himalayan) elements are present in the mountain flora of northern Luzon and a few Australian ones at low altitudes (Merrill, 1923).

In general, the dominant species along the shore, in the settled areas at low elevations, in the large areas of open grassland (derived), and in the secondary forests are identical with widely distributed Indo-Malaysian species. In the primary forests, about 75% of the dominant plant species are endemics.

The Dipterocarpaceae comprise the dominant floral elements of the forests in which the butaan is found. This family is represented by 300 species, distributed from India to Australia. It seems to have arrived in Southeast Asia during the mid-Tertiary, probably from farther west, possibly from India (Ashton, 1963). Borneo has the largest number of dipterocarp species (ca 100). Both the Philippines and Malaysia have about 50 species each. There are fewer in the Celebes and fewer yet in New Guinea. There are no dipterocarp forests in Formosa. Thus the dominant trees of the butaan's habitat have the center of their distribution in the Indo-Malaysian area. It seems likely that it is the same area in which the ancestor of *V. olivaceus* originated. However, the forests in which the butaan is found are comprised of many floral elements, a number of which have completely different geographic distributions. This fact suggests that the evolutionary dispersal of the entire forest cannot be based on the distribution of only one or two families, even if they are dominant.

Table 14-1 shows that, of the major butaan foods, most represent fairly large and widely dispersed families—many being pantropical in distribution. However, at the generic level Southeast Asian endemics emerge as an important group. The table also shows that while many food species are Philippine endemics, those that are not are found in the greater Malaysian area, including the Greater Sunda Islands (Borneo, Java, etc.). The distribution of animal food species is narrower than that of plants; most frequently eaten molluscs are endemics on Luzon or even southeast Luzon.

Table 14-2 shows the large number of plant genera shared by Sundaland, Malaysia, and the Philippines. It seems significant that all of the genera that serve as food plants for *V. olivaceus* are represented in this table, but the plants of other parts of Luzon are not (table 14-3). Many common plant fami-

TABLE 14-1. Geographic distribution of major foods of *V. olivaceus*.

| Family | | | | Worldwide | | Philippines | |
Genus	Geographic range	Range of species eaten	Genera	Species	Genera	Species
PLANTS						
Palmae	Pantropical		150	1350	21	100
Caryota	Southeast Asia, Indo-Australia	Philippines		12		4
Moraceae	Pantropical		67	1000	13	150
Ficus	Pantropical	Philippines		800		100
Burseraceae	Pantropical		16	?	4	?
Canarium	Tropical Asia	Philippines		150		35
Meliaceae	Pantropical	Malaysia, Philippines	47	800	18	90
Sandoricum	India to Malaysia	Introduced		7		2
Aglaia	Southeast Asia, Indo-Australia, Polynesia	Malaysia, Philippines		100		40
Anacardiaceae	Pantropical		60	500	11	36
Spondias	Pantropical	Malaysia, Philippines		6		2
Dracontomelum	Southeast Asia, Indo-Australia	Malaysia, Philippines		5		2
Tiliaceae	Pantropical		45	600	11	45
Grewia	Old World Tropics	Malaysia, Philippines		150		20
ANIMALS						
Helicarionidae	Pantropical		86	900	12	165
Lepidotrichia	Philippines	East Luzon, Catanduanes		25		25
Camaenidae	Pantropical		70	900	7	87
Obba	Borneo, Celebes, Philippines	Northern Philippines		35		28

TABLE 14-2. Native Philippine tree genera common in habitat of *V. olivaceus*, as well as in Borneo, Celebes, and the Malaysian peninsula.

	Celebes	Borneo	Malaysia
Aglaia[a]	●		●
Aleurites			●
Anisoptera		●	
Artocarpus	●	●	●
Averrhoa	●	●	●
Bauhinia	●	●	●
Calophyllum			●
Canarium[a]	●		
Caryota	●	●	●
Clerodendron			
Dillenia	●		●
Diospyros			●
Dipterocarpus	●	●	●
Dracontomelum[a]	●		
Dysoxylon			●
Eugenia			●
Ficus[a]	●		●
Garcinia			
Gmelina			●
Gnetum	●		●
Grewia[a]			●
Macaranga	●		●
Mangifera	●	●	●
Memecylon			●
Pandanus[a]	●	●	●
Pentacne		●	
Pisonia	●	●	●
Pterocarpus	●	●	
Schefflera			
Semecarpus			●
Shorea			●
Sterculia	●	●	●
Terminalia	●	●	●
Tristania			●
Vitex			●

a. *V. olivaceus* food species included.

lies in the butaan's habitat are missing from the mountains of the northern part of the island (including two important families of food plants), in spite of close geographical positions and a common land surface.

In addition, there are few Australian or Formosan plant genera in the forests of southern Luzon, and none of these are food trees. Thus there is no phytogeographical evidence that either the Formosan or Australian area was ever a center of dispersal for *V. olivaceus*. On the other hand, the forests of Borneo, the Celebes, and Malaysia contain many of the common plant genera found in the butaan's habitat, including all the food species. On the basis of present plant distributions in surrounding areas, especially the distributions of butaan food plants, the butaan is more closely associated with the Indo-Malaysian flora than with those from Formosa to the north or from the Australian area to the southeast.

The distribution of plants on Luzon corresponds in a general way with the distribution of *V. olivaceus*. Merrill (1923, 1926) recognized a number of floral subprovinces on the island. None is well defined, except for the Mountain Provinces Subprovince (northern Luzon), clearly a southward extension of part of the Himalayan flora that evidently reached Luzon over Formosa (von Koenigswald, 1935, 1939). This same route is apparently the one followed by some of the Pleistocene fossil proboscideans found in the Philippines and some islands southeastward (Hooijer, 1967, 1972). However, these flora and fauna evidently have little to do with the distribution and dispersion of the butaan.

Merrill also proposed a Sierra Madran Floral Subprovince that extended along the entire eastern seaboard of Luzon from Isabella to Sorsogon, then through eastern Samar, Leyte, and Mindanao. The subprovince is not well defined, but it does reflect a zone of high endemism among closely related plants, suggesting a biologically significant separation from the remainder of Luzon. However, its geographic restriction may be due largely to edaphic conditions, such as its generally higher rainfall (see fig. 4-3). In any event, it is this flora with which the butaan is most intimately associated. However, the range of the butaan as now known does not include the entire north-south extent of the subprovince but only the middle section encompassing southern Luzon and Catanduanes Islands.

In summary, the butaan is closely associated with a rather distinctive lowland flora of eastern Luzon. This flora has little in common with the Himalayan-derived flora of the mountainous district of central and northern Luzon.

TABLE 14-3. Plant families found in the geographic range of the butaan but not in forests of northern Luzon.

Acanthaceae	Dilleniaceae	Myristicaceae
Annonaceae	Dipterocarpaceae	Nyctaginaceae
Apocynaceae	Ebenaceae	Ochnaceae
Bignoniaceae	Flacourtiaceae	Pandanaceae[a]
Capparidaceae	Guttiferae	Sapotaceae
Combretaceae	Lecythidaceae	Sterculiaceae
Connaraceae	Malvaceae	Verbenaceae
Convolvulaceae	Meliaceae[a]	

a. *V. olivaceus* food species included.

Rather, it is restricted to floras of Indo-Malaysian type, particularly to those from the Celebes, Borneo, and the Malay Peninsula (i.e., the geographic area that also contains *V. bengalensis nebulosus*—the form to which *V. olivaceus* is thought to be most closely related). Most important, all of the food plant genera of *V. olivaceus* are found in that area as well. This fact suggests that the butaan's evolutionary history is tied to the development and dispersal of the lowland dipterocarp forest. *V. bengalensis nebulosus* is a member of that association wherever it occurs, especially the drier facies. Thus the butaan ancestor dispersed into the Philippine Archipelago from the south, southwest, or west at some time prior to the establishment of the current geography, probably concomitant with the dispersal of the dipterocarp flora into this area. McGregor (*in* Dickerson, 1928) believes that the hornbill ancestor reached the archipelago in the Pliocene-Pleistocene (perhaps too late an estimate), which may have some connection with the dispersal of the butaan, for locally the hornbill often prefers many of the same food species as the monitor lizard. With *V. olivaceus* and several native rodents, the hornbills are probably one of the few vertebrates with a local diet that regularly includes the chemically protected members of the family Burseriaceae (chap. 11).

Geology

Unfortunately, the geology of the Philippine Islands is not as well known as that of some surrounding areas. Still, within a span of no more than three decades, scientific insight into the geophysiological evolution of continents has undergone profound and revolutionary modifications. With the discoveries of continental drift and suboceanic crustal tectony, almost half of all data accumulated on mountain building had to be revised, including much of what was believed about the Philippines. Data resulting from current petroleum exploration in the area is bound to revise much of what we now believe to be correct. The following account is based on current understandings, which are clearly going to change in the 1990s. However, I do not believe the new information will substantially modify the broader outline of this geologic history. Nor will such changes in detail have much appreciable effect on how I view the paleogeography of the butaan and its immediate ancestors, except in minor ways, unless the range of the present populations is found to be different than now believed.

Though some ancient surface rocks are known in the Philippine Archipelago, the islands have no definitive history older than 30 myBP. (Radioisotope dating of rocks from Luzon [after Samson, 1979] is still in an early stage, and only a few dates are available: Sierra Madre Mts. 13–16 myBP, Mayon Volcano 13 myBP.) There is also no definitive evidence that any part of the archipelago was ever connected to any continent during this entire time, though interconnections between some of the islands making up the archi-

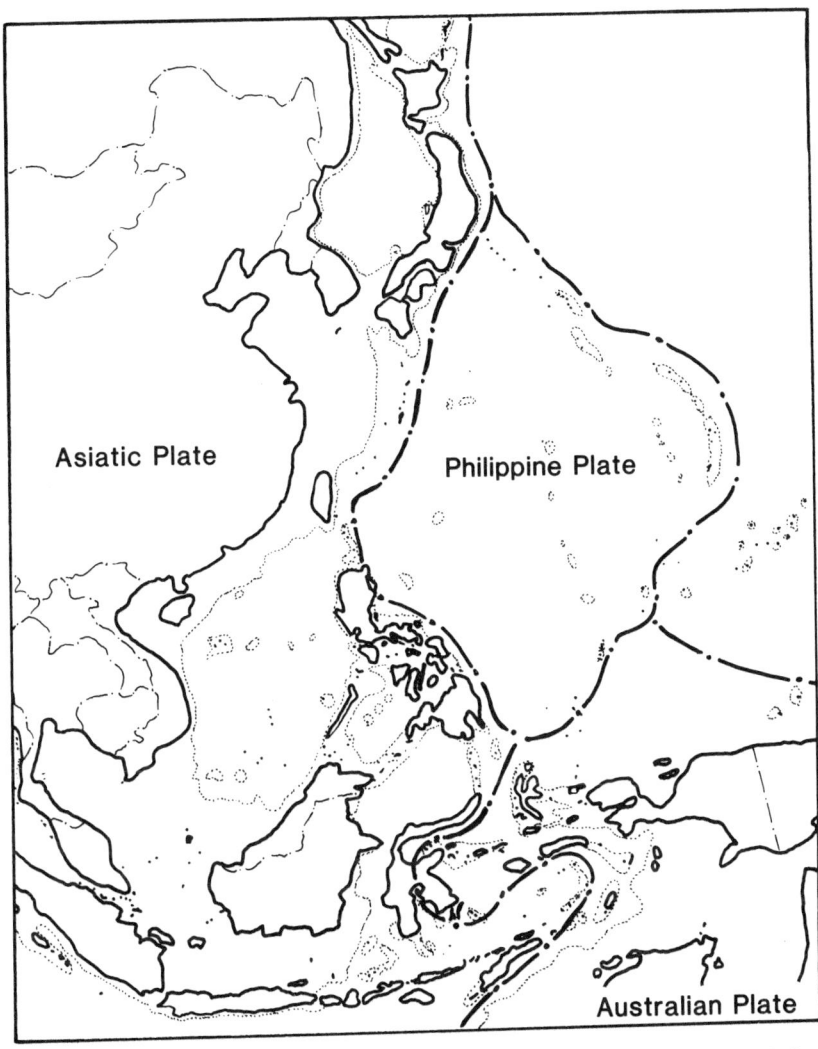

Fig. 14-7. Tectonic plate structures in the western Pacific that affect the distribution and evolution of *V. olivaceus* (largely after Audley-Charles, 1978).

pelago and others near it are known to have been made and broken—often several times.

The butaan is restricted to the islands along the boundary between the Philippine and Asiatic tectonic plates. While these islands are products of many geological factors, the major ones are crustal uplift due to horizontal pressure exerted by the edges of the Pacific, Philippine, and Asiatic plates on one another (fig. 14-7), volcanism, and changing sea levels. All of these three factors are related to the history of the butaan.

Pre-Tertiary rocks.—The basement rocks that make up the core of the Philippine area (40,814 mi^2 and about 7000 islands) are probably Mesozoic (Smith, 1924; Audley-Charles, 1978), though their precise age remains unknown. They are all extrusives forced up from great depths, suggesting diastrophism and volcanism on a grand scale. Most often these rocks have formed narrow ridges as a result of past east-west compression, probably the result of horizontal movements of the western edge of the Pacific plate against the edge of the Philippine plate, pushing it westward against the eastern edge of the Asiatic plate. By the end of the Mesozoic (Cretaceous), some of the main islands of the archipelago, including Luzon, were already formed, though considerably different from their modern counterparts in both shape and position. For the most part, these early islands were narrow strips on one or both sides of the major fault zone that separates the Asiatic from the Philippine plate (Dickerson, 1928; Patanne, 1972; Audley- Charles, 1978). Though this "Eastern Philippine Strip" (of Samson, 1979) had nothing directly to do with the evolution or zoogeography of *V. olivaceus,* it provided the foundation upon which the remainder and more important part of the geological history of the area, from the standpoint of the butaan, is based.

Tertiary and later rocks.—Lying unconformably on this wrinkled basement complex of pre-Tertiary rocks are the Tertiary ones. Most are sedimentary, consisting of extensive beds of clastic and interbedded limestones formed under shallow marine conditions, but a few terrestrial deposits (including lignitic beds) are also present (Dickerson, 1924; Smith, 1924). What is suggested, therefore, is a series of probably narrow, northwest-southeast-oriented, eroding islands with their detrital sediments incorporated in the marine limestones forming in the shallow seas between them. Though some of these rocks are Eocene and Oligocene in age, the main body of the geographic range of the butaan is underlain with Miocene-Pliocene materials, most deposited in down-warped areas between earlier narrow, rocky islands and ridges.

The Miocene rocks are the best known of any age in the Philippine Islands. Most are marine limestones containing a rich invertebrate fauna best studied on the Bondoc Peninsula, near (or within) the butaan's range. In general, these limestones were deposited in shallow seas between earlier more or less linear, ridgelike islands, whose exact position, shape, and number remain unknown. This uncertainty is largely because of extensive folding and erosion, which have destroyed the original form of the sedimentary basins between them (in which the limestone was being deposited). In some areas where erosion has not been as great (such as Samar, where elevation of these beds above the sea was much later), these beds are less disturbed and eroded and form plateaus.

Adjacent to areas of basement rocks, the Miocene beds are often volcanic and form the major mountain masses of Bicol (such as the Caralingan Mountains, which make up the spine of the Caramoan Peninsula) and the southern

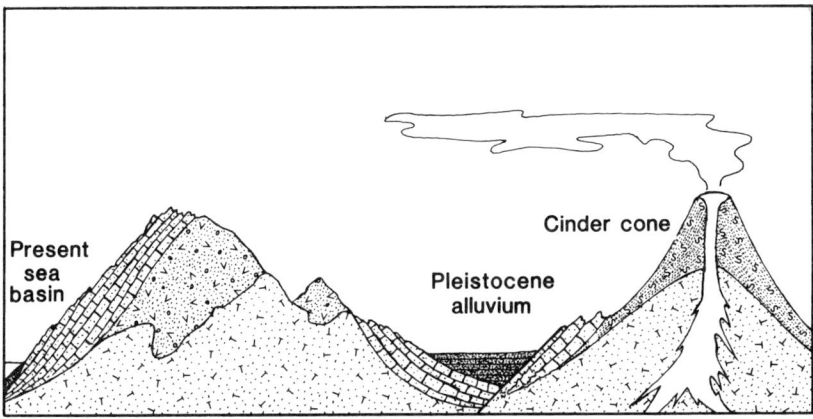

Fig. 14-8. Idealized cross section of geographic area in which *V. olivaceus* is found in southern Luzon.

part of the Sierra Madre Mountains (the part northward not inhabited by the butaan is made up largely of Pleistocene volcanics). However, the prominent features of the foothills in Luzon and Catanduanes are formed of Miocene and Pliocene limestones (fig. 14-8).

Quaternary sediments form most of the plains areas (mostly highly modified by agriculture), such as the Bicol River valley. Quaternary volcanics form great cones that dominate the landscape, particularly in Bicolandia. Peripheral sheets of lavas and tuffs, such as those on the flanks of Mt. Mayon and Mt. Isarog, normally extend for several kilometers in all directions. Some of these Quaternary volcanos are still active. Though important features of the current landscape, their areal extent is considerably limited compared to the Tertiary volcanics and sedimentaries among which they lie. Most volcanos are formed near major fault zones and so tend to be clustered or arranged linearly.

Tectonic features.—The Philippines lie in one of the most mobile belts of the earth and have been unstable for millions of years (Smith, 1924; Audley-Charles, 1978). Their instability makes interpreting the past history of *V. olivaceus*, highly conjectural under even near-perfect conditions, extremely difficult. However, some knowledge of the factors and problems involved in such an analysis is important in appreciating the evolutionary processes that have helped shape this extraordinary reptile.

The major topographic and geologic features within the butaan's range—ridges, mountain ranges, peninsulas, and islands—all have their northwest-southeast main axes (fig. 14-9). A major fault zone (Philippine Fault Zone of Willis, 1937) separates this most eastern area of northwest-southeast trends from another area to the west with landforms showing a northeast-southwest trend (compare trends of Palawan and the Sulu Archipelago with southern Luzon and Leyte, for example, in fig. 14-8). In Bicol the Philippine Fault

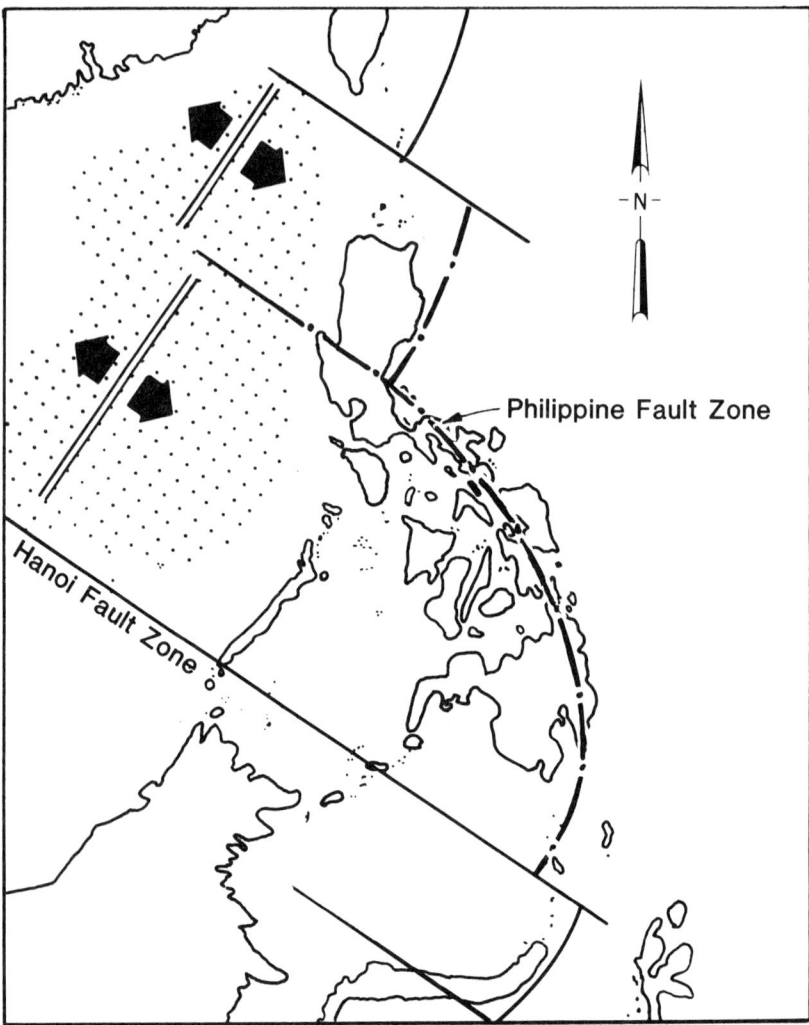

Fig. 14-9. Philippine area, showing major fault zones associated with the contact of the Philippine and Asiatic plates and the seafloor spreading in the China Sea (arrows and dotted area; largely after Willis, 1937, 1939; Hess, 1948; Irving, 1953; Amato, 1965; and Audley-Charles, 1978).

Zone shows a striking northwest-southeast alignment of long, narrow uplands, expressed as ridges and valleys on land and as elongate islands and channels where partly submerged. The Bondoc Peninsula, Ticao Island, and the Bicol River valley are all examples. The misalignment of the Caramoan Peninsula and Catanduanes Island with the rest of this system in Bicol will be discussed. Though less distinctive northwest of Bicol, the same fault zone still forms a low gap in the eastern Sierra Madre Mountains at Dingalan Bay (King and McKee, 1948; fig. 6-1). It is this gap that may be the northern boundary of the geographic range of *V. olivaceus*. If so, then this species is restricted to a zone that has suffered considerable diastrophism, with extensive land changes through time as the result of folding and lateral displacement accompanying plate tectonics. This apparently constant changing of the landscape, particularly in respect to the past formation and disappearance of islands in Bicol, may be part of the reason the species is so restricted geographically (see below).

Drawing on geological data (Saderra and Smith, 1913; Smith, 1915, 1924; Dickerson, 1924; Alvir, 1926; Respetti, 1933; Willis, 1937, 1939, 1944; King and McKeen, 1948; Anon., 1970), it seems clear that these east and west areas, defined partly by their trend lines, were produced during entirely different geological periods. The oldest is the eastern sector of northwest-southeast trends along the Asiatic and Philippine tectonic plates boundary. The land masses found here were apparently produced by earlier northeast-southwest compressional forces. They form the "probable Philippine Miocene Islands" of DuPree (1954) and Samson (1979), based in part on earlier maps by others (shown here slightly modified in fig. 14-10). As shown by all other authors,

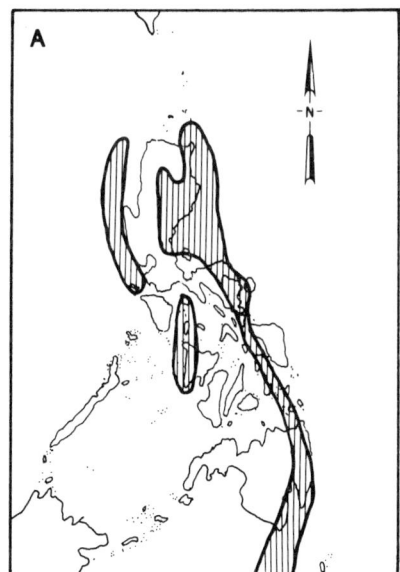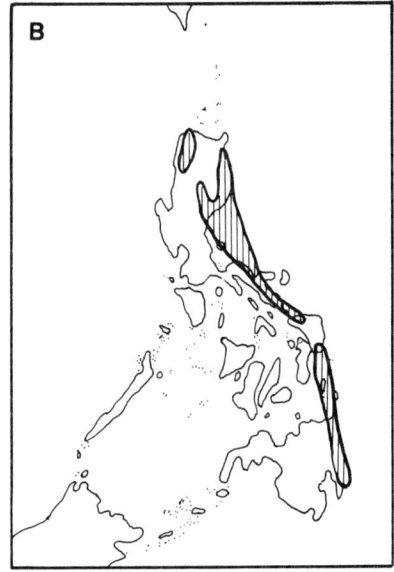

Fig. 14-10. Probable land areas (striped) in Philippine Archipelago during the late Tertiary. *A*, Vigo Miocene (largely after Dickerson, 1924, and DuPree, 1954); *B*, Early to Middle Pliocene (largely after Dickerson, 1924; King and McKee, 1948; DuPree, 1954; and Sampson, 1979).

the northern part of the largest and most eastern island was apparently connected not to Formosa but probably to the Celebes (Dickerson, 1924).

Later in the Miocene, a probably long, narrow island was affected by the collision of the drifting Australian plate with both the Pacific and Asiatic plates (DuPree, 1954), generating massive shocks and diastrophism from Indonesia to Japan. In Indonesia, the collision caused the major collapse and extensive horizontal folding (see Auffenberg, 1980a) of the previously more or less "draped" double ring of islands of the southeastern edge of the Asian plate (Molengraaf and Brouwer, 1929; Barber and Audley-Charles, 1976). These rings now form the Lesser Sundas, including the Celebes.

In the Philippines this north-to-east pressure produced a number of major geological changes of importance to understanding the history of *V. olivaceus*: (1) the buckling of the Philippine East Strip in central Luzon, causing (2) the realigning of the southern part of the Sierra Madre Range in a northeast-southwest direction (rather than N-S), (3) the pulling of the Caramoan Peninsula and Catanduanes out of their originally more N-S orientation into a northwest-southeast one (the present island of Mindoro may be part of the most westward-pushed portion of this buckled part of the East Strip), and (4) the pushing of the entire Philippines plate northward (Samson, 1979), resulting in the massive volcanism that accompanied it along the Philippine Fault Zone (over the common border of the Asian and Philippine plates). South of this giant buckle the narrow East Strip apparently extended all the way to the Celebes (Dickerson, 1924, 1928), and it is along this strip that the butaan must have moved north in company with the dipterocarp forests and particularly with the food species with which it is now so closely associated. (Also probably moving north were the tamarow [*Bubalus mindorensis*] and the Philippine cloud rat [*Phloemys cummingii*]; Dickerson's supposition that all of the mammals were upland types is now known to be incorrect, for the cloud rat is also reported from low elevations on Luzon and nearby islands [Oldfield, 1898; Musser and Gordon, 1981; this study].)

During the Pliocene the area inhabited by the butaan continued to be modified by tectonic movements, apparently caused by diastrophic turmoil emanating from the area of the China Sea. Land areas formed during that time have a northeast-southwest axis, resulting from pressure from the northwest. Palawan, part of Borneo, the Sulu Archipelago, and many of the Visayan Islands in the central part of the archipelago were formed at that time, and the Eastern Strip was splintered in the Samar-Leyte area, isolating the part north of this area from that in Mindanao. As now understood, the butaan is restricted to this northern section. During the same period, the land connection between the Celebes and Mindanao was broken (Dickerson, 1928; Samson, 1979).

It is important to realize that some of the islands, particularly in the Bicol area, were probably connected and disconnected many times, but all were undoubtedly bordered by lowland tropical forests similar to those found today. In fact, a Pliocene fossil flora from Bondoc Peninsula (Merrill, 1923) has only species characteristic of today's lowland dipterocarp forest (*Shorea polysperma, S. guiso, Calophyllum blancoi, Diplodiscus paniculatus, Phoebe sterculioides*, etc.), suggesting that the butaan habitat was well developed at that time within the area of its present range.

In general, the islands of Bicolandia were lower during the Pliocene (to 1500 m; Samson, 1979) than the Miocene, since diastrophism was somewhat reduced. Much of Samar was below sea level during Pliocene (and early Pleistocene) time, as was almost all of Masbate, Bondoc, Ticao, and Burias (fig. 14-10B). It is this period, plus part of the Miocene, that produced most of the limestone found on the flanks of older uplifted deposits, now heavily eroded and forming one of the major refuge areas for butaan populations throughout the entire range. The submergence of much of Samar during this time may explain why the butaan is not found there at present. Although it was probably never connected to Luzon, the straits between them are deep (fig. 14-11). Most of Samar was not lifted above the sea before middle Pleistocene time.

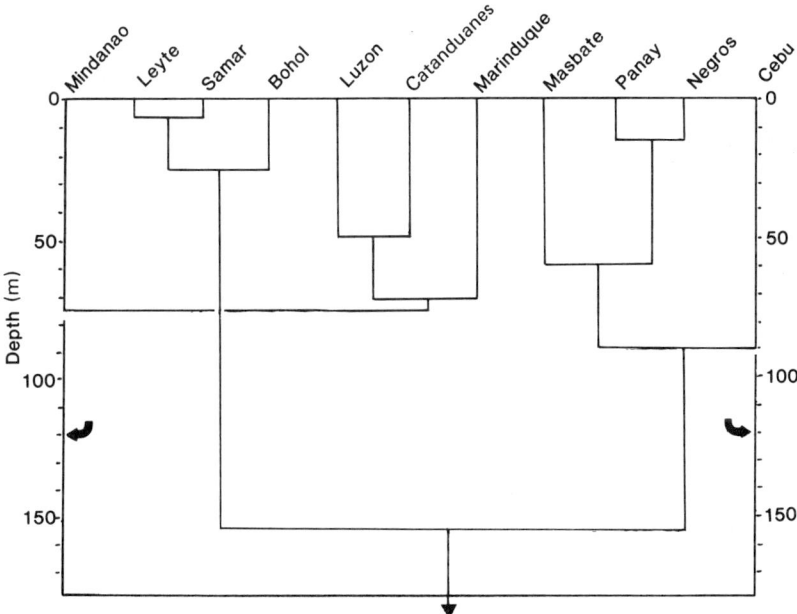

Fig. 14-11. Depths of straits separating Philippine islands near southern Luzon. Arrows represent the lowest sea levels during the Pleistocene.

Bicolandia was also separated from the rest of Luzon by a wide channel between Limon Bay and the Sibuyan Sea (Ragay Gulf). This "Camarines Island" remained distinct throughout much of the Pleistocene as well, for a marine terrace at 9 m exists in the lowlands between Ragay Gulf and Calauog Bay (DuPree, 1954). Other islands probably existed north of this one, including parts of the southern Sierra Madre Range. The Zambalas Mountains in the southwest of Luzon were completely separated from the Sierra Madre Range at that time, as was the extensive Mt. Province area in central and northern Luzon. However, there were broad connections between Borneo, Palawan, and Mindoro during much of the Pliocene and extending into parts of the Pleistocene (fig. 14-12).

During the Pleistocene, most of the Bicol area began to be consolidated into the area we recognize today. The shallow seaway that existed in the Bicol River valley changed to a freshwater swamp, though it was still an important barrier to the east-west dispersion of butaan populations on this part of Luzon Island at that time. More of Samar was exposed. On the other hand, eastern Luzon, Pollilo, Marindugue, Burias, and Catanduanes were all connected—largely as a result of lowered sea levels. Samar was at the same time connected to Leyte, Bohol, and Surigao but never to Luzon. The Zambalas Range in western Luzon was connected to the remainder of Luzon only during the late Pleistocene. Volcanism in Bicol continued (Mayon, Iriga, Isarog, etc.) but played a minor role in the zoogeography of the area.

There is no evidence that Taiwan was ever connected to Luzon during the Pleistocene, though Dickerson (1928) suggested that there is a possibility of island hopping between them during this time by some plants and insects. Both Hooijer (1967, 1972) and von Koenigswald (1956) have suggested that some large land mammals, such as proboscideans, may have crossed the water gap. It is important that any new paleontological sites containing vertebrate remains be thoroughly investigated—particularly on islands that at some time in their geological history were part of the Philippine East Strip. Islands such as Palawan, where some work continues to be done, will probably be found less instructive in this regard—though valuable in their own right.

In this outline of the geological history of southern Luzon, I have tried to show that the area inhabited by the butaan is geologically one of the most ancient in the entire northern part of the Philippines. Its importance biologically is shown by the fact that Dickerson (1928) referred to it as one of the "fundamental" biotic provinces of the archipelago, containing a relatively high level of endemism of plants and animals, of which *V. olivaceus* is one. I have also tried to show through its plant associations that this specialized monitor lizard is part of a Southeast Asian biotic community with clear continental lowland (to 500 m) affinities and that the community is probably one of the oldest extant in the entire Philippines. I believe that *V. olivaceus* (and its

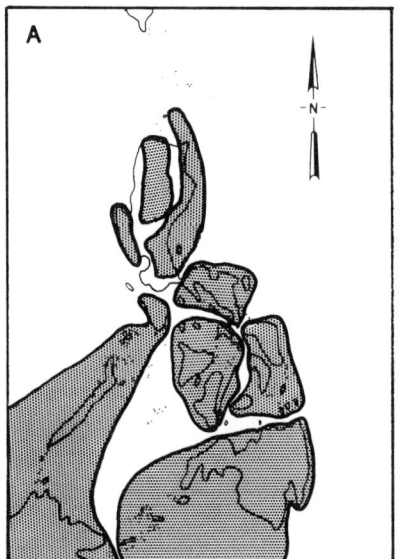 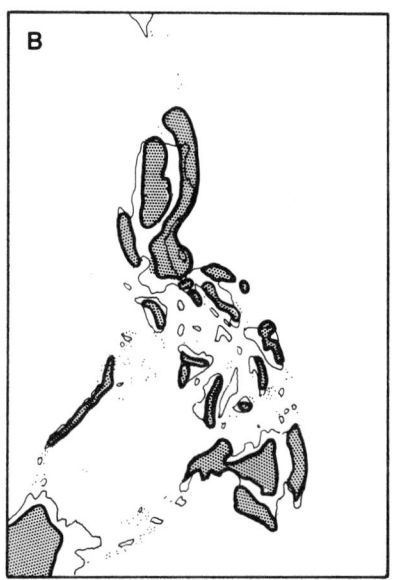

Fig. 14-12. Probable near maximum land exposure (dotted area) during Plio-Pleistocene (*A*, Pliocene), and maximum submergence (*B*, Pleistocene), after Faustino, 1926; Case, 1927; Dickerson, 1928; and Feliciano and Pelaez, 1940. Idealized presentation, pending needed data.

faunal and floral associates) moved into the Philippines during the Miocene—probably from the Celebes—and was later isolated in southeastern Luzon. It remained so due to both climatic and, particularly, tectonic events that created a series of long-lived though constantly changing islands. What this hypothesis does not explain is what happened to the geographically intermediate populations between its present range in Bicol and the island corridor extending southward to the Celebes through which *V. olivaceus* had to move in order to arrive at its present isolated geographic position. While it is clear that large segments of this corridor have disappeared below the sea surface and have only recently risen above it (parts of Samar, for example), the eastern Mindanao strip has apparently remained above the sea's surface since the Miocene. Yet, to my knowledge, the butaan does not occur there. However, if later work shows that it does occur on Samar, Leyte, or Mindanao, the basic tenets of the hypothesis presented here would be strengthened. This hypothesis also falls short of explaining why this lizard has such a restricted range on Luzon, even though it may occur northward of the presently established range along the Sierra Madre Mountains (see chaps. 6, 15). It does point out a correspondence between an important geological fact and the unusual known limits of the species in the Sierra Madre Range. Finally, geographically the hypothesis does not completely link *V. olivaceus* with the geographic range of its presumed closest living relative, *V. bengalensis nebulosus*; the most eastern records for the latter occur in eastern Java (Mertens, 1942), and there is no indication of any land connection between the Celebes and Java at any time in the past. The ecology and distribution of *V. bengalensis* (both subspecies) tend generally to disfavor rafting between islands, unless coastal forests are of a monsoon (seasonally dry) type. Nevertheless, the Lesser Sunda Islands, of which the Celebes seem to be a displaced segment, has scattered populations of both plants and animals characteristic of such dry forests and savannas. Their present distribution is completely relictual and highly spotty. Reptile examples include *Vipera russelli, Python molurus,* and *V. komodoensis* (see Auffenberg, 1980a, for details). In my view, *V. bengalensis* (*nebulosus*) is clearly one of these, found in the drier forests of continental Southeast Asia and only spottily distributed in small pockets of such forests scattered in isolated parts of Sumatra and Java. Other elements of the same semixeric flora are found farther eastward along the island chain (*Python molurus* to Sumbawa, *Vipera russelli* to Flores, etc.). In an earlier paper (1980a) I hypothesized that these fauna and flora were pulled eastward along the island chain as part of the moving monsoon rain shadow cast by the eastward-drifting Australian continent before, during, and after its collision with the Lesser Sunda Islands at the juncture of the Asiatic and Pacific tectonic plates. I presume that *V. bengalensis nebulosus,* or a closely related ancestral form, was pulled eastward along the Sunda Island arc as part of that more xeric community. This move would have placed it closer to the Celebes, making the suggested range shifts from the Sundas to Luzon more likely than present geography would suggest, if in fact the ancestor of the butaan was *not* specifically adapted to slightly drier conditions but rather to more mesic ones. Therefore, it would have been the change to drier conditions brought about by the passing Australian continental mass that resulted in the disappearance of ancestral populations in the Lesser Sunda Islands.

15

Conservation

BEFORE this study was begun, the butaan was considered one of the rarest lizards in the world. It was thought to be either extremely restricted geographically or perhaps extinct (Taylor, 1922). Even after the current work was well under way, individuals remained difficult to find. In spite of the fact that both the ecology and general geographical distribution of the species had become apparent (Auffenberg, 1978a, 1979a), I was still led to describe it as one of the world's rarest species of larger lizards (Auffenberg, 1979c, 1980b). The work accomplished during the last eighteen months of this study clearly shows that, as expected, the butaan is more widely distributed than previously documented (Auffenberg, 1979a, c). Though its geographic range is now believed to have been generally established, later work may show that it occurs all the way to the northeastern corner of Luzon along the Sierra Madre Range (unconfirmed reports from the extreme northeastern part of Luzon). The chances that it occurs on Samar Island, south of its known range, are also good but less than the chance of an extension northward (see chap. 6; figs. 6-1, 6-2). Still, the geographic area known to be inhabited by the butaan is small compared to that of the other larger vertebrates of the world. The total known range is approximately 5000 km^2, only seven times larger than the known range of *V. komodoensis*—considered by many conservationists as the most endangered of all species of monitors (IUCN, 1982). The range as presently understood is in fact so small that, next to the Komodo monitor, it is probably the second smallest for any monitor species. The small range, constant pursuit by hunters, and an alarmingly fast rate of destruction of its habitat dictates a vigilant and aggressive policy of protection for the next several decades. Fortunately, within this small range the species is actually common, so it is in no danger of imminent extirpation.

My previous recommendations regarding conservation of *V. komodoensis* in Indonesia (Auffenberg, 1981a) were made with some optimism, which the present situation of *V. olivaceus* does not permit. I think its situation cannot be improved materially—rather it is bound to deteriorate. The problem facing conservationists in the Philippines is how to slow the rate of destruction until socioeconomic factors begin to change current attitudes regarding not only this species but wildlife species in general.

In view of the rapidly expanding Philippine human population, the objectives of any successful conservation of its wildlife can be realized only by scientific management of the multifaceted environmental factors involved, coupled with a sound and vigorous educational program beginning at the elementary school level.

I am generally pessimistic about the success of conserving wildlife species that require relatively undisturbed habitat anywhere. Continued multiplication of the world's human population and increasing demand for material resources by industrialized and industrializing nations are the ultimate detrimental factors responsible for destruction of natural habitats. In the final analysis, all local conservation problems stem from one or both of these factors, in the Philippines as elsewhere. The human species has not yet learned to cope with massive environmental destruction resulting from burgeoning population growth brought about primarily by advances in medical technology. Unfortunately, even a modicum of success on the local level demands more than recognition of the basic underlying problems. In fact, most conservation proposals openly condemning human population growth and economic expansion are bound to fail from the start—and for cogent reasons. To understand the nature of the underlying problem(s) is fundamental, but to effect a positive change locally in any conservation ethic demands an intimate knowledge and appreciation of the many ecological, sociological, and economic factors that enhance or detract from such a program. The fact that cannot be ignored in dealing with the conservation of the butaan (and most other endangered animals and plants throughout the world) is that it lives in an area in which people are poor and often hungry. Frequently they lack adequate education to cope with the changes in land use patterns, in their culture, and in technological advances characteristic of the world today, and by all portents they will be even less prepared in the next several decades.

Yet the butaan is common in what remains of its natural habitat. Furthermore, it does not compete with human beings for food, though it is sometimes eaten by them. Apparently it can survive and reproduce even in disturbed forested situations. Its foods are not completely restricted to primary evergreen forests (though some of the tree species whose fruits it eats and all of the snails that it eats are destroyed if all the protecting forest cover is removed, even temporarily). So it has at least some characteristics in its favor from the standpoint of conservation. Even if all the lowland evergreen forests within its range were disturbed (not destroyed), it could survive if not regularly hunted by humans and their dogs. The outlook for conservation of the butaan is somewhat analogous to that of the manatees in the southern United States, the batagur freshwater turtle in southern Asia, and hundreds of other species harmless to man, his crops, and his domestic animals. That is, the butaan can coexist peacefully with man as long as its habitats remain relatively intact and it is left undisturbed.

Refuges

Providing wildlife with protection from human disturbance is not a long-standing policy in the Philippines (Grimwood, 1978). Though national parks have existed there for a long time, only recently has there been much effort expended on essential aspects of their development, and modern game management practices are noticeably lacking.

Table 15-1 provides data on the parks and refuges in the presently known butaan range. On the surface the total area devoted to forest lands is impressive (7.3% of total area in the presently known range of the butaan; 11.6% of all Philippine national parks are in its geographic range). However, it must be pointed out that almost all of this area represents mountain forest types; few lowland or hill dipterocarp forests are included. So there is actually little refuge land available for the butaan and other organisms of the same ecosystem.

It is probably completely unrealistic in the Philippines at this time to expect

TABLE 15-1. National parks with suitable habitat in the geographic range of *V. olivaceus*.

Province	Total area of province (km^2)	Park	Total area of park (approx. km^2)	Percent of total province area
Quezon	11,957	Aurora Memorial[a]	57.0	0.3
		Quezon	10.0	0.1
		Banahaw-San	111.0	0.7
Nueva Ecija	5,492	Aurora Memorial (part)		0.4
Bulacan	2,516	Biak-na-Bato I	21.0	0.8
Rizal	2,049	Banahaw-San (part)		0.2
Laguna	1,204	Cristobal Mountains	18.0	0.1
Camarines Norte	21,147	Bicol	52.0	0.2
Camarines Sur	5,336	Bicol (part)	13.0	0.2
		Caramoan	3.0	0.1
		Mt. Isarog	101.0	1.9
Albay	4,001	Mayon Volcano	5.5	1.4
Sorsogon	2,054	Bulusan Volcano	3.7	1.8
Catanduanes	1,448		0.0	0.0

a. Primarily a historical monument but appropriate for habitat.

inviolate reserves of the European and American kind. For example, few local officials attempt to curb hunting in national parks, at least not in the Bicol area, because there is no clear ruling regarding such matters and because local administrative offices lack the personnel necessary to enforce such provisions (Lantican, 1981). Recently the problem has been magnified by the stationing of large numbers of military and paramilitary units near some national parks because there are political dissidents living in the forests. The result is an increase in firearms in just those areas where, from a conservation standpoint, they are least wanted.

Act 2590 of the Philippine legislature is specifically designed to protect game birds, mammals, and fish. It has been amended by Acts 491 and 3730, as well as several Presidential Decrees, whereby the protected species include practically all wildlife of the Philippines. However, these amendments were added mainly to make it illegal to carry firearms in the forest because of spotty but persistent and sometimes serious dissident activities in forested mountainous areas—not to protect wildlife. Still, the laws exist, and prosecutions ensue, with the result that gun hunting is not common among nonmilitary personnel and actually rare among the military.

However, hunting with dogs is not specifically discouraged at the local level, and parties regularly use them to scavenge whatever meat can be obtained. The use of dogs in protected forests must be regulated if there is any hope for wildlife protection, at least in specified forests. Rural people are often protein-starved, and it is clear that families must often supplement their diets with wild flesh. The problem exists where dogs are used.

Much of the archipelago is beginning to experience pressure on "unused" land, and resettlement programs and squatters' rights have become important political realities on the fringes of refuge and park areas (Sajise, 1981). The human population is already starting to press hard on agricultural resources, particularly on Luzon; hence the immigration of Tagalog people to distant islands such as Palawan and Mindanao. (This migration explains some of the political-cultural problems on Mindanao in the last several decades.) Both legal and illegal schemes to settle primitive areas are under way throughout the entire island nation. (The "KKK" program, though, aims to assist rural people in place rather than to resettle them [Benitez, 1981].) It is mainly because of this urgent need for productive land in an area of generally low soil fertility that conservationists must be prepared to compromise between important wild animal and plant resources and equally important and immediate human needs.

Not only are there too few protected lands, generally representing different habitats than those preferred by the butaan; those that are available are not always efficiently administered (Hyman, 1984a) or protected (Grimwood, 1975, 1978; Lantican, 1981). Few personnel are adequately trained for the particularly difficult job facing them. Even when stationed near the area of their responsibility (which is not always possible or preferable due to local political or logistic problems), personnel are often absent. Some absences are clearly justified because of the meetings and reports required of higher level governmental officers, but some result from a general lack of interest in the wilderness or the job or because the daily expense rate is higher in cities than in rural areas. Even when "in station," local wardens have a difficult and sensitive job; there is considerable local pressure to utilize forest products, and success in any conservation program demands the local community's appreciation of its goals and consent and cooperation in its methods. This is not to say that dedicated persons cannot be found in wildlife-related professions in the Philippines. Efficient, well-trained, well-intentioned staff are found at all levels. But the combination of poor salaries and working conditions, ignorance, and apathy regarding wildlife protection, a common desire not to upset routine practices, and the necessity of facing an often hostile, unappreciative citizenry makes the job of the wildlife biologist—not only in the Philippines but in many developing smaller nations throughout the world—a difficult one even under the best local conditions.

In my view (and that of Grimwood, 1975), one final problem is the frequent combining of wildlife departments and forestry units, often with the forestry interests in supervisory positions. Prevailing philosophies in these two areas are somewhat in opposition. Foresters are interested in conservation and management solely for future exploitation; nongame wildlife conservation units are, or should be, interested solely in preservation and research pertaining to it. As a result, parks and wildlife units are under the aegis of departments interested in *using* resources. Thus the emphasis in refuge and park site selection and maintenance tends to be on recreation rather than preservation; the habitat is intended to serve as a local bank for genetic pools, study, enjoyment, or even limited exploitation. The two views can and often do coexist in national parks in some parts of the world, but the danger of a shift in emphasis to a user philosophy is always present (as witnessed in the 1980s in the United States, for example).

Predation by Man

Rural Caramoans eat the butaan for several reasons: it is available meat in a generally protein-starved region, it provides fat for frying and folk medicines of several types, and with python and dog it is a somewhat prestigious "palutan" (finger food) to be eaten at men's drinking parties. Though the water monitor is regularly used in the Philippine leather industry (see Auffenberg, 1982b), the butaan is rarely caught in the traps used by commercial hide hunters. In the large dying vats in Manila, I found that butaan skins accounted for only 0.5% of all monitor skins being processed, and this number may be higher than normal because a large shipment from southern Luzon was being processed at that time.

Varanid lizards have long been used as food by man. Archeological evidence indicates that, in Borneo, monitors were eaten as early as 40,000 YBP in the Niah Cave area (W. King, 1962), and in Java bones representing this family have been interpreted as remains of food eaten by early hominids of middle Pleistocene time (material in Florida State Museum Collections, unpublished). Throughout the range of monitor lizards, these and other data show that these lizards were regularly eaten by hunting and gathering societies from Africa (Fitzsimons, 1943) through Asia (Deraniyagala, 1953) and into Australia (Cogger, 1967) for thousands of years.

The butaan does not eat carrion, so the usual meat-baited traps for capturing monitor lizards (particularly *V. salvator*) cannot be used for this species. Instead, guided by one or two human hunters, two or three trained dogs search out the scent of the monitors. The dogs are released in a likely spot, and the party makes a rough transect through the forest. Places that show some sign of local monitor activity, such as scratches on tree trunks or droppings, are selected for intensive search. Most hunters expect to be in the field by 1100 hrs, though there is considerable superstition regarding the activity periods of these animals: butaans are said to come out only on certain days of the week, never in the rain, only once a month, etc. These presumed limitations to hunting times are for the most part explanations for the infrequency with which individuals are seen, even with trained dogs and experienced human hunters in areas where the species is known to occur and in reasonably high numbers. Infrequent sightings are the reason that the butaan is often described as being extremely shy or, more commonly, "ashamed."

The dogs usually locate whatever butaan scent they find on the ground. When the monitors are flushed, they dash for an available rock crevice or, more often, up a nearby tree. If vines and epiphytic growth are dense on the trunk and lower branches, the butaan usually hides in these; if the trunk and larger branches are bare, the butaan exhibits "squirreling" behavior, by which it tries to keep itself positioned on the side of the trunk away from the hunter (chap. 7). However, when located and even approached, it seldom bolts until the hunter has climbed to within a meter or so. This is why it is called the "blind lizard" (*buta* in the local dialect). However, when it does bolt, it does so quickly, usually throwing itself out of the tree from considerable height (hence the common name of "jumping lizard" in some areas), landing uninjured on the often jagged limestone rocks many meters below, and dashing into the nearest fissure, where it often remains very still, even if only its head is covered.

Of 126 butaans brought into our camp during the study period (1976–83),

61 were injured when captured by the dogs on the ground or in holes. Mortal injuries, though rare, are often internal, caused by dog bites; strangely, many of those bitten become ill and even succumb to sepsis resulting from such wounds. This is unusual because healthy *V. salvator* and *V. komodoensis* are resistant to infection (Auffenberg, 1981a), as are many reptiles (Reichenbach-Klinke and Elkan, 1965).

Male and female butaans are caught in equal numbers so that the technique of capture is not biased on that account. However, of all the butaans first located by dogs, none was less than 1 m total length, probably because the young spend more time in trees (chap. 6).

It is unlikely that all the local butaans could be found even within a period of several years of intensive hunting with dogs (at least on highly karstified surfaces). The dogs not only fail to find the young, but they often miss the scent of one or sometimes even two butaans around a particular tree, in spite of the fact that radio telemetric tracking and later verification proved that the butaans were indeed there.

Most hunting parties spend about 5 hours per day in the field for one or two days before a butaan is caught. However, they rarely go out specifically for them, generally hunting anything of interest (wild boar, cloud rats, water monitors, etc.). The eventual extermination of local butaans in any area is undoubtedly due to a combination of such organized hunts, unrestrained and unattended hunting of village dogs in nearby forested areas (each rural village has 1–3 dogs), and the fact that teenage males are encouraged to bring home whatever meat they can (usually by use of slingshot, at which they are particularly expert).

The energetic costs expended by a man and his two or three dogs when hunting are generally hardly worth the effort. While the sale of all wildlife meat (in any form) is specifically forbidden by law in the Philippines, the practice is nevertheless common within all forested areas in which the butaan occurs. Unlike subsistence hunting, commercial exploitation of the butaan is economically profitable because those who eventually buy it pay a higher price than they would for domestic animal meats. Though the edible portion of a butaan carcass seems reasonably large, it compares poorly to a mean of about 50% for many domestic species (Ensminger, 1977). In a 2-kg butaan, only about 220 g (11%) are edible. At a selling price of about 30 pesos locally, the meat would thus cost about 120 pesos/kg (local pork was 13 pesos/kg and water buffalo 10 pesos/kg in mid-1983). Thus, for even remote houses in the forest, less expensive meats are the rule; furthermore, they are much more generally available.

However, it is not only the food value or taste of the butaan that encourages its continued capture and sale. There are important additional factors that can be called cultural. Some ethnic and social groups in the Philippines are sometimes more interested in eating wild than domestic meat; even native, free-ranging chickens and their eggs are thought to be more nutritious and healthful than chickens raised under supervision and control, a feeling apparently analogous to that of Western organic food cults. In addition, some groups have exploited reptiles as food for many centuries and continue to do so despite obviously rising costs as the species become more difficult to obtain. In many communities throughout Southeast Asia, it is primarily (but not exclusively) the local Chinese who are associated with such preferences, but in southern Luzon the preference exists in many rural communities. Some young men are particularly prone to indulge periodically in the purchase of butaan and similar wild game. Besides liking the taste (butaan is often described by aficionados as "more delicious than chicken"), some think of the butaan as possessing certain medicinal properties. The flesh of monitors (and pythons) is said to "give strength" to the body, to provide "heat." The fat of the butaan particularly is preferred as a cooking oil, for it does not burn when heated to high temperatures. It is also thought by many to cure skin infections. The gall bladder is often given to children sick with fevers and convulsions or chills from several causes (see Auffenberg, 1982b, for similar uses of the water monitor outside the range of the butaan). The body fat of the butaan is sometimes used as a skin balm for babies, especially when mixed with the mother's milk and baby oil. Based on meager evidence indeed, I would place the monthly harvest of butaans throughout the known range at about 20 adults (no juveniles) per month, about 240 per year.

Habitat Destruction

The fact that butaans feed on a remarkably small number of fruiting trees and primarily one species of land snail in at least the Caramoan area suggests that the remaining populations are vulnerable to extirpation. Burning and cutting destroys not only seedling and mature fruit trees but the habitat of the snails as well.

Perhaps the oldest means of controlling unwanted natural vegetation is to burn it. Many rural people throughout the world periodically burn forests. Forests are ignited accidentally by beehunters, charcoal burners, lumbermen, and coffee plantation owners in the Caramoan area, in addition to vandal-set fires. Because of the generally high moisture of surface debris, lightning ignition is rare in this area.

While fire has a definitely beneficial role to play in the maintenance of certain xeric habitats and in some wildlife management programs (see Komarek, 1969), it has no place in evergreen tropical forests. They have no mechanisms to cope with it, so fire is extremely destructive in this type of forest. There were times during the dry season when we could stand on the hill at our base camp and count as many as 32 fires in the primary and secondary forests within view. Though the climatic conditions were particularly conducive to fire at that time, advantage was taken of other dry periods to burn a little more.

However, fire by itself is not an important factor in the ecology and management of *V. olivaceus*. While many fires are ignited, the areas burned are always small, unless the forest has already been destroyed and overgrown with gogon grass. Then the fires are intensely hot and burn off large areas of hillsides. With each burning these areas become larger because of the drying effect both of the fires and of the open character of the habitat. Unfortunately, such fires also destroy whatever seedling trees are there either as part of reforestation projects or from natural germination near the edges. With repeated fire, gogon areas remain self-perpetuating, for they can be destroyed only when overgrown by brush and trees.

Most original lowland dipterocarp forests that once blanketed southern Luzon have been destroyed by urbanization, agriculture, and lumbering, and any appropriate habitat left is found in more dissected upland areas (fig. 15-1) that have so far escaped massive clear-cutting (though even here, in the study area, dramatic changes have occurred since 1976). Within the Bicol area, the

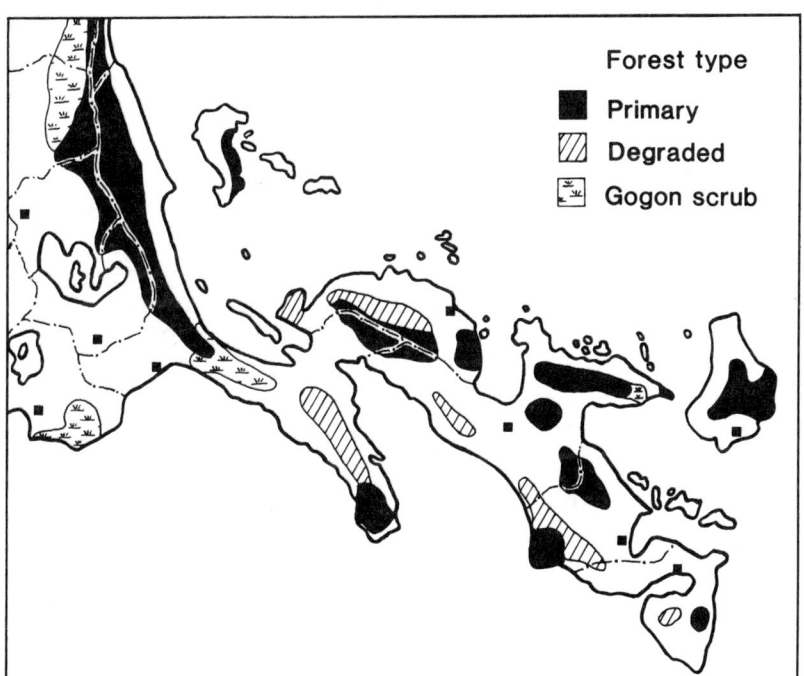

Fig. 15-1. Distribution of generally forested land within the butaan's range (after Anon., 1975).

butaan is restricted to the remaining forests dominated by broad-leafed evergreen trees. It is not known to what extent the butaan occurred in the level lowland forests that have been destroyed. But its occurrence at sea level in areas and habitats that were more widely distributed in the past suggests that populations were more continuous than at present and that they covered a greater area. Known current distributions plotted on a soil map suggest that, in the Bicol region alone, there were once about 617,200 ha of suitable habitat compared with about 218,300 ha today—a reduction of 65%.

Since studies of timber trends and prospects in Southeast Asia only two decades ago (FAO, 1961), there has been a complete change in trade of forest products, resulting in great increases in the need for wood. More recent reviews (FAO, 1976a, b) show that the current supply in the region, in spite of rich indigenous forest resources, may be insufficient to meet future demands, particularly for fuel wood in the Philippines (Knowland and Ulinski, 1979; Hyman, 1981a, b, 1983, 1984b). Unless greater supplies of fuel wood become available to rural Filipinos, much of the remaining lowland primary and secondary forest in which the butaan is found in southern Luzon will be almost completely gone by the year 2000. Not only will the short supply of fuel wood result in the destruction of butaan forests; it will also destroy the ability of the forest to regenerate and perform its role in protecting watersheds and preventing soil erosion and flooding. Enormous importance attaches, therefore, to the projected increases in fuel wood consumption in this area (FAO, 1976b) and the consequent increase in pressure on the remaining forests. To make matters worse, insular Southeast Asia, including the Philippines, shows every sign of remaining the most important source of wood-based products for export to other parts of the world for the next two decades, yet it may be one of the first tropical regions to run out of wood (Guppy, 1983). Though countries in this area with rich forest resources are naturally concerned with forest conservation to ensure steadily high export profits, they are also often short of foreign currency; log exports frequently represent an important source of it. In the Philippines, forest products accounted for about 27% of the annual gross national product in 1970 (Cheetham and Hawkins, 1976), but that amount later dropped to 15% (FAO, 1976b).

Contributing to the general problem is the density and growth of the human population in the Philippines (fig. 15-2). The rate of population growth is estimated at 2.3% (FAO, 1976b)—about average for Asia. With a human population of about 39 million on a small land area, the density is high (130/km^2). Fuel wood consumption per year is about 500 cubic meters/1000 persons, which is equivalent to most other Asian countries of the same economic status. However, the human density of the Philippines combined with this relatively high usage of fuel wood means that, in the Philippines, fuel wood is disappearing at a more rapid rate than in most other Asian countries. Prevent-

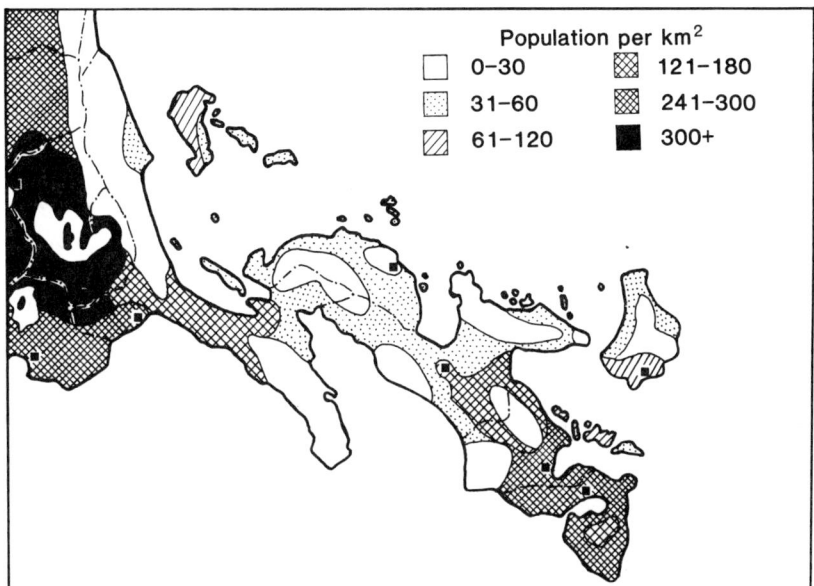

Fig. 15-2. Human population density within the butaan's range (after Anon., 1975).

ing destruction of the forest is critical not only for survival of the butaan but to avoid the more serious consequences of accompanying climatic changes, flooding, and siltation. To make matters worse, such problems may be felt at great distances from the places where the forests are destroyed.

Within insular Southeast Asia it is estimated that about 530 million ha of primary forest remain, of which about 315 million are in the developing countries, 10.9 million in the Philippines (Anon., 1981). Of the 10.9 million, about 9 million represented dipterocarp forests in 1980 (Anon., 1981). Of the 9 million, 25% is regarded as unusable because it is economically inaccessible, or it is not suitable for commercial utilization, or commercial utilization already has been (theoretically) excluded by establishment of forest reserves and national parks. The remaining 75% will certainly be exploited commercially (FAO, 1976b). Some may be converted into plantation-type planting (perhaps not as feasible in tropical as in temperate areas; see Janzen, 1971a, and Halos, 1980, for biological problems, Hyman, 1984a, for technical and administrative problems), but most will be clear-felled prior to agricultural plantings.

Conversion of forest to use by either long-term legal or short-term illegal shifting agricultural interests is not only a critical factor in reducing wildlife resources of the Philippines; it also reduces potential and actual protective effects of local forests (Guppy, 1983). In addition, any change in tropical forests reduces productivity (Goodland, 1975). Even the current legal conversion to agriculture in the Philippines will destroy a large part of the remaining standing forests of the country within 50 years or less (Guppy, 1983; Hyman, 1984b). There is probably an even higher rate of forest destruction because of illegal shifting agriculture. Specific data are available for only a few areas, and no nationwide surveys have yet been completed for the Philippines (Duldulao, 1981a, b). But it is estimated that about 173,000 ha are destroyed each year by shifting cultivators (FAO, 1976b). Unfortunately, continued cropping and burning in the Philippines, particularly of hillside agricultural lands, quickly lead to the formation of degraded gogon (*Imperata* sp.) lands—apparently completely unsuited to agriculture thereafter. These areas can be restored only by shading out the grass with tree planting. Unfortunately, massive efforts by the Bureau of Forestry in the Caramoan area to plant gogon areas with fast-growing trees have met with relatively little success because local rural folk characteristically burn the gogon grass each year, also killing the seedling trees. In Indonesia as much as 30 million ha of gogon land have resulted from overcropping and burning land ill-suited to such manipulation (FAO, 1976b). In the Philippines about 5 million ha have been permanently degraded to gogon (ca 16.7% of total land area; Palaypayon, 1977).

The combination of agriculture clearance (both legal and otherwise), shifting agriculture, and destructive logging practices is expected to reduce the primary forest in the limestone regions of Bicol inhabited by the butaan by about 5% in the next 20 years, perhaps by as much as 20% in the volcanic-soil areas. Since much of the original lowland dipterocarp forest inhabited by the butaan has already been destroyed, the prospects seem poor indeed. The situation is made immeasurably worse when one considers that the continued destruction of forests places more pressure on existing forests to meet future wood needs in the Philippines.

Since any forest reduction affects the butaan adversely, I am led to conclude that the best policy for future conservation is not to stress the importance of the butaan as a valuable resource to people currently unable to recognize its value but to encourage reforestation projects to ease the pressure on existing mixed forests. Even existing forests will suffer severe degradation during the next several decades from legal lumbering in response to the need for processed wood. (As an example, house repair and construction in urban areas of Southeast Asia at present require about 4 million m^3 annually and are estimated to require 15 million m^3 in a few decades [FAO, 1976b].) Therefore, it seems that the best policy is to concentrate on the sections of forest that will not be logged commercially, preferably in highly karstified reserve or national park areas, where there is already a double protection at least against legal commercial exploitation. The Caramoan karst area (only a small part of which is included in the Caramoan National Park) is a good example. Similar karst areas in eastern Catanduanes and western Albay provinces could be designated as protected forests. However, even in the Caramoan karst, squatters, shifting agriculture, and illegal log cutting for lumber and fuel wood steadily reduce the remaining undisturbed forest land. The area demands considerably more protection than that given by local wardens. Forestry officials must be alert to apprehend persons who break the laws regarding violation of public lands. They must send regular patrols into forested lands and prosecute all violators within these small protectable regions. The remaining forests of the Caramoan Peninsula, located on largely igneous soils, will undoubtedly be completely destroyed regardless of any efforts to save them.

The government's schemes to plant trees in gogon areas must succeed; the trees are needed to produce fuel wood for villagers to reduce pressure on protected lands. Such programs have the added advantage of providing the opportunity to improve the location of the wood supply for rural people. Projects demanding the use of large volumes of wood should be discouraged near such protected forests, unless there is every assurance that the local wood supply in plantations (ipil-ipil, etc.) can fill local needs.

While solutions seem easy when problems are viewed from thousands of miles away, it is clear that the conservation of any animal resource is not a simple matter at the local level—the only place where the effectiveness of a program can be measured. In southern Luzon the pressures for wood (especially fuel wood), protein, and agricultural land are the primary threats to the butaan, and all are problems too important to dismiss. Destruction of forest cover will destroy food resources of the butaan; it will also reduce the wood supply and eliminate all of the beneficial effects of tree cover. Too often the agriculture to which the hillside lands, particularly, are converted involves poor techniques that lead to deterioration of already poor soils, to erosion, and to extreme fluctuation in local water levels, and removal of the forest will increase these effects, causing harm beyond the extirpation of the butaan. These effects are aggravated when removal of trees is followed by these destructive agricultural techniques. This succession has already led in some areas to the failure of local agriculture and the reduction of these areas to wasteland.

Recognition of the long-term measures involved in conservative agricultural management and tree planting is problematical for communities faced with immediate and desperate problems of food and fuel supply. The membership of such communities will be even less willing to contribute to conservation measures if they are unsure of their claim to the eventual benefits of community projects (i.e., cooperative fuel wood plantations).

The Philippine government is faced with the decision of investing in the production of fuel wood for the use of its own communities. In certain instances it has decided to invest as heavily as it can, but funds are short. The project becomes even more expensive because of the need to convince rural people of the wisdom of the investment of time, effort, and land in such schemes. Such programs are intricately linked with others concerning agriculture and wildlife conservation, all of which are vital to future environmental quality. Securing the cooperation of the various governmental agencies involved and the local communities will not be easy. It is indeed doubtful that much of the remaining forest land harboring butaans can be saved by any means in the foreseeable future. It is clear that some forested areas can be saved, but decisions must be made soon on the location of these areas and on coordinated programs that must be developed by the public and private sectors to make such efforts feasible. (The Ministry of Human Settlements has the authority to define critical areas, where "any change in land use would negatively affect the inherent characteristics of the areas that are important to the public welfare.")

It is difficult to establish meaningful management policies in the face of local and international pressures to use Philippine forest products as capital. All participants have to agree on a workable plan to limit utilization of national park and refuge lands and then to abide by it. A fundamental ingredient of conservation policies is that free enterprise, which creates the demand for forest products, must be partially replaced by a cooperative policy. The Bureau of Forestry of the Philippine government and the Parks, Range, and Wildlife offices have worked together for some time to control forest and wildlife degradation—a difficult task with often imperfect results. The remainder of the job is partly the responsibility of the international community—to try to preserve at least a part of the remaining forests, to establish new reserves in areas not yet represented, and to assist financially and with contributed personnel in the wardenship of such protected lands. These tasks are clearly within the spheres of interest and influence of such international groups as the Species Survival Commission of the International Union for the Conservation of Nature, whose members voluntarily advise governments on the status and management of wilderness areas and wild species. The World Wildlife Fund, Nature Conservancy, New York Zoological Society, and International Foundation for Conservation of Birds have also shown concern over the past several decades. A perusal of such lists makes it clear that, with few exceptions, the concerns for conservation of endangered habitats and species are held primarily by members of private rather than government organizations and, though often prestigious, they do not carry the political clout of official government policy. The Food and Agriculture Organization of the United Nations has prestige and power, but its interests are chiefly, and understandably (in view of its mission), in the area of manipulation for future exploitation. What is needed is a multimillion-dollar effort by responsible, concerned, and wealthy nations to achieve massive assistance of several types. Such an organization(s) should not replace the private agencies currently operating well on a number of fronts but join them to strengthen the concept of saving now for the future. Only by considerable governmental pressures and assistance can useful conservation be affected on a global scale.

Recommendations

I believe that, in view of the current political, cultural, and economic climate, the Philippine government is probably doing all that it can to protect the butaan. This is not to say that improvements are not possible or that they should not be made, but what is needed is more than the Philippines can do alone. Broad, international, governmentally financed programs are generally prestigious in developing countries, for they represent jobs, new experiences and skills, and interest by other nations in the host country's problems. Together these advantages tend to foster a feeling of the project's importance that cannot be achieved through small, single-species, or single-refuge projects, such as this study of the butaan.

The butaan should be thoroughly protected in at least part of its presently known range (Echano, 1984). To do so requires a broad, multifaceted program involving (1) teaching people to care for future fuel wood and food needs and to educate their children in these matters, and (2) implementing assistance programs relating to the wardenship of the protected land and research programs dealing with the included wildlife. Both parts are necessary to justify the entire venture in the short-term view as well as the long-term goal of preserving local wildlife.

Only long-term assistance programs have any chance of succeeding. Often, partly because of the enormity of the task on even a national scale, private agencies can fund projects only briefly. The effect of a few well-trained persons working for a short period is short lived. That a massive, long-term program will have longer effects is demonstrated in the relative success of human population growth and rural health programs in many parts of the world. Where single communities are selected as "showcase" villages or districts, and much money and effort are poured into a series of interrelated projects in that area, benefits are visible for other communities to recognize and emulate.

Appendix 1
Annotated List of Plants Mentioned in Text

Abutilon indicum — "Malbas," shrub, to 6 m, common, clustered, on forest edges and in openings generally; introduced but widely distributed; Malvaceae.

Aglaia harmsiana — "Matabato," a medium subcanopy tree, to 20 m, common, scattered, on variety of soils; nearly spherical fruit (red) with thin pericarp, 1.7 cm dia., in axillary panicles, available February–May, eaten by many birds and mammals, including *V. olivaceus;* seed oval with three ridges; Meliaceae.

Aleurites (cf. *A. trisperma* and *A. thurifera*) — "Dagang," a large emergent tree, to 45 m, common, scattered, with straight, long, unbranched trunk, canopy dense in the wet season, open in the dry, somewhat restricted to deeper soils; not important shelter tree for *V. olivaceus;* Dipterocarpaceae.

Amorphophallus campanulatus — An herb, to 1 m, thickets and rocky places, often on dry soils, scattered, uncommon; Araceae.

Antidesma leptocladum — "Bignay," a medium shrub of open areas, thickets, ecotones, to 4 m, common, usually clustered; fruit (red to purple) in dense racemes, 5 mm dia., often eaten by birds, mammals, and humans but not by *V. olivaceus;* Euphorbiaceae.

Areca catechu — "Bunga," "betal," a medium subcanopy tree, to 15 m, usually on moist soils, common, clustered; probably introduced; nut chewed by humans; Palmae.

Artocarpus communis — "Rimas," "breadfruit," a subcanopy tree, to 15 m, often on forest edges, where it is frequently planted, on various soils, common, clustered; unquestionably introduced; fruit (green) cauliflorous, to 20 cm long; Moraceae.

Artocarpus elastica — "Gumihan," a canopy tree, to 30 m, on various soils, common, scattered; fruit (green) large, cauliflorous, hairy; Moraceae.

Artocarpus integrifolia — "Lanca," "jak-fruit," a subcanopy tree, on various soils, common, usually scattered; fruits (green) large (to 20 kg and 100 cm long), cauliflorous, commonly cultivated and regularly eaten by humans (only); probably introduced; Moraceae.

Asplenium nidus and *A. macrophyllum* — "Bird's-nest fern," an epiphytic fern, to 0.4 m, on all parts of trunks and large branches, uncommon, scattered; not important arboreal cover for *V. olivaceus;* Polypodiaceae.

Averrhoa bilimbi — "Iba," a small subcanopy tree, to 12 m, common, scattered; introduced but thoroughly naturalized; fruit (green) cauliflorous, eaten by many forest vertebrates, as well as by man, but not by *V. olivaceus;* Oxalidaceae.

Bambusa blumeana — A spiny bamboo, 15 m, common in openings, along stream forests, ecotones generally, often on damp soils (at least seasonally), often densely clustered; Gramineae.

Bambusa vulgaris — A nonspiny bamboo, 12 m, generally same as above but even more common; extensively used by rural people; Gramineae.

Barringtonia asiatica and *B. luzonensis* — "Boton," a small canopy tree of coastal forests and low-lying streamside forests, 15 m, common, usually scattered, on moist soils; fruit (green) 8–12 cm dia., not eaten by many forest vertebrates (if any); Lecythidaceae.

Bauhinia sp.— "Camat-cabog," a vine growing on trees and shrubs in a variety of habitats, to 7 m, common, often clustered; sometimes an important arboreal retreat for *V. olivaceus,* especially when combined with other vines to make a dense thicket; Leguminosae.

Calamus mollis — "Rattan," "climbing palm," a slender, scandant palm growing into trees to 20 m, clustered; not important arboreal retreat for *V. olivaceus* due to open habit but often forming thickets with many other scandants and vines; Palmae.

Callicarpa blancoi — A shrub, to 2 m, common, often clustered along forest ecotones, on a variety of soil types; Verbenaceae.

Calophyllum blancoi — "Palo-maria," "bitanhol," a canopy tree, to 25 m, scattered, often open places and ecotones but also in primary forest; Guttiferae.

Calophyllum sp.— "Dangkalan," a canopy tree to 20 m, uncommon, scattered; fruit (yellow) available May–July; sometimes eaten by mammals and birds but not by *V. olivaceus;* Guttiferae.

Canarium hirsutum — "Milipili," a small resinous subcanopy tree, to 8 m, common, scattered on a variety of soils but often in rocky places; fruits (bluish black) an important food of *V. olivaceus,* triangular, in large terminal panicles but covered with small irritating hairs until ripe, available May–February; seeds pointed on both ends, eaten by some rodents; Burseraceae.

Canarium luzonicum — "Kiramo," a large resinous canopy tree, to 35 m, on a variety of soils, common, scattered; fruit an edible triangular nut (purple-black) on large terminal panicles, available July–November; pericarp often eaten by larger frugivorous birds, omnivorous mammals, and *V. olivaceus,* but seeds not eaten by rodents, perhaps because kernel seed coat causes diarrhea in this and related species; Burseraceae.

Canarium ovatum — "Pili," a resinous canopy tree in both primary and secondary forests, to 20 m, on a variety of soils, common, scattered; edible, single-seeded, triangular nut with stone pointed at both ends, often harvested by rural people and frequently eaten by larger birds and mammals in the forest and by *V. olivaceus;* fruits (blackish purple), 1–2 on each terminal twig, available May–August; Burseraceae.

Canarium villosum — Similar to *C. luzonicum,* but fruit 1 cm and rounded in cross section, probably eaten by *V. olivaceus;* tree to 30 m, rare locally; Burseraceae.

Caryota cumingii — Similar to *C. Rhumphiana,* but smaller tree and fruit (5–7 m), similar ecology, rarer, fruit not eaten by *V. olivaceus,* perhaps because of the stinging calcium oxylate crystals in the pericarp (see Sulit, 1918); Palmae.

Caryota Rhumphiana — "Bagsang," "fish-tail palm," a small subcanopy tree, to 5 m, on generally moist soils of mountain slopes and ravines, common and often clustered; subglobose fruit with thin pericarp (brownish black), produced on large axillary panicles, April–September, an important food of *V. olivaceus;* seed smooth, subglobose; Palmae.

Cassia alata — A medium shrub, to 3 m, in open places and ecotones, common, often clustered; introduced but thoroughly naturalized; Leguminosae.

Appendix 1. Annotated List of Plants

Cerbera odollam — "Lipata," a small tree, to 6 m, common, clustered, a subcanopy type of coastal forests and along floodplains of low-lying rivers; Apocynaceae.

Clerodendron quadriloculare — "Baguae," a small subcanopy tree, to 5 m, uncommon, scattered; Verbenaceae.

Cocos nucifera — "Niog," "coconut," tree, to 20 m, cultivated, often semiwild, common, clustered; large fruits often eaten by rodents but not by *V. olivaceus;* Palmae.

Costus speciosus — Tall herb, to 1.5 m, growing mainly on damp soils in primary forests and streamside woodlands, common, somewhat clustered; Zingiberaceae.

Crinum asiaticum — An herb, to 1 m, on various soils, frequently near coast, common, clustered; Amaryllidaceae.

Cycas circinalis — A shrub, to 2 m, uncommon, scattered, usually on moist slopes; Cycadaceae.

Dendrobium ssp.— An epiphytic orchid, small but growing to tops of largest trees, usually in primary forests, on both trunks and large branches, common, clustered in dense masses; forms important cover for *V. olivaceus* in trees; Orchidaceae.

Dillenia philippinensis — "Katmon," a small tree, to 15 m, common, scattered, often on damp soils; fruit (green) 5–6 cm dia., not commonly eaten by forest vertebrates or by *V. olivaceus;* Dilleniaceae.

Dysoxylum decandrum and *Dysoxylum* sp. — Medium-sized trees, part of main canopy, common, scattered, not on rocky ridges, though on variable soils; fruits (yellow) in axillary panicles; seeds brightly colored, pericarp eaten by bats but not by *V. olivaceus* as far as is known; Meliaceae.

Erythrina indica —"Dap-dap," a small tree, to 15 m, common, often clustered along streams near coast and along floodplains of low-lying rivers; Leguminosae.

Eugenia calukob — "Amhi," a canopy tree, to 25 m, common especially along streams; fruits (purple-red, subglobose) to 5 mm long, available June–August, edible, eaten by many large birds, civet cats, and humans but not by *V. olivaceus;* Myrtaceae.

Eugenia cumingii and *E. polycephaloides* — "Duhat," "amhi," subcanopy trees to 15 m, common, scattered, in open places and secondary forest; fruit (red) to 2 cm long, available June–August, regularly eaten by many large birds and mammals, including man, but not by *V. olivaceus;* Myrtaceae.

Ficus benjaminus, F. concinna, F. forsteni, etc. — "Balete," "strangler figs," medium to large trees, to 20 m, important canopy species with wide, spreading crowns, some deciduous; fruit variable in size, axillary or terminal, often in pairs (yellow, green, purple-black, orange), eaten by many forest vertebrates, including *V. olivaceus,* though seasonal for the latter; fruits at different times of year, often one short pulse per tree but heavy crops; Moraceae.

Ficus hauli — "Siau," a small subcanopy tree, to 8 m, usually in openings, ecotones, common, sometimes in small clusters; fruits not known to be eaten by any vertebrates; twigs and leaves contain poisonous sap (see Sulit, 1918); Moraceae.

Ficus nota, F. variagata, and relatives — "Tabog," "tiago," "tabgon," etc., small to medium trees, almost always subcanopy types, to 10 m, various soils but frequently slightly damp, often at the base of hills and in floodplains; fruits (green, brown, red, or orange) from several mm to several cm in diameter, all cauliflorous, frequently eaten by birds, bats, and rodents but rarely by *V. olivaceus* (not an important food type for this species); fruits all year depending on species, often regularly pulsed; Moraceae.

Ficus pseudopalma — "Niog-niog," a small, unbranched, subcanopy tree, to 6 m, on various soils, common, scattered; fruits (brown) terminal, not an important food of forest vertebrates and apparently not eaten by *V. olivaceus;* Moraceae.

Ficus ulmifolia, F. minahasae, and *F. cumingii* — "Hagupit," shrub-type figs, to 5 m, in thickets, open places, ecotones, dry ridges, common (at least some species), often clustered; fruits (red, orange, brownish red, green, depending on species) usually axillary, single or in pairs, ripen over a long part of year, but few fruits develop at any one time; an important food type for many vertebrates and apparently sometimes eaten by *V. olivaceus*, though these figs are most common outside major butaan habitat; Moraceae.

Garcinia venulosa and *Garcinia* sp. — "Bilukan," subcanopy trees to 15 m, common, sometimes clustered, especially along streambeds; fruit (yellow-orange) to 2 cm long, available May–July, regularly eaten by larger birds and mammals (including man) but rarely by *V. olivaceus*; Guttiferae.

Gnetum gnemon — "Bagu," an emergent tree, to 30 m, common, scattered, usually on damp soils; fruits (red) eaten by many vertebrates but not by *V. olivaceus* as far as is known; Gnetaceae.

Grewia stylocarpa — "Parong," a small subcanopy tree, to 8 m, on various soils, common, scattered; fruit (orange) ca. 2 cm dia., available May–August, borne in large terminal panicles, regularly eaten by birds, bats, mammals, and *V. olivaceus*; seeds hairy, triangular; Tiliaceae.

Gymnartocarpus woodii — "Bitanghol," a canopy tree, to 20 m, uncommon, scattered; fruits (green) 6–8 cm, thin pericarp; large seeds, eaten by rodents when fruit falls to ground; Moraceae.

Hopea basilanica — "Yakal," a large emergent tree of primary forests, to 45 m locally; Dipterocarpaceae.

Hoya (cf. *H. luzonica*) — An epiphytic vine, to 4 m, on trunks of large trees; not important cover for *V. olivaceus*; Asclepiadaceae.

Hymenocallis tenuiflora — "Spider lily," an herb, to 1 m, moist soils, uncommon, clustered; introduced but thoroughly naturalized; not important to *V. olivaceus*; Amaryllidaceae.

Kleinhofia hospita — "Tanag," a canopy tree, to 20 m, common, clustered, often on damp soils, particularly along streams; Sterculiaceae.

Kolowratia elegans — A tall herb, to 4 m, always on damp, usually floodplain soils, common, clustered; Zingiberaceae.

Lantana camara — A shrub, to 2 m, common, sometimes densely clustered, many soil types, almost always on forest edge; introduced but completely naturalized; Verbenaceae.

Laportea gracilipes and *L. subclausa* — "Sagai," herbs, to 1 m, damp soils, forested slopes, common, scattered; irritating stinging hairs on leaves and stems (see Sulit, 1918); Urticaceae.

Laportea meyeniana — "Langatun," a small subcanopy tree, to 5 m, common, somewhat clustered, on various soils; fruits not eaten by vertebrates, leaves and stems with stinging hairs; Urticaceae.

Leucaena glauca — "Ipil-ipil," a small tree, to 6 m, open places and ecotone generally, common, often clustered; introduced but thoroughly naturalized; Leguminosae.

Lygodium japonicum and *L. flexuosum* — "Nito," a vinelike fern, to 6 m, growing thickly on shrubs and small trees, common, clustered; forms dense thickets that are often important arboreal refuges for *V. olivaceus*; Schizaeaceae.

Macaranga tanarius — "Binonga," a small tree, to 8 m, common in openings and ecotones, particularly on damp soils, usually clustered; Euphorbiaceae.

Malaisia scandans — "Malaisis," a climbing shrub (8 m), not common, usually in ecotonal area; fruit deep red receptacle, sweet-smelling, to 34 mm long but usually shorter, tiny seeds in a raspberry-like pericarp, eaten by *V. olivaceus* and other reptiles (scincids) as well as many mammals and birds, presence in digestive tract easily noted on basis of pleasant odor; monotypic genus, widely distributed from southern China to Polynesia; Moraceae.

Appendix 1. Annotated List of Plants

Mangifera altissima — "Paho," a canopy tree, to 15 m, common, scattered, on good, damp soils; fruit (green) eaten by hornbills and monkeys but not by *V. olivaceus;* Anacardiaceae.

Mesoneurum pubescens — "Camat-cabog," a vine, to 5 m, on shrubs and small trees in dry places, common, of local occurrence; not an important arboreal cover for *V. olivaceus;* Leguminosae.

Morinda citrifolia — "Nino," a small subcanopy tree, to 10 m, common, often somewhat clustered, on a variety of soils; Rubiaceae.

Mucuna negricans and *M. sericophylla* — Vines, to 10 m, common, scattered, widely distributed; irritating hairs on dry pods (see Sulit, 1918); Leguminosae.

Onychium siliculosum — A finely divided fern, to 0.5 m, generally restricted to damp soils, including stream banks, often growing in large, dense stands; an important ground cover in appropriate habitats, particularly on rocky substrates, because the land snails upon which *V. olivaceus* feeds are often found in largest numbers in such stands; Polypodiaceae.

Ophioglossum pendulum — "Adder's tongue fern," a long, straplike, epiphytic type, to 1 m, that often grows in dense masses on the large horizontal branches of canopy and emergent trees; an important refuge for *V. olivaceus;* Polypodiaceae.

Pandanus radicans — "Ollano," a small tree, to 5 m, uncommon, scattered, often on rocky mountain slopes; fruit (syncarp) several kg, composed of ca. 200 drupes (yellow), elongate, each ca. 3 cm dia., available January–April, an important food of *V. olivaceus,* apparently also eaten by hornbills; seed fibrous; Pandanaceae.

Pandanus simplex — "Karagumoi," a shrub, to 3 m but usually lower because leaves are constantly harvested for manufacture of household items, common, clustered, usually on moist soils; fruit a large syncarp (to 8 kg) of about 1000 drupes (reddish orange), available January–April, probably an important food of *V. olivaceus* if fruits allowed to mature, also eaten by large rodents; Pandanaceae.

Pandanus tectorius — "Pandan," a small tree, to 5 m, near coast and along tidal streams, common, often densely clustered; fruit a large syncarp of several hundred drupes (red), available January–April, eaten often by parrots and some larger rodents; Pandanaceae.

Pentacne contorta — "Lauaan," Philippine mahogany, a large, emergent tree, to 45 m, usually on deep soils of igneous mountain slopes, scattered, uncommon due to local logging; Dipterocarpaceae.

Phalaenopsis sp. — "Epiphytic orchids," see *Dendrobium;* Orchidaceae.

Pisonia aculeata — "Anuling," a scandant shrub, to 5 m, usually on rocky soils, common, scattered; Nyctaginaceae.

Polygonum barbatum — An herb, to 0.6 m, usually on damp soils, common, clustered, often in dense stands which are an important ground cover hiding the snails and fallen fruits on which *V. olivaceus* feeds; Polygonaceae.

Polypodium sp. — Terrestrial ferns, to 1.5 m; see *Dryopteris;* Polypodiaceae.

Psidium guajava — "Bayabas," "guava," a small tree, to 6 m, common locally, clustered, in thickets, open places, and ecotones, second growth generally; fruit (green) eaten by humans and many forest vertebrates but not by *V. olivaceus;* Myrtaceae.

Pteris ssp. — Terrestrial ferns, to 2.5 m, see *Dryopteris;* Polypodiaceae.

Pterocarpus indica — "Narra," a large, emergent tree, to 45 m, scattered, common; protected as the "national tree" but in demand for furniture, etc.; Dipterocarpaceae.

Pterocymbium tinctorium — "Talo-to," a large emergent tree, to 25 m, common, scattered, often in rocky places; Sterculiaceae.

Rhaphidophora merrillii — "Amlong," a stout vine, to 6 m, on large tree trunks and lowest branches, common, clustered; an important retreat for *V. olivaceus,* particularly on the trunks of large trees, due to the great size of its leaves; fruit not eaten, perhaps because of stinging crystals; Araceae.

Sandoricum koetjape — "Santol," a small canopy and subcanopy tree, common, scattered, on damp soils; fruits edible (yellow), large, available August–September, eaten by man and many forest animals, including *V. olivaceus;* Meliaceae.

Santiria sp. — a subcanopy shrub, to 2.5 m, common, scattered; fruits (red) axillary, to 2 cm, eaten by birds but not by *V. olivaceus;* Burseraceae.

Schefflera sp. — "Galamoi-amo," a small subcanopy tree, to 6 m, common, scattered, often in rocky places; Araliaceae.

Schizostachyum diffusum and *S. dielsianum* — "Bical," climbing bamboo, to 20 m, on large forest trees, common, clustered; often used as arboreal retreat by *V. olivaceus;* Gramineae.

Semecarpus cuneiformis, S. gigantifolia, and other rarer species — "Ligas," "ingas," small tree(s), to 8 m, common, clustered, usually unbranched, sap poisonous (see Sulit, 1918); fruit (purple) edible but not eaten by *V. olivaceus;* Anacardiaceae.

Shorea guiso — "Guiho," a large emergent tree, to 40 m, often found in rocky areas locally, common, scattered; frequently used by *V. olivaceus* for basking; Dipterocarpaceae.

Shorea polysperma — "Tanguile," a large emergent tree, to 45 m, not common locally, scattered, generally more abundant on igneous soils; Dipterocarpaceae.

Sida cordifolia — A shrub, to 1 m, common, clustered, on forest edges and in openings; introduced but widely distributed; Malvaceae.

Spondias pinnata — "Lubas," a small subcanopy tree, to 10 m, common, scattered, often on rocky soils; oval fruits (yellow) ca. 2.6 cm dia., in large terminal panicles, available May–June, eaten by many larger birds and mammals, including *V. olivaceus;* seeds oval, hairy; Anacardiaceae.

Sterculia cuneata and *S. philippinensis* — "Balinad," a canopy tree, to 20 m, common, clustered, usually in rocky places; fruit regularly eaten by hornbills and larger parrots but not by *V. olivaceus;* Sterculiaceae.

Syzygium cumini — A subcanopy tree, to 10 m, common, especially in open places and ecotones, often left standing during clearing around villages; fruit (red) eaten by birds and rodents but not by *V. olivaceus;* Myrtaceae.

Terminalia catappa — "Talisay," a large canopy tree, to 25 m, common, usually scattered, in coastal forests and along low-lying streams; fruit not eaten by forest vertebrates as far as is known; Combretaceae.

Thespesia populnea — "Banago," a small tree, to 10 m, common, often clustered in coastal and low-lying streamside forests; Malvaceae.

Uvaria rufa — "Ulagak," a climbing shrub, to 6 m, on small trees, common, clustered, usually in openings and ecotones; fruit (red), to 5 cm, eaten by birds, probably by mammals but not by *V. olivaceus* as far as is known; Annonaceae.

Uvaria sorsogonensis — "Ulagak," a large vine, to 15 m, uncommon, scattered, growing into tall trees; fruits (yellow) clustered along vine stem, to 10 cm long, eaten by bats, birds, and monkeys, rarely by *V. olivaceus;* Annonaceae.

Vitex parviflora — "Molave," a small subcanopy tree, to 15 m, common, often clustered, especially on rocky soils of mountain slopes near coast; frequently used for basking by *V. olivaceus,* perhaps because it often grows in exposed locations; Verbenaceae.

Zingiber zerumbet and *Zingiber* sp. — "Pine-cone lily," an herb, to 2 m, most common on damp soils, generally clustered; Zingiberaceae.

Appendix 2

Annotated List of Birds and Mammals Whose Food Resources Overlap Those of *Varanus olivaceus* in the Caramoan Area*

BIRDS

    Columbidae: doves and pigeons.

        Treron vernans — "Punay," pink-necked green pigeon; 260 mm; common in secondary forest and other open associations below 300 m; widely distributed but generally near coast; gathers in flocks August–December; feeds on a variety of small fruits.

        Treron ponpodora — Pompador green pigeon; 280 mm; local distribution and less common than *T. vernans;* similar habits; widely distributed outside Philippines.

        Ptilinopus occipitalis — Yellow-breasted fruit dove; 280 mm; common from coast to 1000 m, usually in primary forests; Philippine endemic; feeds on small fruits.

        Phapitreron leucotis — "Bato-batong," white-eared brown fruit dove; 240 mm; deep forests from near coast to 1000 m, more common near sea level; often in large numbers in single fruiting tree, though they arrive singly or in pairs, usually feeds in the lowest canopy and shrub layers; a Philippine endemic.

        Ducula aenea — "Balud," green imperial pigeon; 420 mm; in primary forests, especially near coasts, not above 500 m; inland feeding September–January, coastal otherwise (Rabor, 1977); buds, shoots, and larger fruits (to *Canarium ovatum* size). Feeding habits and large flocks frequently formed suggest that it may be a major competitor of the butaan; widely distributed outside of *V. olivaceus*.

* Sources: Hachisuka, 1931; Taylor, 1934; Delacour, 1946; Walker et al., 1964; Rabor, 1977.

Ducula bicolor — "Balud puti," nutmeg imperial pigeon; 395 mm; common; large flocks in September–October, often feeding with other pigeons and doves; feeds on fruits generally; most common on smaller islands.

Ducula poliocephala — "Balud," pink-bellied imperial pigeon; 425 mm; a bird of higher mountain forests (600–1500 m); usually feeds on small fruits in groups of 1–4 birds, in upper canopy; widely distributed outside Philippines.

Macropygia phasianella — "Batik-lauin," slender-billed cuckoo dove; 400 mm; common, usually feeds in lower canopy, often in pairs, sometimes in mixed flocks; in lowlands to 1600 m, in clearings, bamboo jungles, and second growth; feeds on a variety of small fruits.

Streptopelia bitorquata — "Bato-bato de collar," Philippine turtle dove; 300 mm; common, especially in open forests and ecotones; feeds on small fruits of many types; range extends beyond Philippines.

Gallicolumba luzonica — "Punalada," bleeding heart dove; 300 mm; in thick forests; usually feeds on ground on fruits and insects; a Philippine endemic.

Psittacidae: parrots.

Prioniturus discurus — "Kanauihan," blue-headed racket-tailed parrot; 255 mm; eats fruits, including some large ones, such as pandan and banana; also feeds extensively on seeds, is more a seed predator than a distributor; usually feeds in upper canopy, though variable; common in primary forest and adjacent openings and clearings, frequently in small flocks of 4–6; a Philippine endemic.

Tanygnathus lucionensis — "Loro," blue-naped parrot; 315 mm; common in primary forest to ca 1000 m but usually far from coast; feeds chiefly on fruits, often in small flocks; a Philippine endemic.

Loriculus philippinensis — "Kollasisi," Philippine hanging parakeet; 150 mm; common, seems to prefer deep forests from coast to 1000 m, singly or in pairs; feeds on small soft fruits and blossoms; a Philippine endemic.

Bucerotidae: hornbills.

Buceros hydrocorax — "Kalao," rufous hornbill; 960 mm; a common large bird of lowland forests, to 500 m, sometimes in small flocks whose loud choruses can be heard up to 2 km away; feeds mainly on large fruits, including the most common ones eaten by *V. olivaceus;* a Philippine endemic.

Penelopides panini — "Tariktik," grooved hornbill; 650 mm; common in well-forested areas from sea level to 800 m; feeds mostly on medium to large fruits, several of them important resources of *V. olivaceus* (see chap. 12), but also occasionally on insects and small vertebrates (Rabor, 1977); a Philippine endemic.

Capitonidae: barbets.

Megalaema haemacephala — "Puk-puk," crimson-breasted barbet; 163 mm; fairly common in variety of open-forested situations; feeds on fruits (chiefly *Ficus*) and insects; ranges well beyond Philippines.

Oriolidae: orioles.
> *Oriolus chinensis* — "Kiliaon," black-naped oriole; 280 mm; usually in open forests and clearings, from sea level to 500 m; feeds on fruit (perhaps chiefly; Delacour, 1946) and insects (Rabor, 1977); widely distributed outside Philippines.

Corvidae: crows and jays.
> *Corvus macrorhynchos* — "Uwak," large-billed crow; 500 mm; common, particularly in open situations, clearings, and ecotones; omnivorous, with fruits making up about 25% of diet (Rabor, 1977); widely distributed outside Philippines.

Sturnidae: starlings and relatives.
> *Sarcops calvus* — "Koleto," black-backed coleto; 290 mm; generally in open forests and clearings from coast to 500 m, singly to 12 individuals; feeds on fruits from near ground level to the highest canopy level (Gilliard, 1949); a Philippine endemic.

MAMMALS

Pteropidae: flying foxes and fruit bats.
> *Ptenochirus jagorii* — "Kabag," dog-faced fruit bat; 130 mm; common in mountainous country from coast to 1000 m; feeds on a variety of fruit; roosts in colonies of 1–12; a Philippine endemic (chap. 12).
>
> *Rousettus amplexicaudatus* — "Kabag," Geoffroy's rousette; 130 mm; widely distributed in and outside Philippines; feeds almost entirely on fruits, though Dobson (in Rabor, 1977) reports it feeding on carrion; roosting colonies of 1–12.
>
> *Pteropus vampyrus* — "Paniki," flying fox; 280 mm; a large fruit specialist, flying great distances from roosting colonies (up to several thousand individuals, often near the coast) to the feeding trees; probably the greatest food resource overlap with *V. olivaceus* among local bats; widely distributed outside Philippines (see chap. 12).
>
> *Acerodon jubatus* — "Paniki," yellow-crowned flying fox; 285 mm; habits nearly identical with those of *Pteropus*, often mixes in roosts with the latter; an important food overlap with *V. olivaceus;* range extends much beyond Philippines (see chap. 12).

Cercopithidicae
> *Macaca philippinensis* — "Amo," Philippine macaque; 1000 mm; a common local species generally found in little-settled places, from coast to 1000 m; travels in bands of up to 20; omnivorous, diet including small vertebrates, insects, crabs, leaves, buds, shoots, and fruits, as well as a large variety of human crops; an endemic species, close to *M. fasciolarius;* an important food overlap with *V. olivaceus* but with a wide food spectrum (see chap. 12).

Hominidae
> *Homo sapiens* — Cosmopolitan, opportunistic species, preying on *V. olivaceus* (probably most important of all predators at least on adults)

and competing with it for some larger forest fruits; most important factor is this species' capacity for destroying natural habitats (see chap. 12).

Muridae

Batomys granti — "Keeno," Luzon forest rat; 300 mm; common in forested places from coast to over 2000 m, especially in rocky areas; nocturnal; feeds extensively on fruits of all sizes; among rodents, probably the most important food resource overlap with *V. olivaceus;* a Philippine endemic.

Phloeomys cumingi — "Bog-kon," Luzon cloud rat; 950 mm; common in rocky forested areas from sea level to over 3000 m; feeds on a variety of fruits but mostly large types, such as coconut; feeds on *Canarium* spp. fruits, a major food source for *V. olivaceus,* but most of its food consists of bamboo leaves, particularly of the scandant genus *Schizostachyum;* restricted to Luzon and Catanduanes.

Viverridae

Viverra tangalunga — "Sing-galong," Malay civet; 810 mm; less common than *Paradoxurus* but often seen locally; nocturnal, in open forests and along their edges and in streamside forests; climbs trees but only when it can find a slanting trunk (Rabor, 1977); feeds on a large variety of small animals (chiefly freshwater crabs locally) and some fruits, which it finds on the ground; widely distributed outside range of *V. olivaceus.*

Paradoxurus philippinensis — "Amid," palm civet; 970 mm; common in a wide variety of habitats but usually in open forests; nocturnal; an expert tree-climber, exhibiting great agility in the branches; feeds on a variety of fruits of all sizes and some insects (see Auffenberg, 1981a, for competition with *V. komodoensis*); utilizes many of the food resources of *V. olivaceus;* widely distributed outside range of *V. olivaceus.*

Suidae: pigs.

Sus celebensis — "Baboy ihalas," wild boar; 1350 mm; largest of the important competitors of *V. olivaceus;* common in less populated areas but not in highly karstified limestone hills of Caramoan area; has a wide variety of habitats that offer thick cover from coast to high in mountains; singly and in bands of 20 or more; omnivorous, eating many fallen fruits (see Auffenberg, 1981a, for competition with *V. komodoensis,* and Vogel, 1979, for competition with *V. salvator*).

Appendix 3

Abbreviations Used

MUSEUMS
- ANSP — Academy of Natural Sciences (Philadelphia)
- BMNH — British Museum (Natural History) (London)
- PNM — Philippine National Museum (Manila)
- SMF — Senckenberg Museum (Frankfurt)
- UF/FSM — University of Florida/Florida State Museum (Gainesville)
- USNM — United States National Museum of Natural History (Washington)
- ZMS — Munich Natural History Museum (Munich)

MEASUREMENTS
- bhd — breast-height diameter
- cal — calories
- chap. — chapter
- cm — centimeter
- cm^2 — square centimeter
- cm^3 — cubic centimeter
- CTM — critical thermal maximum
- CV — coefficient of variation
- df — degrees of freedom
- dia. — diameter
- f — frequency
- F — F-Max Test
- f_o — observed frequency
- f_h — hypothetical frequency
- g — gram
- gr — greatest
- ha — hectare
- hr — hour
- ht — height
- kg — kilogram
- km — kilometer

km²	square kilometer	
l	length	
lst	least	
m	meter	
m²	square meter	
m³	cubic meter	
min	minute	
mg	milligram	
MHz	megahertz	
ml	milliliter	
mm	millimeter	
Mo	mode	
mon	month	
μm	micron	
μ, mu	millimicron	
N	number (sample size)	
OR	overall range	
p	probability	
PI	preference index	
pop	population	
r	regression coefficient	
SD	one standard deviation	
SE	standard error	
SE_m	standard error of mean	
SVL	snout-vent length	
t	Student's t-test	
T_a	environmental ambient temperature	
T_b	body temperature	
T_t	transmitter ambient temperature	
tw	total weight	
w	width	
wt	weight	
$\bar{X}$	arithmetic mean	
X^2, Chi²	chi-square test	
YBP	years before present	
yr	year	

SYMBOLS

°C	degrees Celsius (centigrade)
%	percent
Σ, σ	"the sum of"
/	per

Literature Cited

Abdoessoeki, E. n.d. *Varanus komodoensis* Ouwens pada habitatnya di Pulau Komodo. Udayana Univ. Mimeographed. 9 p.

Adams, L., and S. B. Davis. 1967. The internal anatomy of home range. J. Mamm. 48(4): 529–36.

Adamson, M. L. 1986. *Meteterakis vaucheri* new species (Nematoda: Heterakoidea) from *Varanus grayi* (Varanidae) in the Philippines. Canadian J. Zool. 64(4):814–817.

Alexander, C. E., and W. G. Whitford. 1968. Energy requirements of *Uta stansburiana*. Copeia 1968(4):678–83.

Alexandre, D. Y. 1980. Caractère saisonnaire de la fructification dans une forêt hygrophile de Côte-d'Ivoire. Rep. Ecol. (Terr. Vie) 34:335–53.

Ali, S. 1944. Courtship of the monitor lizard (*Varanus monitor*). J. Bombay Nat. Hist. Soc. 44:479.

Allard, P. 1947. Light intensity studies in Canaan Valley, West Virginia. Castanea 12:63–74.

Alvir, A. D. 1926. A theory on the major tectonic structure of the island of Luzon, Philippines. Proc. 3d Pan-Pacific Congress 1:451–54.

Amato, F. L. 1965. Stratigraphic paleontology in the Philippines. Phil. Geol. 19:1–24.

Anastos, G. 1950. The scutate ticks, or Ixodidae, of Indonesia. Entomol. Am., Brooklyn Entomol. Soc. 20:1–144.

Anderson, S. C. 1963. Amphibians and reptiles from Iran. Proc. Calif. Acad. Sci. 31:417–98.

Andrews, R., and A. S. Rand. 1974. Reproductive effort in anoline lizards. Ecology 55: 1317–27.

Anon. 1970. Tectonic framework of the Philippines (maps). Bur. Mines (Manila).

———. 1975. The Philippine Atlas. Vols. I, II. 303, 125 p. Manila.

———. 1979. Caramoan, Philippines (topo. maps). A. F. P. Mapping Center (Manila), Ed. 1, based on 1956 photogrammetric survey.

———. 1981. Policy and issues in natural resources. Manila: Natural Resources Management Center, Ministry of Natural Resources. 182 p.

Ashton, P. S. 1963. The taxonomy and ecology of the Dipterocarpaceae in Brunei State. Ph.D. dissertation, Cambridge Univ. 289 p.

———. 1969. Speciation among tropical forest trees: Some deductions in the light of recent evidence. Biol. J. Linn. Soc. 1:155–96.

Literature Cited

Audley-Charles, M. G. 1978. The Indonesian and Philippine archipelagos. *In* M. Moullade and A. E. M. Nairn (eds.). The Phanerozoic Geology of the World, vol. 2, pp. 163–207. Amsterdam: E. Scientific Publ.

Auffenberg, K., and T. Auffenberg. In press. Biomass, density, spatial distribution, and activity patterns of *Geophorus bothropoma* (Prosobranchia, Helicinidae). Mal. Rev.

Auffenberg, T. In press. *Amblyomma helvolum* (Ixodidae) as a parasite of scincid and varanid reptiles in the Philippines. Internatl. J. Parasitology.

———. MS. Fig fruiting phenology in the Philippines.

Auffenberg, W. 1964. Notes on the courtship of the land tortoise *Geochelone tranvancorica* (Boulenger). J. Bombay Nat. Hist. Soc. 61(21):1–7.

———. 1965. Sex and species discrimination in two sympatric South American tortoises. Copeia 1965(3):335–42.

———. 1976. First description of an adult *Varanus grayi*. Copeia 1976(3):586–88.

———. 1978a. Gray's monitor lizard—status survey. World Wildlife Yearbk. 1977–78:129–31.

———. 1978b. Feeding and social behavior in *Varanus komodoensis*. *In* N. Greenberg and P. McLean (eds.). Neurology and Behavior of Lizards, pp. 301–31. Washington: Natl. Inst. Mental Health, U.S. Dept. HEW Publ. No. 77–491.

———. 1979a. A monitor lizard in the Philippines. Oryx 15(1):38–46.

———. 1979b. Intersexual differences in behavior of captive *Varanus bengalensis* (Reptilia, Lacertilia, Varanidae). J. Herp. 13(3):313–15.

———. 1979c. The butaan: World's rarest lizard? Topic, Jan.-Feb., pp. 14–16.

———. 1980a. The herpetofauna of Komodo with notes on adjacent areas. Bull. Fla. State Mus., Biol. Sci. 25(2):39–156.

———. 1980b. The butaan: World's rarest lizard? *In* N. Sitwell (ed.). Wildlife '80, pp. 70–75. London: Crolier Enterprises Corp.

———. 1981a. The Behavioral Ecology of the Komodo Monitor. Gainesville: Univ. Presses Florida. 406 p.

———. 1981b. Combat behaviour in *Varanus bengalensis* (Sauria: Varanidae). J. Bombay Nat. Hist. Soc. 7(1):54–72.

———. 1982a. Feeding strategy of the Caicos ground iguana, *Cyclura carinata*. *In* G. M. Burghardt and S. Rand (eds.). Iguanas of the World: Their Behavior, Ecology, and Conservation, pp. 84–116. Park Ridge, NJ: Noyes Publ. 516 p.

———. 1982b. Catch a lizard—use a lizard. Internatl. Wildlife. Nov.-Dec. 10:16–19.

———. 1983a. The food and feeding of juvenile Bengal monitor lizards (*Varanus bengalensis*). J. Bombay Nat. Hist. Soc. 80(1):119–124.

———. 1983b. Courtship behavior in *Varanus bengalensis* (Sauria: Varanidae). *In* A. G. J. Rhodin and K. Miyata (eds.). Advances in Herpetology and Evolutionary Biology: Essays in Honor of Ernest E. Williams, pp. 535–51. Cambridge: Mus. Comp. Zool., Harvard Univ.

———. 1984. Notes on feeding behaviour of *Varanus bengalensis* (Sauria: Varanidae). J. Bombay Nat. Hist. Soc., 80(2):286–302.

Auffenberg, W., and T. Auffenberg. 1987. Resource partitioning and reproductive strategies in Philippine skinks. Bull. Fla. State Museum.

Auffenberg, W., and W. G. Weaver. 1969. *Gopherus berlandieri* in southeastern Texas. Bull. Fla. State Mus., Biol. Sci. 13(3):146–91.

Augsberger, C. K. 1982. A case for synchronous flowering. *In* E. G. Leigh, A. S. Rand, and D. M. Windsor (eds.). The Ecology of a Tropical Forest, pp. 133–50. Washington: Smithsonian Inst. Press.

Avery, R. A. 1971. Estimates of food consumption by the lizard *Lacerta vivipara* Jacquin. J. Anim. Ecol. 40:351–65.

Bacon, R. F. 1909. Philippine terpenes and essential oils, pt. III, sec. A. Phil. J. Sci. 4:93–265.
———. 1910. Ibid., pt. IV, sec. A. Phil. J. Sci. 5:1–257.
Bahl, K. N. 1937. Skull of *Varanus monitor* (Linn.). Rec. Indian Mus. 39: 133–74.
Baker, J. S., and I. Baker. 1936. The seasons in a tropical rain-forest (New Hebrides), part 2. Botany. J. Linn. Soc. (Zool). 39:507–19.
Barber, A. J., and M. G. Audley-Charles. 1976. The significance of the metamorphic rocks of Timor in the development of the Banda Arc, Indonesia. Tectonomorphics 30:119–28.
Barbour, T. 1921. Aquatic skinks and arboreal monitors. Copeia 1921(1): 42–44.
———. 1926. Reptiles and Amphibians. Their Habits and Adaptations. Boston: Houghton-Mifflin and Co. 125 p.
Bartels, E. 1964. *Paradoxurus hermaphroditus javanicus* (Horsfield, 1824). Beaufortia 10(124): 193–201.
Bartholomew, G. A. 1966. A field study of temperature relations in the Galapagos marine iguana. Copeia 1966(3):241–50.
Bartholomew, G. A., and V. A. Tucker. 1964. Size, body temperature, thermal conductance, oxygen consumption, and heart rate in Australian varanid lizards. Physiol. Zool. 37: 341–54.
Bell, G. H., J. N. Davidson, and H. Scarborough. 1963. Textbook of Physiology and Biochemistry, 5th ed. Edinburgh: Livingstone. 458 p.
Bellairs, A. d'A. 1961. Apparent failure of tooth replacement in *Varanus niloticus*. British J. Herp. 3:14–15.
———. 1970. The Life of Reptiles. 2 vols. New York: Universe Books. 368, 451 p.
Bellairs, A. d'A., and A. E. W. Miles. 1960. Apparent failure of tooth replacement in monitor lizards, with remarks on loss of teeth in other reptiles. Brit. J. Herp. 1:189–94.
Ben-Avraham, Z., and S. Uyeda. 1973. The evolution of the China Basin and the Mesozoic palaeogeography of Borneo. Earth Planet. Sci. Lett. 18:365–76.
Benitez, J. 1981. Ordinary minor forest products for Kilusang Kabuhayan at Kaunlaran (KKK). Manila: Ministry of Human Settlements. 43 p.
Bennett, A. F. 1971. Oxygen transport and energy metabolism in two species of lizards, *Sauromalus obesus* and *Varanus gouldii*. Ph.D. dissertation, Univ. Michigan, Ann Arbor.
———. 1972a. The effect of activity on oxygen consumption, oxygen debt, and heart rate in the lizards *Varanus gouldii* and *Sauromalus hispidus*. J. Comp. Physiol. 79:259–80.
———. 1972b. A comparison of activities of metabolic enzymes in lizards and rats. Comp. Biochem. Physiol. 42B:637–47.
———. 1973. Blood physiology and oxygen transport during activity in two lizards, *Varanus gouldii* and *Sauromalus hispidus*. Comp. Biochem. Physiol. 46A:673–90.
Berry, A. J. 1966. Population structure and fluctuations in the snail fauna of a Malayan limestone hill. J. Zool. Soc. London 150:11–27.
Berry, K. H. 1974. The ecology and social behavior of the chuckawalla, *Sauromalus obesus obesus* Baird. Univ. Calif. Publ. Zool. 101:1–60.
Bogert, C. M. 1949. Thermoregulation in reptiles, a factor in evolution. Evolution 3:195–211.
Bollard, E. G. 1970. The physiology and nutrition of developing fruits. *In* A. C. Hulme (ed). The Biochemistry of Fruits and Their Products, pp. 387–427. New York: Academic Press.
Bonaccorso, F. J. 1979. Foraging and reproductive ecology in a Panamanian bat community. Bull. Fla. State Mus., Biol. Sci. 24(4):359–408.
Boulenger, G. A. 1885. Catalogue of the lizards in the British Museum (Natural History). Vol. 2. London: Taylor and Francis. 341 p.
Bradbury, J., D. Morrison, E. Stashko, and R. Heithaus. 1979. Radio-tracking methods for bats. Bat Res. News 20(1):9–17.

Brandis, D. 1895. An enumeration of the Dipterocarpaceae. J. Linn. Soc. Bot. 31:1–148.
Brattstrom, B. H. 1968. Heat retention by large Australian monitor lizards. Amer. Zool. 8(4):144.
———. 1971. Social and thermoregulatory behavior of the bearded dragon, *Amphibolurus barbatus*. Copeia 1971(3):484–97.
———. 1973. Rate of heat loss by large Australian monitor lizards. Bull. So. Calif. Acad. Sci. 72(1):52–54.
Braysher, M., and B. Green. 1970. Absorbtion of water and electrolytes from the cloaca of an Australian lizard, *Varanus gouldii* (Gray). Comp. Biochem. Physiol. 35:607–14.
———. 1972. The absorbtion of insulin by cloacas and bladders in reptiles and the chicken. Comp. Biochem. Physiol. 43A:613–19.
Brill, H. C., and F. Agcaoili. 1915. Philippine oil-bearing seeds and their properties, pt. II. Phil. J. Sci., Ser. A, 10:1–110.
Brongersma, I. D. 1932. Über die Eiablage und die Eier von *Varanus komodoensis* Ouwens. Der Zool. Garten (Leipzig) (n.s.) 5(1/3):45–46.
———. 1958. On an extinct species of the genus *Varanus* (Reptilia: Sauria) from the island of Flores. Rijksmus. Nat. Hist. Leiden 36(7):114–25.
Broughton, Lady M. 1936. A modern dragon hunt on Komodo. Natl. Geogr. Mag. 70:321–31.
Brown, W. H., and A. F. Fischer. 1920a. Philippine bamboos. *In* Minor Products of Philippine Forests. Bur. For. Bull. 1(22):250–310.
———. 1920b. Philippine fiber plants. *In* Minor Products of Philippine Forests. Bur. For. Bull. 1(22):311–412.
———. 1921. Wild food plants of the Philippines. *In* Minor Products of Philippine Forests. Bur. For. Bull. 1(23):224–377.
Brown, W. H., and D. M. Mathews. 1914. Philippine dipterocarp forests. Phil. J. Sci. 9(5): 413–553.
Brown, W. H., and E. D. Merrill. 1920. Philippine palms. *In* Minor Products of Philippine Forests. Bur. For. Bull. 1(22):135–240.
Bullet, P. 1942. Beiträge zur Kenntnis des Gebisses von *Varanus salvator* Laur. Vierteljahrsschrift d. Naturf. Gesell. Zurich 87:139–92.
Bullock, J. A. 1963. Notes on the cave faunas of two limestone massifs in the Taman Negara. Malay. Nat. J.:46–52.
Burden, W. D. 1928. Observations on the habits and distribution of *Varanus komodoensis* Ouwens. Amer. Mus. Nov. 316:1–10.
Burger, W. C. 1974. Flowering periodicity at four altitudinal levels in eastern Ethiopia. Biotropica 6:38–42.
Burghardt, G. M. 1970. Chemical perception in reptiles. *In* C. J. Johnston, W. Moulton, and S. Turk (eds.). Communication by Chemical Signals, pp. 241–308. New York: Appleton-Century-Crofts.
Burghardt, G. M., and A. S. Rand (eds.). 1982. Iguanas of the World: Their Behavior, Ecology, and Conservation. Park Ridge, NJ: Noyes Publ. 516 p.
Burrage, B. R. 1973. Comparative ecology and behaviour of *Chamaeleo pumilus* (Gmelin) and *C. namaquensis* A. Smith (Sauria: Chamaeleonidae). Ann. S. Afr. Mus. 61:1–158.
Calhoun, J. B., and J. U. Cosby. 1958. The calculation of home range and density of small mammals. Public Health Manag. No. 55, U.S. Dept. HEW. 24 p.
Camper, A. 1812. Mémoire sur queles parties moins connues du squelette des sauriens fossiles de Maestricht. Ann. Mus. Hist. Nat. 19:215–41.
Canby, T. Y. 1984. El Niño's ill wind. Natl. Geogr. 165(1):144–83.
Carey, W. M. 1975. The rock iguana, *Cyclura pinguis*, on Anagada, British Virgin Islands, with

notes on *Cyclura ricordi* and *Cyclura cornuta* on Hispaniola. Bull. Fla. State Mus., Biol. Sci. 19(4):189–238.
Carpenter, C. C. 1977. Communication and displays of snakes. Amer. Zool. 17:217–33.
Carpenter C. C., and G. W. Ferguson. 1977. Variation and evolution of stereotyped behavior in reptiles. *In* C. Gans and D. W. Tinkle (eds.). Biology of the Reptilia. Pt. 1, A Survey of Patterns, in vol. 7, Ecology and Behavior, pp. 335–518. London: Academic Press.
Carpenter, C. C., J. C. Gillingham, J. B. Murphy, and L. A. Mitchell. 1976. A further analysis of the combat ritual of the pygmy mulga monitor, *Varanus gilleni* (Reptilia: Varanidae). Herpetologica 32(1):35–40.
Case, G. S. 1927. The geographic regions of the Philippine Islands. J. Geogr. 26(2):41–52.
Case, T. J. 1976. Body size differences between populations of the chuckawalla, *Sauromalus obesus*. Ecology 57:313–23.
Charles-Dominique, P. 1978. Ecologie et vie sociale de *Nandinia binotata* (Carnivores, Viverrides): Comparison avec les Prosimiens sympatriques du Gabon. Terre et Vie 32:477–528.
Charles-Dominique, P., M. Atramentowicz, M. Charles-Dominique, H. Gerard, A. Hladik, C. M. Hladik, and M. F. Prevost. 1981. Les mammifères frugivores nocturnes d'une forêt Guyanaise: Inter-relations plants-animaux. Rev. Ecol., Terre et Vie 35:342–435.
Charnov, E. 1976. Optimal feeding: The marginal value theory. Theoret. Pop. Biol. 9:129–36.
Cheetham, R., and E. Hawkins. 1976. The Philippines, priorities, and prospects for development. Washington: World Bank.
Chivers, D. J., and C. M. Hladik. 1980. Morphology of the gastrointestinal tract in Primates: Comparisons with other mammals in relation to diet. J. Morphol. 166:337–86.
Christian, K. A., C. R. Tracy, and W. P. Porter. 1984. Diet, digestion and food preferences of Galapagos land iguanas. Herpetologica 40(2):205–12.
Christiansen, J. L. 1973. Reproduction of *Cnemidophorus inornatus* and *Cnemidophorus neomexicanus* (Sauria: Teiidae). Herpetologica 29(1):195–204.
Cisse, M. 1972. L'Alimentation des Varanides au Senegal. Bull. l'Inst. Fondam. Afr. Noire, sér. A, 39:503–15.
Cloudsley-Thompson, J. L. 1966. Seasonal changes in the daily rhythms of animals. Int. J. Biomet. 10:119–25.
———. 1967. Water-relations and diurnal rhythm of activity in the young Nile monitor. Brit. J. Herp. (2):296–300.
———. 1970. On the biology of the desert tortoise *Testudo sulcata* in Sudan. Proc. Zool. Soc. (London) 160:17–33.
Clutton-Brock, T. H. 1974. Primate social organisation and ecology. Nature (London) 250:539–42.
Clutton-Brock, T. H., and P. H. Harvey. 1977. Species differences in feeding ranging behavior in primates. *In* T. H. Clutton-Brock (ed.). Primate Ecology, pp. 557–84. London: Academic Press.
Cochran, W. W., and R. D. Lord, Jr. 1963. A radio-tracking system for wild animals. J. Wildl. Mgmt. 27:9–24.
Cody, M. L. 1974. Optimization in ecology. Science 183:1156–64.
Cogger, H. 1967. Australian Reptiles in Colour. Sydney: A. H. and A. W. Reed. 52 p.
Cole, C. J. 1966. Femoral glands in lizards: A review. Herpetologica 22(2):199–206.
Collins, P. 1956. My "Komodo dragon" circus. Anim. King. 59(2):34–42.
Coney, C. C., W. A. Tarplay, T. C. Warden, and J. W. Nagel. 1982. Ecological studies of land snails in the Hiwasee River Basin of Tennessee, U.S. A. Malacol. Rev. 15:69–106.
Cook, A. 1977. Mucus trail following by the slug *Limax grossui*. Anim. Beh. 25:774–81.
Corkill, N. L. 1928. Notes on the desert monitor (*Varanus griseus*) and the spiny tailed lizard (*Uromastix microlepis*). J. Bombay Nat. Hist. Soc. 32:608–10.

Corner, E. J. H. 1940. Wayside trees of Malaya. 2 vols. Singapore: Government Printer. 343, 406 p.
———. 1952. Ibid., 2d ed. 356, 467 p.
Coronos, P. 1918. Census of the Philippine Islands, vol. 1. Manila: Phil. Bur. Sci. 331 p.
Cottam, G., and J. Curtis. 1956. The use of distance measures in phytosociological sampling. Ecology 37:451–60.
Cowles, R. B. 1930. The life history of *Varanus niloticus* (Linnaeus) as observed in Natal, South Africa. J. Ent. Zool. 22:1–31.
Cox, C. R., and B. J. LeBoeuf. 1977. Female incitation of male competition: A mechanism in sexual selection. Amer. Nat. 111:317–35.
Crews, D. P. 1973. Coition-induced inhibition of sexual receptivity in female lizards (*Anolis carolinensis*). Physiol. Behav. 11:463–68.
———. 1975. Psychology of reptilian reproduction. Science 189:1059–65.
———. 1978. Integration of internal and external stimuli in the regulation of lizard reproduction. *In* N. Greenberg and P. D. McLean (eds.). Behavior and Neurology of Lizards, pp. 149–71. Washington: Nat. Inst. Mental Health, U.S. Dept. HEW.
Crews, D. P., and E. E. Williams. 1977. Hormones, reproductive behavior and speciation. Amer. Zool. 17:271–86.
Croat, T. B. 1969. Seasonal flowering behavior in central Panama. Ann. Missouri Bot. Gard. 56:295–307.
Croze, H. 1970. Searching image in carrion crows. Zeitschr. Tierpsychol. Beih. 5:1–85.
Cummins, K. W. 1967. Caloric equivalents for studies in ecological energetics. 2d ed. Pittsburgh: Univ. Pittsburgh Press. 208 p.
Curio, E. 1976. The Ethology of Predation. Berlin: Springer-Verlag. 250 p.
Cuvier, G. 1805. Leçons d'Anatomie Comparée. Paris: Crochard Press. Pp. 440–64, 475–77.
d'Abreu, E. A. 1932. Notes on some monitor lizards. J. Bombay Nat. Hist. Soc. 36:369–70.
Darwin, C. 1845. Journal of researches into the natural history and geology of the countries visited during the voyage of the H. M. S. Beagle around the world. 2d ed. London: Ward, Lock and Co. 431 p.
Daubenmire, R. 1970. Phenology and other characteristics of tropical semideciduous forest in northwestern Costa Rica. J. Ecol. 60:147–70.
Daudin, F. M. 1803. Histoire naturelle générale et particulière des reptiles. Vol. 8. Paris. 301 p.
Dawson, T. J. 1977. Kangaroos. Sci. Amer. 237(2):78–87.
de Elera, C. 1895–96. Catalogo systematico de toda la fauna de Filipinas . . . la coleccion zoologica del museo de . . . Sto. Tomas de Manila. 3 vols. Manila: Santo Tomas College. 858 p.
de Jong, J. K. 1927. Die Bewegungen im Schadel von *Varanus komodoensis*. Zool. Anz. 70:65–69.
———. 1937. Ein en onder *Varanus komodoensis*. Natuurwet. Tijdschr. Ned. Indie (Batavia) 97(8):173–208.
Delacour, J. 1946. Birds of the Philippines. New York: Macmillan and Co. 532 p.
Deraniyagala, P. E. P. 1931. Some Ceylon lizards. Spol. Zeylan. 16:139–80.
———. 1953. A Colored Atlas of Some Vertebrates from Ceylon, vol. 2, Tetrapod Reptiles, pp. 86–87. Colombo: Government Press.
———. 1957. Reproduction in the monitor lizard *Varanus bengalensis* (Daudin). Spol. Zeyl. 28(2):161–66.
Deraniyagala, R. Y. 1958. Pseudocombat of the monitor *Varanus bengalensis*. Spol. Zeyl., Bull. Natl. Mus. Ceylon 28:11–13.
Dessam, D. 1985. Ontogenetic changes in the dentition and diet of *Tupinambis* (Lacertilia: Teiidae). Copeia 1985(1):245–47.

deVos, A., and H. S. Mosby. 1963. Evaluation of habitat. *In* Wildlife Investigational Techniques, chap. 4. Ann Arbor: Edwards Bros. 449 p.

Dickerson, R. E. 1924. Tertiary paleogeography of the Philippines. Phil. J. Sci. 25(1):11–50.

———. 1928. Distribution of Life in the Philippines. Manila: Bur. Printing. 322 p.

Distal, H., and J. Veazey. 1982. The behavioral inventory of the green iguana, *Iguana iguana*. *In* G. M. Burghardt and A. S. Rand (eds.). Iguanas of the World: Their Behavior, Ecology, and Conservation, pp. 25–106. Park Ridge, NJ: Noyes Publ.

Ditmars, R. L. 1955. Reptiles of the World. New York: Doubleday & Co. 218 p.

Dixon, R. M. 1894. The strychnine tree. J. Bombay Nat. Hist. Soc. 9:102–5.

Dryden, G. L. 1965. The food and feeding of *Varanus indicus* on Guam. Micronesia 2:72–76.

Dubuis, A., L. Faurel, C. Grenot, and R. Vernet. 1971. Sur le régime alimentaire du lizard Saharien *Uromastix acanthinurus* Bell. C. R. Acad. Sci. Paris (ser. D) 273:500–503.

Dugan, B. A. 1982. A field study of the headbob displays of the male green iguanas (*Iguana iguana*): Variation in form and context. Anim. Beh. 30:327–38.

Duldulao, A. 1981a. The implications of forest occupancy management on natural resources conservation. Likas-Yaman J. Nat. Resour. Manage. Forum. 3:32.

———. 1981b. Final report of the committee on rural development. Baguio (Philippines): Forestry Policy Review Consultative Workshop. December. 18 p.

Dunn, J. E. 1977. The multivariate Ornstein-Uhlenbeck process in studies of home range. Univ. Arkansas, Statistics Lab., Tech. Rept. (2):1–13.

Dupree, L. 1954. A survey of the geography and geology of the Philippines, with emphasis on the Pleistocene. Univ. Manila J. East Asiatic Studies 3(2):183–98.

Echano, B. 1984. Who cares for the butaan? Bacong Sibol. Bur. Forestry Devel. (Quezon City). 2(4):14.

Edmund, A. G. 1960. Tooth replacement phenomena in the lower vertebrates. Royal Ont. Mus., Life Sci. Contr. 52:1–190.

———. 1969. Dentition. *In* C. Gans, A. d'A. Bellairs, and T. S. Parsons (eds.). Biology of the Reptilia, vol. 1, Morphology, pp. 117–200. London: Academic Press.

Ehrlich, P. R., and P. H. Raven. 1964. Butterflies and plants: A study in coevolution. Evolution 18:586–608.

Elias, B. E., and D. H. Nelson. 1978. The evolutionary significance of delayed emergence from the nest by hatching turtles. Evolution 32(2):297–303.

Elicano, V., and L. A. Faustino. 1926. Preliminary geological map of the Philippine Islands. Div. Geol. Mines, Phil. Bur. Sci.

Ellis, J. E., J. A. Wiens, C. F. RoDell, and J. C. Anway. 1976. A conceptual model of diet selection as an ecosystem process. J. Theor. Biol. 60:93–108.

Emlen, J. M. 1966. The role of time and energy in food preference. Amer. Nat. 100:611–17.

———. 1968. Optimal choice in animals. Amer. Nat. 102:385–90.

Ensminger, M. E. 1977. Animal Science. Danville, IL: Interstate Printers and Publ., Inc., 1047 p.

Estabrook, G. F., and A. E. Dunham. 1976. Optimal diet as a function of absolute abundance, relative abundance and relative value of available prey. Amer. Nat. 110(973):401–13.

Estes, R., and E. E. Williams. 1984. The ontogenetic variation in the molariform teeth of lizards. J. Vert. Paleont. 4:96–107.

Evans, L. T. 1936a. A study of social hierarchy in the lizard *Anolis carolinensis*. J. Genet. Psychol. 48:88–111.

Falk, K. 1921. Sudwestafrikanische Reptilien und ihre Heimat. VII, Reptilien bei Omaruru. Wiss. Aquar. Terrar. Kunde. 18:231–32.

Faustino, L. A. 1926. History of the strand line of the Philippine Islands during Pleistocene and post-Pleistocene times. Proc. 3d Pan-Pacific Congr. (Tokyo), vol. 2:1807–11.

———. 1932. The development of karst topography in the Philippine Islands. Phil. J. Sci. 29:203–9.
Fedorov, A. A. 1966. The structure of tropical rain forest and speciation in the humid tropics. J. Ecol. 54:1–11.
Feeny, P. P. 1975. Biochemical coevolution between plants and their insect herbivores. *In* L. E. Gilbert and P. H. Raven (eds.). Coevolution of Animals and Plants, pp. 102–230. Austin: University of Texas. 349 p.
Fejervary, S. J. 1918. Contributions to a monograph on fossil Varanidae and on Megalanidae. Ann. Mus. Natl. Hungarica 16:341–467.
———. 1935. Further contributions to a monograph of the Megalanidae and fossil Varanidae—with notes on recent varanians. Ann. Mus. Natl. Hungarica 29:1–130.
Feliciano, J., and V. Pelaez. 1940. Pleistocene orogenic movements in the Philippine Islands. Proc. 6th Pacific Sci. Congr., vol. 2:811–14.
Fenchel, T. M., C. P. McRoy, J. C. Oyden, P. Parker, and W. E. Rainey. 1979. Symbiotic cellulose degradation in green turtles. *Chelonia mydas* L. Appl. Environ. Microbiol. 37:348–50.
Ferguson, G. W. 1977. Variation and evolution of stereotyped behavior in reptiles: Social Displays. *In* C. Gans and D. W. Tinkle (eds.). Biology of the Reptilia, vol. 7, Ecology and Behavior, pp. 405–508. London: Academic Press.
Ferner, J. W. 1974. Home-range size and overlap in *Sceloporus undulatus erythrocheilus* (Reptilia: Iguanidae). Copeia 1974(2):332–37.
Field, C. R. 1967. Palatability factors and nutrient values of the food of the buffalo (*Syndocerus caffer*) in Uganda. E. Afr. Wildl. J. 14:181–201.
Fitch, H. S. 1954. Life history and ecology of the five-lined skink, *Eumeces fasciatus*. Univ. Kansas Mus. Nat. Hist. Publ. 8:1–156.
———. 1970. Reproductive cycles in lizards and snakes. Univ. Kansas Mus. Nat. Hist., Misc. Publ. 53:1–247.
Fitzsimons, V. F. 1943. The lizards of South Africa. Transvaal Mus. Mem. (1):1–402.
Fleay, R. 1958. How to capture a goanna. Anim. King. 61:73–75.
Fleck, A., and N. H. Munro. 1965. The determination of organic nitrogen in biological materials: A review. Clinical Chem. Acta. 11:2–12.
Fleming, T. H. and E. R. Heithaus. 1981. Frugivorous bats, seed shadows and the structure of tropical forests. Biotropica. 13(Suppl.):45–53.
Flenley, J. 1979. The Equatorial Rain Forest: A Geological History. London: Butterworths. 203 p.
Fogden, M. P. L. 1972. The seasonality and population dynamics of equatorial forest birds of Sarawak. Ibis 114:307–43.
Food and Agricultural Organization (FAO) of the United Nations. 1961. FAO Yearbook of Forest Products. Rome: FAO. 145 p.
———. 1976a. Forest Resources in Asia and the Far East Region. Rome: FAO. 81 p.
———. 1976b. Development and Forest Resources in Asia and the Far East: Trends and Perspectives. Rome: FAO. 88 p.
Foster, R. B. 1973. Seasonality of fruit production and seed fall in a tropical forest ecosystem in Panama. Ph.D. dissertation, Duke Univ., Durham, NC. 156 p.
———. 1982. The seasonal rhythm of fruit fall on Barro Colorado Island. *In* E. G. Leigh, A. S. Rand, and D. N. Windsor (eds.). The Ecology of a Tropical Forest, pp. 151–72. Washington: Smithsonian Inst. Press.
Fox, H. 1977. The urinogenital system of reptiles. *In* C. Gans and T. S. Parsons (eds.). Biology of the Reptilia, vol. 6, Morphology, pp. 1–158. London: Academic Press.
Francaz, J. M., M. Dudemaine, R. Vernet, and C. Grenot. 1976. Etude de l'évolution de la température interne et du rythme cardiaque chez le lézard saharien: *Varanus griseus* Daudin par radio télémétrie. C. R. Acad. Sci. Paris 282:1199–1202.

Frankie, G. W., H. G. Baker, and P. A. Opler. 1974. Comparative phenological studies of trees in tropical wet and dry forests in the lowlands of Costa Rica. J. Ecol. 62:881–919.

Frazetta, T. H. 1962. A functional consideration of cranial kinesis in lizards. J. Morphol. 11(3):287–320.

———. 1983. Adaptation and function of cranial kinesis in reptiles. In A. G. J. Rhodin and K. Miyata (eds.). Advances in Herpetology and Evolutionary Biology: Essays in Honor of Ernest E. Williams, pp. 223–44. Cambridge: Mus. Comp. Zool., Harvard Univ. .

Freeland, W. J., and D. H. Janzen. 1974. Strategies in herbivory by mammals: The role of plant secondary compounds. Amer. Nat. 108:269–89.

Furbringer, M. 1922. Das Zungenbein der Wirbeltiere, ins besondere der Reptilien und Vögel. Abh. Heidelberg. Akad. Wiss. math. nat. Kl. Abt. B. (11):23–46.

Gardarsson, A., and R. Moss. 1970. Selection of food by Icelandic ptarmigan in relation to its availability and nutritive value. In A. Watson (ed.). Animal Populations in Relation to Their Food Resources, pp. 47–71. London: Blackwell.

Gibbons, J. W. 1972. Reproduction, growth and sexual dimorphism in the canebrake rattlesnake (*Crotalus horridus atricaudatus*). Copeia 1972(2): 222–26.

———. 1976. Aging phenomena in reptiles. In M. F. Elias, B. E. Eleftheriou, and P. K. Elias (eds.). Special Review of Experimental Aging Research: Progress in Biology, pp. 454–75. Bar Harbour, ME: EAR, Inc.

Gilliard, E. T. 1949. A study of the coleto or bald starling (*Sarcops calvus*). Amer. Mus. Nat. Hist., Novitates (1429):1–6.

Gillingham, J. C. 1979. Reproductive behavior of the rat snakes of eastern North America, Genus *Elaphe*. Copeia 1979(2):319–31.

Gittings, S. D., and J. J. Raemaekers. 1978. Siamang, lar, and agile gibbons. In D. J. Chivers (ed.). Malayan Forest Primates, pp. 89–105. New York: Plenum Press.

Gloyd, M. 1967. Mean crowding. J. Anim. Ecol. 36(1):1–30.

Gonor, J. J. 1967. Predator-prey reactions between two marine prosobranch gastropods. The Velliger 7:228–32.

Goodland, R. J. A. 1975. Amazon Jungle: Green Hell to Red Desert. Oxford: Elsevier Press. 212 p.

Gorman, G. C., L. Atkins, and T. Holzonger. 1967. New karyotypic data on 15 genera of lizards in the family Iguanidae, with a discussion of taxonomic and cytological implications. Cytogenetics (6):286–99.

Gray, J. E. 1831. A synopsis of the species of the Class Reptilia. In B. Cuvier (arranger). The Class Reptilia: With Specific Descriptions by E. Griffith and E. Pidgeon, pp. 19–21 in Cuvier's Animal Kingdom 9:1–311. London: Whittaker, Treacher, and Co.

———. 1845. Catalogue of the specimens of lizards in the collection of the British Museum. London: Taylor and Francis. 365 p.

Green, B. 1969. Water and electrolyte balance in the sand goanna *Varanus gouldii* (Gray). Ph.D. dissertation, Univ. Adelaide. 203 p.

———. 1972. Water loss of the sand goanna (*Varanus gouldii*) in its natural environments. Ecology 53:452–57.

Green, B., and D. King. 1978. Home range and activity patterns of the sand goanna *Varanus gouldii* (Reptilia: Varanidae). Aust. J. Wildl. Res. 5:417–25.

Greenberg, B., and G. K. Noble. 1944. Social behavior of the American chameleon (*Anolis carolinensis* Voigt). Physiol. Zool. 17:392–439.

Greene, H. W. 1969. Brief notes on the Komodo dragon, *Varanus komodoensis*. Mimeographed. 2 p.

———. 1980. On the diets of snakes and "carnivorous" lizards. Abstract, ASIH 60th Ann. Mtg., p. 62.

Grenot, C. 1967. Etude comparative de la résistance à la chaleur d'*Uromastix acanthinurus* et de *Varanus griseus*. Terre et Vie 1968:390–409.

———. 1976. Ecophysiologie de lézard saharien *Uromastix acanthinurus* Bell, 1825 (Agamidae herbivore). Ecole Norm. Super., Publ. Lab. Zool. 7:1–323.

Griffiths, D. 1980. Foraging costs and relative prey size. Amer. Nat. 116:743–52.

Grijs, P. 1899. Beobachtungen an Reptilien in der Gefangenschaft. Zool. Gart. 40:175–91, 210–26, 236–48, 267–77, 302–15.

Grimwood, I. R. 1975. National parks and wildlife conservation in the Philippines. *In* FAO, Training in multiple use forest management, W3/F6827, pp. 1–24. Rome.

———. 1978. Wildlife conservation in the Philippines. Tiger paper. 5(4):9–11.

Grosenbaugh, L. 1952. Plotless timber estimates. J. Forestry 50:32–37.

Grote, H. 1911. Kurze biologische Notizen über einige Sanger und Reptilien Ostafrikas. Zool. Beob. 52:346–49.

Gunther, A. 1861. On the anatomy of *Regina ocellata*. Proc. Zool. Soc. London 12(1):60–62.

Guppy, N. 1983. Tropical deforestation: A global view. Foreign Affairs (Spring 1984):928–65.

Haas, G. 1973. Muscles of the jaws and associated structures in the Rhynchocephala and Squamata. *In* C. Gans and T. Parsons (eds.). Biology of the Reptilia, vol. 4, Morphology, pp. 285–490. London: Academic Press.

Hachisuka, M. 1931–35. Birds of the Philippine Islands. Vols. 1–2. London: H. F. and G. Witherby. 1078 p.

Hagen, B. 1890. Die Pflanzen und thierwelt von Deli auf der Ostküste Sumatra's. Natuurw. Skitzen Beiträge Tijd. Kon. Ned. Amdrijksk. Genootsch. (2):1–240.

Haggag, G., K. A. Raheem, and F. Khalil. 1965. Hibernation in reptiles: Changes in blood electrolytes. Comp. Biochem. Physiol. 16:457–65.

Hahn, W. E., and D. W. Tinkle. 1965. Fat body cycling and experimental evidence for its adaptive significance to ovarian follicle development in the lizard *Uta stansburiana*. J. Exp. Zool. 158:79–86.

Hallinan, T. 1920. Notes on lizards of the Canal Zone, Isthmus of Panama. Copeia 1920(83):45–49.

Hallowell, E. 1857. Notes on the reptiles in the collection of the Museum of the Academy of Natural Sciences, Philadelphia, pp. 146–153.

Halos, S. 1980. Ipil-ipil needs a second look. Canopy Internatl. 6(10):1–8.

Hamilton, W. J., and C. G. Coetzee. 1969. Thermoregulatory behavior of the vegetarian lizard *Angolosaurus skoogi* on the vegetationless northern Namib Desert dunes. Sci. Pap. Namib Desert Res. Sta. 47:95–103.

Hamilton, W. J., and W. E. Watt. 1970. Refuging. Ann. Rev. Ecol. Syst. 1:263–97.

Hansen, R. M., and C. K. Sylber. MS. Cellulosa digestion by young yellow giant chuckawalla lizards.

Harestad, A. S., and F. L. Bunnell. 1979. Home range and body weight—a reevaluation. Ecology 60(2):389–402.

Harris, V. A. 1964. The life of the rainbow lizard. London: Hutchinson. 203 p.

Harrison, J. L. 1955. Data on the reproduction of some Malayan mammals. Proc. Zool. Soc. London 125:445–60.

———. 1969. The abundance and population density of mammals in Malayan lowland forests. Malay. Nat. J. 22:174–78.

Hathaway, D. E. 1959. Myrobalans: An important tanning material. Tropical Sci. 1:85–106.

Hazlett, B. A., and W. H. Bossert. 1965. A statistical analysis of the aggressive communication systems of some hermit crabs. Anim. Beh. 13:357–73.

Heatwole, H. 1970. Thermal ecology of the desert dragon *Amphibolurus inermi*. Ecol. Monogr. 40:425–57.

Heatwole, H., and A. Heatwole. 1978. Ecology of the Puerto Rican camaenid tree snails. Malacologia 17(2):241–315.
Hediger, H. 1934. Beiträg auf Herpetologie und Zoogeographie Neu Britanniens und einiger umliegender Gebiete. Zool. Jahrb. Syst. 65:44–582.
Heithaus, E. R., P. A. Opler, and H. G. Baker. 1974. Bat activity and pollination of *Bauhinia pauletia:* Plant-pollinator coevolution. Ecology 55:412–19.
Hendry, R. S. 1959. Atlas of the Philippines. Manila: Phil.-ASEAN Publ., Inc. 182 p.
Henne am Rhyne, R. 1903. Einige merkwurdige Kreichtiere der Sunda-Inseln. Zool. Anz. 26:167–72.
Herberer, G. 1930. Das Bodeninterstitum von *Varanus komodoensis,* Ouwens. Zeit. mikro.-anat. Forsch. 20:398–416.
Herrera, C. M. 1982. Seasonal variation in the quality of fruits and diffuse coevolution between plants and avian dispersers. Ecology 63(3):773–85.
Hickman, J. C. 1975. Environmental predictability and plastic energy allocation in the annual *Polygonium cascadense* (Polygonaceae). J. Ecol. 63:689–701.
Hladik, A., and C. M. Hladik. 1969. Rapports trophique entre végétation et primates dans la forêt de Barro Colorado (Panama). Terre et Vie 116:25–117.
Hladik, C. M. 1976. Surface relative du tractus digestif de quelques primates morphologie des villosités intestinales et corrélations avec la régime alimentaire. Mammalia 31(1):120–47.
———. 1978. Adaptive strategies of primates in relation to leaf-eating. *In* G. L. Montgomery (ed.). Ecology of Arboreal Foliovores, pp. 373–96. Washington: Smithsonian Inst.
Hladik, C. M., and A. Hladik. 1972. Disponibilité alimentaires et domaines vitaux des primates à Ceylan. Terre et Vie 26:149–215.
Hofer, H. 1960. Vergleichende Untersuchungen am Schadel von *Tupinambis* und *Varanus* mit besondere Berücksichtigung ihrer Kinetik. Morph. J. 100(4):706–46.
Hoffstetter, R., and J.-P. Gasc. 1969. Vertebrae and ribs of modern reptiles. *In* C. Gans, A. d'A. Bellairs, and T. S. Parsons (eds.). Biology of the Reptilia, vol. 1, Morphology, pp. 201–310. New York: Academic Press.
Holdridge, L. R. 1964. Life Zone Ecology. San Jose, Costa Rica: Tropical Sci. Cent. 124 p.
Holmes, C. H. 1942. Flowering and fruiting of forest trees of Ceylon. Indian For. 68(8):411–20, (9):488–99, (11):580–87.
Holmes, R. S., M. King, and D. King. 1975. Phenetic relationships among varanid lizards based upon comparative electrophoretic data and karyotypic analyses. Biochem. Syst. Ecol. 3:257–62.
Honegger, R. E., and H. Heusser. 1969. Beiträge zum Verhalteninventar des Bindenswaren (*Varanus salvator*). Zool. Garten 36:251–60.
Hoogerwerf, A. 1970. Udjung Kulon: The Land of the Last Javan Rhinoceros. Leiden: Brill. 308 p.
Hoogstraal, H., and A. Aeshlimann. 1982. Tick-host specificity. Mittl. Schweis. Entomol. Gesell. 55:5–32.
Hooijer, D. H. 1967. Indo-Australian insular elephants. Genetics 38:143–62.
———. 1972. *Stegodon trigonocephalus florensis* Hooijer, and *Stegodon timorensis* Sartono. K. Nedl. Akad. Wet. 75(1):12–32.
Horn, H. G. 1980. Bisher unbekante Details zur Kenntnis von *Varanus varius* auf Grund von feld herpetologischen und terraristischen Beobachtungen (Reptilia: Sauria: Varanidae). Salamandra 16:1–18.
Horn, H. G., and G. Petters. 1982. Beiträge zur Biologie des Rauhnackenswarans, *Varanus (Dendrovaranus) rudicollis* Gray. Salamandra 18(1/2):29–40.
Howe, H. F. 1982. Fruit production and animal activity in two tropical trees. *In* E. G. Leigh, Jr., A. S. Rand and D. M. Windsor (eds.). The Ecology of a Tropical Forest, pp. 189–99. Washington: Smithsonian Inst. Press.

Howes, F. N. 1953. Vegetable Tanning Materials. London: Butterworth. 368 p.
Hughes, R. N. 1979. Optimal diets under the energy maximization premise: The effects of recognition time and learning. Amer. Nat. 113:209–21.
Humason, G. L. 1979. Animal tissue techniques. 4th ed. San Francisco: W. H. Freeman Co. 512 p.
Hyman, E. 1981a. Policy options for solving woodfuel energy problems. In Proceedings of the Third International Congress on Energy and Environment, pp. 18–45. Manila: Soc. Engin. Sci.
———. 1981b. Wood and charcoal as direct energy source for consumption by households and rural industries in the Philippines. Likas-Yaman J. Nat. Resour. Mgmt. Forum. 3(28):1–72.
———. 1983. Analysis of woodfuels market: A survey of fuelwood sellers and charcoal makers in the province of Ilocos Norte, Philippines. Biomass 3(3):167–97.
———. 1984a. Forestry administration and policies in the Philippines. Environ. Mgmt. 7(6): 511–24.
———. 1984b. Analysis of the demand for woodfuels by cottage industries in the province of Ilocos Norte, Philippines. Energy 5(3):113–19.
Impey, O. R. 1967. Functional aspects of cranial kineticism in the Lacertilia. Ph.D. dissertation, Oxford Univ. 304 p.
Irving, E. M. 1952. Geological history and petroleum possibilities of the Philippines. Bull. Am. Assoc. Pet. Geol. 36:437–76.
IUCN. 1982. Amphibia-Reptilia Red Data Book. Vol. II. Morges: IUCN. 214 p.
Iverson, J. B. 1977. Behavior and ecology of the rock iguana *Cyclura carinata*. Ph.D. dissertation, Univ. Florida, Gainesville. 303 p.
———. 1979. Behavior and ecology of the rock iguana *Cyclura carinata*. Bull. Fla. State Mus., Biol. Sci. 24(3):175–358.
———. 1980. Colic modifications in iguanine lizards. J. Morph. 163: 79–93.
———. 1982. Adaptations to herbivory in iguanine lizards. In G. M. Burghardt and A. S. Rand (eds.). Iguanas of the World: Their Behavior, Ecology, and Conservation, pp. 60–76. Park Ridge, NJ: Noyes Publ. 516 p.
Jacob, D., and L. S. Ramaswami. 1976. The female reproductive cycle of the Indian monitor lizard, *Varanus monitor*. Copeia 1976(2):256–60.
Janzen, D. H. 1970. Herbivores and the number of tree species in tropical forests. Am. Nat. 104:501–28.
———. 1971a. Seed predation by animals. Ann. Rev. Ecol. Syst. 2:465–92.
———. 1971b. Escape of *Cassia grandis* L. beans from predation in time and space. Ecology 52:964–79.
———. 1973a. Sweep samples of tropical foliage insects: Description of study sites with data on species abundances and size distributions. Ecology 54:659–86.
———. 1973b. Ibid.: Effects of season, vegetation types, elevation, time of day and insularity. Ecology 54:689–708.
———. 1973c. Host plants as islands: Competition in evolutionary and contemporary time. Am. Nat. 107:786–90.
———. 1975a. Ecology of Plants in the Tropics. London: Arnold. 301 p.
———. 1975b. Intra- and interhabitat variations in *Guazuma ulmifolia* (Sterculiaceae) seed predation by *Amblycerus cistelinus* (Bruchidae). Biotropica 6:69–103.
———. 1978. Complications in interpreting the chemical defenses of trees against tropical arboreal plant-eating vertebrates. In G. C. Montgomery (ed.). The Ecology of Arboreal Foliovores, pp. 73–84. Washington: Smithsonian Inst. Press.
———. 1979. How to be a fig. Ann. Rev. Ecol. Syst. 10:13–51.

Literature Cited

Jenkins, R. 1969. Ecology of three species of saltators with special reference to their frugivorous diet. Ph.D. dissertation, Howard Univ., Washington, D. C. 318 p.

Jennich, R. I., and F. B. Turner. 1969. Measurement of non-circular home range. J. Theoret. Biol. 22:227–37.

Johnson, C. R. 1972. Head-body temperature differences in *Varanus gouldii* (Sauria: Varanidae). Comp. Biochem. Physiol. 43A:1025–29.

———. 1976. Some behavioral observations on wild and captive sand monitors, *Varanus gouldii* (Sauria: Varanidae). Zool. J. Linn. Soc. 59:377–80.

Johnson, D. R. 1966. Diet and estimated energy assimilation of three Colorado lizards. Am. Midl. Nat. 76:504–9.

Johnson, R. N., and H. Lillywhite. 1979. Digestive efficiency of the omnivorous lizard *Klauberina riversiana*. Copeia 1979(3):431–37.

Jolly, G. M. 1965. Explicit estimates from capture-recapture data with both death and immigration-stochastic models. Biometrika 52:225–47.

Jorgensen, C. D., and W. W. Tanner. 1963. The application of the density probability function to determine the home ranges of *Uta stansburiana stansburiana* and *Cnemidophorus tigris tigris*. Not. Herp. Chiapasiae IV. Herpetologica 19(2):105–15.

Kamil, A., and T. Sargent. 1981. Foraging behavior: Ecological, ethological, and psychological approaches. New York: Garland STPM. 285 p.

Karr, J. R. 1976. Seasonality, resource availability, and community diversity in tropical bird communities. Am. Nat. 110(976):973–94.

Kaufman, V. 1972. The ixodid tick genus *Aponomma*. Ph.D. dissertation, Univ. Maryland, College Park. 401 p.

Kehl, R., and C. Combescot. 1955. Reproduction in the Reptilia. In The Comparative Endocrinology of Vertebrates. Mem. Soc. Endocrin. (4):57–74.

Kimmich, H. P. 1980. Artifact-free measurement of biological parameters: Biotelemetry, a historical review and layout of modern developments. In C. J. Amlaner, Jr., and D. W. MacDonald (eds.). A Handbook on Biotelemetry and Radio Tracking, pp. 1–120. Oxford: Pergamon Press.

King, D. 1977. Temperature regulation in the sand goanna *Varanus gouldii'* (Gray). Ph.D. dissertation, Univ. Adelaide, Australia. 58 p.

———. 1980. The thermal biology of free-living sand goannas (*Varanus gouldii*) in southern Australia. Copeia 1980(4):755–67.

———. In press. The diet and foraging strategy of *Varanus acanthurus*.

King, D., and B. Green. 1979. Notes on the diet and reproduction of the sand goanna, *Varanus gouldii rosenbergi*. Copeia 1979(1):64–70.

King, D., and L. Rhodes. 1982. The sex ratio and breeding season of *Varanus acanthurus*. Copeia 1982(4):784–87.

King, J., and A. McKee. 1948. Physiographic map of the Philippine Islands. New York: Rand McNally and Sons.

King, M., and D. King. 1975. Chromosomal evolution in the lizard genus *Varanus* (Reptilia). Australian J. Biol. Sci. 28:89–108.

King, W. 1962. Paleolithic reptile and amphibian remains from Niah Great Cave. Sarawak Mus. J. 10(19–20):450–52.

Kleiman, D. G. 1977. Monogamy in mammals. Quart. Rev. Biol. 52:39–69.

Knowland, B., and C. Ulinski. 1979. Traditional fuels: Present data, past experience, and possible strategies. Washington: AID. 79 p.

Koelmeyer, K. O. 1960. The periodicity of leaf change and flowering in the principal forest communities of Ceylon. 1960. Pts. 1–11, Ceylon Forester 4(2):157–89, 308–62.

Komarek, E. V. 1969. A discussion of wildlife management, fire and the wildlife landscape. Proc. 8th Ann. Tall Timbers Fire Ecol. Conf. (Tallahassee, FL):85–125.

Kratzer, H. 1973. Beobachtungen über die Zeitigungsdauer eines Eigeleges von *Varanus salvator.* Salamandra 9(1):27–33.

Krebs, J. R. 1973. Behavioral aspects of predation. *In* P. P. G. Bateson and P. H. Klopfer (eds.). Perspectives in Ethology, pp. 73–111. New York: Plenum Press.

Krishnaswamy, V. S., and A. S. Mathauda. 1954. Phenological behaviour of a few forest species at New Forest, Dehra Dun. Indian For. 80:124–53, 187–206.

Lack, D. 1962. Ecological Adaptations for Breeding in Birds. London: Methuen. 409 p.

Lakjer, T. 1926. Studien über die trigeminusversorgte Kaumuskulatur der Sauropsiden. Copenhagen: C. A. Rietzel. 112 p.

Landsmeer, J. M. F. 1981. Digital morphology in *Varanus* and *Iguana.* J. Morph. 168:289–95.

Lang, H. 1919. *In* H. A. Pilsbry. A review of the land mollusks of the Belgian Congo. Am. Mus. Nat. Hist. Bull. 40(1):54–58.

Lantican, C. 1981. Final report of the committee on support forestry programs. Forestry Policy Review Consultative Workshop, Bagio, December. Mimeographed. 8 p.

Leck, C. F. 1969. Palmae: Hic et ubique. Principes 13:80.

———. 1970. The seasonal ecology of fruit and nectar eating birds in lower Middle America. Ph.D. dissertation, Cornell Univ., Ithaca, NY. 183 p.

———. 1972. Seasonal changes in feeding pressures of fruit- and nectar-eating birds in the neotropics. Condor 74:54–60.

Lecuru, S. 1968. Remarques sur le scapulo-coracoide des lacertiliens. Ann. Sci. Nat., Zool., Paris, 12th ser. 10:475–510.

———. 1969. Etude morphologique de l'humerus des lacertiliens. Ann. Sci. Nat., Zool., Paris, 12th ser., 10:515–58.

Lederer, G. 1929. Ein zahmer Bindenwaran. Wschr. Aquar. Terrar. Kunde 26:19–20.

———. 1933. Beobachtungen an Waranen im Frankfurter Zoo. Zool. Garten (n.s.) 6:118–26.

———. 1942. Der Drachenwaran (*Varanus komodoensis* Ouwens). Zool. Garten (n.s.) 14(5/6):227–44.

Leenhouts, P. W. 1955. Burseraceae. Florae Malaysiana 5(2):1–309.

Lesson, R. P. 1830. Zoologie. *In* L. J. Duperrey (ed.). Voyage autout du monde, execute par ordre du roi. Paris. 403 p.

Leung, W. T. W. 1968. Food composition tables for use in Africa. Bethesda: U.S. Public Health Service. FAO–HEW Publ. (63):1–136.

Licht, P., and G. C. Gorman. 1970. Reproduction and fat bodies in Caribbean *Anolis* lizards. Univ. California Publ. Zool. 95:1–95.

Linz, H. 1939. Das hyobranchialskelett von *Varanus albigularis* Daud. Ph.D. dissertation, Münster Univ., Münster. 21 p.

Lloyd, M. 1967. Mean crowding. J. Anim. Ecol. 36(1):1–30.

Lof, G. O. G., J. A. Duffie, and C. O. Smith. 1966. World distribution of solar radiation. Univ. Wisconsin Eng. Exper. Sta., Solar Energy Lab. Rept. 21. 21 p.

Lonnberg, E. 1902. On the adaptation to a molluscivorous diet in *Varanus niloticus.* Arkiv. f. Zool. 1:32–56.

Loop, M. S. 1974. The effect of relative prey size on the ingestive behavior of the Bengal monitor, *Varanus bengalensis* (Sauria: Varanidae). Herpetologica 30:123–27.

Lorenz, K. 1937. The companion of the birds world. Auk 54:245–73.

Lucas, L. R., F. G. Salazar, and L. Engle. 1947. Soils of Camarines Norte. Manila: Bur. Soils. Map.

Luna, L. G. 1968. Manual of histological staining methods of the Armed Forces Institute of Pathology, 3d ed. New York: McGraw-Hill Book Co. 617 p.

MacArthur, R. H. 1972. Geographical Ecology. New York: Harper and Row. 211 p.
MacArthur, R. H., and E. R. Pianka. 1966. On the optimal use of a patchy habitat. Am. Nat. 100:603–9.
McBee, R. H. 1971. Significance of intestinal microflora in herbivory. Ann. Rev. Ecol. Syst. 2:165–76.
———. 1977. Fermentation in the hindgut. *In* R. Clarke and T. Bauchop (eds.). Microbial Ecology of the Gut, pp. 185–222. London: Academic Press.
McBee, R. H., and V. H. McBee. 1982. The hindgut fermentation in the green iguana, *Iguana iguana*. *In* G. M. Burghardt and A. S. Rand (eds.). Iguanas of the World: Their Behavior, Ecology, and Conservation, pp. 77–82. Park Ridge, NJ: Noyes Publ. 516 p.
McClure, H. E. 1966. Flowering, fruiting and animals in the canopy of a tropical rain forest. Malay For. 29(3):182–202.
MacDonald, D. W. 1980. Rabies and Wildlife: A Biologist's Perspective. Oxford: Oxford Univ. Press. 203 p.
MacDonald, D. W., and C. J. Amlaner, Jr. 1980. A practical guide to radio tracking. *In* C. J. Amlaner, Jr., and D. W. MacDonald (eds.). A Handbook on Biotelemetry and Radio Tracking, pp. 213–48. Oxford: Pergamon Press. 667 p.
MacDonald, D. W., F. G. Ball, and N. G. Hough. 1980. The evaluation of home range size and configuration using radio tracking data. *In* C. J. Amlaner, Jr., and D. W. MacDonald (eds.). A Handbook on Biotelemetry and Radio Tracking, pp. 405–34. Oxford: Pergamon Press. 667 p.
McKay, D. 1974. Adaptive patterns in alkaloid physiology. Am. Nat. 108:305–32.
MacKay, R. S. 1970. Bio-medical Telemetry, 2d ed. New York: John Wiley and Sons. 312 p.
McNab, B. K. 1963. Bioenergetics and the determination of home range size. Am. Nat. 97:133–40.
McNab, B. K., and W. Auffenberg. 1976. The body temperature of the Komodo monitor *Varanus komodoensis*. Comp. Biochem. Physiol. 55:45–50.
Magnan, A. 1912. Le régime alimentaire et la longueur de l'intestine chez les mammifères. C. R. Acad. Sci. 154:129–31.
Mahendra, B. C. 1931. How the monitor lizard sits in its burrow. J. Bombay Nat. Hist. Soc. 34:255–56.
Marfori, R. T. 1947. Soils of Camarines Sur Province. Manila: Bur. Soils. Map.
———. 1948a. Soils of Sorsogon Province. Manila: Bur. Soils. Map.
———. 1948b. Soils of Albay Province. Manila: Bur. Soils. Map.
Marshall, A. J., and R. Hook. 1960. The breeding of equatorial vertebrates: Reproduction of the lizard Agama agama lionotes Boulenger at lat. 0 01'N. Proc. Zool. Soc., London 134:197–205.
Mautz, W. J. 1982. Use of cave resources by a lizard community. *In* N. J. Scott (ed.). Herpetological Communities, pp. 129–34. U.S. Fish Wildl. Res. Rept. 13.
Mayhew, W. W. 1963. Some food preferences of captive *Sauromalus obesus*. Herpetologica 19(1):10–16.
Mazurkiewicz, M. 1969. Elliptical modification of the home range pattern. Bull. Acad. Polish Sci. Ser. II, 17(7):427–31.
Mead, A. R. 1961. The Giant African Snail: A Problem in Economic Malacology. Chicago: Univ. Chicago Press. 257 p.
Medway, Lord. 1969. The Wild Mammals of Malaya. Kuala Lumpur: Oxford Univ. Press. 431 p.
———. 1972. Phenology of a tropical rain forest in Malaya. Biol. J. Linn. Soc. 4:117–46.
Medway, Lord, and D. R. Wells. 1976. The birds of the Malay Peninsula. Kuala Lumpur: Univ. Malaya. 685 p.
Merrill, E. D. 1912. A flora of Manila. Manila: Bur. Sci. Publ. (5): 1–491.

———. 1923. Distribution of Dipterocarpaceae: Origins and relationships of the Philippine flora and causes of the difference between the flora of eastern and western Malaysia. Phil. J. Sci.:1–33.

———. 1926. An enumeration of Philippine plants. Vols. 1–4. Manila: Bur. Sci. 813 p.

Mertens, R. 1937. Zwei Bemerkungen über Warane (Varanidae). Senckenburgiana 19:177–81.

———. 1942. Die Familie der Warane (Varanidae). Abh. Senckb. Naturforsch. Ges., Pts. 1–3(465):1–399.

———. 1958. Bemerkungen über die Warane Australiens. Senck. Biol. 47:229–64.

———. 1962. *Philippinosaurus*, eine neue Untergattung von *Varanus*. Senck. Biol. 43(5): 331–33.

———. 1971. Unerwartete Bananenfressen unter Reptilien. Salamandra 7:39–40.

Miller, M. R. 1969. The cochlear duct of lizards. Proc. Calif. Acad. Sci. 33(11):255–359.

Milstead, W. W. 1961. Observations of the activities of small animals (Reptilia and Mammalia) on a small quadrat in southwestern Texas. Am. Midl. Nat. 65:127–38.

Milton, K. 1979. Factors influencing leaf choice by howler monkeys: A test of some hypotheses of food selection by some generalist herbivores. Am. Nat. 114:362–78.

Minnich, J. E., and J. E. Shoemaker. 1970. Diet, behavior, and water turnover in the desert iguana, *Dipsosaurus dorsalis*. Am. Midl. Nat. 84:476–509.

Minton, S. A. 1966. A contribution to the herpetology of West Pakistan. Bull. Am. Mus. Nat. Hist. 134:27–184.

Moberly, W. R. 1968a. The metabolic responses of the common iguana, *Iguana iguana*, to walking and diving. Comp. Biochem. Physiol. 27:21–32.

———. 1968b. The metabolic responses of the common iguana, *Iguana iguana*, to activity under restraint. Comp. Biochem. Physiol. 21:1–20.

Moehn, L. D. 1984. Courtship and copulation in the Timor monitor, *Varanus timorensis*. Herp. Rev. 15(1):14–16.

Mohr, C. O., and W. A. Stumpf. 1966. Comparison of methods for calculating areas of animal activity. J. Wildl. Mgmt. 30(2):293–304.

Molengraaf, G. A. F., and H. A. Brouwer. 1929. Over de bergvormende in het gebeid der boogvormige eil andenrecksen van het osstelijk dedeelte van den O. I. archipel. Proc. Akad. Wet. (Amsterdam) 24:768–69.

Morris, D. J. 1956. The function and causation of courtship ceremony. *In* R. Grass (ed.). L'Instinct dans le Compartement des Animaux et de l'Homme, pp. 261–87. Paris: Masson and Cie.

———. 1962. The behavior of the acouchi (Myoprocta pratti) with special reference to scatter hoarding. Proc. Zool. Soc. London 139(4):701–32.

Morrison, D. W. 1978. On the optimal searching strategy for refuging predators. Am. Nat. 112(987):925–34.

Morton, E. S. 1973. On the evolutionary advantages and disadvantages of fruit-eating in tropical birds. Am. Nat. 107(953):8–22.

Moss, R. 1977. The digestion of heather by red grouse during the spring. Condor 79:471–77.

Moviser, H. L., Jr. 1944. Early man and Pleistocene stratigraphy in southern and eastern Asia. Pap. Peabody Mus. (Howard Univ.) 19:1–125.

Mueller, D., D. Dombois, and H. Ellenberg. 1974. Aims and Methods of Vegetation Ecology. New York: John Wiley and Sons. 547 p.

Muller, L. 1905. Der westafrikanische Steppenwaran (*Varanus exanthematicus* Bosc.). Bl. Aquar. Terr. Kunde 16:266–68.

———. 1923. Aquarium and Reptilien Haus der Zoologischen Gartens aus Frankfurt. Bl. Aquar. Terr. Kunde 20:289–91, 308–13, 324–29.

Murphy, J. B., and L. A. Mitchell. 1974. Ritualized combat behavior of the pygmy mulga monitor lizard, *Varanus gilleni* (Sauria: Varanidae). Herpetologica 30(1):90–97.

Musser, G. G. and L. K. Gordon. 1981. A new species of *Crateromys* (Muridae) from the Philippines. J. Mamm. 62(3):513–25.

Nagy, K. A. 1973. Behavior, diet, and reproduction in the desert lizard, *Sauromalus obesus*. Copeia 1973(1):93–102.

Ng, F. S. P. 1972. Tree Flora of Malaysia. Vol. 1. Kuala Lumpur: Longman. 336 p.

Ng, F. S. P., and H. S. Loh. 1975. Flowering-to-fruiting periods of Malaysian trees. Malay. For. 37(2):127–32.

Noble, G. K. 1937. Sense organs involved in courtship of *Storeria, Thamnophis,* and other snakes. Bull. Amer. Mus. Nat. Hist. 73:673–725.

Norris, K. S. 1953. The ecology of the desert iguana *Dipsosaurus dorsalis*. Ecology 34(2):265–87.

Oden, N. 1977. Partitioning dependence in non-stationary behavioral sequences. *In* B. A. Hazlett (ed.). Quantitative Methods in the Study of Animal Behavior, pp. 203–20. New York: Academic Press.

Oesman, H. 1967. Breeding behavior of *V. komodoensis* at the Jogjakarta Zoo. Int. Zoo. Yearbk. 7:10–11.

Oldfield, T. 1898. On the mammals obtained by Dr. John Whitehead during his recent expedition to the Philippines. Trans. Zool. Soc. London, 14:337–412.

Oosting, P. 1958. A Study of Plant Communities. San Francisco: W. H. Truman and Co. 440 p.

Oppenheimer, J. R. 1982. *Cebus capucinus*: Homing, population diagnosis, and interspecific relationships. *In* E. G. Leigh, Jr., A. S. Rand, D. M. Windsor (eds.). The Ecology of a Tropical Forest, pp. 253–72. Washington: Smithsonian Inst. Press.

Owen, R. 1840–45. Odontography; or a treatise on the comparative anatomy of the teeth . . . in . . . vertebrate animals. London: Taylor and Co. 872 p.

Palaypayon, W. 1977. A biological perspective on deforestation. Canopy 3(1977):13–14.

Panigel, M. 1956. Contribution a l'étude de l'ovoviviparité chez les reptiles: Gestation et parturition chez le lézard vivipare *Zootoca vivipara*. Ann. Sci. Nat. 18:569–668.

Parker, G. A. 1974. Courtship persistence and female-guarding on male time investment strategies. Behavior 48:157–84.

Parker, W. S., and E. R. Pianka. 1975. Comparative ecology of populations of the lizard *Uta stansburiana*. Copeia 1975(4):615–32.

Parmentier, C. J. 1976. The natural history of the Australian freshwater turtle *Chelodina longicollis* Shaw (Testudinata: Chelidae). Ph.D. dissertation, Univ. New England, Perth, Australia. 210 p.

Parr, M. J., T. J. Gaskell, and B. J. George. 1968. Capture-recapture methods of estimating animal numbers. J. Biol. Educ. 2:95–117.

Parry, N. E. 1932. Some notes on the water monitors in the Garo Hills, Assam. J. Bombay Nat. Hist. Soc. 35:901–5.

Patanne, E. D. 1972. The Philippines in the World of Southeast Asia. Quezon City: Enterprise Publ. 218 p.

Payne, J. B., 1978. Competitors. *In* D. J. Chivers (ed.). Malayan Forest Primates, pp. 261–330. New York: Plenum Press.

Peters, L. L. 1969. Zum usten mal ingefanganschaft einblage und Schlupf von *Varanus spenceri*. Aquar. Ten. 16:306–7.

Peyer, B. 1929. Das Gebiss von *Varanus niloticus* L. und *Dracaena guianensis* Daud. Rev. Suisse Zool. 36:112–45.

Phillips, E. A. 1959. Methods of vegetation study. New York: Holt-Dryden Press. 258 p.

———. 1964. Field Ecology. Boston: D. C. Heath and Co. 100 p.
Pianka, E. R. 1968. Notes on the biology of *Varanus eremius*. W. Aust. Nat. 11(3):39–44.
———. 1969a. Notes on the biology of *Varanus caudolineatus* and *Varanus gilleni*. W. Aust. Nat. 11(3):76–82.
———. 1969b. Sympatry of desert lizards (*Ctenotus*) in western Australia. Ecology 50: 1012–30.
———. 1969c. Notes on the biology of *Varanus gouldi flavirufus*. W. Aust. Nat. 11(6):141–44.
———. 1970. Comparative autecology of the lizard *Cnemidophorus tigris* in different parts of its geographic range. Ecology 51:703–20.
———. 1971. Notes on the biology of *Varanus tristis*. W. Aust. Nat. 11(7):180–83.
———. 1978. Evolutionary Ecology, 2d ed. New York: Harper and Row. 434 p.
Pianka, E. R., and W. S. Parker. 1975. Age-specific reproductive tactics. Amer. Nat. 109(968): 453–64.
Pianka, E. R., and H. D. Pianka. 1970. The ecology of *Moloch horridus* (Lacertilia: Agamidae) in western Australia. Copeia 1970(2):90–103.
Piazzini, G. 1960. The Children of Lilith. New York: E. P. Dutton and Co. Translation. 192 p.
Pijl, L. van der. 1957. The dispersal of plants by bats (Chiropterochary). Acta. bot. nedrl. 6:291–315.
Pilsbry, H. A., and J. C. Bequaert. 1927. The aquatic mollusks of the Belgian Congo. Bull. Am. Mus. Nat. Hist. 52(2):69–602.
Pitelka, F. A. 1942. Territory and related problems in North American hummingbirds. Condor 44:189–204.
Poore, M. E. D. 1964. Integration in the plant community. J. Ecol. 52(Suppl.):213–16.
———. 1968. Studies in Malaysian rain forest. I, The forest on Triassic sediments in Jengka Forest Reserve. J. Ecol. 56:143–96.
Pough, F. H. 1973. Lizard energetics and diet. Ecology 54:837–44.
Prater, S. H. 1965. The Book of Indian Animals, rev. ed. Bombay: Bombay Nat. Hist. Soc. 323 p.
Pregill, G. 1984. Durophagous feeding adaptations in an amphisbaenid. J. Herp. 18(2):186–91.
Presst, I. 1971. An ecological study of the viper *Vipera berus* in southern Britain. J. Zool., London, 164:373–418.
Procter, J. B. 1929. On a living Komodo dragon (*Varanus komodoensis* Ouwens) exhibited at the scientific meeting, October 23, 1928. Proc. Zool. Soc. London 23:1017–19.
Pulliam, H. R. 1975. On the theory of optimal diets. Am. Nat. 108:59–79.
Rabor, D. 1977. Philippine birds and mammals. Quezon City: Univ. Philippines Press. 283 p.
Raemaekers, J. J., F. P. G. Aldrich-Blake, and J. B. Payne. 1978. The forest. *In* D. J. Chivers (ed.). Malayan Forest Primates, pp. 29–38. New York: Plenum Press.
Rand, A. S. 1967a. The adaptive significance of territoriality in iguanid lizards. *In* W. W. Milstead (ed.). Lizard Ecology: A Symposium, pp. 106–15. Austin: Univ. Texas Press.
———. 1967b. Predator-prey interactions and the evolution of aspect diversity. Atlas Simposio soli a Biotica Amazonica 5:73–83.
———. 1978. Reptilian arboreal foliovores. *In* G. C. Montgomery (ed.). The Ecology of Arboreal Foliovores, pp. 115–22. Washington: Smithsonian Inst. Press.
Rapport, D. J. 1971. An optimization model of food selection. Am. Nat. 105:575–87.
Real, L. A. 1980. On uncertainty and the law of diminishing returns in evolution and behavior. *In* J. E. R. Staddon (ed.). Limits to Action: The Allocation of Individual Behavior, pp. 37–64. New York: Academic Press.
Redford, K. H., G. A. Bouchardet da Fonseca, and T. E. Lacher. 1984. The relationship between frugivory and insectivory in primates. Primates. 25(4):433–40.
Regal, P. J. 1978. Behavioral differences between reptiles and mammals: An analysis of activity and mental capabilities. *In* N. Greenberg and P. D. McLean (eds.). Behavior and Neurology

of Lizards, pp. 183–202. Washington: Natl. Inst. Mental Health, U.S. Dept. HEW No. 77-491.

Reichenbach-Klinke, H., and E. Elkan. 1965. Principal Diseases of Lower Vertebrates. Vol. 3, Diseases of Reptiles. Hong Kong: T. F. H. Publ. Inc. 600 p.

Reinholz, H. 1923. Über die Befestigen der Zähne von *Varanus niloticus*, ein Beiträg zur Frage nach der Herkunft des Zementes. Jen. Z. Naturweisse 59:53–64.

Respetti, W. E. 1933. Structure of the earth's outer crustal layer in the region of the Philippine Islands. Proc. 5th Pacific Sci. Congr. 3:2527–32.

Richards, P. W. 1952. The Tropical Rain Forest. Cambridge: Cambridge Univ. Press. 401 p.

Richter, H. 1933. Das Zungenbein und seine Muskulatur bei den *Lacertilia vera*. Jen. Z. Naturwiss. 66:395–480.

Rick, C. M., and R. I. Bowman. 1961. Galapagos tomatoes and tortoises. Evolution. 15:407–17.

Ridley, H. N. 1930. The dispersal of plants throughout the world. Kent: Reeve Publ., Inc. 744 p.

Rieppel, O., and L. Labhardt. 1979. Mandibular mechanics in *Varanus niloticus* (Reptilia: Lacertilia). Herpetologica 35:158–63.

Rixford, L. N. 1912. Fructification of the fig by *Blastophaga*. J. Econ. Entomol. 5:349–55.

Robinson, M. H., and B. Robinson. 1970. Prey caught by a sample population of the spider *Argiope argentata* (Araneae: Araneidae) in Panama: A year's census data. Zool. J. Linn. Soc. 49:345–58.

Rose, F. L., and F. W. Judd. 1982. Biology and status of Berlandier's tortoise (*Gopherus berlandieri*). *In* R. B. Bury (ed.). North American Tortoises: Conservation and Ecology, pp. 57–70. U.S. Dept. Inter., Fish Wildl. Serv., Rept. 12.

Roux, R. 1936. Les grandes Sauriens en captivité. Terre et Vie (Nov.–Dec.):1–11.

Rowe, J., and J. Janulaw. 1971. A noteworthy conservation achievement. Internatl. Turtle Tort. Soc. J. (Aug.–Oct.):20–30.

Rudran, R. 1978. Sociology of the blue monkeys (*Cercopithecus mitis stuhlmanni*) in the Kibale Forest, Uganda. Smith. Contrib. Zool. (249):1–32.

Rumney, G. R. 1968. Climatology and the World's Climate. New York: Macmillan and Co. 318 p.

Saderra, M., and M. Smith. 1913. The relation of seismic disturbances in the Philippines to the geologic structure. Phil. J. Sci. 8:199–252.

St. Girons, H., and M. C. St. Girons. 1959. Escape vital domaine et territoire chez les vertebres terrestres (Reptiles et Mammifères). Mammalia 23:448–76.

Sajise, P. 1981. Final report of the committee on environmental aspects of forestry policies. Forestry Policy Review Consultative Workshop, Baguio. Mimeographed. 15 p.

Salvoza, F. M., and M. Lagrimas. 1940. Checklist of the trees of the Philippines, pts. 1–4. Phil. J. Forestry 3(4):477–547, 4(1):63–104, 4(2):191–218, 4(3):285–313.

Samson, J. A. 1979. The Geo-physical History and Geomorphology of the Philippines Archipelago. Caonatan City, Philippines: Dorman Press. 62 p.

Sanderson, G. C. 1966. The study of mammal movements—a review. J. Wildl. Mgmt. 30(2):215–35.

Scherer, J. 1903. Der Nilwaren (*Varanus niloticus* L.). Natur u. Haus 11:249–51.

Schimper, A. F. W. 1903. Plant Geography Upon a Physiological Basis. Oxford: Oxford Univ. Press. 239 p.

Schmidt, K. P. 1927. The reptiles of Hainan. Bull. Am. Mus. Nat. Hist. 54:395–465.

Schmidt-Nielson, K. 1964. Desert Animals. Oxford: Oxford Univ. Press. 326 p.

———. 1975. Animal Physiology: Adaptation and Environment. London: Cambridge Univ. Press.

Schnell, R. 1971. Introduction à la Phytogéographie des pays Tropicaux. 2 vols. Paris: Gauthier-Villars. 951 p.

Schoener, T. W. 1969a. Models of optimal size for solitary predators. Am. Nat. 103:277–313.
―――. 1969b. Optimal size and specialization in constant and fluctuating environments: An energy-time approach. Brookhaven Symp. Biol. 22:103–14.
―――. 1971. The theory of feeding strategies. Ann. Rev. Ecol. Syst. 2:369–404.
Schumacher, G-H. 1973. The head muscles and hyolaryngeal skeleton of turtles and crocodilians. In C. Gans and T. Parsons (eds.). Biology of the Reptilia, vol. 4, pp. 101–99. London: Academic Press.
Scott, J. P. 1950. Methodology and techniques for the study of animal societies. Ann. N. Y. Acad. Sci. 51(6):1001–1122.
Semlitsch, R. D., and J. W. Gibbons. 1978. Reproductive allocation in the brown water snake, *Natrix taxispilota*. Copeia 1978(4):721–23.
Sexton, O. J., E. P. Ortleb, L. M. Hathaway, R. E. Ballinger, and P. Licht. 1971. Reproductive cycles of three species of anoline lizards from the Isthmus of Panama. Ecology 52:201–15.
Simon, C. A. 1975. The influence of food abundance on territory size in the iguanid lizard *Sceloporus jarrovi*. Ecology 56:993–98.
Siniff, D. B., and J. R. Tester. 1965. Computer analysis of animal movement data obtained by telemetry. BioScience 15(2):104–8.
Slaven, J. L., and H. A. Marlett. 1980. Effect of refined cellulose on apparent energy, fat and nitrogen digestibilities. J. Nutrition 110:287–97.
Smith, H. C. 1931. The monitor lizards of Burma. J. Bombay Nat. Hist. Soc. 34:367–73.
Smith, K. K. 1982. An electromyographic study of the function of the jaw adducting muscles in *Varanus exanthematicus* (Varanidae). J. Morph. 173:137–58.
Smith, K. K., and W. L. Hylander. 1985. Strain gauge measurement of mesokinetic movement in the lizard *Varanus exanthematicus*. J. Exp. Biol. 114:53–70.
Smith, M. A. 1932. Some notes on the monitors. J. Bombay Nat. Hist. Soc. 35:615–19.
Smith, W. D. 1915. Notes on a geological reconnaissance of the Mountain Provinces, Luzon. Phil. J. Sci. 4:1–15.
―――. 1924. Geology and mineral resources of the Philippine Islands. Phil. J. Sci. 8:1–215.
Smythe, N. 1970. Relationships between fruiting seasons and seed dispersal methods in a Neotropical forest. Am. Nat. 104(935):25–35.
―――. 1974. Biological monitoring data—insects. In R. W. Rubinoff (ed.). 1973 Environmental Monitoring and Base Line Data, pp. 70–115. Washington: Smithsonian Inst. Press.
Smythe, N., E. W. Glanz, and E. G. Leigh, Jr. 1982. Population regulation— some terrestrial frugivores. In E. G. Leigh, Jr., A. S. Rand, and D. M. Windsor (eds.). The Ecology of a Tropical Forest, pp. 227–38. Washington: Smithsonian Inst. Press.
Snow, D. W. 1962. The natural history of the oil-bird *Steatornis caripensis*. Zoologica 47:199–221.
―――. 1965. A possible selective factor in evolution of fruiting seasons in a tropical forest. Oikos 15:274–81.
―――. 1971. Evolutionary aspects of fruit-eating by birds. Ibis 113: 194–202.
Snow, D. W., and B. K. Snow. 1964. Breeding seasons and annual cycles of Trinidad land-birds. Zoologica 49(1):1–39.
Sokol, O. M. 1967. Herbivory in lizards. Evolution 21:192–94.
―――. 1971. Lithography and geography in reptiles. J. Herp. 5:69–71.
Sokolov, V. E., V. P. Sukhov, and Y. M. Chernyshou. 1975. A radiotelemetric study of diurnal fluctuations of body temperature in the desert monitor (*Varanus griseus*). Zool. Zh. 54: 1347–56.
Spatz, P. 1931. Wüstenwaran (*Varanus griseus* Daudin). Das Aquar. 10: 160–62.
Spears, G. 1977. Predation on colubrid snakes. Ph.D. dissertation, Univ. Florida, Gainesville. 50 p.

Literature Cited

Spellerberg, I. F. 1971. Thermoregulation and temperature tolerances in the *Sphenomorphus cuoyi* complex (Lacertilia: Scincidae) of south-east Australia. Ph.D. dissertation, La Trobe Univ., Canberra, Australia. 218 p.
Stamps, J. A. 1973. Displays and social organization in female *Anolis aeneus* (Sauria: Iguanidae). Copeia 1973(2):264–71.
Stebbins, R. C., and R. E. Barwick. 1968. Radiotelemetric study of thermoregulation in a lace monitor. Copeia 1968(3):541–47.
Stirling, E. C. 1912. Observations on the habits of the large central Australian monitor (*Varanus giganteus*), with a note on the "fat bodies" of this species. Trans. Proc. Soc. (So. Australia) 36:26–33.
Stiven, A. E. 1970. Respiration in the snail *Caracolus caracolla* and an estimate of the relative density and biomass of litter snails. *In* H. T. Odum and R. F. Pigeon (eds.). A Tropical Rain Forest. U.S. Atom. Energy Comm. 3(1):I-65–I-67.
Styles, E. W. 1980. Patterns of fruit presentation and seed dispersal in bird-disseminated woody plants in the eastern deciduous forest. Am. Nat. 116(5):670–713.
Sulit, M. D. 1918. Plants to be avoided or handled with care in the forest. Phil. J. Fores. 3(2):177–89.
Swain, T. 1976. Angiosperm-reptile co-evolution. *In* A. d'A. Bellairs and C. B. Cox (eds.). Morphology and Biology of Reptiles, pp. 107–22. London: Academic Press.
Symington, C. F. 1943. Foresters manual of dipterocarps. Malay. For. Rec. 16:1–41.
Symons, C. T. 1912. Note on the arboreal habits of the Kabaragoya (*Varanus salvator*) and the Talagoya (*V. bengalensis*). Spol. Zeyl. 8:65–66.
Taylor, E. H. 1922. The lizards of the Philippine Islands. Phil. Bur. Sci. Monogr. (17):13–269.
———. 1934. Philippine land mammals. Phil. Bur. Sci. Monogr. (30):1–548.
Telford, S. R. 1969. The ovarian cycle, reproductive potential and structure in a population of the Japanese lacertid *Tachydromus tachydromoides*. Copeia 1969(3):548–67.
Terbough, J., and J. M. Diamond. 1970. Niche overlap in feeding assemblages of New Guinea birds. Wilson Bull. 82:29–52.
Thilenius, G. 1897. Der Farbenwechsel von *Varanus griseus* bei *Uromastix acanthurus* und *Agama inermis*. Morph. Jarb. 7:515–45.
Throckmorton, G. 1973. Digestive efficiency in the herbivorous lizard *Ctenosaura pectinata*. Copeia 1973(3):431–35.
Tinkle, D. W. 1967. The life and demography of the side-blotched lizard, *Uta stansburiana*. Misc. Publ. Mus. Zool., Univ. Michigan 132:1–182.
———. 1969. The concept of reproductive effort and its relation to the evolution of the life histories of lizards. Am. Nat. 103(933):501–16.
Tinkle, D. W., and N. F. Hadley. 1973. Reproductive effort and winter activity in the viviparous montane lizard, *Sceloporus jarrovi*. Copeia 1973(2):272–77.
Tinkle, D. W., D. McGregor, and D. Sumner. 1962. Home range ecology of *Uta stansburiana stejnegeri*. Ecology 43(2):223–29.
Tinkle, D. W., H. M. Wilbur, and S. G. Tilley. 1970. Evolutionary strategies in lizard reproduction. Evolution 24(1):55–74.
Tomlinson, P. B., and M. H. Zimmermann. 1978. Tropical Trees as Living Systems. Cambridge: Cambridge Univ. Press. 128 p.
Townsend, C. R. 1974. Mucus trail following by the snail *Biomphalaria glabrata* (Say). Anim. Beh. 22:170–77.
Troyer, K. 1982. Transfer of fermatative microbes between generations in a herbivorous lizard. Science 216:540–42.
———. 1983. The biology of iguanine lizards: present status and future directions. Herpetology 39(3): 317–28.

Tubbs, A. A. 1972. Non-specific stress in juvenile lizards, *Sceloporus undulatus garmani*. M. S. thesis, Kansas State Univ., Manhattan. 203 p.
Tullock, G. 1970. Switching in general predators. Bull. Ecol. Soc. Am. 51:21–23.
———. 1971. The coal tit as a careful shopper. Am. Nat. 105:77–80.
Turner, F. B., G. A. Hoddenbach, P. A. Medica, and J. R. Lannom. 1969. The demography of the lizard *Uta stansburiana* Baird and Girard, in southern Nevada. J. Anim. Ecol. 39: 505–19.
Underwood, E. E. 1970. Quantitative stereology. Reading, MA: Addison-Wesley Publ. Co., Inc. 238 p.
Upadhyay, S. N., and S. S. Gukaya. 1972. Histochemical observation on the interstitial gland (or Leydig) cells of a lizard testes. Gen. Comp. Endocr. 19:88–95.
Vallois, H. 1955. Ordre des Primates. *In* P. P. Grasse (ed.). Traite de Zoologie 17, Mammifères, pp. 1873–75, 1945–47. Paris: Masson.
VanDevender, R. W. 1982. Growth and ecology of spiny-tailed and green iguanas in Costa Rica, with comments on the evolution of herbivory and large body size. *In* G. M. Burghardt and A. S. Rand (eds.). Iguanas of the World: Their Behavior, Ecology, and Conservation, pp. 162–83. Park Ridge, NJ: Noyes Publ.
Vellayan, S. 1981. The nutritive value of *Ficus* in the diet of Lar gibbon (*Hylobates lar*). Malay. Appl. Biol. 10(2):177–81.
Vernet, R. 1977. Recherches sur l'on écologie de *Varanus griseus* Daudin (Reptilia: Sauria: Varanidae), dans les ecosystèmes sableux du Sahara nord-occidental (Algerie). Ph.D. dissertation, Univ. Pierre et Marie Curie, Paris. 117 p.
Versluys, J. 1910. Streptostylie bei Dinosaurieren. Zool. J. Abt. Anat. 30:177–258.
Vitt, L. J., and J. D. Congdon. 1978. Body shape, reproductive effort, and relative clutch mass in lizards: Resolution of a paradox. Am. Nat. 112(985):595–608.
Vitt, L. J., and H. J. Price. 1982. Ecological and evolutionary determinants of relative clutch size in lizards. Herpetologica 38(1):237–55.
Vogel, P. 1979. Zur Biologie des Bindeswarens (*Varanus salvator*) in Westjavanischen Naturschutsgebiet Ujung Kulon. Ph.D. dissertation, Univ. Basel, Switzerland. 139 p.
Vogel, Z. 1929. Aus dem Leben der Reptilien. Prague. 116 p.
Volsoe, H. 1944. Structure and seasonal variation of the male reproductive organs of *Vipera berus* (L.). Spolia Zool. Mus. Haunensis 5:1–171.
Von Koenigswald, G. H. R. 1935. Die fossile Säugetierfauna Javas. Proc. Kon. Akad. Wetensch. (Amsterdam) 38:188–98, 872–79.
———. 1939. The relationship between the fossil mammalian fauna of Java and China, with special reference to early man. Peking Nat. Hist. Bull. 13:293–98.
———. 1956. Fossil mammals from the Philippines. Far East. Pre-hist. Anthrop. Cong.; Nat. Res. Comm. (Phil.), Quezon City: Univ. Philippines Press. 156 p.
Waite, E. R. 1927. The Fauna of Kangaroo Island, South Australia. Part 3, The Reptiles and Amphibians. Trans. Soc. So. Australia 51:326–29.
———. 1929. The Reptiles and Amphibians of South Australia. Adelaide: Government Printer. 183 p.
Walker, E. P. 1964. Mammals of the World. 2d ed. 2 vols. Baltimore: Johns Hopkins Press. 1500 p.
Walter, H. 1971. Ecology of Tropical and Subtropical Vegetation. Edinburgh: Oliver and Boyd. 539 p.
———. 1973. Vegetation of the Earth. New York: Springer-Verlag, Inc. Translation. 228 p.
Weaver, W. J. 1970. Courtship and combat behavior in *Gopherus berlandieri*. Bull. Fla. State Mus., Biol. Sci. 15(1):1–43.

Webb, G., H. Heatwole, and J. DeBavay. 1971. Comparative cardiac anatomy of the Reptilia, 1. The chambers and septa of the varanid ventricle. J. Morph. 134:335–50.

Wegner, R. N. 1922. Der Stutzknochen, Os nariae, in der Nasenhöhle bei den Gürteltieren, Dasypodidae, and seine homologen Gebilde bei Amphibien, Reptilien und Monotremen. Morphol. Jb. 51(43):413–92.

Wells, M. J., and S. K. L. Buckley. 1972. Snails and ticks. Anim. Behav. 20:345–55.

Werner, F. 1893. Bemerkungen über Reptilien und Batracheir aus dem tropischen Asien und von der Sinae-Halbinsel. Verh. Zool.-Bot. Ges. Wien 43:349–59.

———. 1908. Ergibnisse der mit Subvention aus der Erbschaft Treitl unternommenen zoologischen Forschungsreise Dr. Fran Werner's nach dem Aegyptischen Sudan und Nord-Uganda. XII. Die Reptilien und Amphibien. Akad. Wis. Wien, Math.-Nat. 116(1):1–213.

West, A. P., and W. H. Brown. 1921. Philippine resins, gums, seed oils and essential oils. In Minor Products of Philippine Forests, vol. 1. Bur. For. (Manila) Bull. 22:5–224.

Westoby, M. 1974. An analysis of the diet selection by large generalized herbivores. Am. Nat. 108:290–304.

Wheatley, B. P. 1976. The ecology strategy of the long-tailed macaque, *Macaca fascicularis,* in the Kutai Nature Reserve, Kalimantan. Frontier 5:27–32.

———. 1978. Foraging patterns in a group of long-tailed macaques in Kalimantan Timur, Indonesia. In D. J. Chivers and J. Herbett (eds.). Recent Advances in Primatology, vol. 1, Behaviour, pp. 347–49. London: Academic Press.

Whitaker, A. H. 1968. The lizards of Poor Knights Island, New Zealand. New Zealand J. Sci. 11(4):623–51.

Whitford, H. N. 1906. The vegetation of the Lamao forest reserve. Phil. J. Sci. 1(1):373.

———. 1911. The forests of the Philippines, pts. 1, 2. Bur. For. (Manila) Bull. 10:1–401.

Whitmore, T. C. 1972. Burseraceae. In Tree Flora of Malaya: A Manual for Foresters, vol. 1, pp. 124–36. Kuala Lumpur: Longman.

———. 1975. Tropical Rain Forests of the Far East. Oxford: Clarendon. 381 p.

Wiewandt, T. A. 1977. Ecology, behavior, and management of the Mona Island ground iguana, *Cyclura stejnegeri.* Ph.D. dissertation, Cornell Univ., Ithaca, NY. 398 p.

Wight, N. 1938. Field and Laboratory Techniques in Wildlife Management. Ann Arbor: Univ. Michigan Press. 126 p.

Willey, A. 1909. Ambalantota to Hambegamuwa. Spol. Zeyl. 6:39–53.

Willis, B. 1937. Geological observations in the Philippines Archipelago. Phil. Isl. Nat. Res. Counc. Bull. B 13:1–18.

———. 1939. The Philippine Archipelago: An illustration of continental growth. Proc. 6th Pacific Sci. Congr. 1:125–200.

———. 1944. Philippine earthquakes and structure. Bull. Seismol. Soc. Am. 34:669–81.

Wilson, E. O. 1975. Sociobiology. Cambridge: Harvard Univ. Press. 697 p.

Wilson, E. O., and W. H. Bossert. 1963. Chemical communication among animals. Rec. Prog. Hormone Res. 19(4):673–716.

Wilson, V. J. 1968. The leopard tortoise, *Testudo pardalis babcocki* in eastern Zambia. Arnoldia 40(3):1–11.

Wong, Y. K. 1967. Some indications of total volume of wood per acre in lowland dipterocarp forest. Malay. For. Dept., Res. Pamph. (53):1–16.

Wyatt-Smith, J. 1953. Malaya forest types. Malay. Nat. J. 7:45–55, 91–98.

Zander, A. 1895. Einige trankapische Reptilien, *Varanus griseus.* Zool. Gart. 36:298–301.

Zavattari, E. 1908. Matererioli per lo studio dell'osso ioide dei sauri. Atti. Acc. Sci. Torino 43:1138–45.

Index

Abbreviations, App. 3
Activity
  body temperature and, 89, 107, 109–11, 244, figs. 5–12, 5–13, 5–17
  diurnal (daily) pattern, 93, 97 ff., 105, 109 ff., 143, 254, figs. 5–3, 5–4, 5–6, 5–7
  effect of rain on, 114, 116, 143, 145, figs. 5–15, 9–12
  effect of temperature on, 89, 106 ff., 244, figs. 5–12, 5–13, 5–17
  effect of wind on, 114, 145
  and food, 191–92, 219 ff., 224, 256, 288
  light and, 81, 105 ff., 109, fig. 5–7
  nocturnal, 142
  patterns of, 109 (*see also* Activity, diurnal; Activity, seasonal; Cycles; Peak activity periods)
  range, home, 134 ff., 142, 244, 252
    compared to theoretical size, 135
    for other species, 135
    shape, 142, fig. 6–4
    size, 135, fig. 6–4
    technique to estimate, 135
    vertical dimension, 135
  seasonal cycles, 105, 135, 257
  sex and, 105, 116
  substrate and, 88, 114–15, 143–45
  See also Movement
Adaptations
  arboreal, 20, 22, 156, 330
  climbing, 20, 22, 156, 330
  crushing snails, 193, 237, 334
  feeding, 42 ff., 47, 179 ff., 236 ff., 334
  gastrointestinal, 58 ff., 260
  swallowing, 42, 334, fig. 3–4
Age
  at maturity, 159
  longevity, 151
Aggression
  in combat, 149, 294, 299
  in courtship, 146, 294, 310, 313
  in protection, 152
Aglaia harmsiana, 221, 227 ff., 247, 258, 276, App. 1
Agonistic behavior. *See* Aggression
Agriculture, and butaan conservation, 364 ff.
Air
  temperature, 78, 89, 91, 244
  humidity, 88
Albay (Province, Luzon), 125, 131
Altitude
  and range, 133
  and vegetation changes, 81 ff., 133, fig. 4–7
Amblyomma, 279 ff.
Amblyrhinchus, 179, 239
Amplexus, 310 ff., fig. 13–2
Anatomy
  body form, 17–20
  feet, 24–27, 155–56
  gastrointestinal, 51 ff., 58 ff., fig. 3–7
  head, 33 ff., 39 ff., fig. 3–3
  musculature, 39 ff., 193, fig. 3–3
  skeleton, 42 ff., 51, 334, figs. 3–2, 3–6
  skull, 33 ff., 42–43, 334, fig. 3–2
  teeth, 47 ff., 237, 334, fig. 3–5

Anglosaurus, 178
Anolis, 310
Appendages, use of
  in combat, 149–50, fig. 7-2
  in courtship, 309 ff., fig. 13-2
Arbohabitat, 87 ff., 142, 243–44, 254, figs. 6-6, 7-4
Arboreal habits, 23, 155–56. *See also* Activity, substrate and
Area-concentrated search, 246–47, 253
Area tenure patterns. *See* Home range
Asplenium, 142, App. 1
Aurora (Province, Luzon), 128

Barometric pressure, 79
Barriers, geographic, 125 ff., 133
Basking, 94, 105, 254, fig. 7-4
Bats, as possible fruit competitors, 186, 238, 268, 272, App. 2
Beetles, as food, 207, 261
Behavior
  in combat, 293, 298–99, figs. 2-8, 2-9, 14-1
  in refuges, 142 ff.
  in reproduction, 169 ff., 309 ff., figs. 13-1, 13-2, 13-4
  when approached, 152, 293
Biomass, estimates of
  butaans, 10–11, 158
  fruit in different habitats, 7, 9, 233 ff., 248 (*see also* Fruits)
Bipedal
  movement, 155–56
  position, 155–56, 298
Birds
  as food, 187, 241
  as frugivores, 239, 258, 286, 288, App. 2
  as predators, 268, 280
Biting, 294
Blood parasites, 11, 281
Boar, 280
  as possible fruit competitors, 269 ff., App. 2
Body fat, figs. 3-11, 3-12
  and food, 63 ff., 74, 256–57
  and reproduction, 170, 175
Body part proportions, 18 ff., 333, fig. 2-4
Body positions
  in basking, 94, 105, fig. 7-4
  in combat, 298 ff.
  in courtship, 298 ff., fig. 13-2
  in hiding, 147, 154, 300
Body temperature, 106 ff.
  and activity, 89, 106 ff., 244, figs. 5-12, 5-13, 5-17

  and ambient differentials, 109–10, fig. 5-10
Brachial embrace, 299
Breeding. *See* Courtship
Bulacan (Province, Luzon), 128
Bullimus (rodent), as prey, 187, 209
Burrows, not excavated, 143
Butaans
  biomass, 10–11, 158
  blood parasites of, 11, 281
  color and pattern, 27 ff., 326, 330, pl. 1, figs. 2-8, 2-9, 14-1
  combat in, 149, 294–99
  as common name, 1
  density, 158, 233, 266
  diagnosis, 16
  discovery and history of, 2 ff.
  as energy maximizers, 266
  food requirements of, 182, 186, 233
  form of, 10, 18 ff., 333
  hatchling, size of, 17–18, 175
  intestinal parasites of, 11, 278 ff.
  length, 11, 16–17
  movement, 266 (*see also* Activity; Movement)
  musculature, 39 ff., 193, fig. 3-3
  origin, 329 ff.
  osteoderms, 334
  and prey, 178 ff. (*see also* specific types)
  scalation, 11, 25 ff., 332, figs. 2-6, 2-7
  scale anatomy, 333
  signals and displays, 292 ff.
  size, 11, 16–17
  skeleton, 42 ff., 51, 334, figs. 3-2, 3-4
  skull, 33 ff., 42–43, 334, figs. 3-2, 3-4
  sympatry with water monitor, 15
  teeth, 47 ff., 334, fig. 3-5
  testes, 11, 165
  ticks of, 278 ff.
  as time maximizers, 266
  and water monitors, 15, 18 ff., 179 ff. (*see also* Water monitor)
  viscera, size of, 11, 51 ff., 60 ff., 261, fig. 3-7
  weight of, 16–17, figs. 2-1, 2-2, 2-3
  weights of body parts, compared to water monitor, 20
  young, as predators, 261, 266

Caecum, 56 ff.
Calories, of food, 224–25
Camarines Norte (Province, Luzon), 126, 131
Camarines Sur (Province, Luzon), 125, 131
Camouflage, 31, 280. *See also* Coloration
Canarium

Index

distribution in forest, 133, 220
competition for, 258, 274
as food, 182, 191, 204, 238, 285 ff.
fruit phenology, 229–30, 284
fruit shadow, 230, fig. 10–4
hirsutum, 133, 143, 187, 230 ff., 285, 287, App. 2
ovatum, 204
pattern of fruits on ground, 230
seed germination, 290–91
toxicity, 204, 238, 243
vrieseanum, 221, 229, 232
See also App. 1
Cannibalism, none, 182
Captives, methods of study, 11
Carrion, as food, 264, 268
Caryota
  cumingii, 205, 248, 258, 285–87
  as food, 182, 191–92, 204, 258, 276
  fruit phenology, 227 ff.
  pattern of fruits on ground, 226, 230, 246–47
  Rhumphiana, 258
  See also App. 1
Casting behavior, 253
Catanduanes Island, 73, 122, 131
Cerastes cerastes, 281
Chamaeleo pumilis, 64
Characteristics, individual, 147 ff.
Chemical signals, 333
Chemistry of foods, 192, 201, 224–25, 238, 242–43, 286–87
Civet cat (*Paradoxus hermaphroditus*)
  as competitor, 238, 258, 269, 274 ff.
  density, 276
  food selection, 187, 199, 201, 236, 274 ff.
  habits, 187, 244, 275
  See also App. 2
Claws
  morphology, 24
  use in climbing, 24, 156–57
  use in feeding, 155, 209
Clearance rates, 209, 211, 285 ff.
Climate, methods of study of, 32
Climatic factors, 75
  and feeding, 255 ff.
  and movement, 105 ff., figs. 5–3, 5–4, 5–6, 5–7, 5–15, 9–12
Climbing, use of claws in, 24, 156–57
Cloacal scents, 333
Clutches, 163, 167, 171–72
  See also Eggs
Coactions
  competition for fruit, 258–59, 269 ff., fig. 12–1

interspecific, 269 ff. (*see* specific organism)
intraspecific, 258–59 (*see also* Combat; Courtship)
  with man, 274, 357 ff., 364 ff.
  with *V. salvator,* 258, 264, 267, 277–78
Cochlosyla pithogaster, 216
Coevolution, with plants, 226, 258
Coloration
  of butaans, 27 ff., 326, 330, pl. 1, figs. 2–8, 2–9, 14–1
  of foods, 199
Combat
  behavior, 292, 298–99, fig. 13–3
  and spermatogenesis, 170
  typical sequences, 307–8
  and wounds, 149–50, fig. 7–2
Communities, 81 ff., fig. 4–7
  density of, 253
  See also specific habitats
Commuting distance, 252
Competition, 258 ff., 269 ff., App. 2
Competitors, 258 ff. *See also* specific types
Conservation, 357 ff.
Coprophagy, 239
Copulation, 311
Core area, 134 ff., 142, 244, 247
Corpora lutea, 11–12, 162–63, 171, figs. 8–2, 8–3
Courtship, figs. 13–2, 13–4
  aggressive phases of, 310
  biting in, 310, 313
  cues during, 310 ff., 323, fig. 13–2
  and egg laying, 310, 323
  primers, 323
  releasers, 320
  seasonality of, 310 ff.
  success of, 310
  typical sequences, 312, 313 ff., fig. 13–2
  See also Sexual behavior
Cover density of vegetation, 7, 253
Crabs, as food, 179, 182, 191, 207
Crevices
  temperatures of, 92 ff., 113, fig. 4–11
  use as shelter, 113, 114, 142–43, figs. 6–7, 6–8
Critical thermal maximum (CTM), 106 ff., 254–55, fig. 5–9
  effect on behavior, 109 ff.
Cryptic coloration, 32, 326, 330
Ctenosauria, 239
Cues. *See* Courtship, cues during
Currents, air. *See* Air
Cycles
  daily, 94, 101, 109, 143, 254

Cycles (*continued*)
  fruiting, 227, 289, fig. 10-1
  reproductive, 159 ff., 167-68, figs. 8-1, 8-4, 8-5
  seasonal, 187 ff., 224 ff., 289, fig. 10-1
Cyclophorus, 207, 270
Cyclura carinata
  as carnivore, 233
  as herbivore, 179, 186, 204, 233, 239, 242, 244, 252, 260, 268, 288
Cyclura cornuta, 233
Cyclura cychlura, 186

Daily foraging pattern, 254. *See also* Activity, diurnal
Day, length of, 81, 173
Death, butaan, from heat, 107, 254, fig. 5-9
Deer, as possible fruit competitors, App. 2
Defecation, 212
Defense, 152, 293
Delayed emergence, of hatchlings, 172
Dendrovaranus, 329, 339
Density
  of butaans, 158
  dependency, 253
  methods of estimation, 7 ff.
  of plants, 205, 234, 253, 266
  of prey, 224, 253, 264
Diel activity cycles, 94, 101, 109, 143, 254
Digestion, 210 ff. *See also* Clearance rates
Digging, 155
Dinochloa, 142
Dipterocarp forest, 81 ff., 342, pls. 2, 3. *See also* Mixed dipterocarp forest
Dipterocarpus, 82, App. 1
Diseases, 150-51
Display behavior. *See* Combat; Courtship
Distribution
  of butaans, 81, 88 ff., 122 ff., 342, fig. 6-3
  of food
    animal, 133
    plant, 81 ff., 133, 207, 233, 253, 266
Dominance behavior. *See* Aggression
Dorsal arch, flattening, 294
Doves, as possible fruit competitors, App. 2
Dracaena guianensis, 42, 182, 208
Dracontomelium
  as food, 192, 204, 224-25, 238, 258, 277, 286
  fruit phenology, 229
  fruit shadow, 224-25, 247
Drinking, 68
Drynaria, 142
Dry season, 75, 256. *See also* Season

Eating habits. *See* Feeding; Foods; Predation
Ecological barriers, 88 ff., 133
Ecology and phylogeny, 339
Ecosystems. *See* Communities
Ecotone, 90
Ectoparasites, 278 ff.
  distribution on host, 279-80
  species, 278
Edaphic factors important to butaan, 68 ff.
Eggs, butaan, 167 ff., 170-71, 172 ff.
  of birds as butaan food, 187
Empagusia, as subgenus, 37
Endoparasites, 11, 280 ff.
Enemies, 280
Energetics of food search, 253, 266
Environment, alteration by man, 357 ff., 364 ff.
Escape behavior, 154
Evolution, butaan, 329
Extinction, presumed, of the butaan, 3

Face-off. *See* Aggression
Factors affecting behavior
  environment, 140 ff., figs. 5-3, 5-4, 5-6, 5-7, 5-15, 9-12
  food availability, 205, 233-34, 252, 264, 266 (*see also* Fruits)
  season, 138 ff., 191 ff., 234, 255, 288
  sex (*see* Combat; Courtship)
  social position, 292 (*see also* Aggression)
  temperature, 92 ff., 107 ff., 140 ff., 244, figs. 5-12, 5-13, 5-17
Facultative monogamy, 268
Fat, storage of, 63 ff.
  related to food, 64-65, 224, 257, 288
  and reproductive cycles, 170, 176
  seasonal, 65, figs. 3-11, 3-12
  in sexes, 64-65
Fecal pellets, 212
Feeding
  and egg laying, 171, 191
  habits, 201, 207, 244 ff.
  and ontogeny, 184
  rate, 208
  specializations, 42-43, 196, 239
  technique, 42-43, 196 ff., 244, 252-53
  time spent, 208, 244, 268
  See also Foods; Predation
Feet and claws, use of
  in climbing, 24, 156-57
  in combat, 149-50
  in courtship, 309 ff.
  in foraging, 155, 207
Female reproductive cycle, 160 ff. *See also* Reproduction

Index

Ferns, 142
Ficus, 6, 274, App. 1
　as butaan food, 191, 196, 201, 205, 258, 276–77
　distribution in forest, 82–83
　fruiting phenology, 199 ff., 255, 284–85
Fighting behavior. *See* Aggression
Fire and habitat destruction, 88, 364 ff.
Floral diversity, 81, 233–34, fig. 10-2
Folivory, general, 178 ff., 263
Food(s)
　acceptability, 186–87, 201, 240, 242–43
　amounts eaten, 182 ff., 208
　and body fat, 256–57
　calcium-rich, 192
　calories, 224, 241
　cannibalism, 182
　in captivity, 187, 241
　carrion, 264, 268
　characteristics, 193 ff., 224, 238, 284–86
　chemistry, 192, 201, 224, 238, 242–43, 284–86
　feast or famine, pattern, 187
　item size, 182, 193, 238
　item weight, 182, 193, 238, 240
　as limiting factor, 234, 264
　and lizard size, 182, 238, 240, 241
　local mix important, 233, 242, 257
　location (spacing), 205 ff., 234, 245, figs. 10-3, 10-4
　methods of study, 7
　number of items/gut, 182
　number of species/gut, 182
　oily, 192, 231–32, 238, 240, 247, fig. 9-3
　patches, 253, 266
　percentage of total flora, 234, 238, 240, 244, fig. 10-3
　piracy, 188
　position on parent tree, 199
　protein-rich, 192, fig. 9-3
　rarely ingested, 186–87, 191, 240
　requirements, 182, 186, 233
　retention in gut, 285, fig. 12-2
　ripeness, 187, 204, 242
　seasonal use, 175, 188 ff., 224, 289, figs. 9-1, 9-2, 10-1
　selection, 187–88, 201, 236, 238, 240 ff., 266
　and sex, 182, 188
　species, 179 ff.
　spectrum, 238–40
　state, 187, 266
　sugary, 192, 230, 238, 241–42, 247, 252, fig. 9-3
　toxic, 192, 201, 238, 243, 263, 286 ff.
　utilization ratios, 188, 238
　See also Feeding; Predation
Foraging
　area, 201, 204, 207, 241, 244, 247, figs. 10-3, 10-4
　efficiency, 199, 254 ff.
　periodicity, 254
　persistent patterns of, 240
　speed, 204, 253
　strategy, 204–7, 235 ff.
　use of feet in, 204
　use of snout in, 25, 207, 237
Forbs, App. 1,
Forest communities, 81 ff., fig. 4-8
　butaans, density in, 233
　destruction of, 358, 364 ff., fig. 15-1
　See also Mixed dipterocarp forest
Forest stratification, 81 ff., figs. 4-8, 5-10
Form, ontogenetic. *See* Ontogenetic differences
Frugivory, general, 178 ff., 259 ff.
Fruits
　as food, 179, 201, 207, 232 ff., 241, 259, 262 ff.
　below surface, 207
　chemical analyses, 7, 204
　estimating technique, 9
　fruit shadows, 225 ff., 241, 247, fig. 10-4
　fruit shape and distribution on ground, 225, 241
　patchy distribution, 233 ff., 248–49, figs. 10-6, 10-7
　pulse, 255, fig. 10-6
　seasonally available, 192, 227, 232, 245, 254 ff., 264, 288, fig. 10-1

Gall bladder, 57 ff.
Gaping
　in defense, 293
　at high temperatures, 107, 255
Gastric parasites. *See* Endoparasites
Gastric pellet, 213
Gastrointestinal morphology, fig. 3-7
　and diet, 59 ff.
　lining, 51 ff.
　size, 51 ff., 61
Geckos, 277
Geochelone species, delayed hatching, 172
Geographical origin, 329 ff.
Geographic distribution, 68 ff., 88 ff., 122, 131–32, fig. 6-3
Geologic subsidence, 68 ff., fig. 14-10
Geology, general, 6, 68 ff., figs. 14-7, 14-8,

Geology (*continued*)
    14–9, 14–10. *See also* Miocene;
      Pliocene
Geophorus (mollusc)
    habitat, 207
    as prey, 196
Germination of seeds, 290–91
Gogan grass (*Imperata cylindrica*), 88,
    App. 1
Gopherus, 252
Granivory, 178
Grasslands, caused by fire, 88
Grewia stylocarpa, App. 1
    distribution in forest, 133, 229 ff.
    as food, 182, 191, 224, 238, 277, 286
    fruit phenology, 229
    pattern of fruits on ground, 224
Groundwater, 68
Growth, 151, 260
Gular expansion,
    as display, 296
    in thermoregulation, 107

Habitat
    modification of, 358, 364
    of study area, general, 6, pls. 1–3, fig.
      4–8
Habits, butaans affected by
    plants, 81 ff., 182, 227 ff.
    rain, 114, 116, 143, 145, fig. 5–15
    sounds, 152
    temperature, 89, 106 ff., 244, figs. 5–12,
      5–13, 5–19
    wind, 114, 145
Hatching. *See* Eggs, hatching of
Hatchlings
    olivaceous
      feeding, 240, 266
      weight, 17–18
    salvator, 17–18, 174
Haul-in time, 100, 143, fig. 5–9
Haul-out time, 99–100, 143, fig. 5–12
Head positions, 294 ff.
Hearing. *See* Senses
Heating rates, 106, fig. 5–9
Hemiglypha (mollusc), habitat, 207
Herbivory, general
    in lizards, 178, 263
    in varanids, 179
Hermit crabs (*Coenobitas*), as food, 182
Hierarchy. *See* Social behavior
Hiss, 293
History, of names *V. grayi* and *V. olivaceus,*
    2, 327
Holotypes of *grayi* and *olivaceus,* 327
Home range, 133 ff., 142, 244, 246, fig.
    6–4. *See also* Activity, range; Core area;
      Foraging, area
Hopea, 82, 86, App. 1
Horizontal distribution, 122 ff., fig. 4–7
Hornbills, 258, 288
    as possible fruit competitors, 207, 269 ff.,
      285, App. 2
Humidity, relative
    vertical distribution in forest, 88
    See also Air
Hunger, 188
Hunting. *See* Foraging
Hyoid apparatus, 46–47, 334, fig. 14–3
Hydrosaurus pistulosus, 51 ff., 211, 243,
    261, 263, 267

Iguana, 21, 239, 243, 252
Imperata cylindrica. See Gogan grass
Incubation. *See* Eggs, incubation
Indovaranus, 329, 339
Ingestive behavior
    rate, 268
    and specialization of skull, 42 ff., 196,
      334, fig. 3–4
Injuries
    and behavior, 149, fig. 7–2
    and body parts, 147 ff., figs. 7-1, 7–2
    and sex, 150, fig. 7–1
Insects
    as competitors, 269
    as food, 184, 240, 261, 266
Integumentary scents, 310, 323, fig. 13–5
Interspecific behavior
    coactions (*see* Combat; Courtship)
    competition, 258–59
Intestinal parasites, 11, 56, 269
Intraspecific behavior. *See* Combat; Courtship
Investigative behavior, 155, fig. 13-5

Jumping, 154
Juveniles
    as energy maximizers, 266
    as insectivores, 184, 261, 266

Karyotypes, 329

Lacerta vivipara, 210
Laguna (Province, Luzon), 127, 132
Large intestine, 54–56
Lateral orientation, 294
Learning, food sites, 248
Leaves, as food, 192
Legs, proportions compared with *V. salvator,*
    18 ff., fig. 2–4
Length, 11, 16–17. *See also* Size; Snout-vent
      length

Index

Lepidophyma, 240
Lepidotrichia semisculpta
  as food, 191, 196, 206, 277
  habitat of, 133, 207, 216 ff., 277
Licking, 266, 295, 321, fig. 13-5
Life span. *See* Age
Light, 81, 104-6, fig. 5-7
Limiting factors, 235, 264
Live v. dead prey, 187, 241, 264, 268
Liver, 57, 288, figs. 3-8, 3-9
Local mix of fruits, 233
Locomotion
  bipedal, 155
  escape, 154
  spoor, 153-54
  See also Movement
Longevity. *See* Age
Lygodium, 142

Macaca philippinensis. See Monkey, macaque
Macroecological factors, 68 ff.
Malasia, as food, 192, 238, 258, 285
Male reproductive cycle, 165 ff. *See also* Reproduction
Mammals
  as food, 274
  as potential competitors, 186, 258, 274 ff., App. 2
  predation, 281, App. 2
Man
  as competitor, 274 ff., App. 2
  effect of
    on butaan density, 361
    on habitat, 358, 364 ff., figs. 15-1, 15-2
  as predator, 361 ff., App. 2
Management, 357 ff., 371
Maps, mapping, 15
Mating behavior. *See* Courtship; Reproduction
Maturity, 159.
Maxillary teeth
  number, 48-49, figs. 3-5, 14-4
  shape, 47-50, fig. 3-5
Metabolic heat, 108
Methods, study, 7 ff.
Microbial fermentation, 239, 260-61
Microclimates, 91 ff., 110
Miocene, 348 ff., fig. 14-10
Mixed dipterocarp forest, 81 ff., 342 ff., pls. 2, 3, figs. 4-8, 4-9, 4-10
Molluscs
  as food, 42, 181 ff., 190, 191, 196, 206, 214-15, 256, 264, 269
  habitats of, 206

pressure needed to crush shells, 196, 237, 334
Monitor, as genus, 18
Monkey, macaque (*Macaca philippinensis*), 186, 269, 274, App. 2
Monsoon, 75, 256
  effect on butaan movements, 116-19, 143, 145
Morphology
  claws and feet, 24-27, 155-56
  ear opening, 24
  gastrointestinal, 51 ff., 58 ff., fig. 3-7
  hatchlings, 17
  head, 33 ff., 39 ff., figs. 3-3, 3-4
  hyoid, 46-47, 334, fig. 14-3
  length, 11, 16-17
  muscles, 39 ff., 193
  narial opening, 18, 25, 331-32, fig. 14-2
  proportions, 18 ff., 20, 333, fig. 2-4
  scalation, 11, 25 ff., 332-33, figs. 2-6, 2-7
  size, form, and mass, 11, 17-20
  skeleton, 42 ff., 51, 334, figs. 3-2, 3-4, 3-6
  tail, 18-19, figs. 2-4, 2-5
  teeth, 47 ff., 334, figs. 3-5, 14-4
  weight, 17
Mortality, 151.
Mounting, 298, fig. 13-2
Movement, 96 ff.
  patterns, 93 ff., 97 ff., 106, 109-11, 143, 254, figs. 5-3, 5-4, 5-6, 5-7, 5-16
  telemetry techniques to determine, 96, figs. 5-1, 5-2, 5-5
  See also Digging; Haul-in time; Haul-out time; Locomotion; Swimming; Walking
Mucous trails, 205
Musculature, 39 ff., 196, fig. 3-3

Narial opening and feeding, 25, 207, 237
Neck bite, 298
Nematodes, 280-81
Nests, nesting, 171
Nocturnal behavior, 142
Nueva Ejica (Province, Luzon), 128
Nutrient mix, 233, 242, 257

Obba (mollusc)
  habitat, 207
  as prey, 192, 196, 206
Object-concentrated search, 253
Oily fruits, 192, 231-32, 238, 240, 247, 258, 289, fig. 9-3
Ondatria, as subgenus, 37
Ontogenetic differences, in food, 240, 266
Ophiophagus, as predator, 280

Optimum diet, 236 ff.
Osteoderms, 334
Osteology. See Anatomy
Ovary, 11
  development of, 160 ff.
  See also Reproductive cycle
Oviduct, 160 ff.
Oviposition and feeding, 171

Pair bonding, 139, 268
Palm civet, 186–87, 238, 244, 258, 269, 274 ff., App. 2
Pandanus
  as food, 184, 191, 196, 239, 258, 263, 286
  fruit shadow, 282
  radicans, 133, 184, 286
  simplex, App. 1
  tectorius, App. 1
Panting, 107
Paradoxurus hermaphroditus. See Civet cat
Parasites. See Ectoparasites; Endoparasites
Parrots, as possible competitors, 274, 285
Peak activity periods, 92 ff., 104 ff., 142, 254–55, fig. 5–16
  diurnal, 254
  seasonal, 254
Pens, activity, 11
Pentacne, 82, 85, App. 1
Perceptual field, 253
Pheromones, 153
Philippinosaurus, as subgenus, 3, 329, 339
Phloemys (cloud rat), 285, 329, 352, App. 2
Phylogeny, 342 ff.
Pigeons, as possible fruit competitors, 272, App. 2
Pigs. See Boar
Plant foods, methods of study, 7
Plants
  and butaans, 81 ff., 182 ff., 226 ff. (*see also* Coevolution; Foods)
  ecology and importance of, 81 ff., 339 ff.
  methods of study, 7
  shelter, 90 ff., 142–43, 244, 278
  temperature, 92 ff., fig. 4–12
  vertical stratification of, 82 ff., figs. 4–7, 4–8
Pleistocene, 133, 346, 348 ff., figs. 14–11, 14–12
Pliocene, 346, 348 ff., figs. 14–10, 14–12
Populations
  characteristics, 157 ff., fig. 2–3
  density, 158
  estimates for study area, 7
  limits, 81 ff.
  See also Competition; Predation
Precipitation. See Rainfall

Predation
  by butaans, 187, 240, 260–61
  on juveniles, 261
  pressure on butaans, 261, 280, 361 (*see also* App. 2)
  See also Feeding; Foods
Predators, 268, 280, 361
Prey
  birds, 182, 188, 240
  density, 246 ff., 253
  insects, 182 ff., 260 ff., fig. 10–1
  molluscs, 182 ff., 267
  number eaten, 182 ff., 187–88, 288
  pursuit, 240
  recognition, 253
  rodents, 187, 240
  selection, 188, 191, 199, 201, 236 ff., 266
  size, 184, 186, 196, 237
  See also specific types
Primates. See Monkey, macaque
Proportions
  compared to other species, 18 ff., 333, fig. 2–4
  of *olivaceus*, 18 ff., fig. 2–4
Protein foods, 191–92, 240 ff., 284, fig. 9–3
Python molurus, 356
Python reticulatus
  as competitor, 187, App. 2
  as predator, 281, App. 2

Quezon (Province, Luzon), 127, 132

Radiation, solar, 79
Rainfall, 75, 172, 256, figs. 4–3, 4-4, 4–5, 4–6, 5–15, 9–12, 10–5
Range. See Core area; Home range
Raphidophora merrillii, 142, App. 2
Rats
  as competitors, 201, 258, 284, App. 2
  as food resources in different habitats, 187, 209
Refuging behavior, 252
Relative humidity, 88
Reproduction, 159 ff.
  adult size and clutch size, 170–71
  and body fat, 175
  control of cycle, 173–74
  strategy, 174 ff.
Reproductive behavior, 170 ff.
Reproductive cycle, 159 ff., figs. 8–1, 8–4, 8–5
  methods of study, 11
Reproductive effort, 159, 175–76, fig. 8–6
Reproductive female cohort, 165
Reptiles. See specific type
Risk, of different sexes, 244–45

Rizal (Province, Luzon), 128, 132
Rodents
  as possible fruit competitors, 199, 258, 285, App. 2
  as prey, 187, 240
Ryssota (mollusc)
  habitat, 133, 207, 277
  as prey, 196, 201, 206, 277, App. 2

Salivation at high temperatures, 107, 255
Sandoricum koetjape, 224, 233, 238, 258, App. 1
  as food competitors, 186
Satiation, 184. *See also* Food
Scalation, 11, 25 ff., 332, figs. 2-6, 2-7
Scent
  of fruit, 187, 205, 238, 242, 263, 266
  of interacting butaans, 154, 333, fig. 13-5
Schizostachyum, 142, App. 1
Scincids
  feed on fruits, 259
  feed on molluscs, 182
Scratching
  for food, 155, 204, 207
  one another, 149-50, 309 ff.
Searching efficiency, 243 ff.
Search path, 244-46, 253
Search time, 247, 253
Season
  and behavioral cycle, 256
  and butaan movement, 104 ff., 135, 256 ff.
  dry, 75, 256
  and food, 256 ff., fig. 10-1
  wet, 75, 256
Secondary forest, 86-87
Secondary frugivory, 261
Secondary sexual characters, 153
Seed destroyers, 258, 274
Seed dispersal, 224, 258, 285, figs. 10-5, 12-2
Seed shape, 258, 285, fig. 12-2
Selection of food, 188, 190, 199, 201, 236, 239, 241 ff., 266
Senses, 152 ff.
  hearing, 152
  sight, 152
  smell, 152, 187 ff., 205, 242, 263, 266, 333
Sex ratio, 120, 156
Sexual behavior
  characteristics, 159 ff.
  See also Courtship
Sexual maturity, 159
Shedding, 151
Shelter, 142 ff.
  and activity of butaans, 142, 254
  arboreal types, 90-91, 142, 254, fig. 6-6
  crevice types, 92 ff., 113-14, 142-43, figs. 4-1, 4-11, 6-7, 6-8
  hollow trees, 92, 254
  and rain, 92 (*see also* Activity; Rainfall)
  seasonal use, 92 (*see also* Activity; Season)
  sharing with other animals, 277
  and temperature, 92 ff., 142
Shorea, 82, 86, App. 1
Shrews, 258, 264, 270, App. 2
Sight. *See* Senses
Signals and displays. *See* Combat; Courtship
Site familiarity, 240 ff., 246
Size
  and body temperature, 106 ff., 254-55
  range, 17 ff.
Skinks, as frugivores, 259, fig. 3-10
Skull, 33 ff., 39 ff., 334, figs. 3-2, 3-4
  kinetics, 42-43
  modification for feeding, 42-43, 196
Slash and burn agriculture, 364 ff.
Sleep, 97-99, 114, 142
Slugs, as food, 182, 207, 258
Small intestine, 55 ff.
Smell. *See* Senses
Snails. *See* Molluscs
Snakes
  as competitors, 187, App. 2
  as predators, 280
    cobra (*Ophiophagus hannah*), 280
    reticulated python (*Python reticulatus*), 280
Snout-vent length, 16-17
Social behavior, 292 ff.
Social organization, 292
Soil, 72-73
Solar radiation, 79
Sorsogon (Province, Luzon), 125, 132
Sounds made, 152
Spatial overlap, 138 ff.
Spatial relations, 122 ff.
Spermatogenesis. *See* Spermatogenic cycle
Spermatogenic cycle, 11, 165 ff.
Spermatozoa. *See* Spermatogenic cycle
Spiders, as food, 207, 258
Spondias pinnata, App. 1
  distribution in forest, 133, 214
  as food, 201, 214, 224, 238, 258, 276, 286
  fruit phenology of, 230
  pattern of fruit on ground, 225, 246
Spoor, 153-54
Stomach contents, 179 ff.
Stratification of temperature, 92 ff., figs. 4-11, 4-12
Stress reactions, 296

Study area, description of, 6, fig. 1-1
Sugary fruit, 192, 230, 241-42, 247, 252, 277, fig. 9-3
Surface water. *See* Water
Swallowing, 42, 237, fig. 3-4
Swimming, 155
Synchronous fruiting, 230, 258
Systematics, 32 ff.

Tactile cues (signals)
  raking behavior of males, (=scratching), 298
Tail
  lift, 298, fig. 13-2
  positions, 154, 293, 300, fig. 7-3
  shape of, 19, fig. 2-5
  use of
    in defense, 293
    in locomotion, 155
Tanqua (intestinal parasite), 280
Tectonic features, 349, fig. 14-7
Tectovaranus, 339
Teeth, 47 ff., 237, 334, figs. 3-5, 14-4
Telemetry, methods of use and analysis, 12, 96
Temperament, 152, 293
Temperatures
  air, 78, 89, 92 ff., 243-44, figs. 4-4, 4-6, 4-12
  of butaan, 106-7, figs. 5-8, 5-10, 5-11
  and butaan behavior, 89, 106 ff., 244, figs. 5-12, 5-13, 5-17
  diel pattern in habitat, 92 ff.
  and egg laying, 172
  gradients, 91, 244, figs. 4-11, 4-12
  in vegetation, 92, 244, fig. 4-12
  See also Death, butaan
Territoriality, 268. *See also* Social organization
Tertiary geology, 348 ff.
Testes, 11, 165
Thermal stratification, 91 ff., 255, figs. 4-11, 4-12, 5-10
Thermoregulation, 254-55, 266
Ticks. *See* Ectoparasites
Tongue, use of, figs. 13-1, 13-5
  in combat, courtship, 295, 321
  in food search, 205
Topography, figs. 4-1, 4-2
  and food availability, 72
  and wind, 75
Topping, 296
Transects, methods of plant study, 8
Transient butaans, 266
Trees
  climbing in, 20, 24, 89, 142, 155-56, 330, fig. 7-4
  hollow, 92, 254
    as nests, 171
    as shelters, 90-91, 142, 254, fig. 6-6
Trochomorpha (mollusc), 196, 206
Typhoons, and butaan activity, 114 ff., 143, 145

Uaranus, as genus, 18
Uaranus ornatus (=*Varanus ornatus*), 3, 326
Uromastix, as herbivore, 178
Uvaria, as food, 238, 258, App. 1

Varanus
  acanthurus, 63-64, 157, 172, 199, 240
  bengalensis, 32, 39, 42, 46, 47, 50, 65, 103, 150 ff., 159-61, 170, 172, 175, 188, 207, 209, 240, 253, 280, 293 ff., 305 ff., 329, 330 ff., 346
  brachyurus, 37
  brevicauda, 63
  dumerilii, 4, 24, 32, 47, 156, 206, 329 ff.
  eremius, 24
  exanthematicus, 17, 24, 32, 37, 42-43, 46-47, 63, 156, 181, 208, 214, 309, 331 ff.
  flavescens, 32, 37, 42-43, 330-31, 339, 341
  giganteus, 63, 334
  gilleni, 299, 304
  gouldii, 100-101, 107, 109, 111, 119, 136, 138, 154, 244, 252, 299
  grayi (=*olivaceus*) (*see* specific part or function wanted)
  griseus, 24, 32, 46, 136, 151, 156, 172, 244, 309, 331
  hooijeri, 42
  indicus, 32, 42, 181
  komodoensis, 2, 23-24, 33, 37, 44, 50, 63, 90, 101, 107, 109, 135, 145, 151, 155-56, 159, 165, 172, 173, 176, 179, 181, 187, 237, 242, 244, 254, 266, 268, 279, 280, 288, 293 ff., 329 ff.
  mertensi, 294
  niloticus, 2, 37, 42, 46, 47, 50, 151, 172, 175, 181, 208, 294, 331 ff.
    convergence with butaan, 39, 329
  olivaceus (*see* Butaan)
    history, 2, 327 ff.
    believed extinct, 3
  prasinus, 23, 32, 37, 47, 156, 182
  rudicollis, 23, 24, 32, 37, 128, 155, 182, 207, 239, 330 ff., 341
    in Philippine Islands, 4

salvadori, 294
salvator, 15, 17 ff., 22–24, 31, 37, 42, 50 ff., 63, 66, 147, 152, 171–72 (*see also* Water monitor)
spenceri, 172, 294, 299
timorensis, 309
varius, 39, 294, 299, 329
Vegetative cover, pls. 2, 3, figs. 4–8, 4–9, 4–10
  density of, 7, 253
Vegetative factors important to butaan, 81 ff., fig. 4–8
Vertical limits
  of plants, 81 ff., 133, fig. 4–7
  of snails, 133
  of *V. olivaceus,* 133, fig. 4–7
  of *V. salvator,* 133, fig. 4–7
Vipera berus, 64
Vipera russelli, 356
Vision. *See* Senses
Visual cues. *See* Courtship
Vitellogenic cycle, 161
Vitex, 82–83, 85, App.1
Viverra, 276, 280, App. 2

Waking, 99
Walking, by butaan, 155
Warning behaviors, 293
Water
  drinking, 68 ff.
  loss, 109
  subsurface, 73–74
  surface, 73
Water monitor (*V. salvator*)
  combat, 300 ff.
  energetics of foraging, 267
  feeding on carrion, 187, 268
  gall bladder, 57
  habitat, 88, 279
  injuries, 147
  interactions with butaans, 15, 258 ff., 264 ff., 269, 279
  liver, 57–58, 286 ff., fig. 3–9
  parasites, 278 ff.
  prey, 179 ff., 188, 236 ff., 266
  proportions of body, 18 ff., 237, fig. 2–4
  proportions of skull, 39, 42–43
  reproductive allocation, 176
Weather, methods of study, 14
Weight
  of butaans, 17
  of fruits, 184, 196, 199
Wet season. *See* Season
Wildlife refuges, 359 ff.
Wind, 75, 78–79, 244
  and butaan movement, 116 ff., 145, 244
  and seed dispersal, 284–85
Wounds, 147 ff., figs. 7–1, 7–2
Wrestle, 299